Gamma-Ray Bursts

Diego Götz • Maurizio Falanga • Zigao Dai •
Emeric Le Floc'h • Nial Tanvir • Bing Zhang
Editors

Gamma-Ray Bursts

A Tool to Explore the Young Universe

Chapters 2–10 were previously published in *Space Science Reviews* Volume 202, Issues 1–4, 2016

Springer

Editors
Diego Götz
Servide d'Astrophysique
CEA
Gif-sur-Yvette, France

Maurizio Falanga
International Space Science Institute
Bern, Switzerland

Zigao Dai
School of Astronomy and Space Science
Nanjing University
Nanjing, China

Emeric Le Floc'h
Irfu-Departement d'Astrophysique
CEA
Gif-sur-Yvette, France

Nial Tanvir
Department of Physics and Astronomy
University of Leicester
Leicester, UK

Bing Zhang
University of Nevada
Las Vegas, NV, USA

ISSN 1385-7525   Space Sciences Series of ISSI
ISBN 978-94-024-1278-9

Library of Congress Control Number: 2018937109

Cover Image: Credit: Gemini Observatory/AURA, artwork by Lynette Cook

Printed on acid-free paper

This Springer imprint is published by the registered company Springer Science+Business Media B.V. part of Springer Nature.
The registered company address is: Van Godewijckstraat 30, 3311 GX Dordrecht, The Netherlands

Contents

Preface

Diego Götz[1] · Maurizio Falanga[2]

This book reviews a comprehensive set of problems raised by the use of Gamma-Ray Bursts (GRBs) as cosmological probes, as they were addressed and debated during the 2015 ISSI-Beijing workshop "Gamma Ray Bursts: a tool to explore the young Universe," which took place between April 13–17 in Beijing. Particularly original about this workshop was that the invited scientists represented both the GRB community *stricto sensu* and the community studying high-redshift galaxies and star formation history.

Despite the recent progresses in Gamma-Ray Burst science, obtained in particular thanks to the *Swift* and *Fermi* satellites, there are still many open questions in the field. One concerns the mechanisms that power these extreme explosions (in a handful of seconds the isotropic equivalent energy emitted by GRBs spans from 10^{50} to 10^{54} erg, making them the most luminous events in the Universe), which is still unclear after more than four decades since their discovery. In particular, the content of the relativistic flow that produces the GRBs remains to be investigated, especially in terms of its bulk Lorentz factor, its magnetization, its baryon loading and their consequences on the possibility of GRBs being the sources of Ultra High Energy Cosmic Rays (UHECRs).

Another open issue concerns the nature of GRBs progenitors: while thanks to a handful of spectroscopic associations with type I c Supernovae, long GRBs are currently considered the endpoint of very massive (>30–50 solar masses) stars (but their potential binary nature is an open issue), the situation is less clear for what concerns the short GRBs. The most popular models involve the possibility of a coalescence of two compact objects (black holes or neutron stars), and the first direct proof may be represented by GRB/GW 170817, for which a gravitational wave signal from two coalescent neutron stars has been associated with a short flash of gamma ray radiation, but the derived isotropic emitted energy is far below the one measured for previously known short GRBs.

✉ D. Götz
diego.gotz@cea.fr

1 CEA Service d'Astrophysique, Gif-sur-Yvette, France

2 International Space Science Institute, Bern, Switzerland

In addition to particle acceleration, radiation physics, and stellar evolution, GRBs are also pertinent to other branches of astrophysics, such as cosmology. Indeed, observers are able to pinpoint and study the most distant galaxies thanks to the imprinting on afterglow data of the GRB close environment and the intervening matter between the GRB and the Earth. Absorption lines contain a wealth of information on the structure and physical state of the gas in the Universe.

The aim of this ISSI-Beijing workshop was to elucidate the connection between GRBs and the deep Universe. In recent years much effort has been put into the ground based follow-up of GRB afterglows: the median measured redshift of long GRBs is $\sim$2 (with the highest value reported to date of $\sim$9.4), which means that GRBs are among the best candidates to study the deep Universe. Ten years after the launch of the NASA *Swift* satellite, statistics on GRBs above $z = 3$ have been constantly improving, reaching more than 40 events to date. It was hence timely to coherently review and assess the current knowledge on Gamma-Ray Bursts, their progenitors and their environment in order to understand their nature and origin. Understanding the GRB phenomenon globally is of paramount importance for the use of GRBs as beacons of the deep Universe.

At the same time, we felt it would be wise to start thinking about how to develop world-wide synergies between the high-redshift GRB studies and the surveys of the deep Universe, especially in the framework of the future GRB-dedicated facilities in which China is deeply involved. This workshop was also labeled as a scientific event celebrating the 50th anniversary of the diplomatic relationship between France and China and is related to the more general framework of the upcoming Sino-French space mission *SVOM* (Space based Variable astronomical Object Monitor), dedicated to GRBs and time domain astronomy studies.

The chapters by Zhang et al. and Dai et al. contain valuable introductory material about the GRBs observational properties and the associated theoretical models and provide the basis for the following sections. The review by Levan et al. deals with the GRB progenitors, covering the different theories about what parent star(s) could be at the origin of these peculiar objects. The chapter by Nagamine et al. summarizes our current knowledge on galaxy formation and stellar evolution especially at early times. Perley et al. review the characteristics of the galaxies hosting GRBs. The chapter by Cucchiara et al. provides some examples about the use of GRBs as cosmological probes, using them as powerful background lighthouses shining through the Universe. Toma et al. investigate the possibility of producing and detecting GRBs from the first (so-called population III) stars. The chapter by Chary et al. deals with the possible use of GRBs as (unbiased?) tracers of the early star formation. Wang et al. review the progress on the possibility of probing fundamental cosmology and physics using GRBs. Finally, Yuan et al. provide an interesting perspective on the future facilities that could tackle the observation of highly redshifted GRBs such as *SVOM*, *THESEUS*, *Einstein Probe*, *SKA*, *WFIRST*, *JWST*, *ELTs*, etc.

It is our honor to warmly compliment the co-editors and conveners (Z. Dai, N. Gehrels, E. Le Floc'h, N. Tanvir, B. Zhang) and the organizers of the workshop; they conducted their job of producing this extensive review volume with great enthusiasm and dedication. The fast-growing space research community in China and elsewhere in Asia is eager to increase its scientific exchanges with colleagues from other parts of the world, and, at the same time, scientists from outside Asia will find new opportunities in Asia/China through ISSI-Beijing. Thus, 46 highly respected scientists from all over the world representing different scientific communities could be gathered to work on these review papers. We thank all who participated and who contributed to the papers. This excellent set of review papers represents an important outcome of ISSI-Beijing: congratulation to all.

Finally, we would like to dedicate this book to Neil Gehrels, who passed away during the editing process. He has been an example for all of us: an outstanding scientist full of curiosity and always ready to embark enthusiastically on new projects. In addition, he had exceptional human skills, being ready to listen and discuss with all the members of our community. We will all miss him.

Space Sci Rev (2016) 202:5–34
DOI 10.1007/s11214-016-0305-9

GRB Observational Properties

Bing Zhang[1] · **Hou-Jun Lü**[2,3] · **En-Wei Liang**[2,3]

Received: 23 August 2016 / Accepted: 26 October 2016 / Published online: 10 November 2016

Abstract We summarize the basic observational properties of gamma-ray bursts (GRBs), including prompt emission properties, afterglow properties, and classification schemes. We also briefly comment on the current physical understanding of these properties.

Keywords Gamma-ray bursts

1 Introduction

Gamma-ray bursts (GRBs) are energetic bursts of γ-rays from deep space. They are the most luminous objects in the universe, signifying births of stellar-mass black holes or millisecond magnetars in most violent explosions, such as core collapse of rapidly spinning massive stars or mergers of two compact stars (NS-NS or NS-BH). Many in-depth reviews have been written on the subject: e.g. (Fishman and Meegan 1995; Piran 1999; van Paradijs et al. 2000; Mészáros 2002; Zhang and Mészáros 2004; Piran 2004; Mészáros 2006; Zhang 2007; Gehrels et al. 2009; Kumar and Zhang 2015). This chapter summarizes the basic observational properties of GRBs, including prompt emission (Sect. 2), afterglow (Sect. 3), and classification schemes (Sect. 4). Besides outlining the observational facts, a brief discussion on the current physical understanding of the data is also presented.

✉ B. Zhang
zhang@physics.unlv.edu

H.-J. Lü
lhj@gxu.edu.cn

E.-W. Liang
lew@gxu.edu.cn

1 Department of Physics and Astronomy, University of Nevada Las Vegas, NV 89154, USA

2 GXU-NAOC Center for Astrophysics and Space Sciences, Department of Physics, Guangxi University, Nanning 530004, China

3 Guangxi Key Laboratory for Relativistic Astrophysics, Nanning, Guangxi 530004, China

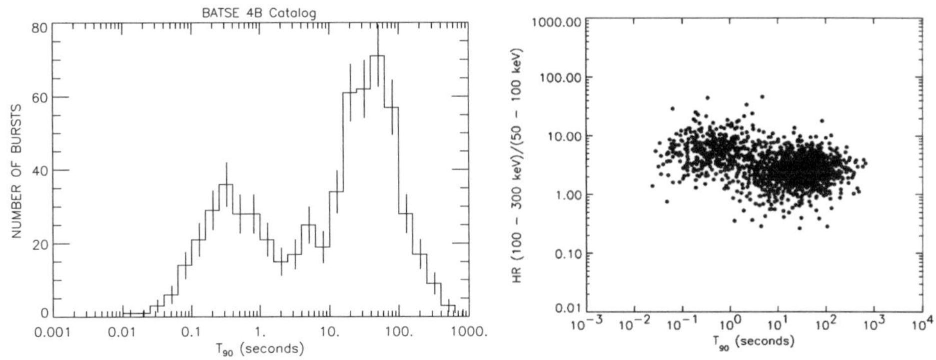

Fig. 1 *Left*: The bimodal distribution of GRB durations. *Right*: The two dimensional distribution of GRBs in the T_{90}–HR domain. From BATSE GRB catalogs (http://gammaray.msfc.nasa.gov/batse/grb/catalog/)

2 Prompt Emission

GRB prompt emission is usually defined as the soft γ-ray/hard X-ray emission detected by the GRB triggering detectors (such as *Swift*/BAT and *Fermi*/GBM). Some GRBs are detected in higher (e.g. GeV) or lower (e.g. optical) energies during the prompt emission phase.

2.1 Temporal Properties

2.1.1 Duration Distribution

GRB duration is usually quantified as T_{90}, the duration during which 5 % to 95 % of the total fluence is detected by the triggering detector. The definition of T_{90} depends on the energy band and sensitivity of the detector. A same burst may have a longer T_{90} if the detector is more sensitive or has a softer energy band.

In the CGRO/BATSE band (25–350 keV), GRBs showed a bimodal distribution with a rough separation at 2 s (Kouveliotou et al. 1993). This was the foundation of classifying GRBs into long ($T_{90} > 2$ s) and short ($T_{90} < 2$ s) categories (Fig. 1a). The bimodal distribution was confirmed by other later detectors (e.g., Sakamoto et al. 2011; Paciesas et al. 2012; Zhang et al. 2012; Lien et al. 2016), even though the relative fractions of the two types vary for different detectors and also for different energy bands in the same detector (Qin et al. 2013). The bimodal classification is more evident in the two-dimensional T_{90}–HR domain, where HR is the hardness ratio of a burst. Long GRBs are typically softer than short GRBs (Fishman and Meegan 1995, see Fig. 1b). Broad-band observations now show that the two duration classes roughly correspond to two types of progenitor systems: long GRBs are related to deaths of massive stars, and short GRBs are related to mergers of compact objects (see Sect. 4 for more discussion).

2.1.2 Lightcurves

GRB lightcurves are erratic, usually display overlapping emission episodes with a wide range of variability time scales (Fishman and Meegan 1995). Some bursts have one or several clearly separated "pulses", which typically show a fast-rising and exponential-decay

(FRED) shape. The duration of these pulses is typically seconds. Some GRBs have a "precursor" followed by a quiescent gap before the main burst comes out. The emission properties of the precursor is not too different from the main burst (e.g., Lazzati 2005; Burlon et al. 2008; Hu et al. 2014). A power density spectrum (PDS) analysis does not reveal quasi-periodic oscillations, but rather show a featureless power law (Beloborodov et al. 2000; Guidorzi et al. 2012). However, there is evidence of the superposition of a "fast" variability component on top of a "slow" variability component (Gao et al. 2012).

2.2 Spectral Properties

2.2.1 *Spectral Models and Multiple Components*

When enough photons are detected (large enough of GRB fluence), the GRB spectra are usually fit by a phenomenological model called the "Band" function (Band et al. 1993). It is essentially a broken power law model with a smooth (exponential) transition between the two segments. The typical low-energy and high-energy photon indices are $\alpha \sim -1 \pm 1$ and $\beta \sim -2^{+1}_{-2}$ for the GRBs observed with BATSE in the 20–2000 keV band (Preece et al. 2000), and confirmed for GRBs detected by *Fermi*/GBM and Integral (Zhang et al. 2011; Nava et al. 2011; Gruber et al. 2014). The peak energy of a GRB (when $\beta < -2$ is satisfied) is defined by $E_p = (2+\alpha)E_0$, where E_0 is the break energy in the photon spectrum of the Band function. The distribution of E_p is about several hundred keV for bright GRBs, but can vary from several keV for GRB 060218 (Campana et al. 2006) to 15 MeV for GRB 110721A (Axelsson et al. 2012). The spectra of some GRBs may be fit with a cut-off power law, and some others with a single power-law function if the GRB is not bright enough or it is observed with a narrow band instrument such as *Swift* BAT, even though the true spectral shape may be Band-like (Zhang et al. 2007a).

In some GRBs, the superposition of multiple spectral components was observed. A thermal like spectral component claimed in the pre-*Fermi* era (Ryde 2005) was confirmed in the *Fermi* era. First, GRB 090902B was found to have a narrow Band component superposed on a power law component (Abdo et al. 2009a). The narrow Band component could be also fit as a multi-color blackbody (Ryde et al. 2010). When the time bin is small enough, the data could be even fit with a blackbody plus power law model (Zhang et al. 2011). A similar case may be made to the short GRB 090510 (Ackermann et al. 2010). Later, a weak thermal component was suggested to exist in the low-energy shoulder of the Band component of some GRBs (e.g., Guiriec et al. 2011, 2013, 2015; Axelsson et al. 2012). Some GRBs, on the other hand, show featureless Band-like spectra without the need of a thermal component (e.g., GRB 130606B, Zhang et al. 2016). Another spectral component is the power law component as seen in some GRBs, e.g. GRB 090902B and GRB 090510. This component extends several orders of magnitude from keV range sometimes all the way to the GeV range. It usually has a positive slope in νF_ν spectral representation, which means that there should be a high-energy E_p in the GeV range. Indeed, GRB 090926 was found to have such a component with a high-energy cutoff (Ackermann et al. 2011). Some example GRB spectra are presented in Fig. 2. Overall, GRB prompt emission spectra may have three distinct components (Zhang et al. 2011, Fig. 2d): I a Band component; II a thermal component; and III a power law component. Different spectral components may dominate in different GRBs. Indeed GRB spectral with different combinations of these three components have been observed. Guiriec et al. (2015) showed that all three components may exist in at least some bright GRBs.

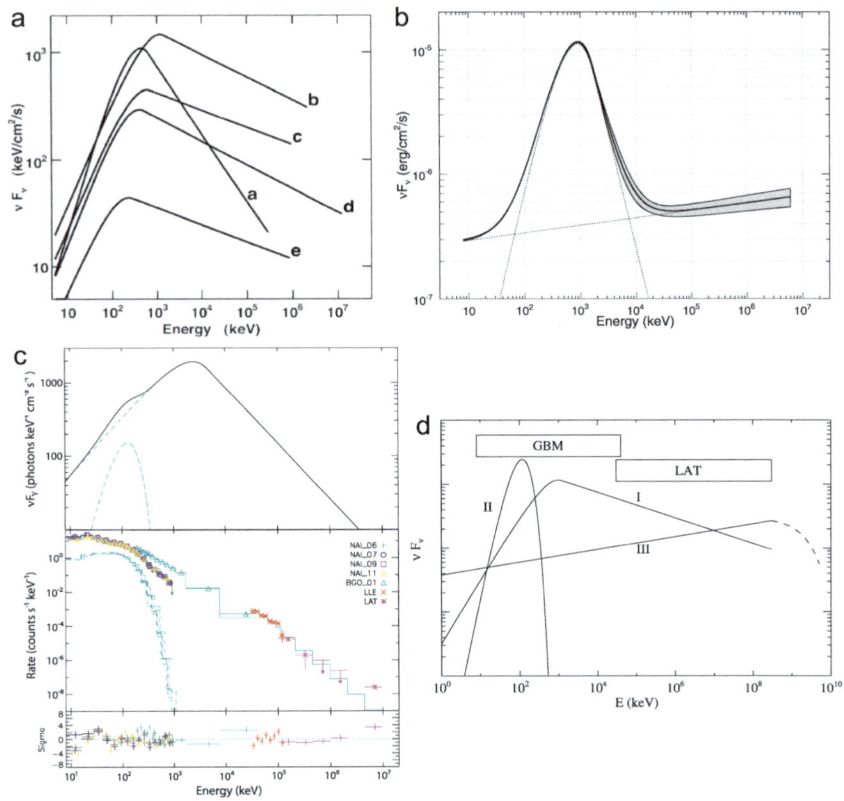

Fig. 2 Some example GRB spectra: (**a**) the Band function in GRB 080916c (Abdo et al. 2009b); (**b**) a quasi-thermal spectrum superposed on a power law component in GRB 090902B (Abdo et al. 2009a); (**c**) a thermal component superposed on a Band component in GRB 110721 (Axelsson et al. 2012); (**d**) three possible elemental spectral components that shape GRB spectra (Zhang et al. 2011)

2.2.2 Spectral Evolution

Prompt emission spectra rapidly evolve with time. This can be seen from two different aspects. First, if one displays lightcurves in different energy bands, one would see significant "spectral lags", i.e. a pulse observed in a softer band is usually broader than and lagged behind the same pulse observed in a harder band (Norris et al. 2000; Liang et al. 2006a). Second, if one displays time-dependent spectra, one would see that E_p rapidly evolves with time. There are two types of evolution patterns (Norris et al. 1986; Golenetskii et al. 1983; Lu et al. 2010, 2012): hard-to-soft evolution and intensity tracking.

2.3 Prompt Emission in Other Wavelengths

During the prompt emission phase, some GRBs are also detected in other energy bands. In the high-energy regime, GeV emission was detected in the prompt emission phase of several GRBs. In general, GeV emission is delayed with respect to the sub-MeV emission (e.g., Abdo et al. 2009a,b). Sometimes the GeV emission peak seems to align with some sub-MeV emission peaks, but in most cases, the GeV emission lasts longer than the MeV

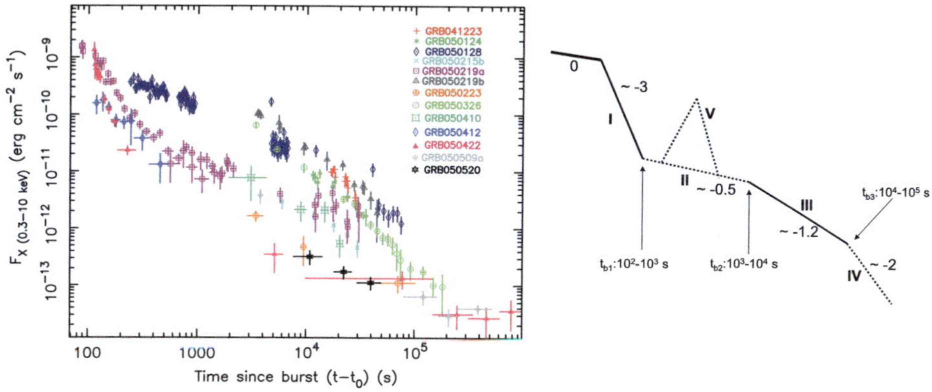

Fig. 3 *Left*: Examples of observed XRT lightcurves (Nousek et al. 2006). *Right*: A canonical XRT lightcurve (Zhang et al. 2006)

component (Ghisellini et al. 2010; Zhang et al. 2011), suggesting the emergence of a new emission component.

Prompt optical emission was detected in a few cases. Some GRBs show a tracking behavior between the optical and sub-MeV emission (e.g., the famous "naked-eye" GRB 080319B Racusin et al. 2008), while some others show a distinct peak near the end of prompt emission which is offset from the sub-MeV peaks (e.g., in GRB 990123 Akerlof et al. 1999). The combination of these two components was also seen in some GRBs (e.g. in GRB 050820A Vestrand et al. 2006).

On the other hand, extensive searches of high-energy neutrinos associated with GRBs during prompt emission phase have been carried out, but no detection has been made so far. An increasingly stringent upper limit of the prompt neutrino flux has been placed (Abbasi et al. 2012; Aartsen et al. 2016).

2.4 Physical Understanding

No general consensus is reached in the community regarding how prompt emission properties are interpreted. This is because there are some uncertainties in the physical models of GRB prompt emission (Zhang 2011, for a review): jet composition (matter-dominated fireball or Poynting-flux-dominated outflow), energy dissipation mechanism and particle acceleration mechanism (internal shocks or magnetic reconnection), and radiation mechanism (synchrotron radiation or comptonization of quasi-thermal photons). As a result, many models have been proposed to interpret GRB prompt emission. According to the location of emission site, these models may be grouped into three types: 1 the dissipative photosphere models (e.g., Rees and Mészáros 2005; Pe'er et al. 2006; Giannios and Spruit 2007; Beloborodov 2010; Lazzati and Begelman 2010; Toma et al. 2011; Murase et al. 2012, which interpret prompt emission as comptonized quasi-thermal photons from the photosphere); 2 the internal shock models (e.g., Rees and Meszaros 1994; Daigne and Mochkovitch 1998; Daigne et al. 2011, which interpret prompt emission as synchrotron radiation of electrons accelerated in the internal shocks); and 3 the forced magnetic dissipation models with large emission radii (e.g., the internal collision-induced magnetic reconnection and turbulence, ICMART, model of Zhang and Yan 2011, which interpret prompt emission as synchrotron radiation of electrons accelerated in the magnetic dissipation region). These models have different magnetization parameter σ in the emission region. The first two models have $\sigma \ll 1$, while the third model has $\sigma \geq 1$ in the emission region.[1]

The erratic lightcurves of GRB prompt emission are interpreted differently in the three models. In the photosphere models, all the variabilities are directly produced by the generalized central engine (including the central black hole and the stellar envelope through which the jet propagates). The rapid variabilities may directly reflect the intrinsic fluctuation of the jet power from the central black hole, while the slow component may be caused by the modulation of the stellar envelope as the jet emerges from the central engine (Morsony et al. 2010). In the internal shock models, the observed emission is from many internal shocks, some at small radii (the fast component), and some at large radii (the slow component) (Hascoët et al. 2012). In the ICMART model, the slow component reflects the history of central engine activity. Each broad pulse is from one single emission unit (one ICMART event), which gives a seconds-long duration pulse as the emission region moves forward. The rapid variabilities may be caused by the mini-jets produced in the local reconnection regions within the jet (Zhang and Zhang 2014).

It is agreed among the modelers that the thermal component observed in the spectra of some GRBs is emission from the outflow photosphere. The origin of the most common Band component is, however, subject to debate. One view is that it is also emission from the photosphere. The other view is that it is synchrotron emission from an optically thin region (internal shocks or ICMART regions). The simplest version of both models have trouble to interpret the typical low-energy photon index $\alpha \sim -1$: The photosphere model typically predicts $\alpha \sim +1.4$ (Beloborodov 2010; Deng and Zhang 2014), whereas the synchrotron model typically predicts $\alpha \sim -1.5$ due to fast cooling of the electrons. In order achieve $\alpha \sim -1$, the photosphere model needs to introduce a special type of structured jet model (Lundman et al. 2013). Within the synchrotron model, $\alpha \sim -1$ (or an even harder spectrum) could be achieved if one considers the global decay of the magnetic field strength in the emission region, as is naturally expected in an expanding shell (Uhm and Zhang 2014a). The superposition of a thermal component on top of a Band component poses a challenge to the photosphere model.[2] Within the internal shock and ICMART models, the Band component can be interpreted as the synchrotron radiation from an optically thin emission region (internal shocks or ICMART region), while the thermal component is the emission from the photosphere. Different bursts have different relative strength between the thermal and non-thermal components. This could be understood within the framework of a hybrid central engine (Gao and Zhang 2015) characterized by different dimensionless entropy η (for the fireball component) and magnetization parameter σ_0 (the Poynting flux component).

The power law component extending from low energy to very high energy is not easy to explain. One possibility is that it actually includes two emission components, i.e. a low energy synchrotron component and a high-energy inverse Compton component. With certain parameters they may form an effective power law by coincidence (Pe'Er et al. 2012).

Spectral evolution carries important clues regarding the origin of prompt emission. Spectral lags and E_p evolution patterns all concern the broad pulses in GRB lightcurves rather than the rapid variabilities. This suggests that each broad pulse reflects emission from one single emitting shell which travels in space. Uhm and Zhang (2016a) showed that the properties of spectral lags can be reproduced within this picture given that the emission radius

[1]The so-called electromagnetic model (Lyutikov and Blandford 2003) conjectures $\sigma \gg 1$ in the emission region which is close to the deceleration radius. It is hard to maintain such a huge σ at the deceleration radius. The detections of the thermal components also disfavor this possibility.

[2]Vurm et al. (2011) interprets the Band component as the photosphere emission while the "thermal" bump as the synchrotron radiation.

is large, the magnetic field strength in the emission region is decaying (as expected), and the emission region is undergoing bulk acceleration. The last requirement is consistent with dissipation of magnetic fields in a Poynting-flux-dominated outflow. The same theoretical framework can naturally explain E_p hard-to-soft evolution (Uhm and Zhang 2014a) and intensity tracking (Uhm and Zhang 2016b, in preparation). It is very difficult to interpret broad-pulse spectral lags and E_p evolution patterns within the photosphere models and internal shock models. The stringent upper limits of neutrino flux during prompt emission phase is also consistent with the picture that most GRBs have a Poynting-flux-dominated jet composition (Zhang and Kumar 2013; Aartsen et al. 2016).

3 Afterglow

GRB afterglow is the broad-band emission detected after the prompt emission phase. It is characterized by a multi-segment broken power-law spectrum at a given instant, and a multi-segment broken power law decaying lightcurve in a given observational band. The afterglow was theoretically predicted (Mészáros and Rees 1997) before the first discoveries (Costa et al. 1997; van Paradijs et al. 1997; Frail et al. 1997). After two decades of observations, broad-band afterglows are well studied phenomenologically and physically.

3.1 X-Ray Afterglow

GRB X-ray afterglow was first discovered with BeppoSAX for GRB 970228 (Costa et al. 1997). A systematic study of X-ray afterglows started in the *Swift* era, when *Swift*/XRT regularly observe the GRB source starting from tens of seconds after the trigger. A canonical lightcurve can describe the X-ray afterglow lightcurves of most GRBs, which include five components (Zhang et al. 2006; Nousek et al. 2006): a steep decay phase, a shallow decay phase (or plateau), a normal decay phase, a late jet-break-like steepening, and erratic X-ray flares (Fig. 3). A small fraction of XRT lightcurves show a monotonic decay with a single power law (Liang et al. 2009). Empirically, O'Brien et al. (2006) and Willingale et al. (2007) showed that the X-ray afterglow data can be fit by the superposition of a prompt emission component and an afterglow component, even though no physical model predicts the specific mathematical forms of the model. Some correlations between X-ray afterglow properties and prompt γ-ray properties have been found (Grupe et al. 2013).

The observational properties of the five afterglow components may be summarized as follows:

- Steep decay phase: The temporal decay slope of this segment is usually steeper than -3, sometimes even as steep as -10. This segment is usually smoothly connected to the last prompt emission pulse, suggesting that it is the tail emission of prompt emission (Barthelmy et al. 2005). Strong spectral softening is usually observed during the steep decay phase (Zhang et al. 2007; Campana et al. 2006; Mangano et al. 2007; Butler and Kocevski 2007). The origin of the steep decay phase is usually attributed to the high-latitude "curvature effect", with a proper zero time adopted (Kumar and Panaitescu 2000; Zhang et al. 2006; Liang et al. 2006b). The rapid spectral evolution requires a curved spectrum at the end of prompt emission (Zhang et al. 2007, 2009).
- Shallow decay phase (plateau): This segment is usually seen in the early XRT lightcurves following the steep decay phase. Its decay slope ranges from 0 to -0.5 and the typical duration of this segment is $\sim 10^4$ seconds with a fluence of $\sim 3 \times 10^{-7}$ erg/s in the XRT band (Liang et al. 2007). The typical photon index is ~ 2.1 (O'Brien et al. 2006; Liang

et al. 2007), and there is no spectral evolution seen across the break from this segment to the normal segment afterwards (Liang et al. 2007). The luminosity at the break time is anti-correlated with the break time in the burst frame (Dainotti et al. 2010; Li et al. 2012). In most cases, the shallow decay phase is followed by a normal decay with flux decreasing with time as $\sim t^{-1}$. These plateaus can be understood within the standard external shock model with continuous energy injection into the blastwave (Zhang et al. 2006; Nousek et al. 2006; Rees and Mészáros 1998; Dai and Lu 1998c,a; Zhang and Yan 2011; Yu and Dai 2007). In a few cases, the shallow decay segment appears as a plateau followed by a very steep decay (with index around -9) in both long and short GRBs (Troja et al. 2007; Liang et al. 2007; Lyons et al. 2010; Rowlinson et al. 2010, 2013). These cases cannot be interpreted within the external shock models, and therefore are termed as "internal plateaus". They demand direct energy dissipation from a long-lasting central engine, likely a millisecond magnetar (Troja et al. 2007; Lyons et al. 2010; Rowlinson et al. 2010, 2013; Lü and Zhang 2014; Lü et al. 2015).

- Normal decay and jet break: The normal decay phase has a typical decay index ~ -1, consistent with the prediction of the standard external forward shock model (Mészáros and Rees 1997; Sari et al. 1998). In some GRBs, there is an additional steepening break at later times, after which the decay slope may reach ~ -2 (Liang et al. 2008; Racusin et al. 2009). This is consistent with the so-called jet break (Rhoads 1999; Sari et al. 1999). Some bursts were observed with *Swift*/XRT and/or *Chandra* weeks and even months after the GRB triggers, with no evidence of detecting a jet break in their X-ray lightcurves (e.g., Grupe et al. 2006).
- X-ray flares: X-ray flares are detected in less than half of *Swift* GRBs. Their lightcurves show rapid rise and fall (Burrows et al. 2005; Falcone et al. 2006; Romano et al. 2006) with strong spectral evolution (Lazzati and Perna 2007; Chincarini et al. 2007; Margutti et al. 2011). The fluence of a flare is usually about 1–10 % of that of prompt emission, but occasionally may be comparable. Temporal and spectral analyses of X-ray flares reveal many properties analogous to prompt emission (Burrows et al. 2005; Chincarini et al. 2010; Margutti et al. 2011). They signify the reactivation of the central engine at later times (Burrows et al. 2005; Fan and Wei 2005; Zhang et al. 2006; Lazzati and Perna 2007).

3.2 Optical Afterglow

In the pre-*Swift* era, afterglow observations were mostly made in the optical bands after hours of GRB triggers. The late optical light curves usually show a single power law decay with a decay index of $\sim t^{-1}$. Occasionally, a two-segment broken power law from $\sim t^{-1}$ to $\sim t^{-2}$ was observed, with the temporal break defined as a "jet break". In some cases, an early optical flash was detected, which is characterized by a steep decay with $\sim t^{-2}$ (e.g., GRB 990123 Akerlof et al. 1999) and interpreted as emission from the reverse shock (Mészáros and Rees 1997, 1999; Sari and Piran 1999b,a). Kann et al. (2010, 2011) presented optical afterglow lightcurves in both the observer frame and a common rest-frame at $z = 1$ for a large sample of bursts (Fig. 4a). Similar to the X-ray afterglow lightcurves, the optical afterglow lightcurves are also composed of several different emission components (Li et al. 2012, Fig. 4b): Ia prompt optical flares; Ib an early optical flare from the reverse shock; II early shallow decay segment; III the standard afterglow component (sometimes led by an afterglow onset hump due to deceleration); IV the post jet break phase; V optical flares; VI rebrightening humps; VII late supernova (SN) bumps. The components II–V can find their counterparts in the canonical X-ray lightcurve.

One interesting feature in some optical lightcurves is an early hump that marks the deceleration of the ejecta (e.g., Rees and Meszaros 1992; Meszaros and Rees 1993; Sari and Piran

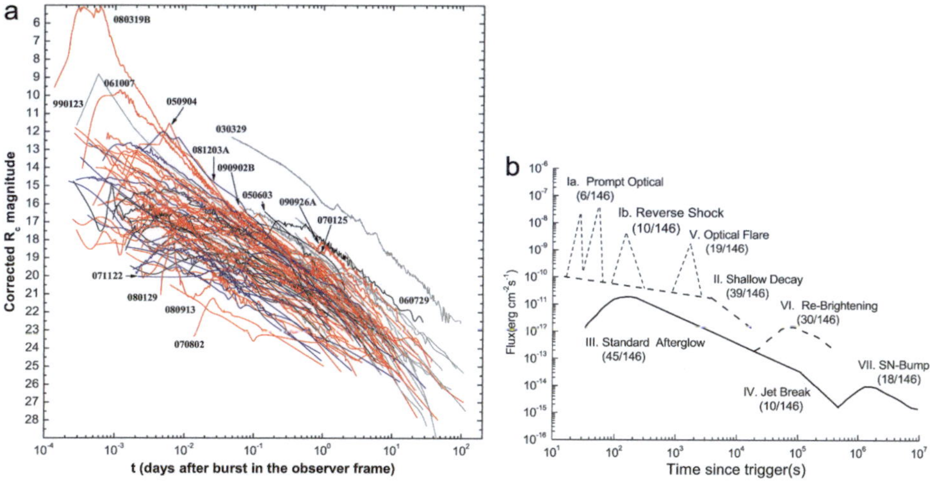

Fig. 4 (**a**) Observer-frame optical lightcurves for a sample of GRBs (Kann et al. 2010); (**b**) Synthetic schematic light curve of multiple optical emission components, from Li et al. (2012)

1999a, for the thin shell case, and Kobayashi et al. 1999; Kobayashi and Zhang 2007, for the thick shell case). The peak time in the lightcurve can be used to infer the initial Lorentz factor of the ejecta during the afterglow phase (e.g. Molinari et al. 2007; Xue et al. 2009; Melandri et al. 2010; Liang et al. 2010). A statistical analysis (Liang et al. 2013) suggests that the typical rising and decaying slopes of the onset hump are $\sim$1.5 and ~ -1.15, respectively. The peak luminosity is anti-correlated with the peak time, $L_p \propto t_p^{-1.81\pm0.32}$. Both L_p and the isotropic energy release of the onset bumps are correlated with $E_{\gamma,iso}$. Liang et al. (2010) discovered a correlation between Γ_0 and $E_{\gamma,iso}$ for a constant density medium. A similar relation between Γ_0 and γ-ray isotropic luminosity was also claimed (Ghirlanda et al. 2011; Lü et al. 2012).

One important subject is the "chromaticity" of the afterglows. Within the standard external forward shock models, optical and X-ray emission comes from the same synchrotron emission component. As a result, the hydrodynamical or geometrical temporal breaks (e.g. the energy injection break and jet break) in the lightcurves should be achromatic, i.e. a break should occur simultaneously in the optical and X-ray bands. However, multi-wavelength observations suggested that some GRBs showed chromatic behaviors in the two bands (e.g., Panaitescu et al. 2006; Fan and Piran 2006; Liang et al. 2007, 2008, Fig. 5). A systematic study (Wang et al. 2015) suggested that the majority of afterglows are consistent with being roughly achromatic, suggesting that the standard external shock models can account for most of the afterglow observations.

3.3 Radio Afterglow

Compared with X-ray and optical afterglows, radio afterglow emission is more difficult to be detected with the available telescopes. About 30 % of GRBs were detected in radio, and this rate did not change before or after *Swift* was launched. Typically radio afterglow light curves at 8.5 GHz show initial rising and a peak time around 3–6 weeks in the rest frame (Fig. 6, Chandra and Frail 2012). This is consistent with the standard external forward shock model, with the peak corresponding to the crossing of the typical synchrotron frequency ν_m

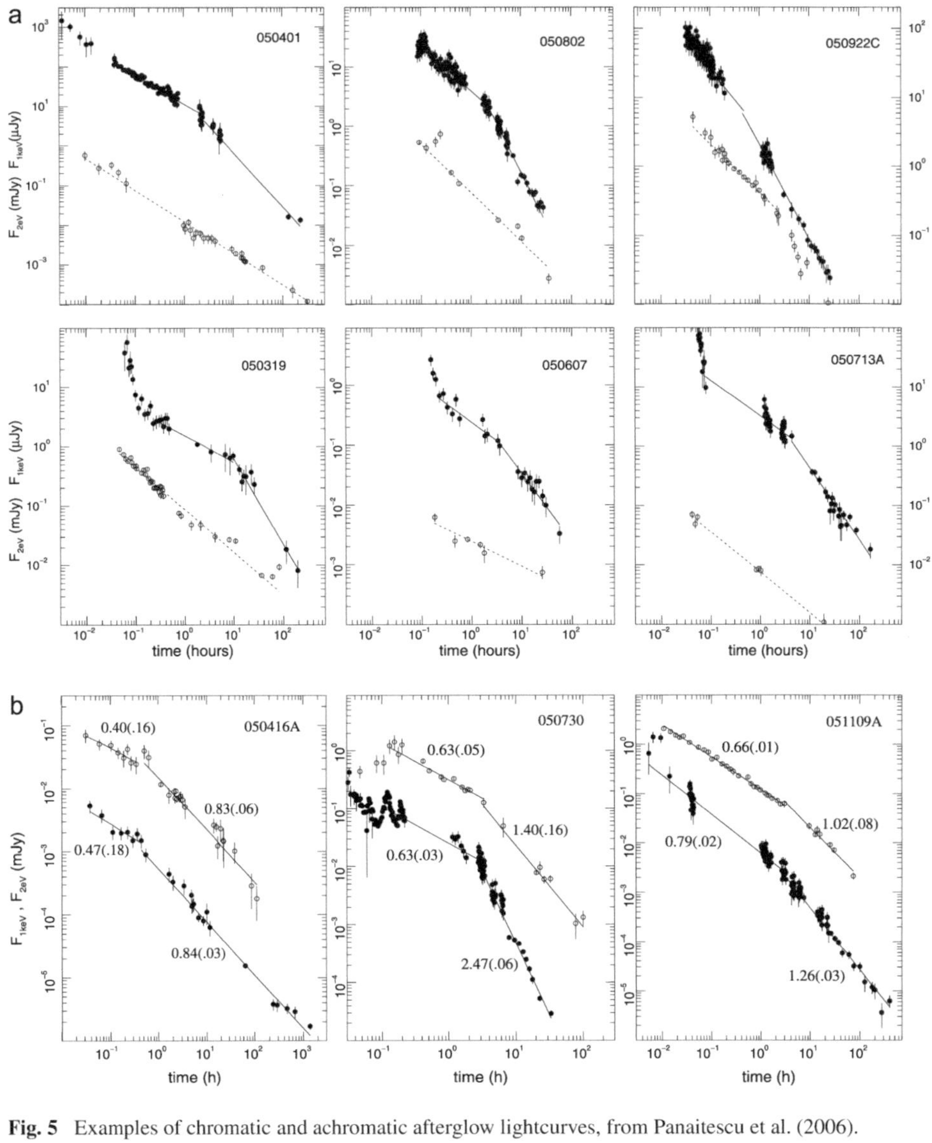

Fig. 5 Examples of chromatic and achromatic afterglow lightcurves, from Panaitescu et al. (2006).

or the self-absorption frequency ν_a in the radio band. There is a clear relationship between the detectability of the radio afterglow and the fluence of a GRB (Chandra and Frail 2012). Some well-monitored bright GRBs show an early radio flare (e.g. GRB 990123, Kulkarni et al. 1999 and GRB 130427A, Anderson et al. 2014), which is usually attributed to the emission from the reverse shock (Sari and Piran 1999b; Kobayashi and Zhang 2003a).

3.4 High Energy Afterglow

Conventionally high-energy emission refers to photons above 100 MeV. The first detection of high-energy afterglow was in the Compton Gamma-Ray Observatory (CGRO) era,

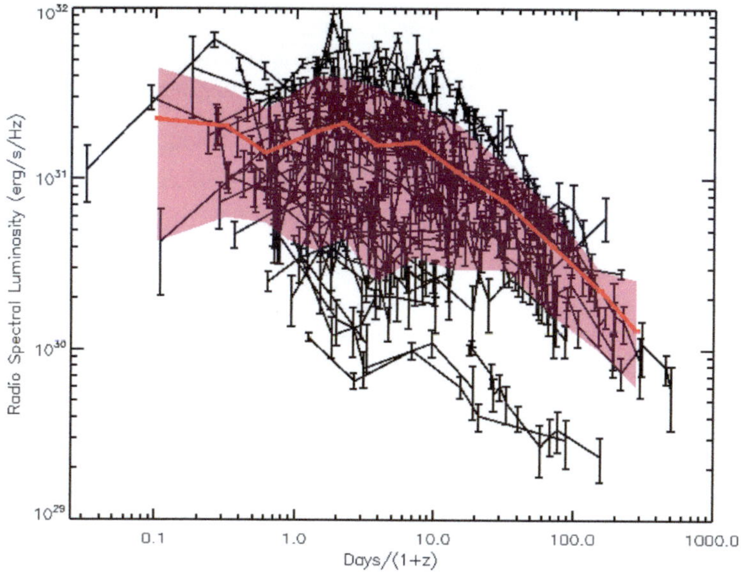

Fig. 6 The radio light curves at 8.5 GHz for the long GRBs in the rest frame time. The *red solid line* represents the mean light curve. The *pink shaded area* represents the 75 % confidence band, from Chandra and Frail (2012)

when GRB 940217 was detected in GeV energies hours after the burst was over (Hurley et al. 1994). *Fermi*/LAT detected a small fraction of GRBs in the GeV range. Most of these GeV-detected GRBs have their high-energy emission lasting longer than the prompt emission itself, suggesting an afterglow origin (e.g., Abdo et al. 2009a,b; Ackermann et al. 2010, 2011). Systematic studies suggest that the lightcurves typically show a power law decay, with the possibility of an early steep-to-shallow break transition (Ghisellini et al. 2010; Zhang et al. 2011; Ackermann et al. 2013).

3.5 Physical Understanding

The standard theoretical framework to understand GRB afterglows is the external shock model. As the relativistic ejecta (also termed as fireball) is decelerated by the circumburst medium, either a constant density ISM or a stratified stellar wind, a pair of shocks are developed: one long-lasting forward shock propagating into the medium and a short-lived reverse shock propagating into the ejecta (Meszaros and Rees 1993; Sari and Piran 1995). Electrons are accelerated from both shocks, which radiate broad-band synchrotron radiation with characteristic frequencies evolving with time. The forward shock gives rise to the broad-band long-lasting afterglow (Mészáros and Rees 1997; Sari et al. 1998; Dai and Lu 1998b; Chevalier and Li 2000; Huang et al. 2000) and a short-lived reverse shock may give additional significant optical and radio emission early on (Mészáros and Rees 1997, 1999; Sari and Piran 1999b,a; Kobayashi 2000; Zhang et al. 2003; Kobayashi and Zhang 2003a,b; Wu et al. 2003; Fan et al. 2004; Zhang and Kobayashi 2005; Resmi and Zhang 2016). For a comprehensive review of all the external shock models see Gao et al. (2013a). If the central engine is long lasting (Dai and Lu 1998c,a; Zhang and Mészáros 2001, 2002) or if the ejecta has a stratified Lorentz factor distribution (Rees and Mészáros 1998; Sari and Mészáros 2000), continuous energy injection into the blastwave (defined as the region between the forward shock and reverse shock) is possible, so that the lightcurve decay slopes are shallower. In these cases, the reverse shock is long lived. If the reverse shock is more magnetized, these long-lasting reverse shocks may be the dominant emitter of afterglow emission (Uhm and Beloborodov 2007; Genet et al. 2007; Uhm et al. 2012).

Over the years, GRB afterglow data have been extensively modeled using the external shock model. In the pre-*Swift* era, afterglow observations were carried out hours after the bursts, and the data are in general well-interpreted by the models (e.g., Panaitescu and Kumar 2001, 2002; Yost et al. 2003). *Swift* unveiled the early afterglow phase and discovered some unpredicted features (e.g. the steep decay phase, shallow decay phase, and X-ray flares). The theoretical framework in the post-*Swift* era invokes emission from both the external shock and internal energy dissipation regions (internal shocks or magnetic dissipation regions) due to the late central engine activities. Shortly after *Swift* started to detect early afterglow regularly, Zhang et al. (2006) suggested that the steep decay phase and the X-ray flares are of an internal origin, and that the shallow decay, normal decay and late steepening phases are from the external shock with the shallow decay phase being due to continuous energy injection into the blastwave. Such a simple paradigm remains valid to interpret most of GRB afterglows after a decade of observations.

In the optical band, the contamination of internal emission from late central engine activities is less, and most lightcurves can be interpreted within the standard forward shock model. Some bursts have early reverse shock component, but some others do not have evidence of reverse shock emission. The diversity may be attributed to different magnetization degree σ of the ejecta. A moderately magnetized reverse shock ($\sigma < 0.1$) is favorable to produce a bright optical flash due to the enhanced synchrotron emission in the reverse shock region (Fan et al. 2002; Zhang et al. 2003; Kumar and Panaitescu 2003), but not necessarily a bright radio flash due to synchrotron self-absorption in the forward shock region (Resmi and Zhang 2016). An ejecta with $\sigma \ll 1$ (weak synchrotron emission) or $\sigma \geq 1$ (weak or no reverse shock) is not expected to produce bright reverse shock emission in optical (Zhang and Kobayashi 2005). The early optical afterglow data of a large sample of GRBs are consistent with a simple picture that GRB ejecta have a distribution of the magnetization parameter σ among different bursts (Gao et al. 2015).

The chromatic nature of afterglow observed in some GRBs casts doubts on interpreting all the afterglow data using the standard forward shock model. At least two different emission components are needed to interpret the data. One possibility is that long-lasting central engine activities are responsible to the entire observed X-ray afterglow emission (e.g., Ghisellini et al. 2007; Kumar et al. 2008a,b), while the optical emission is from the external forward shock. A second possibility is that a long-lasting reverse shock and the forward shock are contributing emission in different energy bands (Uhm et al. 2012; Uhm and Zhang 2014b). Another possibility is to invoke a two-component jet with each component dominating emission in one observational band (e.g., de Pasquale et al. 2009). A recent systematic study (Wang et al. 2015) suggested that the fraction of GRB afterglows that are consistent with being achromatic is at least 50 % and may be as high as 90 %. This suggests that these unconventional models may not be needed to interpret the majority of GRBs.

It is the consensus in the community (Kumar and Barniol Duran 2009, 2010; Zhang et al. 2011; He et al. 2011; Liu and Wang 2011; Maxham et al. 2011) that GeV afterglow after the prompt emission phase is of an external forward shock origin, while GeV emission during the prompt emission phase may have a contribution from the internal dissipation region. The radiation mechanism is believed to be dominated by synchrotron radiation, although synchrotron self-Compton may have been detected in GRB 130427A (Fan et al. 2013; Liu et al. 2013; Tam et al. 2013; Ackermann et al. 2014).

4 Classification Schemes

Diverse GRBs with different observational properties have been observed. Classifying GRBs has been a challenging task, and many suggestions have been made. In general, these proposed classification schemes can be divided in two categories: phenomenological classification schemes and physical classification schemes.

4.1 Phenomenological Classification Schemes

The following classification schemes have been proposed in the literature:

- *Duration-hardness classification scheme:* This is the traditional long/soft vs. short/hard classification scheme (Kouveliotou et al. 1993). The main criterion is T_{90}, and hardness ratio is regarded as a supplementary criterion. The measured duration T_{90} depends on the energy band and sensitivity of the detector (Qin et al. 2013), which may bring in ambiguity of classification in some bursts. For example, in the *Swift* era, some short GRBs are detected to be followed by soft extended emission lasting 10's to ~ 100 s (Norris and Bonnell 2006). Based on T_{90} distribution information, some authors suggested a third, intermediate-duration class (e.g., Mukherjee et al. 1998; Horváth 1998; Hakkila et al. 2003). An ultra-long population was recently suggested (Gendre et al. 2013; Levan et al. 2014), even though whether it indeed form a distinct category is subject to debate (Virgili et al. 2013; Zhang et al. 2014; Boër et al. 2015; Gao and Mészáros 2015).
- *High-luminosity (HL) vs. low-luminosity (LL):* Long GRBs are sometimes classified into two sub-categories based on their luminosities, with a separation line roughly at $(10^{48}\text{–}10^{49})\ \mathrm{erg\,s^{-1}}$. The LL-GRBs typically have long durations and smooth lightcurves, and have a much larger local event rate density than the traditional HL-GRBs (Campana et al. 2006; Soderberg et al. 2006; Liang et al. 2007; Virgili et al. 2009; Sun et al. 2015).
- *GRBs, X-ray rich GRBs, and X-ray flashes:* Based on spectral properties, long GRBs are sometimes further grouped into GRBs, X-ray rich GRBs, and X-ray flashes. The separation line is arbitrary since there is no clear bimodal distribution in the hardness distribution. Studies showed that these events form a continuum in their properties (e.g., Sakamoto et al. 2005). Most LL-GRBs are also X-ray flashes.
- *Supplementary criteria:* Some prompt emission properties can be regarded as supplementary criteria for the duration classification scheme. Besides hardness ratio, spectral lags are often used to define whether a GRB is short (short GRBs typically have negligible spectral lags, e.g. Gehrels et al. 2006). A combination of isotropic energy and rest-frame E_p (the so-called $\epsilon = E_{\gamma,iso}/E_{p,z}^{5/3}$ parameter) serves as a good separator between the long and short GRBs (Lü et al. 2010). Also, the amplitude parameters (f and f_{eff}) can be very good criteria to distinguish between long and short (Lü et al. 2014).
- *Afterglow-based classification schemes:* Based on optical afterglow data, GRBs can be classified as optically bright and optically dark GRBs (Jakobsson et al. 2004; Rol et al. 2005). Based on the X-ray afterglow data one may classify GRBs into those with X-ray flares and those without, or those following a canonical lightcurve and those following a single power law decay.

4.2 Physical Classification Schemes

The ultimate goal of studying astrophysical phenomenology is to uncover the physical nature of the sources. Based on observational properties and theoretical modeling of GRBs over the years, GRBs are physically classified into different categories.

- *Massive star (Type II) vs. Compact star (Type I) GRBs:* The two dominant categories of GRBs are the ones associated with deaths of massive stars and the ones not. The former category roughly correspond to the long duration GRBs, and latter roughly correspond to the short duration GRBs. However, counter examples do exist, e.g. SN-less long duration GRBs (GRB 060614 and GRB 060505) that are likely associated with compact stars (Gehrels et al. 2006; Gal-Yam et al. 2006; Fynbo et al. 2006; Della Valle et al. 2006), and short GRBs that are likely associated with massive stars (e.g. GRB 090426, Levesque et al. 2010; Antonelli et al. 2009; Xin et al. 2011; Thöne et al. 2011). Zhang et al. (2007b) and Zhang (2006) proposed the Type I vs. Type II classification scheme in parallel with the short vs. long scheme, and Zhang et al. (2009) elaborated on the multiple observational criteria to identify the physical category of a certain GRB. Observationally, Type I and Type II GRBs show overlapping behaviors in all individual observational criteria (Li et al. 2016b), suggesting that multi-criteria are needed to perform classification. Zhang et al. (2009) proposed a flowchart to apply multi-criteria to classify GRBs, which was successfully applied to GRBs before 2011 (Kann et al. 2010, 2011).
- *Successful vs. choked jets:* Within long (Type II) GRBs, bursts may be further classified into successful and choked jets. The former apply to the traditional HL-GRBs, while the latter may apply to some LL-GRBs such as GRB 060218 (e.g. Nakar and Sari 2012; Bromberg et al. 2012).
- *Different progenitor stars within Type II GRBs:* For GRBs related to massive stars, there might exist different progenitor star systems. The leading type of progenitor star is Wolf-Rayet stars, rapidly rotating massive stars with hydrogen and helium envelope lost before core collapse (Woosley 1993; MacFadyen and Woosley 1999). The association of long GRBs with Type Ic SNe (no H and He lines) is consistent with such a progenitor picture. Ultra-long GRBs, on the other hand, may call for a different progenitor star system with a larger size, e.g. blue supergiants (Mészáros and Rees 2001; Kashiyama et al. 2013; Greiner et al. 2015), but a strong case of a new type of progenitor is not established. Besides single star collapsars (Woosley and Bloom 2006), a variety of binary progenitor star systems have been also discussed (Fryer et al. 1999; Ruffini et al. 2016). No smoking-gun signature has been identified to judge whether long GRB progenitors are in single or binary star systems.
- *Different progenitor stars within Type I GRBs:* For GRBs not related to massive stars, the leading models invoke mergers of neutron star (NS) binaries (Paczynski 1986; Eichler et al. 1989; Paczynski 1991; Narayan et al. 1992; Rosswog et al. 2003; Rezzolla et al. 2011), either two NSs (NS-NS), or a NS and a black hole (NS-BH). Other types of progenitors (e.g. accretion-induced-collapse, AIC, of NSs, Qin et al. 1998; Dermer and Atoyan 2006) are also possible. A merger-origin of short GRBs may be verified in the future when gravitational wave chirp signals are discovered to be associated with short GRBs. It is interesting to mention the putative short GRB signal detected with *Fermi*/GBM that lagged behind the first gravitational wave event GW 150914 by 0.4 s (Connaughton et al. 2016). If this is the case, BH-BH mergers may also make short GRBs, which may require unconventional mechanisms to produce short GRBs (e.g. Zhang 2016; Loeb 2016; Perna et al. 2016; Li et al. 2016a; Murase et al. 2016; Yamazaki et al. 2016; Veres et al. 2016).
- *GRBs with different central engines:* For both long and short GRBs, two types of central engines have been widely discussed in the literature: hyper-accreting black holes (Narayan et al. 1992; Woosley 1993; Popham et al. 1999) and rapidly rotating magnetars (Usov 1992; Dai and Lu 1998c,a; Zhang and Mészáros 2001; Metzger et al. 2011). Signatures of magnetar spindown have been seen in both long and short GRBs (Lü and Zhang 2014; Lü et al. 2015, and references therein). In particular, a possible magnetar

central engine for short GRBs (Dai et al. 2006; Gao and Fan 2006; Fan and Xu 2006; Metzger et al. 2008), if verified by future gravitational wave data, would have profound implications in constraining poorly known neutron star equation of state (Lasky et al. 2014; Lü et al. 2015; Gao et al. 2016; Li et al. 2016) and in searching for electromagnetic counterparts of gravitational wave events (Zhang 2013; Gao et al. 2013b; Yu et al. 2013; Gao et al. 2016).
- *Other bursts no longer defined as GRBs:* Historically several phenomena were initially confused as GRBs and later found to be other distinct phenomena that also emit bursts of γ-rays. An early example was the so-called soft-gamma-ray repeaters (SGRs), now known as bursts from Galactic magnetars. Some SGRBs emit giant flares characterized with a short hard spike followed by oscillating X-ray pulses (e.g. Palmer et al. 2005). The short hard spikes of SGR giant flares can be detectable up to 80 Mpc, which may be confused as short GRBs. A systematic search suggested that these giant-flare-origin short GRBs may comprise less than 5 % of the short GRB population (Tanvir et al. 2005). One candidate of such events, GRB 051103, has been discovered (Frederiks et al. 2007). Another example was "GRB 110328" which was later identified as a tidal disruption event that launched a jet beaming towards earth (Bloom et al. 2011; Burrows et al. 2011) and renamed as Sw 1644+57. Finally, it has been long proposed that evaporation of primordial black holes with mass $\sim 10^{17}$ g (Hawking 1975) may be able to produce ultra-short GRBs. However, no robust evidence for the existence of these bursts has been collected so far (Ukwatta et al. 2016, and references therein).

5 Summary

After nearly half century of observations, rich data have been accumulated for GRB prompt emission and afterglow. Our physical understanding of GRBs, including their progenitor systems, central engines, jet properties (composition, Lorentz factor), energy dissipation and radiation mechanisms, as well as the interaction between jet and ambient medium, has been greatly advanced. Yet, many open questions remain, which call for more dedicated multi-wavelength, multi-messenger observations that cover the full temporal and spectral windows in the prompt emission and afterglow phases.

Acknowledgements BZ acknowledges the NASA ADAP (NNX14AF85G) and ATP (NNX15AK85G) programs for supports. The research at Guangxi University was supported by the National Basic Research Program (973 Program) of China 2014CB845800, the National Natural Science Foundation of China (Grant Nos. 11533003, 11603006, U1331202), One-Hundred-Talents Program of Guangxi colleges, Scientific Research Foundation of GuangXi University (Grant No. XGZ150299), and Guangxi Science Foundation (Grant Nos. 2013GXNSFFA019001, 2016GXNSFCB380005).

References

M.G. Aartsen, K. Abraham, M. Ackermann, J. Adams, J.A. Aguilar, M. Ahlers, M. Ahrens, D. Altmann, T. Anderson, I. Ansseau et al., An all-sky search for three flavors of neutrinos from gamma-ray bursts with the IceCube Neutrino Observatory. Astrophys. J. **824**, 115 (2016). doi:10.3847/0004-637X/824/2/115

R. Abbasi, Y. Abdou, T. Abu-Zayyad, M. Ackermann, J. Adams, J.A. Aguilar, M. Ahlers, D. Altmann, K. Andeen, J. Auffenberg, et al., An absence of neutrinos associated with cosmic-ray acceleration in γ-ray bursts. Nature **484**, 351–354 (2012). doi:10.1038/nature11068

A.A. Abdo, M. Ackermann, M. Ajello, K. Asano, W.B. Atwood, M. Axelsson, L. Baldini, J. Ballet, G. Barbiellini, M.G. Baring, D. Bastieri, K. Bechtol, R. Bellazzini, B. Berenji, P.N. Bhat, E. Bissaldi, R.D. Blandford, E.D. Bloom, E. Bonamente, A.W. Borgland, A. Bouvier, J. Bregeon, A. Brez, M.S. Briggs, M. Brigida, P. Bruel, J.M. Burgess, D.N. Burrows, S. Buson, G.A. Caliandro, R.A. Cameron, P.A. Caraveo, J.M. Casandjian, C. Cecchi, Ö. Çelik, A. Chekhtman, C.C. Cheung, J. Chiang, S. Ciprini, R. Claus, J. Cohen-Tanugi, L.R. Cominsky, V. Connaughton, J. Conrad, S. Cutini, V. d'Elia, C.D. Dermer, A. de Angelis, F. de Palma, S.W. Digel, B.L. Dingus, E.d.C.e. Silva, P.S. Drell, R. Dubois, D. Dumora, C. Farnier, C. Favuzzi, S.J. Fegan, J. Finke, G. Fishman, W.B. Focke, P. Fortin, M. Frailis, Y. Fukazawa, S. Funk, P. Fusco, F. Gargano, N. Gehrels, S. Germani, G. Giavitto, B. Giebels, N. Giglietto, F. Giordano, T. Glanzman, G. Godfrey, A. Goldstein, J. Granot, J. Greiner, I.A. Grenier, J.E. Grove, L. Guillemot, S. Guiriec, Y. Hanabata, A.K. Harding, M. Hayashida, E. Hays, D. Horan, R.E. Hughes, M.S. Jackson, G. Jóhannesson, A.S. Johnson, R.P. Johnson, W.N. Johnson, T. Kamae, H. Katagiri, J. Kataoka, N. Kawai, M. Kerr, R.M. Kippen, J. Knödlseder, D. Kocevski, N. Komin, C. Kouveliotou, M. Kuss, J. Lande, L. Latronico, M. Lemoine-Goumard, F. Longo, F. Loparco, B. Lott, M.N. Lovellette, P. Lubrano, G.M. Madejski, A. Makeev, M.N. Mazziotta, S. McBreen, J.E. McEnery, S. McGlynn, C. Meegan, P. Mészáros, C. Meurer, P.F. Michelson, W. Mitthumsiri, T. Mizuno, A.A. Moiseev, C. Monte, M.E. Monzani, E. Moretti, A. Morselli, I.V. Moskalenko, S. Murgia, T. Nakamori, P.L. Nolan, J.P. Norris, E. Nuss, M. Ohno, T. Ohsugi, N. Omodei, E. Orlando, J.F. Ormes, W.S. Paciesas, D. Paneque, J.H. Panetta, V. Pelassa, M. Pepe, M. Pesce-Rollins, V. Petrosian, F. Piron, T.A. Porter, R. Preece, S. Rainò, R. Rando, A. Rau, M. Razzano, S. Razzaque, A. Reimer, O. Reimer, T. Reposeur, S. Ritz, L.S. Rochester, A.Y. Rodriguez, P.W.A. Roming, M. Roth, F. Ryde, H.F.-W. Sadrozinski, D. Sanchez, A. Sander, P.M. Saz Parkinson, J.D. Scargle, T.L. Schalk, C. Sgrò, E.J. Siskind, P.D. Smith, P. Spinelli, M. Stamatikos, F.W. Stecker, G. Stratta, M.S. Strickman, D.J. Suson, C.A. Swenson, H. Tajima, H. Takahashi, T. Tanaka, J.B. Thayer, J.G. Thayer, D.J. Thompson, L. Tibaldo, D.F. Torres, G. Tosti, A. Tramacere, Y. Uchiyama, T. Uehara, T.L. Usher, A.J. van der Horst, V. Vasileiou, N. Vilchez, V. Vitale, A. von Kienlin, A.P. Waite, P. Wang, C. Wilson-Hodge, B.L. Winer, K.S. Wood, R. Yamazaki, T. Ylinen, M. Ziegler, Fermi observations of GRB 090902B: a distinct spectral component in the prompt and delayed emission. Astrophys. J. Lett. **706**, 138–144 (2009a). doi:10.1088/0004-637X/706/1/L138

A.A. Abdo, M. Ackermann, M. Arimoto, K. Asano, W.B. Atwood, M. Axelsson, L. Baldini, J. Ballet, D.L. Band, G. Barbiellini et al., Fermi observations of high-energy gamma-ray emission from GRB 080916C. Science **323**, 1688 (2009b). doi:10.1126/science.1169101

M. Ackermann, K. Asano, W.B. Atwood, M. Axelsson, L. Baldini, J. Ballet, G. Barbiellini, M.G. Baring, D. Bastieri, K. Bechtol, R. Bellazzini, B. Berenji, P.N. Bhat, E. Bissaldi, R.D. Blandford, E.D. Bloom, E. Bonamente, A.W. Borgland, A. Bouvier, J. Bregeon, A. Brez, M.S. Briggs, M. Brigida, P. Bruel, S. Buson, G.A. Caliandro, R.A. Cameron, P.A. Caraveo, S. Carrigan, J.M. Casandjian, C. Cecchi, Ö. Çelik, E. Charles, J. Chiang, S. Ciprini, R. Claus, J. Cohen-Tanugi, V. Connaughton, J. Conrad, C.D. Dermer, F. de Palma, B.L. Dingus, E.d.C.e. Silva, P.S. Drell, R. Dubois, D. Dumora, C. Farnier, C. Favuzzi, S.J. Fegan, J. Finke, W.B. Focke, M. Frailis, Y. Fukazawa, P. Fusco, F. Gargano, D. Gasparrini, N. Gehrels, S. Germani, N. Giglietto, F. Giordano, T. Glanzman, G. Godfrey, J. Granot, I.A. Grenier, M.-H. Grondin, J.E. Grove, S. Guiriec, D. Hadasch, A.K. Harding, E. Hays, D. Horan, R.E. Hughes, G. Jóhannesson, W.N. Johnson, T. Kamae, H. Katagiri, J. Kataoka, N. Kawai, R.M. Kippen, J. Knödlseder, D. Kocevski, C. Kouveliotou, M. Kuss, J. Lande, L. Latronico, M. Lemoine-Goumard, M. Llena Garde, F. Longo, F. Loparco, B. Lott, M.N. Lovellette, P. Lubrano, A. Makeev, M.N. Mazziotta, J.E. McEnery, S. McGlynn, C. Meegan, P. Mészáros, P.F. Michelson, W. Mitthumsiri, T. Mizuno, A.A. Moiseev, C. Monte, M.E. Monzani, E. Moretti, A. Morselli, I.V. Moskalenko, S. Murgia, H. Nakajima, T. Nakamori, P.L. Nolan, J.P. Norris, E. Nuss, M. Ohno, T. Ohsugi, N. Omodei, E. Orlando, J.F. Ormes, M. Ozaki, W.S. Paciesas, D. Paneque, J.H. Panetta, D. Parent, V. Pelassa, M. Pepe, M. Pesce-Rollins, F. Piron, R. Preece, S. Rainò, R. Rando, M. Razzano, S. Razzaque, A. Reimer, S. Ritz, A.Y. Rodriguez, M. Roth, F. Ryde, H.F.-W. Sadrozinski, A. Sander, J.D. Scargle, T.L. Schalk, C. Sgrò, E.J. Siskind, P.D. Smith, G. Spandre, P. Spinelli, M. Stamatikos, F.W. Stecker, M.S. Strickman, D.J. Suson, H. Tajima, H. Takahashi, T. Takahashi, T. Tanaka, J.B. Thayer, J.G. Thayer, D.J. Thompson, L. Tibaldo, K. Toma, D.F. Torres, G. Tosti, A. Tramacere, Y. Uchiyama, T. Uehara, T.L. Usher, A.J. van der Horst, V. Vasileiou, N. Vilchez, V. Vitale, A. von Kienlin, A.P. Waite, P. Wang, C. Wilson-Hodge, B.L. Winer, X.F. Wu, R. Yamazaki, Z. Yang, T. Ylinen, M. Ziegler, Fermi observations of GRB 090510: a short-hard gamma-ray burst with an additional, hard power-law component from 10 keV to GeV energies. Astrophys. J. **716**, 1178–1190 (2010). doi:10.1088/0004-637X/716/2/1178

M. Ackermann, M. Ajello, K. Asano, M. Axelsson, L. Baldini, J. Ballet, G. Barbiellini, M.G. Baring, D. Bastieri, K. Bechtol, R. Bellazzini, B. Berenji, P.N. Bhat, E. Bissaldi, R.D. Blandford, E. Bonamente, A.W. Borgland, A. Bouvier, J. Bregeon, A. Brez, M.S. Briggs, M. Brigida, P. Bruel, R. Buehler, S. Buson, G.A. Caliandro, R.A. Cameron, P.A. Caraveo, S. Carrigan, J.M. Casandjian, C. Cecchi, Ö. Çelik, V. Chaplin, E. Charles, A. Chekhtman, J. Chiang, S. Ciprini, R. Claus, J. Cohen-Tanugi, V. Connaughton, J. Conrad, S. Cutini, C.D. Dermer, A. de Angelis, F. de Palma, B.L. Dingus, E.d.C.e. Silva, P.S. Drell, R. Dubois, C. Favuzzi, S.J. Fegan, E.C. Ferrara, W.B. Focke, M. Frailis, Y. Fukazawa, S. Funk, P. Fusco, F. Gargano, D. Gasparrini, N. Gehrels, S. Germani, N. Giglietto, F. Giordano, M. Giroletti, T. Glanzman, G. Godfrey, A. Goldstein, J. Granot, J. Greiner, I.A. Grenier, J.E. Grove, S. Guiriec, D. Hadasch, Y. Hanabata, A.K. Harding, K. Hayashi, M. Hayashida, E. Hays, D. Horan, R.E. Hughes, R. Itoh, G. Jóhannesson, A.S. Johnson, W.N. Johnson, T. Kamae, H. Katagiri, J. Kataoka, R.M. Kippen, J. Knödlseder, D. Kocevski, C. Kouveliotou, M. Kuss, J. Lande, L. Latronico, S.-H. Lee, M. Llena Garde, F. Longo, F. Loparco, M.N. Lovellette, P. Lubrano, A. Makeev, M.N. Mazziotta, S. McBreen, J.E. McEnery, S. McGlynn, C. Meegan, J. Mehault, P. Mészáros, P.F. Michelson, T. Mizuno, C. Monte, M.E. Monzani, E. Moretti, A. Morselli, I.V. Moskalenko, S. Murgia, H. Nakajima, T. Nakamori, M. Naumann-Godo, S. Nishino, P.L. Nolan, J.P. Norris, E. Nuss, M. Ohno, T. Ohsugi, A. Okumura, N. Omodei, E. Orlando, J.F. Ormes, M. Ozaki, W.S. Paciesas, D. Paneque, J.H. Panetta, D. Parent, V. Pelassa, M. Pepe, M. Pesce-Rollins, V. Petrosian, F. Piron, T.A. Porter, R. Preece, J.L. Racusin, S. Rainò, R. Rando, A. Rau, M. Razzano, S. Razzaque, A. Reimer, O. Reimer, T. Reposeur, L.C. Reyes, J. Ripken, S. Ritz, M. Roth, F. Ryde, H.F.-W. Sadrozinski, A. Sander, J.D. Scargle, T.L. Schalk, C. Sgrò, E.J. Siskind, P.D. Smith, G. Spandre, P. Spinelli, M. Stamatikos, F.W. Stecker, M.S. Strickman, D.J. Suson, H. Tajima, H. Takahashi, T. Tanaka, Y. Tanaka, J.B. Thayer, J.G. Thayer, L. Tibaldo, D. Tierney, K. Toma, D.F. Torres, G. Tosti, A. Tramacere, Y. Uchiyama, T. Uehara, T.L. Usher, J. Vandenbroucke, A.J. van der Horst, V. Vasileiou, N. Vilchez, V. Vitale, A. von Kienlin, A.P. Waite, P. Wang, C. Wilson-Hodge, B.L. Winer, K.S. Wood, X.F. Wu, R. Yamazaki, Z. Yang, T. Ylinen, M. Ziegler, Detection of a spectral break in the extra hard component of GRB 090926A. Astrophys. J. **729**, 114 (2011). doi:10.1088/0004-637X/729/2/114

M. Ackermann, M. Ajello, K. Asano, L. Baldini, G. Barbiellini, M.G. Baring, D. Bastieri, R. Bellazzini, R.D. Blandford, E. Bonamente, A.W. Borgland, E. Bottacini, J. Bregeon, M. Brigida, P. Bruel, R. Buehler, S. Buson, G.A. Caliandro, R.A. Cameron, P.A. Caraveo, C. Cecchi, E. Charles, R.C.G. Chaves, A. Chekhtman, J. Chiang, S. Ciprini, R. Claus, J. Cohen-Tanugi, J. Conrad, S. Cutini, F. D'Ammando, A. de Angelis, F. de Palma, C.D. Dermer, E.d.C.e. Silva, P.S. Drell, A. Drlica-Wagner, C. Favuzzi, S.J. Fegan, W.B. Focke, A. Franckowiak, Y. Fukazawa, P. Fusco, F. Gargano, D. Gasparrini, N. Gehrels, N. Giglietto, F. Giordano, M. Giroletti, T. Glanzman, G. Godfrey, J. Granot, J. Greiner, I.A. Grenier, J.E. Grove, S. Guiriec, D. Hadasch, Y. Hanabata, M. Hayashida, E. Hays, R.E. Hughes, M.S. Jackson, T. Jogler, G. Jóhannesson, A.S. Johnson, J. Knödlseder, D. Kocevski, M. Kuss, J. Lande, S. Larsson, L. Latronico, F. Longo, F. Loparco, M.N. Lovellette, P. Lubrano, M.N. Mazziotta, J.E. McEnery, J. Mehault, P. Mészáros, P.F. Michelson, W. Mitthumsiri, T. Mizuno, C. Monte, M.E. Monzani, E. Moretti, A. Morselli, I.V. Moskalenko, S. Murgia, M. Naumann-Godo, J.P. Norris, E. Nuss, T. Nymark, M. Ohno, T. Ohsugi, N. Omodei, M. Orienti, E. Orlando, D. Paneque, J.S. Perkins, M. Pesce-Rollins, F. Piron, G. Pivato, J.L. Racusin, S. Rainò, R. Rando, M. Razzano, S. Razzaque, A. Reimer, O. Reimer, C. Romoli, M. Roth, F. Ryde, D.A. Sanchez, C. Sgrò, E.J. Siskind, E. Sonbas, P. Spinelli, M. Stamatikos, H. Takahashi, T. Tanaka, J.G. Thayer, J.B. Thayer, L. Tibaldo, M. Tinivella, G. Tosti, E. Troja, T.L. Usher, J. Vandenbroucke, V. Vasileiou, G. Vianello, V. Vitale, A.P. Waite, B.L. Winer, K.S. Wood, Z. Yang, D. Gruber, P.N. Bhat, E. Bissaldi, M.S. Briggs, J.M. Burgess, V. Connaughton, S. Foley, R.M. Kippen, C. Kouveliotou, S. McBreen, S. McGlynn, W.S. Paciesas, V. Pelassa, R. Preece, A. Rau, A.J. van der Horst, A. von Kienlin, D.A. Kann, R. Filgas, S. Klose, T. Krühler, A. Fukui, T. Sako, P.J. Tristram, S.R. Oates, T.N. Ukwatta, O. Littlejohns, Multiwavelength observations of GRB 110731A: GeV emission from onset to afterglow. Astrophys. J. **763**, 71 (2013). doi:10.1088/0004-637X/763/2/71

M. Ackermann, M. Ajello, K. Asano, W.B. Atwood, M. Axelsson, L. Baldini, J. Ballet, G. Barbiellini, M.G. Baring, D. Bastieri, K. Bechtol, R. Bellazzini, E. Bissaldi, E. Bonamente, J. Bregeon, M. Brigida, P. Bruel, R. Buehler, J.M. Burgess, S. Buson, G.A. Caliandro, R.A. Cameron, P.A. Caraveo, C. Cecchi, V. Chaplin, E. Charles, A. Chekhtman, C.C. Cheung, J. Chiang, G. Chiaro, S. Ciprini, R. Claus, W. Cleveland, J. Cohen-Tanugi, A. Collazzi, L.R. Cominsky, V. Connaughton, J. Conrad, S. Cutini, F. D'Ammando, A. de Angelis, M. DeKlotz, F. de Palma, C.D. Dermer, R. Desiante, A. Diekmann, L. Di Venere, P.S. Drell, A. Drlica-Wagner, C. Favuzzi, S.J. Fegan, E.C. Ferrara, J. Finke, G. Fitzpatrick, W.B. Focke, A. Franckowiak, Y. Fukazawa, S. Funk, P. Fusco, F. Gargano, N. Gehrels, S. Germani, M. Gibby, N. Giglietto, M. Giles, F. Giordano, M. Giroletti, G. Godfrey, J. Granot, I.A. Grenier, J.E. Grove, D. Gruber, S. Guiriec, D. Hadasch, Y. Hanabata, A.K. Harding, M. Hayashida, E. Hays, D. Horan, R.E. Hughes, Y. Inoue, T. Jogler, G. Jóhannesson, W.N. Johnson, T. Kawano, J. Knödlseder, D. Kocevski, M. Kuss, J. Lande, S. Larsson, L. Latronico, F. Longo, F. Loparco, M.N. Lovellette, P. Lubrano, M. Mayer, M.N. Mazziotta, J.E. McEnery, P.F. Michelson, T. Mizuno, A.A. Moiseev, M.E. Monzani, E. Moretti, A. Morselli, I.V. Moskalenko, S. Murgia, R. Nemmen, E. Nuss, M. Ohno, T. Ohsugi, A. Okumura, N. Omodei, M. Orienti, D. Paneque, V. Pelassa, J.S. Perkins, M. Pesce-Rollins,

V. Petrosian, F. Piron, G. Pivato, T.A. Porter, J.L. Racusin, S. Rainò, R. Rando, M. Razzano, S. Razzaque, A. Reimer, O. Reimer, S. Ritz, M. Roth, F. Ryde, A. Sartori, P.M.S. Parkinson, J.D. Scargle, A. Schulz, C. Sgrò, E.J. Siskind, E. Sonbas, G. Spandre, P. Spinelli, H. Tajima, H. Takahashi, J.G. Thayer, J.B. Thayer, D.J. Thompson, L. Tibaldo, M. Tinivella, D.F. Torres, G. Tosti, E. Troja, T.L. Usher, J. Vandenbroucke, V. Vasileiou, G. Vianello, V. Vitale, B.L. Winer, K.S. Wood, R. Yamazaki, G. Younes, H.-F. Yu, S.J. Zhu, P.N. Bhat, M.S. Briggs, D. Byrne, S. Foley, A. Goldstein, P. Jenke, R.M. Kippen, C. Kouveliotou, S. McBreen, C. Meegan, W.S. Paciesas, R. Preece, A. Rau, D. Tierney, A.J. van der Horst, A. von Kienlin, C. Wilson-Hodge, S. Xiong, G. Cusumano, V. La Parola, J.R. Cummings, Fermi-LAT observations of the gamma-ray burst GRB 130427A. Science **343**, 42–47 (2014). doi:10.1126/science.1242353

C. Akerlof, R. Balsano, S. Barthelmy, J. Bloch, P. Butterworth, D. Casperson, T. Cline, S. Fletcher, F. Frontera, G. Gisler, J. Heise, J. Hills, R. Kehoe, B. Lee, S. Marshall, T. McKay, R. Miller, L. Piro, W. Priedhorsky, J. Szymanski, J. Wren, Observation of contemporaneous optical radiation from a γ-ray burst. Nature **398**, 400–402 (1999). doi:10.1038/18837

G.E. Anderson, A.J. van der Horst, T.D. Staley, R.P. Fender, R.A.M.J. Wijers, A.M.M. Scaife, C. Rumsey, D.J. Titterington, A. Rowlinson, R.D.E. Saunders, Probing the bright radio flare and afterglow of GRB 130427A with the Arcminute Microkelvin Imager. Mon. Not. R. Astron. Soc. Lett. **440**, 2059–2065 (2014). doi:10.1093/mnras/stu478

L.A. Antonelli, P. D'Avanzo, R. Perna, L. Amati, S. Covino, S. Cutini, V. D'Elia, S. Gallozzi, A. Grazian, E. Palazzi, S. Piranomonte, A. Rossi, S. Spiro, L. Stella, V. Testa, G. Chincarini, A. di Paola, F. Fiore, D. Fugazza, E. Giallongo, E. Maiorano, N. Masetti, F. Pedichini, R. Salvaterra, G. Tagliaferri, S. Vergani, GRB 090426: the farthest short gamma-ray burst? Astron. Astrophys. **507**, 45–48 (2009). doi:10.1051/0004-6361/200913062

M. Axelsson, L. Baldini, G. Barbiellini, M.G. Baring, R. Bellazzini, J. Bregeon, M. Brigida, P. Bruel, R. Buehler, G.A. Caliandro, R.A. Cameron, P.A. Caraveo, C. Cecchi, R.C.G. Chaves, A. Chekhtman, J. Chiang, R. Claus, J. Conrad, S. Cutini, F. D'Ammando, F. de Palma, C.D. Dermer, E.d.C.e. Silva, P.S. Drell, C. Favuzzi, S.J. Fegan, E.C. Ferrara, W.B. Focke, Y. Fukazawa, P. Fusco, F. Gargano, D. Gasparrini, N. Gehrels, S. Germani, N. Giglietto, M. Giroletti, G. Godfrey, S. Guiriec, D. Hadasch, Y. Hanabata, M. Hayashida, X. Hou, S. Iyyani, M.S. Jackson, D. Kocevski, M. Kuss, J. Larsson, S. Larsson, F. Longo, F. Loparco, C. Lundman, M.N. Mazziotta, J.E. McEnery, T. Mizuno, M.E. Monzani, E. Moretti, A. Morselli, S. Murgia, E. Nuss, T. Nymark, M. Ohno, N. Omodei, M. Pesce-Rollins, F. Piron, G. Pivato, J.L. Racusin, S. Rainò, M. Razzano, S. Razzaque, A. Reimer, M. Roth, F. Ryde, D.A. Sanchez, C. Sgrò, E.J. Siskind, G. Spandre, P. Spinelli, M. Stamatikos, L. Tibaldo, M. Tinivella, T.L. Usher, J. Vandenbroucke, V. Vasileiou, G. Vianello, V. Vitale, A.P. Waite, B.L. Winer, K.S. Wood, J.M. Burgess, P.N. Bhat, E. Bissaldi, M.S. Briggs, V. Connaughton, G. Fishman, G. Fitzpatrick, S. Foley, D. Gruber, R.M. Kippen, C. Kouveliotou, P. Jenke, S. McBreen, S. McGlynn, C. Meegan, W.S. Paciesas, V. Pelassa, R. Preece, D. Tierney, A. von Kienlin, C. Wilson-Hodge, S. Xiong, A. Pe'er, GRB110721A: an extreme peak energy and signatures of the photosphere. Astrophys. J. Lett. **757**, 31 (2012). doi:10.1088/2041-8205/757/2/L31

D. Band, J. Matteson, L. Ford, B. Schaefer, D. Palmer, B. Teegarden, T. Cline, M. Briggs, W. Paciesas, G. Pendleton, G. Fishman, C. Kouveliotou, C. Meegan, R. Wilson, P. Lestrade, BATSE observations of gamma-ray burst spectra. I—Spectral diversity. Astrophys. J. **413**, 281–292 (1993). doi:10.1086/172995

S.D. Barthelmy, J.K. Cannizzo, N. Gehrels, G. Cusumano, V. Mangano, P.T. O'Brien, S. Vaughan, B. Zhang, D.N. Burrows, S. Campana, G. Chincarini, M.R. Goad, C. Kouveliotou, P. Kumar, P. Mészáros, J.A. Nousek, J.P. Osborne, A. Panaitescu, J.N. Reeves, T. Sakamoto, G. Tagliaferri, R.A.M.J. Wijers, Discovery of an afterglow extension of the prompt phase of two gamma-ray bursts observed by Swift. Astrophys. J. Lett. **635**, 133–136 (2005). doi:10.1086/499432

A.M. Beloborodov, Collisional mechanism for gamma-ray burst emission. Mon. Not. R. Astron. Soc. Lett. **407**, 1033–1047 (2010). doi:10.1111/j.1365-2966.2010.16770.x

A.M. Beloborodov, B.E. Stern, R. Svensson, Power density spectra of gamma-ray bursts. Astrophys. J. **535**, 158–166 (2000). doi:10.1086/308836

J.S. Bloom, D. Giannios, B.D. Metzger, S.B. Cenko, D.A. Perley, N.R. Butler, N.R. Tanvir, A.J. Levan, P.T. O'Brien, L.E. Strubbe, F. De Colle, E. Ramirez-Ruiz, W.H. Lee, S. Nayakshin, E. Quataert, A.R. King, A. Cucchiara, J. Guillochon, G.C. Bower, A.S. Fruchter, A.N. Morgan, A.J. van der Horst, A possible relativistic jetted outburst from a massive black hole fed by a tidally disrupted star. Science **333**, 203 (2011). doi:10.1126/science.1207150

M. Boër, B. Gendre, G. Stratta, Are ultra-long gamma-ray bursts different? Astrophys. J. **800**, 16 (2015). doi:10.1088/0004-637X/800/1/16

O. Bromberg, E. Nakar, T. Piran, R. Sari, An observational imprint of the collapsar model of long gamma-ray bursts. Astrophys. J. **749**, 110 (2012). doi:10.1088/0004-637X/749/2/110

D. Burlon, G. Ghirlanda, G. Ghisellini, D. Lazzati, L. Nava, M. Nardini, A. Celotti, Precursors in Swift gamma ray bursts with redshift. Astrophys. J. Lett. **685**, 19 (2008). doi:10.1086/592350

D.N. Burrows, P. Romano, A. Falcone, S. Kobayashi, B. Zhang, A. Moretti, P.T. O'Brien, M.R. Goad, S. Campana, K.L. Page, L. Angelini, S. Barthelmy, A.P. Beardmore, M. Capalbi, G. Chincarini, J. Cummings, G. Cusumano, D. Fox, P. Giommi, J.E. Hill, J.A. Kennea, H. Krimm, V. Mangano, F. Marshall, P. Mészáros, D.C. Morris, J.A. Nousek, J.P. Osborne, C. Pagani, M. Perri, G. Tagliaferri, A.A. Wells, S. Woosley, N. Gehrels, Bright X-ray flares in gamma-ray burst afterglows. Science **309**, 1833–1835 (2005). doi:10.1126/science.1116168

D.N. Burrows, J.A. Kennea, G. Ghisellini, V. Mangano, B. Zhang, K.L. Page, M. Eracleous, P. Romano, T. Sakamoto, A.D. Falcone, J.P. Osborne, S. Campana, A.P. Beardmore, A.A. Breeveld, M.M. Chester, R. Corbet, S. Covino, J.R. Cummings, P. D'Avanzo, V. D'Elia, P. Esposito, P.A. Evans, D. Fugazza, J.M. Gelbord, K. Hiroi, S.T. Holland, K.Y. Huang, M. Im, G. Israel, Y. Jeon, Y.-B. Jeon, H.D. Jun, N. Kawai, J.H. Kim, H.A. Krimm, F.E. Marshall, P. Mészáros, H. Negoro, N. Omodei, W.-K. Park, J.S. Perkins, M. Sugizaki, H.-I. Sung, G. Tagliaferri, E. Troja, Y. Ueda, Y. Urata, R. Usui, L.A. Antonelli, S.D. Barthelmy, G. Cusumano, P. Giommi, A. Melandri, M. Perri, J.L. Racusin, B. Sbarufatti, M.H. Siegel, N. Gehrels, Relativistic jet activity from the tidal disruption of a star by a massive black hole. Nature **476**, 421–424 (2011). doi:10.1038/nature10374

N.R. Butler, D. Kocevski, X-ray hardness evolution in GRB afterglows and flares: late-time GRB activity without N_H variations. Astrophys. J. **663**, 407–419 (2007). doi:10.1086/518023

S. Campana, V. Mangano, A.J. Blustin, P. Brown, D.N. Burrows, G. Chincarini, J.R. Cummings, G. Cusumano, M. Della Valle, D. Malesani, P. Mészáros, J.A. Nousek, M. Page, T. Sakamoto, E. Waxman, B. Zhang, Z.G. Dai, N. Gehrels, S. Immler, F.E. Marshall, K.O. Mason, A. Moretti, P.T. O'Brien, J.P. Osborne, K.L. Page, P. Romano, P.W.A. Roming, G. Tagliaferri, L.R. Cominsky, P. Giommi, O. Godet, J.A. Kennea, H. Krimm, L. Angelini, S.D. Barthelmy, P.T. Boyd, D.M. Palmer, A.A. Wells, N.E. White, The association of GRB 060218 with a supernova and the evolution of the shock wave. Nature **442**, 1008–1010 (2006). doi:10.1038/nature04892

P. Chandra, D.A. Frail, A radio-selected sample of gamma-ray burst afterglows. Astrophys. J. **746**, 156 (2012). doi:10.1088/0004-637X/746/2/156

R.A. Chevalier, Z.-Y. Li, Wind interaction models for gamma-ray burst afterglows: the case for two types of progenitors. Astrophys. J. **536**, 195–212 (2000). doi:10.1086/308914

G. Chincarini, A. Moretti, P. Romano, A.D. Falcone, D. Morris, J. Racusin, S. Campana, S. Covino, C. Guidorzi, G. Tagliaferri, D.N. Burrows, C. Pagani, M. Stroh, D. Grupe, M. Capalbi, G. Cusumano, N. Gehrels, P. Giommi, V. La Parola, V. Mangano, T. Mineo, J.A. Nousek, P.T. O'Brien, K.L. Page, M. Perri, E. Troja, R. Willingale, B. Zhang, The first survey of X-ray flares from gamma-ray bursts observed by Swift: temporal properties and morphology. Astrophys. J. **671**, 1903–1920 (2007). doi:10.1086/521591

G. Chincarini, J. Mao, R. Margutti, M.G. Bernardini, C. Guidorzi, F. Pasotti, D. Giannios, M. Della Valle, A. Moretti, P. Romano, P. D'Avanzo, G. Cusumano, P. Giommi, Unveiling the origin of X-ray flares in gamma-ray bursts. Mon. Not. R. Astron. Soc. Lett. **406**, 2113–2148 (2010). doi:10.1111/j.1365-2966.2010.17037.x

V. Connaughton, E. Burns, A. Goldstein, L. Blackburn, M.S. Briggs, B.-B. Zhang, J. Camp, N. Christensen, C.M. Hui, P. Jenke, T. Littenberg, J.E. McEnery, J. Racusin, P. Shawhan, L. Singer, J. Veitch, C.A. Wilson-Hodge, P.N. Bhat, E. Bissaldi, W. Cleveland, G. Fitzpatrick, M.M. Giles, M.H. Gibby, A. von Kienlin, R.M. Kippen, S. McBreen, B. Mailyan, C.A. Meegan, W.S. Paciesas, R.D. Preece, O.J. Roberts, L. Sparke, M. Stanbro, K. Toelge, P. Veres, Fermi GBM observations of LIGO gravitational-wave event GW150914. Astrophys. J. Lett. **826**, 6 (2016). doi:10.3847/2041-8205/826/1/L6

E. Costa, F. Frontera, J. Heise, M. Feroci, J. in't Zand, F. Fiore, M.N. Cinti, D. Dal Fiume, L. Nicastro, M. Orlandini, E. Palazzi, M. Rapisarda#, G. Zavattini, R. Jager, A. Parmar, A. Owens, S. Molendi, G. Cusumano, M.C. Maccarone, S. Giarrusso, A. Coletta, L.A. Antonelli, P. Giommi, J.M. Muller, L. Piro, R.C. Butler, Discovery of an X-ray afterglow associated with the γ-ray burst of 28 February 1997. Nature **387**, 783–785 (1997). doi:10.1038/42885

Z.G. Dai, T. Lu, Gamma-ray burst afterglows and evolution of postburst fireballs with energy injection from strongly magnetic millisecond pulsars. Astron. Astrophys. **333**, 87–90 (1998a)

Z.G. Dai, T. Lu, Gamma-ray burst afterglows: effects of radiative corrections and non-uniformity of the surrounding medium. Mon. Not. R. Astron. Soc. Lett. **298**, 87–92 (1998b). doi:10.1046/j.1365-8711.1998.01681.x

Z.G. Dai, T. Lu, γ-ray bursts and afterglows from rotating strange stars and neutron stars. Phys. Rev. Lett. **81**, 4301–4304 (1998c). doi:10.1103/PhysRevLett.81.4301

Z.G. Dai, X.Y. Wang, X.F. Wu, B. Zhang, X-ray flares from postmerger millisecond pulsars. Science **311**, 1127–1129 (2006). doi:10.1126/science.1123606

F. Daigne, R. Mochkovitch, Gamma-ray bursts from internal shocks in a relativistic wind: temporal and spectral properties. Mon. Not. R. Astron. Soc. Lett. **296**, 275–286 (1998). doi:10.1046/j.1365-8711.1998.01305.x

F. Daigne, Ž. Bošnjak, G. Dubus, Reconciling observed gamma-ray burst prompt spectra with synchrotron radiation? Astron. Astrophys. **526**, 110 (2011). doi:10.1051/0004-6361/201015457

M.G. Dainotti, R. Willingale, S. Capozziello, V. Fabrizio Cardone, M. Ostrowski, Discovery of a tight correlation for gamma-ray burst afterglows with "canonical" light curves. Astrophys. J. Lett. **722**, 215–219 (2010). doi:10.1088/2041-8205/722/2/L215

M. de Pasquale, P. Evans, S. Oates, M. Page, S. Zane, P. Schady, A. Breeveld, S. Holland, P. Kuin, M. Still, P. Roming, P. Ward, Jet breaks at the end of the slow decline phase of Swift GRB light curves. Mon. Not. R. Astron. Soc. Lett. **392**, 153–169 (2009). doi:10.1111/j.1365-2966.2008.13990.x

M. Della Valle, G. Chincarini, N. Panagia, G. Tagliaferri, D. Malesani, V. Testa, D. Fugazza, S. Campana, S. Covino, V. Mangano, L.A. Antonelli, P. D'Avanzo, K. Hurley, I.F. Mirabel, L.J. Pellizza, S. Piranomonte, L. Stella, An enigmatic long-lasting γ-ray burst not accompanied by a bright supernova. Nature **444**, 1050–1052 (2006). doi:10.1038/nature05374

W. Deng, B. Zhang, Low energy spectral index and E_p evolution of quasi-thermal photosphere emission of gamma-ray bursts. Astrophys. J. **785**, 112 (2014). doi:10.1088/0004-637X/785/2/112

C.D. Dermer, A. Atoyan, Collapse of neutron stars to black holes in binary systems: a model for short gamma-ray bursts. Astrophys. J. Lett. **643**, 13–16 (2006). doi:10.1086/504895

D. Eichler, M. Livio, T. Piran, D.N. Schramm, Nucleosynthesis, neutrino bursts and gamma-rays from coalescing neutron stars. Nature **340**, 126–128 (1989). doi:10.1038/340126a0

A.D. Falcone, D.N. Burrows, D. Lazzati, S. Campana, S. Kobayashi, B. Zhang, P. Mészáros, K.L. Page, J.A. Kennea, P. Romano, C. Pagani, L. Angelini, A.P. Beardmore, M. Capalbi, G. Chincarini, G. Cusumano, P. Giommi, M.R. Goad, O. Godet, D. Grupe, J.E. Hill, V. La Parola, V. Mangano, A. Moretti, J.A. Nousek, P.T. O'Brien, J.P. Osborne, M. Perri, G. Tagliaferri, A.A. Wells, N. Gehrels, The giant X-ray flare of GRB 050502B: evidence for late-time internal engine activity. Astrophys. J. **641**, 1010–1017 (2006). doi:10.1086/500655

Y. Fan, T. Piran, Gamma-ray burst efficiency and possible physical processes shaping the early afterglow. Mon. Not. R. Astron. Soc. Lett. **369**, 197–206 (2006). doi:10.1111/j.1365-2966.2006.10280.x

Y.Z. Fan, D.M. Wei, Late internal-shock model for bright X-ray flares in gamma-ray burst afterglows and GRB 011121. Mon. Not. R. Astron. Soc. Lett. **364**, 42–46 (2005). doi:10.1111/j.1745-3933.2005.00102.x

Y.-Z. Fan, D. Xu, The X-ray afterglow flat segment in short GRB 051221A: energy injection from a millisecond magnetar? Mon. Not. R. Astron. Soc. Lett. **372**, 19–22 (2006). doi:10.1111/j.1745-3933.2006.00217.x

Y.Z. Fan, D.M. Wei, C.F. Wang, The very early afterglow powered by ultra-relativistic mildly magnetized outflows. Astron. Astrophys. **424**, 477–484 (2004). doi:10.1051/0004-6361:20041115

Y.-Z. Fan, Z.-G. Dai, Y.-F. Huang, T. Lu, Optical flash of GRB 990123: constraints on the physical parameters of the reverse shock. Chin. J. Astron. Astrophys. **2**, 449–453 (2002)

Y.-Z. Fan, P.H.T. Tam, F.-W. Zhang, Y.-F. Liang, H.-N. He, B. Zhou, R.-Z. Yang, Z.-P. Jin, D.-M. Wei, High-energy emission of GRB 130427A: evidence for inverse Compton radiation. Astrophys. J. **776**, 95 (2013). doi:10.1088/0004-637X/776/2/95

G.J. Fishman, C.A. Meegan, Gamma-ray bursts. Annu. Rev. Astron. Astrophys. **33**, 415–458 (1995). doi:10.1146/annurev.aa.33.090195.002215

D.A. Frail, S.R. Kulkarni, L. Nicastro, M. Feroci, G.B. Taylor, The radio afterglow from the γ-ray burst of 8 May 1997. Nature **389**, 261–263 (1997). doi:10.1038/38451

D.D. Frederiks, V.D. Palshin, R.L. Aptekar, S.V. Golenetskii, T.L. Cline, E.P. Mazets, On the possibility of identifying the short hard burst GRB 051103 with a giant flare from a soft gamma repeater in the M81 group of galaxies. Astron. Lett. **33**, 19–24 (2007). doi:10.1134/S1063773707010021

C.L. Fryer, S.E. Woosley, D.H. Hartmann, Formation rates of black hole accretion disk gamma-ray bursts. Astrophys. J. **526**, 152–177 (1999). doi:10.1086/307992

J.P.U. Fynbo, D. Watson, C.C. Thöne, J. Sollerman, J.S. Bloom, T.M. Davis, J. Hjorth, P. Jakobsson, U.G. Jørgensen, J.F. Graham, A.S. Fruchter, D. Bersier, L. Kewley, A. Cassan, J.M. Castro Cerón, S. Foley, J. Gorosabel, T.C. Hinse, K.D. Horne, B.L. Jensen, S. Klose, D. Kocevski, J.-B. Marquette, D. Perley, E. Ramirez-Ruiz, M.D. Stritzinger, P.M. Vreeswijk, R.A.M. Wijers, K.G. Woller, D. Xu, M. Zub, No supernovae associated with two long-duration γ-ray bursts. Nature **444**, 1047–1049 (2006). doi:10.1038/nature05375

A. Gal-Yam, D.B. Fox, P.A. Price, E.O. Ofek, M.R. Davis, D.C. Leonard, A.M. Soderberg, B.P. Schmidt, K.M. Lewis, B.A. Peterson, S.R. Kulkarni, E. Berger, S.B. Cenko, R. Sari, K. Sharon, D. Frail, D.-S. Moon, P.J. Brown, A. Cucchiara, F. Harrison, T. Piran, S.E. Persson, P.J. McCarthy, B.E. Penprase,

R.A. Chevalier, A.I. MacFadyen, A novel explosive process is required for the γ-ray burst GRB 060614. Nature **444**, 1053–1055 (2006). doi:10.1038/nature05373

H. Gao, P. Mészáros, Relation between the intrinsic and observed central engine activity time: implications for ultra-long GRBs. Astrophys. J. **802**, 90 (2015). doi:10.1088/0004-637X/802/2/90

H. Gao, B. Zhang, Photosphere emission from a hybrid relativistic outflow with arbitrary dimensionless entropy and magnetization in GRBs. Astrophys. J. **801**, 103 (2015). doi:10.1088/0004-637X/801/2/103

H. Gao, B. Zhang, H.-J. Lü, Constraints on binary neutron star merger product from short GRB observations. Phys. Rev. D **93**(4), 044065 (2016). doi:10.1103/PhysRevD.93.044065

H. Gao, B.-B. Zhang, B. Zhang, Stepwise filter correlation method and evidence of superposed variability components in gamma-ray burst prompt emission light curves. Astrophys. J. **748**, 134 (2012). doi:10.1088/0004-637X/748/2/134

H. Gao, W.-H. Lei, Y.-C. Zou, X.-F. Wu, B. Zhang, A complete reference of the analytical synchrotron external shock models of gamma-ray bursts. New Astron. Rev. **57**, 141–190 (2013a). doi:10.1016/j.newar.2013.10.001

H. Gao, X. Ding, X.-F. Wu, B. Zhang, Z.-G. Dai, Bright broadband afterglows of gravitational wave bursts from mergers of binary neutron stars. Astrophys. J. **771**, 86 (2013b). doi:10.1088/0004-637X/771/2/86

H. Gao, X.-G. Wang, P. Mészáros, B. Zhang, A morphological analysis of gamma-ray burst early-optical afterglows. Astrophys. J. **810**, 160 (2015). doi:10.1088/0004-637X/810/2/160

H. Gao, B. Zhang, H.-J. Lü, Y. Li, Searching for magnetar powered merger-novae from short GRBs. Astrophys. J. (2016, submitted). arXiv:1608.03375

W.-H. Gao, Y.-Z. Fan, Short-living supermassive magnetar model for the early X-ray flares following short GRBs. Chin. J. Astron. Astrophys. **6**, 513–516 (2006). doi:10.1088/1009-9271/6/5/01

N. Gehrels, E. Ramirez-Ruiz, D.B. Fox, Gamma-ray bursts in the Swift era. Annu. Rev. Astron. Astrophys. **47**, 567–617 (2009). doi:10.1146/annurev.astro.46.060407.145147

N. Gehrels, J.P. Norris, S.D. Barthelmy, J. Granot, Y. Kaneko, C. Kouveliotou, C.B. Markwardt, P. Mészáros, E. Nakar, J.A. Nousek, P.T. O'Brien, M. Page, D.M. Palmer, A.M. Parsons, P.W.A. Roming, T. Sakamoto, C.L. Sarazin, P. Schady, M. Stamatikos, S.E. Woosley, A new γ-ray burst classification scheme from GRB060614. Nature **444**, 1044–1046 (2006). doi:10.1038/nature05376

B. Gendre, G. Stratta, J.L. Atteia, S. Basa, M. Boër, D.M. Coward, S. Cutini, V. D'Elia, E.J. Howell, A. Klotz, L. Piro, The ultra-long gamma-ray burst 111209A: the collapse of a blue supergiant? Astrophys. J. **766**, 30 (2013). doi:10.1088/0004-637X/766/1/30

F. Genet, F. Daigne, R. Mochkovitch, Can the early X-ray afterglow of gamma-ray bursts be explained by a contribution from the reverse shock? Mon. Not. R. Astron. Soc. **381**, 732–740 (2007). doi:10.1111/j.1365-2966.2007.12243.x

G. Ghirlanda, G. Ghisellini, L. Nava, D. Burlon, Spectral evolution of Fermi/GBM short gamma-ray bursts. Mon. Not. R. Astron. Soc. **410**, 47–51 (2011). doi:10.1111/j.1745-3933.2010.00977.x

G. Ghisellini, G. Ghirlanda, L. Nava, C. Firmani, "Late Prompt" emission in gamma-ray bursts? Astrophys. J. Lett. **658**, 75–78 (2007). doi:10.1086/515570

G. Ghisellini, G. Ghirlanda, L. Nava, A. Celotti, GeV emission from gamma-ray bursts: a radiative fireball? Mon. Not. R. Astron. Soc. **403**, 926–937 (2010). doi:10.1111/j.1365-2966.2009.16171.x

D. Giannios, H.C. Spruit, Spectral and timing properties of a dissipative γ-ray burst photosphere. Astron. Astrophys. **469**, 1–9 (2007). doi:10.1051/0004-6361:20066739

S.V. Golenetskii, E.P. Mazets, R.L. Aptekar, V.N. Ilinskii, Correlation between luminosity and temperature in gamma-ray burst sources. Nature **306**, 451–453 (1983). doi:10.1038/306451a0

J. Greiner, P.A. Mazzali, D.A. Kann, T. Krühler, E. Pian, S. Prentice, F. Olivares E., A. Rossi, S. Klose, S. Taubenberger, F. Knust, P.M.J. Afonso, C. Ashall, J. Bolmer, C. Delvaux, R. Diehl, J. Elliott, R. Filgas, J.P.U. Fynbo, J.F. Graham, A.N. Guelbenzu, S. Kobayashi, G. Leloudas, S. Savaglio, P. Schady, S. Schmidl, T. Schweyer, V. Sudilovsky, M. Tanga, A.C. Updike, H. van Eerten, K. Varela, A very luminous magnetar-powered supernova associated with an ultra-long γ-ray burst. Nature **523**, 189–192 (2015). doi:10.1038/nature14579

D. Gruber, A. Goldstein, V. Weller von Ahlefeld, P. Narayana Bhat, E. Bissaldi, M.S. Briggs, D. Byrne, W.H. Cleveland, V. Connaughton, R. Diehl, G.J. Fishman, G. Fitzpatrick, S. Foley, M. Gibby, M.M. Giles, J. Greiner, S. Guiriec, A.J. van der Horst, A. von Kienlin, C. Kouveliotou, E. Layden, L. Lin, C.A. Meegan, S. McGlynn, W.S. Paciesas, V. Pelassa, R.D. Preece, A. Rau, C.A. Wilson-Hodge, S. Xiong, G. Younes, H.-F. Yu, The Fermi GBM gamma-ray burst spectral catalog: four years of data. Astron. Astrophys. Suppl. Ser. **211**, 12 (2014). doi:10.1088/0067-0049/211/1/12

D. Grupe, D.N. Burrows, S.K. Patel, C. Kouveliotou, B. Zhang, P. Mészáros, R.A.M. Wijers, N. Gehrels, Jet breaks in short gamma-ray bursts. I. The uncollimated afterglow of GRB 050724. Astrophys. J. **653**, 462–467 (2006). doi:10.1086/508739

D. Grupe, J.A. Nousek, P. Veres, B.-B. Zhang, N. Gehrels, Evidence for new relations between gamma-ray burst prompt and X-ray afterglow emission from 9 years of Swift. Astron. Astrophys. Suppl. Ser. **209**, 20 (2013). doi:10.1088/0067-0049/209/2/20

C. Guidorzi, R. Margutti, L. Amati, S. Campana, M. Orlandini, P. Romano, M. Stamatikos, G. Tagliaferri, Average power density spectrum of Swift long gamma-ray bursts in the observer and in the source-rest frames. Mon. Not. R. Astron. Soc. **422**, 1785–1803 (2012). doi:10.1111/j.1365-2966.2012.20758.x

S. Guiriec, V. Connaughton, M.S. Briggs, M. Burgess, F. Ryde, F. Daigne, P. Mészáros, A. Goldstein, J. McEnery, N. Omodei, P.N. Bhat, E. Bissaldi, A. Camero-Arranz, V. Chaplin, R. Diehl, G. Fishman, S. Foley, M. Gibby, M.M. Giles, J. Greiner, D. Gruber, A. von Kienlin, M. Kippen, C. Kouveliotou, S. McBreen, C.A. Meegan, W. Paciesas, R. Preece, A. Rau, D. Tierney, A.J. van der Horst, C. Wilson-Hodge, Detection of a thermal spectral component in the prompt emission of GRB 100724B. Astrophys. J. Lett. **727**, 33 (2011). doi:10.1088/2041-8205/727/2/L33

S. Guiriec, F. Daigne, R. Hascoët, G. Vianello, F. Ryde, R. Mochkovitch, C. Kouveliotou, S. Xiong, P.N. Bhat, S. Foley, D. Gruber, J.M. Burgess, S. McGlynn, J. McEnery, N. Gehrels, Evidence for a photospheric component in the prompt emission of the short GRB 120323A and its effects on the GRB hardness-luminosity relation. Astrophys. J. **770**, 32 (2013). doi:10.1088/0004-637X/770/1/32

S. Guiriec, R. Mochkovitch, T. Piran, F. Daigne, C. Kouveliotou, J. Racusin, N. Gehrels, J. McEnery, GRB 131014A: a laboratory for studying the thermal-like and non-thermal emissions in gamma-ray bursts, and the new l_i^{nTh}-$e_{peak,i}^{nTh,rest}$ relation. Astrophys. J. **814**, 10 (2015). doi:10.1088/0004-637X/814/1/10

J. Hakkila, T.W. Giblin, R.J. Roiger, D.J. Haglin, W.S. Paciesas, C.A. Meegan, How sample completeness affects gamma-ray burst classification. Astrophys. J. **582**, 320–329 (2003). doi:10.1086/344568

R. Hascoët, F. Daigne, R. Mochkovitch, Accounting for the XRT early steep decay in models of the prompt gamma-ray burst emission. Astron. Astrophys. **542**, 29 (2012). doi:10.1051/0004-6361/201219339

S.W. Hawking, Particle creation by black holes. Commun. Math. Phys. **43**, 199–220 (1975). doi:10.1007/BF02345020

H.-N. He, X.-F. Wu, K. Toma, X.-Y. Wang, P. Mészáros, On the high-energy emission of the short GRB 090510. Astrophys. J. **733**, 22 (2011). doi:10.1088/0004-637X/733/1/22

I. Horváth, A third class of gamma-ray bursts? Astrophys. J. **508**, 757–759 (1998). doi:10.1086/306416

Y.-D. Hu, E.-W. Liang, S.-Q. Xi, F.-K. Peng, R.-J. Lu, L.-Z. Lü, B. Zhang, Internal energy dissipation of gamma-ray bursts observed with Swift: precursors, prompt gamma-rays, extended emission, and late X-ray flares. Astrophys. J. **789**, 145 (2014). doi:10.1088/0004-637X/789/2/145

Y.F. Huang, Z.G. Dai, T. Lu, On the optical light curves of afterglows from jetted gamma-ray burst ejecta: effects of parameters. Mon. Not. R. Astron. Soc. **316**, 943–949 (2000). doi:10.1046/j.1365-8711.2000.03683.x

K. Hurley, B.L. Dingus, R. Mukherjee, P. Sreekumar, C. Kouveliotou, C. Meegan, G.J. Fishman, D. Band, L. Ford, D. Bertsch, T. Cline, C. Fichtel, R. Hartman, S. Hunter, D.J. Thompson, G. Kanbach, H. Mayer-Hasselwander, C. von Eerten, M. Sommer, Y. Lin, P. Nolan, P. Michelson, D. Kniffen, J. Mattox, E. Schneid, M. Boer, M. Niel, Detection of a γ-ray burst of very long duration and very high energy. Nature **372**, 652–654 (1994). doi:10.1038/372652a0

P. Jakobsson, J. Hjorth, J.P.U. Fynbo, D. Watson, K. Pedersen, G. Björnsson, J. Gorosabel, Swift identification of dark gamma-ray bursts. Astrophys. J. Lett. **617**, 21–24 (2004). doi:10.1086/427089

D.A. Kann, S. Klose, B. Zhang, D. Malesani, E. Nakar, A. Pozanenko, A.C. Wilson, N.R. Butler, P. Jakobsson, S. Schulze, M. Andreev, L.A. Antonelli, I.F. Bikmaev, V. Biryukov, M. Böttcher, R.A. Burenin, J.M. Castro Cerón, A.J. Castro-Tirado, G. Chincarini, B.E. Cobb, S. Covino, P. D'Avanzo, V. D'Elia, M. Della Valle, A. de Ugarte Postigo, Y. Efimov, P. Ferrero, D. Fugazza, J.P.U. Fynbo, M. Gålfalk, F. Grundahl, J. Gorosabel, S. Gupta, S. Guziy, B. Hafizov, J. Hjorth, K. Holhjem, M. Ibrahimov, M. Im, G.L. Israel, M. Jelínek, B.L. Jensen, R. Karimov, I.M. Khamitov, Ü. Kiziloğlu, E. Klunko, P. Kubánek, A.S. Kutyrev, P. Laursen, A.J. Levan, F. Mannucci, C.M. Martin, A. Mescheryakov, N. Mirabal, J.P. Norris, J.-E. Ovaldsen, D. Paraficz, E. Pavlenko, S. Piranomonte, A. Rossi, V. Rumyantsev, R. Salinas, A. Sergeev, D. Sharapov, J. Sollerman, B. Stecklum, L. Stella, G. Tagliaferri, N.R. Tanvir, J. Telting, V. Testa, A.C. Updike, A. Volnova, D. Watson, K. Wiersema, D. Xu, The afterglows of Swift-era gamma-ray bursts. I. Comparing pre-Swift and Swift-era long/soft (Type II) GRB optical afterglows. Astrophys. J. **720**, 1513–1558 (2010). doi:10.1088/0004-637X/720/2/1513

D.A. Kann, S. Klose, B. Zhang, S. Covino, N.R. Butler, D. Malesani, E. Nakar, A.C. Wilson, L.A. Antonelli, G. Chincarini, B.E. Cobb, P. D'Avanzo, V. D'Elia, M. Della Valle, P. Ferrero, D. Fugazza, J. Gorosabel, G.L. Israel, F. Mannucci, S. Piranomonte, S. Schulze, L. Stella, G. Tagliaferri, K. Wiersema, The afterglows of Swift-era gamma-ray bursts. II. Type I GRB versus Type II GRB optical afterglows. Astrophys. J. **734**, 96 (2011). doi:10.1088/0004-637X/734/2/96

K. Kashiyama, D. Nakauchi, Y. Suwa, H. Yajima, T. Nakamura, Luminous supernova-like UV/optical/infrared transients associated with ultra-long gamma-ray bursts from metal-poor blue supergiants. Astrophys. J. **770**, 8 (2013). doi:10.1088/0004-637X/770/1/8

S. Kobayashi, Light curves of gamma-ray burst optical flashes. Astrophys. J. **545**, 807–812 (2000). doi:10.1086/317869

S. Kobayashi, B. Zhang, Early optical afterglows from wind-type gamma-ray bursts. Astrophys. J. **597**, 455–458 (2003a). doi:10.1086/378283

S. Kobayashi, B. Zhang, GRB 021004: reverse shock emission. Astrophys. J. Lett. **582**, 75–78 (2003b). doi:10.1086/367691

S. Kobayashi, B. Zhang, The onset of gamma-ray burst afterglow. Astrophys. J. **655**, 973–979 (2007). doi:10.1086/510203

S. Kobayashi, T. Piran, R. Sari, Hydrodynamics of a relativistic fireball: the complete evolution. Astrophys. J. **513**, 669–678 (1999). doi:10.1086/306868

C. Kouveliotou, C.A. Meegan, G.J. Fishman, N.P. Bhat, M.S. Briggs, T.M. Koshut, W.S. Paciesas, G.N. Pendleton, Identification of two classes of gamma-ray bursts. Astrophys. J. Lett. **413**, 101–104 (1993). doi:10.1086/186969

S.R. Kulkarni, D.A. Frail, R. Sari, G.H. Moriarty-Schieven, D.S. Shepherd, P. Udomprasert, A.C.S. Readhead, J.S. Bloom, M. Feroci, E. Costa, Discovery of a radio flare from GRB 990123. Astrophys. J. Lett. **522**, 97–100 (1999). doi:10.1086/312227

P. Kumar, R. Barniol Duran, On the generation of high-energy photons detected by the Fermi Satellite from gamma-ray bursts. Mon. Not. R. Astron. Soc. **400**, 75–79 (2009). doi:10.1111/j.1745-3933.2009.00766.x

P. Kumar, R. Barniol Duran, External forward shock origin of high-energy emission for three gamma-ray bursts detected by Fermi. Mon. Not. R. Astron. Soc. **409**, 226–236 (2010). doi:10.1111/j.1365-2966.2010.17274.x

P. Kumar, A. Panaitescu, Steepening of afterglow decay for jets interacting with stratified media. Astrophys. J. Lett. **541**, 9–12 (2000). doi:10.1086/312888

P. Kumar, A. Panaitescu, A unified treatment of the gamma-ray burst 021211 and its afterglow. Mon. Not. R. Astron. Soc. **346**, 905–914 (2003). doi:10.1111/j.1365-2966.2003.07138.x

P. Kumar, B. Zhang, The physics of gamma-ray bursts: relativistic jets. Phys. Rep. **561**, 1–109 (2015). doi:10.1016/j.physrep.2014.09.008

P. Kumar, R. Narayan, J.L. Johnson, Mass fall-back and accretion in the central engine of gamma-ray bursts. Mon. Not. R. Astron. Soc. **388**, 1729–1742 (2008a). doi:10.1111/j.1365-2966.2008.13493.x

P. Kumar, R. Narayan, J.L. Johnson, Properties of gamma-ray burst progenitor stars. Science **321**, 376 (2008b). doi:10.1126/science.1159003

P.D. Lasky, B. Haskell, V. Ravi, E.J. Howell, D.M. Coward, Nuclear equation of state from observations of short gamma-ray burst remnants. Phys. Rev. D **89**(4), 047302 (2014). doi:10.1103/PhysRevD.89.047302

D. Lazzati, Precursor activity in bright, long BATSE gamma-ray bursts. Mon. Not. R. Astron. Soc. **357**, 722–731 (2005). doi:10.1111/j.1365-2966.2005.08687.x

D. Lazzati, M.C. Begelman, Non-thermal emission from the photospheres of gamma-ray burst outflows. I. High-frequency tails. Astrophys. J. **725**, 1137–1145 (2010). doi:10.1088/0004-637X/725/1/1137

D. Lazzati, R. Perna, X-ray flares and the duration of engine activity in gamma-ray bursts. Mon. Not. R. Astron. Soc. **375**, 46–50 (2007). doi:10.1111/j.1745-3933.2006.00273.x

A.J. Levan, N.R. Tanvir, R.L.C. Starling, K. Wiersema, K.L. Page, D.A. Perley, S. Schulze, G.A. Wynn, R. Chornock, J. Hjorth, S.B. Cenko, A.S. Fruchter, P.T. O'Brien, G.C. Brown, R.L. Tunnicliffe, D. Malesani, P. Jakobsson, D. Watson, E. Berger, D. Bersier, B.E. Cobb, S. Covino, A. Cucchiara, A. de Ugarte Postigo, D.B. Fox, A. Gal-Yam, P. Goldoni, J. Gorosabel, L. Kaper, T. Krühler, R. Karjalainen, J.P. Osborne, E. Pian, R. Sánchez-Ramírez, B. Schmidt, I. Skillen, G. Tagliaferri, C. Thöne, O. Vaduvescu, R.A.M.J. Wijers, B.A. Zauderer, A new population of ultra-long duration gamma-ray bursts. Astrophys. J. **781**, 13 (2014). doi:10.1088/0004-637X/781/1/13

E.M. Levesque, J.S. Bloom, N.R. Butler, D.A. Perley, S.B. Cenko, J.X. Prochaska, L.J. Kewley, A. Bunker, H.-W. Chen, R. Chornock, A.V. Filippenko, K. Glazebrook, S. Lopez, J. Masiero, M. Modjaz, A. Morgan, D. Poznanski, GRB090426: the environment of a rest-frame 0.35-s gamma-ray burst at a redshift of 2.609. Mon. Not. R. Astron. Soc. **401**, 963–972 (2010). doi:10.1111/j.1365-2966.2009.15733.x

A. Li, B. Zhang, N.B. Zhang, H. Gao, B. Qi, T. Liu, Internal X-ray plateau in short GRBs: signature of supramassive fast rotating quark stars? Phys. Rev. D **94**, 083010 (2016). arXiv:1606.02934

L. Li, E.-W. Liang, Q.-W. Tang, J.-M. Chen, S.-Q. Xi, H.-J. Lü, H. Gao, B. Zhang, J. Zhang, S.-X. Yi, R.-J. Lu, L.-Z. Lü, J.-Y. Wei, A comprehensive study of gamma-ray burst optical emission. I. Flares and early shallow-decay component. Astrophys. J. **758**, 27 (2012). doi:10.1088/0004-637X/758/1/27

X. Li, F.-W. Zhang, Q. Yuan, Z.-P. Jin, Y.-Z. Fan, S.-M. Liu, D.-M. Wei, Implications of the tentative association between GW150914 and a Fermi-GBM transient. Astrophys. J. Lett. **827**, 16 (2016a). doi:10.3847/2041-8205/827/1/L16

Y. Li, B. Zhang, H.-J. Lü, A comparative study of long and short GRBs. I. Overlapping properties. Astrophys. J. Suppl. Ser. (2016b, in press). arXiv:1608.03383

E.-W. Liang, B.-B. Zhang, B. Zhang, A comprehensive analysis of Swift XRT data. II. Diverse physical origins of the shallow decay segment. Astrophys. J. **670**, 565–583 (2007). doi:10.1086/521870

E.-W. Liang, B.-B. Zhang, M. Stamatikos, B. Zhang, J. Norris, N. Gehrels, J. Zhang, Z.G. Dai, Temporal profiles and spectral lags of XRF 060218. Astrophys. J. Lett. **653**, 81–84 (2006a). doi:10.1086/510516

E.W. Liang, B. Zhang, P.T. O'Brien, R. Willingale, L. Angelini, D.N. Burrows, S. Campana, G. Chincarini, A. Falcone, N. Gehrels, M.R. Goad, D. Grupe, S. Kobayashi, P. Mészáros, J.A. Nousek, J.P. Osborne, K.L. Page, G. Tagliaferri, Testing the curvature effect and internal origin of gamma-ray burst prompt emissions and X-ray flares with Swift data. Astrophys. J. **646**, 351–357 (2006b). doi:10.1086/504684

E.-W. Liang, J.L. Racusin, B. Zhang, B.-B. Zhang, D.N. Burrows, A comprehensive analysis of Swift XRT data. III. Jet break candidates in X-ray and optical afterglow light curves. Astrophys. J. **675**, 528–552 (2008). doi:10.1086/524701

E.-W. Liang, H.-J. Lü, S.-J. Hou, B.-B. Zhang, B. Zhang, A comprehensive analysis of Swift/X-ray telescope data. IV. Single power-law decaying light curves versus canonical light curves and implications for a unified origin of X-rays. Astrophys. J. **707**, 328–342 (2009). doi:10.1088/0004-637X/707/1/328

E.-W. Liang, S.-X. Yi, J. Zhang, H.-J. Lü, B.-B. Zhang, B. Zhang, Constraining gamma-ray burst initial Lorentz factor with the afterglow onset feature and discovery of a tight Γ_0-$E_{\gamma,iso}$ correlation. Astrophys. J. **725**, 2209–2224 (2010). doi:10.1088/0004-637X/725/2/2209

E.-W. Liang, L. Li, H. Gao, B. Zhang, Y.-F. Liang, X.-F. Wu, S.-X. Yi, Z.-G. Dai, Q.-W. Tang, J.-M. Chen, H.-J. Lü, J. Zhang, R.-J. Lu, L.-Z. Lü, J.-Y. Wei, A comprehensive study of gamma-ray burst optical emission. II. afterglow onset and late re-brightening components. Astrophys. J. **774**, 13 (2013). doi:10.1088/0004-637X/774/1/13

E. Liang, B. Zhang, F. Virgili, Z.G. Dai, Low-luminosity gamma-ray bursts as a unique population: luminosity function, local rate, and beaming factor. Astrophys. J. **662**, 1111–1118 (2007). doi:10.1086/517959

A. Lien, T. Sakamoto, S.D. Barthelmy, W.H. Baumgartner, J.K. Cannizzo, K. Chen, N.R. Collins, J.R. Cummings, N. Gehrels, H.A. Krimm, C.B. Markwardt, D.M. Palmer, M. Stamatikos, E. Troja, T.N. Ukwatta, The third Swift burst alert telescope gamma-ray burst catalog. Astrophys. J. **829**, 7 (2016). doi:10.3847/0004-637X/829/1/7

R.-Y. Liu, X.-Y. Wang, Modeling the broadband emission of GRB 090902B. Astrophys. J. **730**, 1 (2011). doi:10.1088/0004-637X/730/1/1

R.-Y. Liu, X.-Y. Wang, X.-F. Wu, Interpretation of the unprecedentedly long-lived high-energy emission of GRB 130427A. Astrophys. J. Lett. **773**, 20 (2013). doi:10.1088/2041-8205/773/2/L20

A. Loeb, Electromagnetic counterparts to black hole mergers detected by LIGO. Astrophys. J. Lett. **819**, 21 (2016). doi:10.3847/2041-8205/819/2/L21

H.-J. Lü, B. Zhang, A test of the millisecond magnetar central engine model of gamma-ray bursts with Swift data. Astrophys. J. **785**, 74 (2014). doi:10.1088/0004-637X/785/1/74

H.-J. Lü, E.-W. Liang, B.-B. Zhang, B. Zhang, A new classification method for gamma-ray bursts. Astrophys. J. **725**, 1965–1970 (2010). doi:10.1088/0004-637X/725/2/1965

H.-J. Lü, B. Zhang, E.-W. Liang, B.-B. Zhang, T. Sakamoto, The 'amplitude' parameter of gamma-ray bursts and its implications for GRB classification. Mon. Not. R. Astron. Soc. **442**, 1922–1929 (2014). doi:10.1093/mnras/stu982

H.-J. Lü, B. Zhang, W.-H. Lei, Y. Li, P.D. Lasky, The millisecond magnetar central engine in short GRBs. Astrophys. J. **805**, 89 (2015). doi:10.1088/0004-637X/805/2/89

J. Lü, Y.-C. Zou, W.-H. Lei, B. Zhang, Q. Wu, D.-X. Wang, E.-W. Liang, H.-J. Lü, Lorentz-factor-isotropic-luminosity/energy correlations of gamma-ray bursts and their interpretation. Astrophys. J. **751**, 49 (2012). doi:10.1088/0004-637X/751/1/49

R.-J. Lu, S.-J. Hou, E.-W. Liang, The E_p-flux correlation in the rising and decaying phases of gamma-ray burst pulses: evidence for viewing angle effect? Astrophys. J. **720**, 1146–1154 (2010). doi:10.1088/0004-637X/720/2/1146

R.-J. Lu, J.-J. Wei, E.-W. Liang, B.-B. Zhang, H.-J. Lü, L.-Z. Lü, W.-H. Lei, B. Zhang, A comprehensive analysis of Fermi gamma-ray burst data. II. E_p evolution patterns and implications for the observed spectrum-luminosity relations. Astrophys. J. **756**, 112 (2012). doi:10.1088/0004-637X/756/2/112

C. Lundman, A. Pe'er, F. Ryde, A theory of photospheric emission from relativistic, collimated outflows. Mon. Not. R. Astron. Soc. **428**, 2430–2442 (2013). doi:10.1093/mnras/sts219

N. Lyons, P.T. O'Brien, B. Zhang, R. Willingale, E. Troja, R.L.C. Starling, Can X-ray emission powered by a spinning-down magnetar explain some gamma-ray burst light-curve features? Mon. Not. R. Astron. Soc. **402**, 705–712 (2010). doi:10.1111/j.1365-2966.2009.15538.x

M. Lyutikov, R. Blandford, Gamma ray bursts as electromagnetic outflows. ArXiv Astrophysics e-prints (2003). arXiv:astro-ph/0312347

A.I. MacFadyen, S.E. Woosley, Collapsars: gamma-ray bursts and explosions in "failed supernovae". Astrophys. J. **524**, 262–289 (1999). doi:10.1086/307790

V. Mangano, V. La Parola, G. Cusumano, T. Mineo, D. Malesani, J. Dyks, S. Campana, M. Capalbi, G. Chincarini, P. Giommi, A. Moretti, M. Perri, P. Romano, G. Tagliaferri, D.N. Burrows, N. Gehrels, O. Godet, S.T. Holland, J.A. Kennea, K.L. Page, J.L. Racusin, P.W.A. Roming, B. Zhang, Swift XRT observations of the afterglow of XRF 050416A. Astrophys. J. **654**, 403–412 (2007). doi:10.1086/509100

R. Margutti, G. Bernardini, R. Barniol Duran, C. Guidorzi, R.F. Shen, G. Chincarini, On the average gamma-ray burst X-ray flaring activity. Mon. Not. R. Astron. Soc. **410**, 1064–1075 (2011). doi:10.1111/j.1365-2966.2010.17504.x

A. Maxham, B.-B. Zhang, B. Zhang, Is GeV emission from gamma-ray bursts of external shock origin? Mon. Not. R. Astron. Soc. **415**, 77–82 (2011). doi:10.1111/j.1365-2966.2011.18648.x

A. Melandri, S. Kobayashi, C.G. Mundell, C. Guidorzi, A. de Ugarte Postigo, G. Pooley, M. Yoshida, D. Bersier, A.J. Castro-Tirado, M. Jelínek, A. Gomboc, J. Gorosabel, P. Kubánek, M. Bremer, J.M. Winters, I.A. Steele, I. de Gregorio-Monsalvo, R.J. Smith, D. García-Appadoo, A. Sota, A. Lundgren, GRB 090313 and the origin of optical peaks in gamma-ray burst light curves: implications for Lorentz factors and radio flares. Astrophys. J. **723**, 1331–1342 (2010). doi:10.1088/0004-637X/723/2/1331

P. Mészáros, Theories of gamma-ray bursts. Annu. Rev. Astron. Astrophys. **40**, 137–169 (2002). doi:10.1146/annurev.astro.40.060401.093821

P. Mészáros, Gamma-ray bursts. Rep. Prog. Phys. **69**, 2259–2321 (2006). doi:10.1088/0034-4885/69/8/R01

P. Meszaros, M.J. Rees, Relativistic fireballs and their impact on external matter—models for cosmological gamma-ray bursts. Astrophys. J. **405**, 278–284 (1993). doi:10.1086/172360

P. Mészáros, M.J. Rees, Optical and long-wavelength afterglow from gamma-ray bursts. Astrophys. J. **476**, 232–237 (1997)

P. Mészáros, M.J. Rees, GRB 990123: reverse and internal shock flashes and late afterglow behaviour. Mon. Not. R. Astron. Soc. **306**, 39–43 (1999). doi:10.1046/j.1365-8711.1999.02800.x

P. Mészáros, M.J. Rees, Collapsar jets, bubbles, and Fe lines. Astrophys. J. Lett. **556**, 37–40 (2001). doi:10.1086/322934

B.D. Metzger, E. Quataert, T.A. Thompson, Short-duration gamma-ray bursts with extended emission from protomagnetar spin-down. Mon. Not. R. Astron. Soc. **385**, 1455–1460 (2008). doi:10.1111/j.1365-2966.2008.12923.x

B.D. Metzger, D. Giannios, T.A. Thompson, N. Bucciantini, E. Quataert, The protomagnetar model for gamma-ray bursts. Mon. Not. R. Astron. Soc. **413**, 2031–2056 (2011). doi:10.1111/j.1365-2966.2011.18280.x

E. Molinari, S.D. Vergani, D. Malesani, S. Covino, P. D'Avanzo, G. Chincarini, F.M. Zerbi, L.A. Antonelli, P. Conconi, V. Testa, G. Tosti, F. Vitali, F. D'Alessio, G. Malaspina, L. Nicastro, E. Palazzi, D. Guetta, S. Campana, P. Goldoni, N. Masetti, E.J.A. Meurs, A. Monfardini, L. Norci, E. Pian, S. Piranomonte, D. Rizzuto, M. Stefanon, L. Stella, G. Tagliaferri, P.A. Ward, G. Ihle, L. Gonzalez, A. Pizarro, P. Sinclaire, J. Valenzuela, REM observations of GRB 060418 and GRB 060607A: the onset of the afterglow and the initial fireball Lorentz factor determination. Astron. Astrophys. **469**, 13–16 (2007). doi:10.1051/0004-6361:20077388

B.J. Morsony, D. Lazzati, M.C. Begelman, The origin and propagation of variability in the outflows of long-duration gamma-ray bursts. Astrophys. J. **723**, 267–276 (2010). doi:10.1088/0004-637X/723/1/267

S. Mukherjee, E.D. Feigelson, G. Jogesh Babu, F. Murtagh, C. Fraley, A. Raftery, Three types of gamma-ray bursts. Astrophys. J. **508**, 314–327 (1998). doi:10.1086/306386

K. Murase, K. Asano, T. Terasawa, P. Mészáros, The role of stochastic acceleration in the prompt emission of gamma-ray bursts: application to hadronic injection. Astrophys. J. **746**, 164 (2012). doi:10.1088/0004-637X/746/2/164

K. Murase, K. Kashiyama, P. Mészáros, I. Shoemaker, N. Senno, Ultrafast outflows from black hole mergers with a minidisk. Astrophys. J. Lett. **822**, 9 (2016). doi:10.3847/2041-8205/822/1/L9

E. Nakar, R. Sari, Relativistic shock breakouts—a variety of gamma-ray flares: from low-luminosity gamma-ray bursts to Type Ia supernovae. Astrophys. J. **747**, 88 (2012). doi:10.1088/0004-637X/747/2/88

R. Narayan, B. Paczynski, T. Piran, Gamma-ray bursts as the death throes of massive binary stars. Astrophys. J. Lett. **395**, 83–86 (1992). doi:10.1086/186493

L. Nava, G. Ghirlanda, G. Ghisellini, A. Celotti, Spectral properties of 438 GRBs detected by Fermi/GBM. Astron. Astrophys. **530**, 21 (2011). doi:10.1051/0004-6361/201016270

J.P. Norris, J.T. Bonnell, Short gamma-ray bursts with extended emission. Astrophys. J. **643**, 266–275 (2006). doi:10.1086/502796

J.P. Norris, G.F. Marani, J.T. Bonnell, Connection between energy-dependent lags and peak luminosity in gamma-ray bursts. Astrophys. J. **534**, 248–257 (2000). doi:10.1086/308725

J.P. Norris, G.H. Share, D.C. Messina, B.R. Dennis, U.D. Desai, T.L. Cline, S.M. Matz, E.L. Chupp, Spectral evolution of pulse structures in gamma-ray bursts. Astrophys. J. **301**, 213–219 (1986). doi:10.1086/163889

J.A. Nousek, C. Kouveliotou, D. Grupe, K.L. Page, J. Granot, E. Ramirez-Ruiz, S.K. Patel, D.N. Burrows, V. Mangano, S. Barthelmy, A.P. Beardmore, S. Campana, M. Capalbi, G. Chincarini, G. Cusumano, A.D. Falcone, N. Gehrels, P. Giommi, M.R. Goad, O. Godet, C.P. Hurkett, J.A. Kennea, A. Moretti, P.T. O'Brien, J.P. Osborne, P. Romano, G. Tagliaferri, A.A. Wells, Evidence for a canonical gamma-ray burst afterglow light curve in the Swift XRT data. Astrophys. J. **642**, 389–400 (2006). doi:10.1086/500724

P.T. O'Brien, R. Willingale, J. Osborne, M.R. Goad, K.L. Page, S. Vaughan, E. Rol, A. Beardmore, O. Godet, C.P. Hurkett, A. Wells, B. Zhang, S. Kobayashi, D.N. Burrows, J.A. Nousek, J.A. Kennea, A. Falcone, D. Grupe, N. Gehrels, S. Barthelmy, J. Cannizzo, J. Cummings, J.E. Hill, H. Krimm, G. Chincarini, G. Tagliaferri, S. Campana, A. Moretti, P. Giommi, M. Perri, V. Mangano, V. LaParola, The early X-ray emission from GRBs. Astrophys. J. **647**, 1213–1237 (2006). doi:10.1086/505457

W.S. Paciesas, C.A. Meegan, A. von Kienlin, P.N. Bhat, E. Bissaldi, M.S. Briggs, J.M. Burgess, V. Chaplin, V. Connaughton, R. Diehl, G.J. Fishman, G. Fitzpatrick, S. Foley, M. Gibby, M. Giles, A. Goldstein, J. Greiner, D. Gruber, S. Guiriec, A.J. van der Horst, R.M. Kippen, C. Kouveliotou, G. Lichti, L. Lin, S. McBreen, R.D. Preece, A. Rau, D. Tierney, C. Wilson-Hodge, The Fermi GBM gamma-ray burst catalog: the first two years. Astron. Astrophys. Suppl. Ser. **199**, 18 (2012). doi:10.1088/0067-0049/199/1/18

B. Paczynski, Gamma-ray bursters at cosmological distances. Astrophys. J. Lett. **308**, 43–46 (1986). doi:10.1086/184740

B. Paczynski, Cosmological gamma-ray bursts. Acta Astron. **41**, 257–267 (1991)

D.M. Palmer, S. Barthelmy, N. Gehrels, R.M. Kippen, T. Cayton, C. Kouveliotou, D. Eichler, R.A.M.J. Wijers, P.M. Woods, J. Granot, Y.E. Lyubarsky, E. Ramirez-Ruiz, L. Barbier, M. Chester, J. Cummings, E.E. Fenimore, M.H. Finger, B.M. Gaensler, D. Hullinger, H. Krimm, C.B. Markwardt, J.A. Nousek, A. Parsons, S. Patel, T. Sakamoto, G. Sato, M. Suzuki, J. Tueller, A giant γ-ray flare from the magnetar SGR 1806-20. Nature **434**, 1107–1109 (2005). doi:10.1038/nature03525

A. Panaitescu, P. Kumar, Fundamental physical parameters of collimated gamma-ray burst afterglows. Astrophys. J. Lett. **560**, 49–53 (2001). doi:10.1086/324061

A. Panaitescu, P. Kumar, Properties of relativistic jets in gamma-ray burst afterglows. Astrophys. J. **571**, 779–789 (2002). doi:10.1086/340094

A. Panaitescu, P. Mészáros, D. Burrows, J. Nousek, N. Gehrels, P. O'Brien, R. Willingale, Evidence for chromatic X-ray light-curve breaks in Swift gamma-ray burst afterglows and their theoretical implications. Mon. Not. R. Astron. Soc. **369**, 2059–2064 (2006). doi:10.1111/j.1365-2966.2006.10453.x

A. Pe'er, P. Mészáros, M.J. Rees, The observable effects of a photospheric component on GRB and XRF prompt emission spectrum. Astrophys. J. **642**, 995–1003 (2006). doi:10.1086/501424

A. Pe'Er, B.-B. Zhang, F. Ryde, S. McGlynn, B. Zhang, R.D. Preece, C. Kouveliotou, The connection between thermal and non-thermal emission in gamma-ray bursts: general considerations and GRB 090902B as a case study. Mon. Not. R. Astron. Soc. **420**, 468–482 (2012). doi:10.1111/j.1365-2966.2011.20052.x

R. Perna, D. Lazzati, B. Giacomazzo, Short gamma-ray bursts from the merger of two black holes. Astrophys. J. Lett. **821**, 18 (2016). doi:10.3847/2041-8205/821/1/L18

T. Piran, Gamma-ray bursts and the fireball model. Phys. Rep. **314**, 575–667 (1999). doi:10.1016/S0370-1573(98)00127-6

T. Piran, The physics of gamma-ray bursts. Rev. Mod. Phys. **76**, 1143–1210 (2004). doi:10.1103/RevModPhys.76.1143

R. Popham, S.E. Woosley, C. Fryer, Hyperaccreting black holes and gamma-ray bursts. Astrophys. J. **518**, 356–374 (1999). doi:10.1086/307259

R.D. Preece, M.S. Briggs, R.S. Mallozzi, G.N. Pendleton, W.S. Paciesas, D.L. Band, The BATSE gamma-ray burst spectral catalog. I. High time resolution spectroscopy of bright bursts using high energy resolution data. Astron. Astrophys. Suppl. Ser. **126**, 19–36 (2000). doi:10.1086/313289

B. Qin, X.-P. Wu, M.-C. Chu, L.-Z. Fang, J.-Y. Hu, The collapse of neutron stars in high-mass binaries as the energy source for the gamma-ray bursts. Astrophys. J. Lett. **494**, 57–59 (1998). doi:10.1086/311152

Y. Qin, E.-W. Liang, Y.-F. Liang, S.-X. Yi, L. Lin, B.-B. Zhang, J. Zhang, H.-J. Lü, R.-J. Lu, L.-Z. Lü, B. Zhang, A comprehensive analysis of Fermi gamma-ray burst data. III. Energy-dependent t_{90} distributions of GBM GRBs and instrumental selection effect on duration classification. Astrophys. J. **763**, 15 (2013). doi:10.1088/0004-637X/763/1/15

J.L. Racusin, S.V. Karpov, M. Sokolowski, J. Granot, X.F. Wu, V. Pal'Shin, S. Covino, A.J. van der Horst, S.R. Oates, P. Schady, R.J. Smith, J. Cummings, R.L.C. Starling, L.W. Piotrowski, B. Zhang, P.A. Evans, S.T. Holland, K. Malek, M.T. Page, L. Vetere, R. Margutti, C. Guidorzi, A.P. Kamble, P.A. Curran, A. Beardmore, C. Kouveliotou, L. Mankiewicz, A. Melandri, P.T. O'Brien, K.L. Page, T. Piran,

N.R. Tanvir, G. Wrochna, R.L. Aptekar, S. Barthelmy, C. Bartolini, G.M. Beskin, S. Bondar, M. Bremer, S. Campana, A. Castro-Tirado, A. Cucchiara, M. Cwiok, P. D'Avanzo, V. D'Elia, M. Della Valle, A. de Ugarte Postigo, W. Dominik, A. Falcone, F. Fiore, D.B. Fox, D.D. Frederiks, A.S. Fruchter, D. Fugazza, M.A. Garrett, N. Gehrels, S. Golenetskii, A. Gomboc, J. Gorosabel, G. Greco, A. Guarnieri, S. Immler, M. Jelinek, G. Kasprowicz, V. La Parola, A.J. Levan, V. Mangano, E.P. Mazets, E. Molinari, A. Moretti, K. Nawrocki, P.P. Oleynik, J.P. Osborne, C. Pagani, S.B. Pandey, Z. Paragi, M. Perri, A. Piccioni, E. Ramirez-Ruiz, P.W.A. Roming, I.A. Steele, R.G. Strom, V. Testa, G. Tosti, M.V. Ulanov, K. Wiersema, R.A.M.J. Wijers, J.M. Winters, A.F. Zarnecki, F. Zerbi, P. Mészáros, G. Chincarini, D.N. Burrows, Broadband observations of the naked-eye γ-ray burst GRB080319B. Nature **455**, 183–188 (2008). doi:10.1038/nature07270

J.L. Racusin, E.W. Liang, D.N. Burrows, A. Falcone, T. Sakamoto, B.B. Zhang, B. Zhang, P. Evans, J. Osborne, Jet breaks and energetics of Swift gamma-ray burst X-ray afterglows. Astrophys. J. **698**, 43–74 (2009). doi:10.1088/0004-637X/698/1/43

M.J. Rees, P. Meszaros, Relativistic fireballs-energy conversion and time-scales. Mon. Not. R. Astron. Soc. **258**, 41–43 (1992). doi:10.1093/mnras/258.1.41P

M.J. Rees, P. Meszaros, Unsteady outflow models for cosmological gamma-ray bursts. Astrophys. J. Lett. **430**, 93–96 (1994). doi:10.1086/187446

M.J. Rees, P. Mészáros, Refreshed shocks and afterglow longevity in gamma-ray bursts. Astrophys. J. Lett. **496**, 1–4 (1998). doi:10.1086/311244

M.J. Rees, P. Mészáros, Dissipative photosphere models of gamma-ray bursts and X-ray flashes. Astrophys. J. **628**, 847–852 (2005). doi:10.1086/430818

L. Resmi, B. Zhang, Gamma-ray burst reverse shock emission in early radio afterglows. Astrophys. J. **825**, 48 (2016). doi:10.3847/0004-637X/825/1/48

L. Rezzolla, B. Giacomazzo, L. Baiotti, J. Granot, C. Kouveliotou, M.A. Aloy, The missing link: merging neutron stars naturally produce jet-like structures and can power short gamma-ray bursts. Astrophys. J. Lett. **732**, 6 (2011). doi:10.1088/2041-8205/732/1/L6

J.E. Rhoads, The dynamics and light curves of beamed gamma-ray burst afterglows. Astrophys. J. **525**, 737–749 (1999). doi:10.1086/307907

E. Rol, R.A.M.J. Wijers, C. Kouveliotou, L. Kaper, Y. Kaneko, How special are dark gamma-ray bursts: a diagnostic tool. Astrophys. J. **624**, 868–879 (2005). doi:10.1086/429082

P. Romano, A. Moretti, P.L. Banat, D.N. Burrows, S. Campana, G. Chincarini, S. Covino, D. Malesani, G. Tagliaferri, S. Kobayashi, B. Zhang, A.D. Falcone, L. Angelini, S. Barthelmy, A.P. Beardmore, M. Capalbi, G. Cusumano, P. Giommi, M.R. Goad, O. Godet, D. Grupe, J.E. Hill, J.A. Kennea, V. La Parola, V. Mangano, P. Mészáros, D.C. Morris, J.A. Nousek, P.T. O'Brien, J.P. Osborne, A. Parsons, M. Perri, C. Pagani, K.L. Page, A.A. Wells, N. Gehrels, X-ray flare in XRF 050406: evidence for prolonged engine activity. Astron. Astrophys. **450**, 59–68 (2006). doi:10.1051/0004-6361:20054172

S. Rosswog, E. Ramirez-Ruiz, M.B. Davies, High-resolution calculations of merging neutron stars—III. Gamma-ray bursts. Mon. Not. R. Astron. Soc. **345**, 1077–1090 (2003). doi:10.1046/j.1365-2966.2003.07032.x

A. Rowlinson, P.T. O'Brien, N.R. Tanvir, B. Zhang, P.A. Evans, N. Lyons, A.J. Levan, R. Willingale, K.L. Page, O. Onal, D.N. Burrows, A.P. Beardmore, T.N. Ukwatta, E. Berger, J. Hjorth, A.S. Fruchter, R.L. Tunnicliffe, D.B. Fox, A. Cucchiara, The unusual X-ray emission of the short Swift GRB 090515: evidence for the formation of a magnetar? Mon. Not. R. Astron. Soc. **409**, 531–540 (2010). doi:10.1111/j.1365-2966.2010.17354.x

A. Rowlinson, P.T. O'Brien, B.D. Metzger, N.R. Tanvir, A.J. Levan, Signatures of magnetar central engines in short GRB light curves. Mon. Not. R. Astron. Soc. **430**, 1061–1087 (2013). doi:10.1093/mnras/sts683

R. Ruffini, J.A. Rueda, M. Muccino, Y. Aimuratov, L.M. Becerra, C.L. Bianco, M. Kovacevic, R. Moradi, F.G. Oliveira, G.B. Pisani, Y. Wang, On the classification of GRBs and their occurrence rates. Astrophys. J. (2016, in press). arXiv:1602.02732

F. Ryde, Is thermal emission in gamma-ray bursts ubiquitous? Astrophys. J. Lett. **625**, 95–98 (2005). doi:10.1086/431239

F. Ryde, M. Axelsson, B.B. Zhang, S. McGlynn, A. Pe'er, C. Lundman, S. Larsson, M. Battelino, B. Zhang, E. Bissaldi, J. Bregeon, M.S. Briggs, J. Chiang, F. de Palma, S. Guiriec, J. Larsson, F. Longo, S. McBreen, N. Omodei, V. Petrosian, R. Preece, A.J. van der Horst, Identification and properties of the photospheric emission in GRB090902B. Astrophys. J. Lett. **709**, 172–177 (2010). doi:10.1088/2041-8205/709/2/L172

T. Sakamoto, D.Q. Lamb, N. Kawai, A. Yoshida, C. Graziani, E.E. Fenimore, T.Q. Donaghy, M. Matsuoka, M. Suzuki, G. Ricker, J.-L. Atteia, Y. Shirasaki, T. Tamagawa, K. Torii, M. Galassi, J. Doty, R. Vanderspek, G.B. Crew, J. Villasenor, N. Butler, G. Prigozhin, J.G. Jernigan, C. Barraud, M. Boer, J.-P. Dezalay, J.-F. Olive, K. Hurley, A. Levine, G. Monnelly, F. Martel, E. Morgan, S.E. Woosley, T. Cline, J. Braga, R. Manchanda, G. Pizzichini, K. Takagishi, M. Yamauchi, Global characteristics of X-ray flashes and X-ray-rich gamma-ray bursts observed by HETE-2. Astrophys. J. **629**, 311–327 (2005). doi:10.1086/431235

T. Sakamoto, S.D. Barthelmy, W.H. Baumgartner, J.R. Cummings, E.E. Fenimore, N. Gehrels, H.A. Krimm, C.B. Markwardt, D.M. Palmer, A.M. Parsons, G. Sato, M. Stamatikos, J. Tueller, T.N. Ukwatta, B. Zhang, The second Swift burst alert telescope gamma-ray burst catalog. Astron. Astrophys. Suppl. Ser. **195**, 2 (2011). doi:10.1088/0067-0049/195/1/2

R. Sari, P. Mészáros, Impulsive and varying injection in gamma-ray burst afterglows. Astrophys. J. Lett. **535**, 33–37 (2000). doi:10.1086/312689

R. Sari, T. Piran, Hydrodynamic timescales and temporal structure of gamma-ray bursts. Astrophys. J. Lett. **455**, 143 (1995). doi:10.1086/309835

R. Sari, T. Piran, GRB 990123: the optical flash and the fireball model. Astrophys. J. Lett. **517**, 109–112 (1999a). doi:10.1086/312039

R. Sari, T. Piran, Predictions for the very early afterglow and the optical flash. Astrophys. J. **520**, 641–649 (1999b). doi:10.1086/307508

R. Sari, T. Piran, J.P. Halpern, Jets in gamma-ray bursts. Astrophys. J. Lett. **519**, 17–20 (1999). doi:10.1086/312109

R. Sari, T. Piran, R. Narayan, Spectra and light curves of gamma-ray burst afterglows. Astrophys. J. Lett. **497**, 17–20 (1998). doi:10.1086/311269

A.M. Soderberg, S.R. Kulkarni, E. Nakar, E. Berger, P.B. Cameron, D.B. Fox, D. Frail, A. Gal-Yam, R. Sari, S.B. Cenko, M. Kasliwal, R.A. Chevalier, T. Piran, P.A. Price, B.P. Schmidt, G. Pooley, D.-S. Moon, B.E. Penprase, E. Ofek, A. Rau, N. Gehrels, J.A. Nousek, D.N. Burrows, S.E. Persson, P.J. McCarthy, Relativistic ejecta from X-ray flash XRF 060218 and the rate of cosmic explosions. Nature **442**, 1014–1017 (2006). doi:10.1038/nature05087

H. Sun, B. Zhang, Z. Li, Extragalactic high-energy transients: event rate densities and luminosity functions. Astrophys. J. **812**, 33 (2015). doi:10.1088/0004-637X/812/1/33

P.-H.T. Tam, Q.-W. Tang, S.-J. Hou, R.-Y. Liu, X.-Y. Wang, Discovery of an extra hard spectral component in the high-energy afterglow emission of GRB 130427A. Astrophys. J. Lett. **771**, 13 (2013). doi:10.1088/2041-8205/771/1/L13

N.R. Tanvir, R. Chapman, A.J. Levan, R.S. Priddey, An origin in the local Universe for some short γ-ray bursts. Nature **438**, 991–993 (2005). doi:10.1038/nature04310

C.C. Thöne, A. de Ugarte Postigo, C.L. Fryer, K.L. Page, J. Gorosabel, M.A. Aloy, D.A. Perley, C. Kouveliotou, H.T. Janka, P. Mimica, J.L. Racusin, H. Krimm, J. Cummings, S.R. Oates, S.T. Holland, M.H. Siegel, M. de Pasquale, E. Sonbas, M. Im, W.-K. Park, D.A. Kann, S. Guziy, L.H. García, A. Llorente, K. Bundy, C. Choi, H. Jeong, H. Korhonen, P. Kubànek, J. Lim, A. Moskvitin, T. Muñoz-Darias, S. Pak, I. Parrish, The unusual γ-ray burst GRB 101225A from a helium star/neutron star merger at redshift 0.33. Nature **480**, 72–74 (2011). doi:10.1038/nature10611

K. Toma, X.-F. Wu, P. Mészáros, Photosphere-internal shock model of gamma-ray bursts: case studies of Fermi/LAT bursts. Mon. Not. R. Astron. Soc. **415**, 1663–1680 (2011). doi:10.1111/j.1365-2966.2011.18807.x

E. Troja, G. Cusumano, P.T. O'Brien, B. Zhang, B. Sbarufatti, V. Mangano, R. Willingale, G. Chincarini, J.P. Osborne, F.E. Marshall, D.N. Burrows, S. Campana, N. Gehrels, C. Guidorzi, H.A. Krimm, V. La Parola, E.W. Liang, T. Mineo, A. Moretti, K.L. Page, P. Romano, G. Tagliaferri, B.B. Zhang, M.J. Page, P. Schady, Swift observations of GRB 070110: an extraordinary X-ray afterglow powered by the central engine. Astrophys. J. **665**, 599–607 (2007). doi:10.1086/519450

Z.L. Uhm, A.M. Beloborodov, On the mechanism of gamma-ray burst afterglows. Astrophys. J. Lett. **665**, 93–96 (2007). doi:10.1086/519837

Z.L. Uhm, B. Zhang, Dynamics and afterglow light curves of gamma-ray burst blast waves encountering a density bump or void. Astrophys. J. **789**, 39 (2014a). doi:10.1088/0004-637X/789/1/39

Z.L. Uhm, B. Zhang, Fast-cooling synchrotron radiation in a decaying magnetic field and γ-ray burst emission mechanism. Nat. Phys. **10**, 351–356 (2014b). doi:10.1038/nphys2932

Z.L. Uhm, B. Zhang, Toward an understanding of GRB prompt emission mechanism. I. The origin of spectral lags. Astrophys. J. **825**, 97 (2016a). doi:10.3847/0004-637X/825/2/97

Uhm, Zhang (2016b, in preparation)

Z.L. Uhm, B. Zhang, R. Hascoët, F. Daigne, R. Mochkovitch, I.H. Park, Dynamics and afterglow light curves of gamma-ray burst blast waves with a long-lived reverse shock. Astrophys. J. **761**, 147 (2012). doi:10.1088/0004-637X/761/2/147

T.N. Ukwatta, D.R. Stump, J.T. Linnemann, J.H. MacGibbon, S.S. Marinelli, T. Yapici, K. Tollefson, Primordial black holes: observational characteristics of the final evaporation. Astropart. Phys. **80**, 90–114 (2016). doi:10.1016/j.astropartphys.2016.03.007

V.V. Usov, Millisecond pulsars with extremely strong magnetic fields as a cosmological source of gamma-ray bursts. Nature **357**, 472–474 (1992). doi:10.1038/357472a0

J. van Paradijs, C. Kouveliotou, R.A.M.J. Wijers, Gamma-ray burst afterglows. Annu. Rev. Astron. Astrophys. **38**, 379–425 (2000). doi:10.1146/annurev.astro.38.1.379

J. van Paradijs, P.J. Groot, T. Galama, C. Kouveliotou, R.G. Strom, J. Telting, R.G.M. Rutten, G.J. Fishman, C.A. Meegan, M. Pettini, N. Tanvir, J. Bloom, H. Pedersen, H.U. Nørdgaard-Nielsen, M. Linden-Vørnle, J. Melnick, G. van der Steene, M. Bremer, R. Naber, J. Heise, J. in't Zand, E. Costa, M. Feroci, L. Piro, F. Frontera, G. Zavattini, L. Nicastro, E. Palazzi, K. Bennett, L. Hanlon, A. Parmar, Transient optical emission from the error box of the γ-ray burst of 28 February 1997. Nature **386**, 686–689 (1997). doi:10.1038/386686a0

P. Veres, R.D. Preece, A. Goldstein, P. Mészáros, E. Burns, V. Connaughton, Gravitational-wave observations may constrain gamma-ray burst models: the case of GW150914-GBM. Astrophys. J. Lett. **827**, 34 (2016). doi:10.3847/2041-8205/827/2/L34

W.T. Vestrand, J.A. Wren, P.R. Wozniak, R. Aptekar, S. Golentskii, V. Pal'Shin, T. Sakamoto, R.R. White, S. Evans, D. Casperson, E. Fenimore, Energy input and response from prompt and early optical afterglow emission in γ-ray bursts. Nature **442**, 172–175 (2006). doi:10.1038/nature04913

F.J. Virgili, E.-W. Liang, B. Zhang, Low-luminosity gamma-ray bursts as a distinct GRB population: a firmer case from multiple criteria constraints. Mon. Not. R. Astron. Soc. **392**, 91–103 (2009). doi:10.1111/j.1365-2966.2008.14063.x

F.J. Virgili, C.G. Mundell, V. Pal'shin, C. Guidorzi, R. Margutti, A. Melandri, R. Harrison, S. Kobayashi, R. Chornock, A. Henden, A.C. Updike, S.B. Cenko, N.R. Tanvir, I.A. Steele, A. Cucchiara, A. Gomboc, A. Levan, Z. Cano, C.J. Mottram, N.R. Clay, D. Bersier, D. Kopač, J. Japelj, A.V. Filippenko, W. Li, D. Svinkin, S. Golenetskii, D.H. Hartmann, P.A. Milne, G. Williams, P.T. O'Brien, D.B. Fox, E. Berger, GRB 091024A and the nature of ultra-long gamma-ray bursts. Astrophys. J. **778**, 54 (2013). doi:10.1088/0004-637X/778/1/54

I. Vurm, A.M. Beloborodov, J. Poutanen, Gamma-ray bursts from magnetized collisionally heated jets. Astrophys. J. **738**, 77 (2011). doi:10.1088/0004-637X/738/1/77

X.-G. Wang, B. Zhang, E.-W. Liang, H. Gao, L. Li, C.-M. Deng, S.-M. Qin, Q.-W. Tang, D.A. Kann, F. Ryde, P. Kumar, How bad or good are the external forward shock afterglow models of gamma-ray bursts? Astron. Astrophys. Suppl. Ser. **219**, 9 (2015). doi:10.1088/0067-0049/219/1/9

R. Willingale, P.T. O'Brien, J.P. Osborne, O. Godet, K.L. Page, M.R. Goad, D.N. Burrows, B. Zhang, E. Rol, N. Gehrels, G. Chincarini, Testing the standard fireball model of gamma-ray bursts using late X-ray afterglows measured by Swift. Astrophys. J. **662**, 1093–1110 (2007). doi:10.1086/517989

S.E. Woosley, Gamma-ray bursts from stellar mass accretion disks around black holes. Astrophys. J. **405**, 273–277 (1993). doi:10.1086/172359

S.E. Woosley, J.S. Bloom, The supernova gamma-ray burst connection. Annu. Rev. Astron. Astrophys. **44**, 507–556 (2006). doi:10.1146/annurev.astro.43.072103.150558

X.F. Wu, Z.G. Dai, Y.F. Huang, T. Lu, Optical flashes and very early afterglows in wind environments. Mon. Not. R. Astron. Soc. **342**, 1131–1138 (2003). doi:10.1046/j.1365-8711.2003.06602.x

L.-P. Xin, E.-W. Liang, J.-Y. Wei, B. Zhang, H.-J. Lv, W.-K. Zheng, Y. Urata, M. Im, J. Wang, Y.-L. Qiu, J.-S. Deng, K.-Y. Huang, J.-Y. Hu, Y. Jeon, H.-L. Li, X.-H. Han, Probing the nature of high-z short GRB 090426 with its early optical and X-ray afterglows. Mon. Not. R. Astron. Soc. **410**, 27–32 (2011). doi:10.1111/j.1365-2966.2010.17419.x

R.-R. Xue, Y.-Z. Fan, D.-M. Wei, The initial Lorentz factors of fireballs inferred from the early X-ray data of Swift GRBs. Astron. Astrophys. **498**, 671–676 (2009). doi:10.1051/0004-6361/200811361

R. Yamazaki, K. Asano, Y. Ohira, Electromagnetic afterglows associated with gamma-ray emission coincident with binary black hole merger event GW150914. Prog. Theor. Exp. Phys. **2016**(5), 051E01 (2016). doi:10.1093/ptep/ptw042

S.A. Yost, F.A. Harrison, R. Sari, D.A. Frail, A study of the afterglows of four gamma-ray bursts: constraining the explosion and fireball model. Astrophys. J. **597**, 459–473 (2003). doi:10.1086/378288

Y.W. Yu, Z.G. Dai, Shallow decay phase of GRB X-ray afterglows from relativistic wind bubbles. Astron. Astrophys. **470**, 119–122 (2007). doi:10.1051/0004-6361:20077053

Y.-W. Yu, B. Zhang, H. Gao, Bright "Merger-nova" from the remnant of a neutron star binary merger: a signature of a newly born, massive, millisecond magnetar. Astrophys. J. Lett. **776**, 40 (2013). doi:10.1088/2041-8205/776/2/L40

B. Zhang, Astrophysics: a burst of new ideas. Nature **444**, 1010–1011 (2006). doi:10.1038/4441010a

B. Zhang, Gamma-ray bursts in the Swift era. Chin. J. Astron. Astrophys. **7**, 1–50 (2007). doi:10.1088/1009-9271/7/1/01

B. Zhang, Open questions in GRB physics. C. R. Phys. **12**, 206–225 (2011). doi:10.1016/j.crhy.2011.03.004

B. Zhang, Early X-ray and optical afterglow of gravitational wave bursts from mergers of binary neutron stars. Astrophys. J. Lett. **763**, 22 (2013). doi:10.1088/2041-8205/763/1/L22

B. Zhang, Mergers of charged black holes: gravitational-wave events, short gamma-ray bursts, and fast radio bursts. Astrophys. J. Lett. **827**, 31 (2016). doi:10.3847/2041-8205/827/2/L31

B. Zhang, S. Kobayashi, Gamma-ray burst early afterglows: reverse shock emission from an arbitrarily magnetized ejecta. Astrophys. J. **628**, 315–334 (2005). doi:10.1086/429787

B. Zhang, P. Kumar, Model-dependent high-energy neutrino flux from gamma-ray bursts. Phys. Rev. Lett. **110**(12), 121101 (2013). doi:10.1103/PhysRevLett.110.121101

B. Zhang, P. Mészáros, Gamma-ray burst afterglow with continuous energy injection: signature of a highly magnetized millisecond pulsar. Astrophys. J. Lett. **552**, 35–38 (2001). doi:10.1086/320255

B. Zhang, P. Mészáros, Gamma-ray bursts with continuous energy injection and their afterglow signature. Astrophys. J. **566**, 712–722 (2002). doi:10.1086/338247

B. Zhang, P. Mészáros, Gamma-ray bursts: progress, problems, prospects. Int. J. Mod. Phys. A **19**, 2385–2472 (2004). doi:10.1142/S0217751X0401746X

B. Zhang, H. Yan, The internal-collision-induced magnetic reconnection and turbulence (ICMART) model of gamma-ray bursts. Astrophys. J. **726**, 90 (2011). doi:10.1088/0004-637X/726/2/90

B. Zhang, B. Zhang, Gamma-ray burst prompt emission light curves and power density spectra in the ICMART model. Astrophys. J. **782**, 92 (2014). doi:10.1088/0004-637X/782/2/92

B. Zhang, S. Kobayashi, P. Mészáros, Gamma-ray burst early optical afterglows: implications for the initial Lorentz factor and the central engine. Astrophys. J. **595**, 950–954 (2003). doi:10.1086/377363

B.-B. Zhang, E.-W. Liang, B. Zhang, A comprehensive analysis of Swift XRT data. I. apparent spectral evolution of gamma-ray burst X-ray tails. Astrophys. J. **666**, 1002–1011 (2007). doi:10.1086/519548

B.-B. Zhang, B. Zhang, E.-W. Liang, X.-Y. Wang, Curvature effect of a non-power-law spectrum and spectral evolution of GRB X-ray tails. Astrophys. J. Lett. **690**, 10–13 (2009). doi:10.1088/0004-637X/690/1/L10

B.-B. Zhang, B. Zhang, E.-W. Liang, Y.-Z. Fan, X.-F. Wu, A. Pe'er, A. Maxham, H. Gao, Y.-M. Dong, A comprehensive analysis of Fermi gamma-ray burst data. I. Spectral components and the possible physical origins of LAT/GBM GRBs. Astrophys. J. **730**, 141 (2011). doi:10.1088/0004-637X/730/2/141

B.-B. Zhang, B. Zhang, K. Murase, V. Connaughton, M.S. Briggs, How long does a burst burst? Astrophys. J. **787**, 66 (2014). doi:10.1088/0004-637X/787/1/66

B.-B. Zhang, Z.L. Uhm, V. Connaughton, M.S. Briggs, B. Zhang, Synchrotron origin of the typical GRB Band function: a case study of GRB 130606B. Astrophys. J. **816**, 72 (2016). doi:10.3847/0004-637X/816/2/72

B. Zhang, Y.Z. Fan, J. Dyks, S. Kobayashi, P. Mészáros, D.N. Burrows, J.A. Nousek, N. Gehrels, Physical processes shaping gamma-ray burst X-ray afterglow light curves: theoretical implications from the Swift X-ray telescope observations. Astrophys. J. **642**, 354–370 (2006). doi:10.1086/500723

B. Zhang, E. Liang, K.L. Page, D. Grupe, B.-B. Zhang, S.D. Barthelmy, D.N. Burrows, S. Campana, G. Chincarini, N. Gehrels, S. Kobayashi, P. Mészáros, A. Moretti, J.A. Nousek, P.T. O'Brien, J.P. Osborne, P.W.A. Roming, T. Sakamoto, P. Schady, R. Willingale, GRB radiative efficiencies derived from the Swift data: GRBs versus XRFs, long versus short. Astrophys. J. **655**, 989–1001 (2007a). doi:10.1086/510110

B. Zhang, B.-B. Zhang, E.-W. Liang, N. Gehrels, D.N. Burrows, P. Mészáros, Making a short gamma-ray burst from a long one: implications for the nature of GRB 060614. Astrophys. J. Lett. **655**, 25–28 (2007b). doi:10.1086/511781

B. Zhang, B.-B. Zhang, F.J. Virgili, E.-W. Liang, D.A. Kann, X.-F. Wu, D. Proga, H.-J. Lv, K. Toma, P. Mészáros, D.N. Burrows, P.W.A. Roming, N. Gehrels, Discerning the physical origins of cosmological gamma-ray bursts based on multiple observational criteria: the cases of $z = 6.7$ GRB 080913, $z = 8.2$ GRB 090423, and some short/hard GRBs. Astrophys. J. **703**, 1696–1724 (2009). doi:10.1088/0004-637X/703/2/1696

F.-W. Zhang, L. Shao, J.-Z. Yan, D.-M. Wei, Revisiting the long/soft-short/hard classification of gamma-ray bursts in the Fermi era. Astrophys. J. **750**, 88 (2012). doi:10.1088/0004-637X/750/2/88

Space Sci Rev (2016) 202:35–80
DOI 10.1007/s11214-016-0312-x

Gamma-Ray Burst Progenitors

Andrew Levan[1] · Paul Crowther[2] · Richard de Grijs[3,4] · Norbert Langer[5] · Dong Xu[6,7,8] · Sung-Chul Yoon[9]

Received: 26 August 2016 / Accepted: 8 November 2016 / Published online: 21 November 2016

Abstract We review our current understanding of the progenitors of both long and short duration gamma-ray bursts (GRBs). Constraints can be derived from multiple directions, and we use three distinct strands; (i) direct observations of GRBs and their host galaxies, (ii) parameters derived from modelling, both via population synthesis and direct numerical simulation and (iii) our understanding of plausible analog progenitor systems observed in the local Universe. From these joint constraints, we describe the likely routes that can drive

✉ A. Levan
a.j.levan@warwick.ac.uk

P. Crowther
Paul.crowther@sheffield.ac.uk

R. de Grijs
grijs@pku.edu.cn

N. Langer
nlanger@astro.uni-bonn.de

D. Xu
dxu@nao.cas.cn

S.-C. Yoon
yoon@astro.snu.ac.kr

1 Department of Physics, University of Warwick, Coventry, CV4 7AL, UK

2 Department of Physics and Astronomy, University of Sheffield, Sheffield S3 7RH, UK

3 Kavli Institute for Astronomy & Astrophysics and Department of Astronomy, Peking University, Yi He Yuan Lu 5, Hai Dian District, Beijing 100871, China

4 International Space Science Institute–Beijing, 1 Nanertiao, Zhongguancun, Hai Dian District, Beijing 100190, China

5 Argelander-Institut für Astronomie der Universität Bonn, Auf dem Hügel 71, 53121, Bonn, Germany

6 Dark Cosmology Centre, Niels-Bohr-Institute, University of Copenhagen, Juliane Maries Vej 30, 2100 Copenhagen, Denmark

7 National Astronomical Observatories, Chinese Academy of Sciences, Beijing 100012, China

massive stars to the creation of long GRBs, and our best estimates of the scenarios that can create compact object binaries which will ultimately form short GRBs, as well as the associated rates of both long and short GRBs. We further discuss how different the progenitors may be in the case of black hole engine or millisecond-magnetar models for the production of GRBs, and how central engines may provide a unifying theme between many classes of extremely luminous transient, from luminous and super-luminous supernovae to long and short GRBs.

Keywords Gamma-ray burst: general · Supernovae: general

1 Introduction

Since their discovery in the late 1960s (Klebesadel et al. 1973), unveiling the origin of gamma-ray bursts (GRBs) has been a central goal of contemporary astrophysics. While at one point the number of proposed models was only modestly smaller than the number of detected GRBs (e.g. Nemiroff 1994),[1] over the past ∼20 years we have finally narrowed down this progenitor list. The step change in our ability to study GRBs arose from the discovery of afterglow emission in 1997—these panchromatic afterglows (Costa et al. 1997; Frail et al. 1997; van Paradijs et al. 1997) precisely pinpointed GRBs on the sky, enabling their cosmological origin to be secured via observations of both afterglows and host galaxies. It is now clear that GRBs are exceptionally luminous cosmological explosions, with energies (if considered isotropic) of up to 10^{54} erg (e.g. Maselli et al. 2014; Perley et al. 2014), and redshifts ranging from $z = 0.0085$ (35 Mpc) to $z > 8$ (Tanvir et al. 2009), possibly $z > 9$ (Cucchiara et al. 2011).

It became apparent in the early years of GRB observation that the distribution of durations was not a smooth single population, but consisted of at least two peaks (e.g. Mazets et al. 1981, 1982). This difference, secured by observations with BATSE in the 1990s (Kouveliotou et al. 1993), led to the identification of short and long duration GRBs, one with a typical duration (normally defined as t_{90}, the time over which 90 % of the total energy release in γ-rays is recorded) of around 1 s, and the other with a characteristic duration of about a minute. While further observations have identified additional possible sub-classes (see Fig. 1) at low luminosity (Soderberg et al. 2004, 2006; Liang et al. 2007a), or at intermediate (Mukherjee et al. 1998; de Ugarte Postigo et al. 2011) or ultra-long duration (Levan et al. 2011, 2014a), it is the short and long GRBs that make up the vast majority of the *observed* GRB population,[2] and whose origin has been most intensely sought.

[1] A time of launch of the Compton Gamma-Ray observatory (GCRO) and its BATSE instrument there were ∼ 800 GRBs observed from a variety of missions, while approximately 110 models had been proposed. Interestingly, despite this progress, the now favoured collapsar model for long GRBs was not on that list.

[2] The apparent distribution of GRB luminosities is an excellent example of Malmquist bias, where the brightest events are visible over a much larger volume. Hence, while the observed population is dominated by high luminosity events, the volumetric rates are dominated by much lower luminosity systems that generally escape detection.

8 Key Laboratory of Space Astronomy and Technology, National Astronomical Observatories, Chinese Academy of Sciences, 20A Datun Road, Beijing 100012, China

9 Department of Physics and Astronomy, Seoul National University, Gwanak-ro 1, Gwanak-gu, Seoul 151-742, Republic of Korea

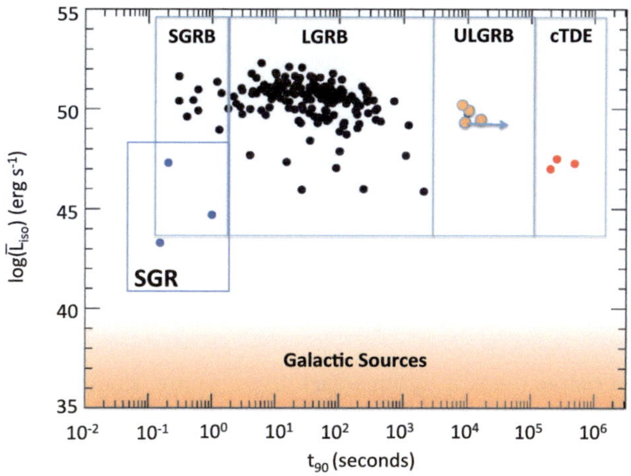

Fig. 1 High energy phase space for gamma-ray bursts, adapted from Levan et al. (2014a). The duration of the bursts is shown, compared to their mean luminosity over that duration. This demonstrates that the majority of observed bursts arise from the long GRB (LGRB) or short GRB (SGRB) population, while outliers are clearly present both at extreme durations, visible as the ultra-long GRBs (ULGRBs) and candidate tidal disruption events (TDEs) and at low luminosities. Indeed, it is the low luminosity long GRBs that provide the best studied associated supernovae. A group of the short GRBs are also likely to arise from giant flares from soft gamma-repeaters (SGRs) in external galaxies (e.g. Palmer et al. 2005; Hurley et al. 2005; Tanvir et al. 2005; Levan et al. 2008), although these have yet to be firmly identified

In this review, we discuss progress towards GRB progenitors that can be made from three distinct strands. The first is direct observations of the GRBs and the host galaxies themselves. From these observations, we can determine the nature of any additional sources of energy in the GRB, be they supernovae (SNe) signatures in long GRBs (e.g. Hjorth et al. 2003), or radioactively powered kilonovae (KNe) created by nucleosynthesis in the neutron-rich disc or ejecta formed in short GRBs (Barnes and Kasen 2013; Berger et al. 2013; Tanvir et al. 2013). These additional sources are frequently observed as late photometric bumps, interrupting the smooth decay of the afterglow light. Their study enables information about the energy and chemical make-up of the GRB explosion to be extracted, and has been very important in pinpointing GRBs progenitors. Studies of the afterglow light also provide the potential to measure beaming angles in GRBs, and hence to convert the observed rate of GRBs to the volumetric rate of GRB-like explosions. Finally, studies of the host galaxies themselves provide information about the stellar population from which the GRB is born, and the dynamics of the progenitors. Taken together these observations provide a significant, but still incomplete view of the stellar systems that create GRBs.

The second strand of our consideration comes from theoretical modelling of both GRB progenitors, and the pathways that lead to their creation. The observed GRB energetics, and the presence of photons from GRBs well above the pair production limit, directly implies relativistic outflows (e.g. Cavallo and Rees 1978), which must somehow pierce their progenitors. This presents immediate and significant constraints on the nature of GRB central engines that must be able to release a significant fraction of a solar rest-mass rapidly, and into a baryon-free environment (the presence of baryons would entrain any ejecta, and make achieving relativistic velocity extremely difficult e.g. Lei et al. 2013). In assessing the progenitors of GRBs from a theoretical perspective it is necessary to both model the details of the proposed progenitor (for example its rotation, the mass of the compact remnant formed, the baryon loading in its immediate environment etc.) and also understand the routes to obtaining these progenitors. Through this route, it is possible to determine both the types of star that might create GRBs, and whether these systems are born at a necessary rate to match the observed GRB population.

The third and final element of this work, and one which is often overlooked is to consider how local populations can inform our studies. Long GRBs are created from massive, and most likely rapidly rotating stars, and there is increasing evidence that these are drawn from low, but not excessively low metallicity (e.g. Fruchter et al. 2006; Graham and Fruchter 2013; Krühler et al. 2015). Indeed, most estimates have a rapid drop-off in the GRB rate somewhere between solar (Krühler et al. 2015; Perley et al. 2015) and 1/3 solar (Graham and Schady 2015b)—in other words, between the Milky Way (roughly solar) and the Small Magellanic Cloud (approximately 1/5th solar). In that sense, studies of the massive stellar populations within the Local Group could place strong constraints on evolutionary pathways that are viable based on star formation in differing environments. Similarly, for short GRBs, the local population of compact binaries, both in the field and clusters can be used as a route to informing the rate and pathways to their production, although again this is a challenging prospect. Such observations are likely incomplete, for example, most NS–NS binaries are found via the radio emission from a spun-up (recycled) pulsar, and there is a clear observational bias against binaries that merge very quickly after their initial formation (i.e. they have already merged and cannot be observed). However, despite these issues, local populations can provide unique diagnostics and constraints and may be able to directly identify plausible GRB progenitors.

By tackling the issue of GRB progenitors from these three routes it is possible to begin to constrain not only the basic properties of the progenitors, but also details about their likely rate, the necessary environmental conditions for their production, the possible presence of their remnants in the local Universe, or the likelihood that any stars identified today may ultimately be GRB progenitors. It is also vital if we wish to place GRBs among the full range of transients produced in the deaths of stars and to understand the links between them. Perhaps most import here are the links between GRBs and super-luminous supernovae (SLSNe, e.g. Gal-Yam 2012). These SNe peak a factor of 100 brighter than most SNe, but interestingly a rather similar set of models are invoked to explain their origin (Gal-Yam et al. 2009; Kasen and Bildsten 2010), and intriguing similarities are present in their environments (Lunnan et al. 2014, 2015; Leloudas et al. 2015; Angus et al. 2016). All of this work has substantial implications in its own right but is also a necessary first step if we hope to use GRBs as increasingly precise cosmological tools, for mapping the history of star formation, the build up of metals, or even as signatures of the collapse of the first stars.

2 What Do Observations Tell Us About Long GRB Progenitors?

Progress towards understanding the nature of long-GRB progenitors has been one of the major success stories of the field and followed rapidly after the first precise locations became available. It is now clear that at least the majority of long GRBs arise from the core collapse of massive stars and are associated with hydrogen poor, high-velocity type Ic supernovae (e.g. Hjorth et al. 2003; Cano 2013). However, beyond this there remain central questions about the progenitors that have yet to be answered; just how massive are the stars creating GRBs, and are they classical Wolf–Rayet stars or something more exotic? Are binary channels important? What is the role of metallicity in creating GRB progenitors and what does this mean about their utility as cosmic probes? Exactly what central engines are created? Are the systems creating the long GRBs similar to those that are seen in the brightest supernovae?

2.1 Building the GRB-SNe Connection

The discovery of the extremely unusual, low luminosity GRB 980425 associated with a broad line type Ic supernova (SN 1998bw) marked the first strong hints as to the progenitors (Galama et al. 2000), although the total energy (if assumed isotropically released) of GRB 980425 was 4–6 orders of magnitude lower than seen in other GRBs, and it had little afterglow (AG) emission (Kouveliotou et al. 2004). Indeed, while GRB 980425 was amongst the first handful of events to be identified, it remains the closest known event of more than 1000 bursts with afterglows to date.

Despite these disparate observational properties, searches for similar supernova signatures in more distant GRBs gradually began, and possible examples were soon found as photometric bumps, causing a re-brightening of the optical counterpart on timescales of 20–30 days after the burst (e.g. Bloom et al. 1999). These photometric bumps apparently had broadly similar peak luminosities to SN 1998bw and importantly were a factor ten or more brighter than the mean peak luminosity of core-collapse SNe. A number of examples were found in the following years, predominantly in the GRB population observed at relatively low redshift ($z < 0.5$) (e.g. Price et al. 2002, 2003a; Garnavich et al. 2003; Bersier et al. 2006), although in some cases with *HST* and 8-m telescopes at somewhat larger distances ($z < 1$) (Masetti et al. 2003; Price et al. 2003b), and in a few cases inferred from bumps in the absence of a redshift (Bloom et al. 1999; Gorosabel et al. 2005; Levan et al. 2005). These observations gradually built a consensus that some form of supernova was present, in at least the majority of long GRBs. However, detailed spectroscopic study, necessary, for example, to identify if hydrogen was present, remained challenging, as the redshifts to the majority of GRBs detected at the time lay beyond the capabilities of available spectroscopy.

Unsurprisingly, the search was on for a low-redshift, but intrinsically energetic GRB that would enable a direct comparison with the local, low-energy GRB 980425. Such an opportunity was presented by GRB 030329, at $z = 0.17$ a redshift at which a supernova signature could be readily visible (SN2003dh). The light curve of GRB 030329 itself was extremely complex (Lipkin et al. 2004), consisting of many re-brightenings due to ongoing energy injection (e.g. Price et al. 2003c; Willingale et al. 2004). However, despite this, the optical light clearly transitioned from the featureless GRB-afterglow power-law into a spectrum with marked similarities to that of SN 1998bw (Hjorth et al. 2003; Stanek et al. 2003), cementing the association of GRBs with broad-lined SN Ic.

2.2 Distribution of Supernova Bulk Properties

A large sample of GRB-SNe or candidate GRB-SNe has now been accrued. The majority of these arise from photometric bumps in the late time light curve, but an increasing number also have direct spectroscopic detections of the SNe in question, although often through noisy spectra at a single epoch.[3] Inevitably this means that the quality of the spectra, and

[3]The GRB/SNe pairs with some spectroscopic evidence include GRB 980425/SN 1998bw, GRB 021211/SN 2002lt, GRB 030329/SN2003dh, GRB031203/SN2003lw, GRB 060218/SN2006aj, GRB100316D/SN2010bh, GRB 10129B/SN2010ma, GRB 111209A/SN 2011kl, GRB120422A/SN2012bz, GRB 130427A/SN2013cq, GRB 130215A/SN2013ez, GRB 130702A/SN 2013dx, GRB 140606B/iPTF14bfu.

decisions of different authors about the merits of including a given burst vary. However, excepting GRB 111209A/SN 2011kl (Levan et al. 2014a; Greiner et al. 2015a) the spectra of all of these events bear marked similarities to that of SN 1998bw, showing broad lines consistent with high-velocity expansion $>20{,}000\ \mathrm{km\,s^{-1}}$, and no sign of hydrogen or helium emission features. All of these SNe are therefore spectroscopically classified as broad-lined type Ic supernovae (SN Ic-BL).

The task of obtaining broad-band spectral shapes and peak luminosities from light curves of these SNe is made challenging by their large luminosity distances, by the contribution of the host galaxy, and by the afterglow contribution. Indeed, at any given time the observed flux depends on the contribution of all three, e.g.

$$F_{\mathrm{obs}}(\nu,t) = F_{\mathrm{AG}}(\nu,t) + F_{\mathrm{SN}}(\nu,t) + F_{\mathrm{host}}(\nu). \tag{1}$$

F_{host} can in principle be obtained from late-time observations and subtraction from early data, although in practice this is often complicated by slightly different filters or instrument combinations, as well as the difficulties introduced by matching ground-based seeing. A typical GRB host is of comparable (or perhaps slightly lower) luminosity than its SNe (Savaglio et al. 2009; Svensson et al. 2010; Perley et al. 2015) and so can create a significant uncertainty, especially for higher redshift bursts, and even more so when spectroscopy is used (since the spectroscopy is inevitably noisier than imaging). The afterglow can normally be treated as a power-law in both frequency (or wavelength) and time e.g.,

$$F_{\mathrm{AG}}(\nu,t) = t^{-\alpha}\nu^{-\beta}. \tag{2}$$

Measurements of the afterglow can be complicated by unseen temporal breaks (e.g. the jet-break due to lateral spreading of the GRB-jet, which typically steepens α to ~ 2 (e.g. Rhoads 1999)) or spectral breaks (e.g. the cooling break, which imparts $\Delta\beta = 0.5$ (e.g. Sari et al. 1998)).

Given the spectral similarities to SN 1998bw, a common approach is to assume that the light curve of the SN evolves in time, t, like that of SN 1998bw but modified by a stretch parameter s, and luminosity scaling,[4] h.

$$F_{\mathrm{SN}} = h F_{\mathrm{98bw}}(t/s), \tag{3}$$

where F_{98bw} is the specific flux of SN 1998bw as observed at the same rest-frame wavelength and time as that of the burst under consideration (i.e. taking into account a k-correction, and cosmological time dilation).

One can fit for these things simultaneously in well-sampled data (e.g. Zeh et al. 2004; Kann et al. 2010; Cano 2013; Cano et al. 2014), although degeneracies can exist (e.g. between h and α), and it is also common to attempt to extrapolate the afterglow contribution, either from earlier observations, or from data in the UV or IR where the SNe contribution is small (e.g. Levan et al. 2014b).

In practice, this is a relatively crude approach, since the SNe light curve may not be a good approximation to a stretched and scaled SN 1998bw. However, in many cases it appears to work well, at least given the data available. More recent attempts have moved beyond simple scaling factors and have attempted to infer bolometric properties, either from the scaled SN 1998bw light curves (e.g. Cano 2013) or using other SNe as templates, enabling single colours to be used for bolometric corrections (e.g. Lyman et al. 2014a, 2016a).

[4]Here we use h to determine the luminosity scaling to avoid confusion with the k-correction, however in other works it is common to see the luminosity scaling expressed as k.

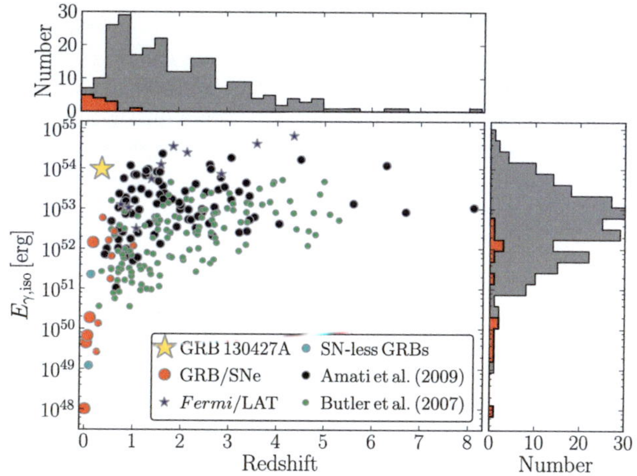

Fig. 2 The distribution of isotropic equivalent energy with redshift for long GRBs from Xu et al. (2013), with the background points determined by Butler et al. (2007) and Amati et al. (2009). Those with spectroscopic detections of a supernova are marked in *red*. As can be seen these are typically low redshift events with a low luminosity (these bursts are likely intrinsically common, but only visible at low-z). Only a handful of events are typical of the more energetic GRBs that make up the bulk of the observed GRB population, with only GRB 130427A lying at the brightest end of the distribution

One thing which is clearly apparent in these observations is that the classical long duration, highly energetic GRBs that comprise the bulk of the *observed* GRB population are highly under-represented in the GRB-SNe sample (Fig. 2). In itself, this is not surprising, since SNe are most readily seen in low redshift examples, and when not outshone by the afterglow (whose luminosity broadly scales with that of the γ-rays, (e.g. Nysewander et al. 2009a; Gehrels et al. 2008)). Indeed, many of the GRBs arise from a potentially distinct population of low-luminosity GRB (LLGRB), while even the Rosetta Stone of GRB 030329/SN 2003dh is of rather intermediate luminosity. Only GRB 130427A/SN 2013cq appears to arise from a highly luminous GRB (Xu et al. 2013; Levan et al. 2014b; Melandri et al. 2014).

While this may not be problematic, it is notable that in some cases these local, low luminosity bursts appear very different from their higher energy cousins. In particular, some are extremely long and dominated by thermal emission (Campana et al. 2006; Starling et al. 2011), features often not seen in other GRBs. It this sense it may be that they are not indicative of GRB-SNe in general, although it may also be that similar components in more distant, luminous GRBs escape detection, and some thermal components have been found in careful searches (Starling et al. 2012; Sparre and Starling 2012), perhaps suggesting similarities. Indeed, the strongest link of the similarities between these very different energies of high energy transient is actually the properties of their supernovae, which are sufficiently similar that it is likely they all arise from the same physical mechanism.

2.3 Test-Bed Examples

An alternative to the large scale samples accrued with data of variable quality is to attempt detailed modelling of well-studied examples, where time series spectra and excellent photometric coverage is available. In these cases it is possible to go beyond simple analytical fits to light curves, or snapshot velocities from spectroscopy, and build detailed models of the explosions of a range of stars that map the nickel releases, velocities, and kinetic energies, providing high-quality spectrophotometric predictions that can be compared with observations. The principle here is to conduct detailed spectral synthesis, in which energy injection from the core is coupled with radiation transport to predict the observed spectrum at given epochs as a function of the various input parameters. In principle, these models can also include detailed geometry, for example, ejecta which includes anisotropies in order to match both the rise/decay times of the SNe. This avoids the need to use scaling relations to obtain SNe properties and instead obtains them from ab initio approaches. It has been successful in several GRBs, including GRB 980425/SN 1998bw (e.g. Höflich et al. 1999; Woosley et al. 1999; Maeda et al. 2002), GRB 030329/SN 2003dh (e.g. Deng et al. 2005) and GRB 060218/SN 2006aj (Pian et al. 2006b; Mazzali et al. 2006). Results from such approaches are reassuringly often similar to those from light curves alone in terms of bulk properties (e.g. nickel mass, Cano 2013), although also provide evidence at times for anisotropies in the explosion, and stronger constraints on pre-explosion core mass and ejecta mass. However, the observational requirements for such work are frequently extreme, and so for the majority of GRBs such detailed information is either too expensive or impossible to obtain via current instrumentation.

2.4 Pushing the Boundaries: From no Supernovae to the Most Luminous Supernovae

While broad-lined type Ic supernovae have been discovered in the vast majority of cases where such a search was plausible there are a handful of cases where such searches have been unsuccessful. In particular, two local long GRBs, 060505 and 060614 (at $z = 0.09$ and $z = 0.125$) lie at redshifts where SNe similar to SN 1998bw should be readily identified, with peak apparent magnitudes in the range $18.5 < m_V < 19.5$. However, deep searches failed to identify any such signatures to limits not only much fainter than GRB-SNe but also to other core-collapse events. Indeed, any SNe in these cases must have been a factor >100 fainter than SN 1998bw (Della Valle et al. 2006; Fynbo et al. 2006; Gal-Yam et al. 2006; Gehrels et al. 2006). It has been noted that GRB 060505 was of relatively short duration $\sim$4 s, while GRB 060614 has subsequently been suggested to be an example of a short burst with extended, softer emission (Gehrels et al. 2006). This scenario seems likely, since further examples of SN-less, but unambiguously long GRBs have not been uncovered. However, other suggestions, for example, that these GRBs may arise in cases where no outward supernova shock is launched and the star collapses directly to a black hole (Fynbo et al. 2006; Fryer et al. 2007) have also been made. This may be particularly relevant given the apparent absence of very massive progenitors to local SN II-P (Smartt et al. 2009), which has led to renewed interest in the prospect of disappearing stars (Kochanek et al. 2008; Kochanek 2014), with recent surveys beginning to find candidate examples (Reynolds et al. 2015). Surprisingly, despite the decade since these discoveries, further examples have not been found, and so their nature remains mysterious.

At the other end of the scale, the recent discovery of an SN-GRB (SN 2011kl/GRB 111209A) which was a magnitude more luminous than SN 1998bw (Greiner et al. 2015a; Kann et al. 2016), challenges the picture of SN homogeneity that has been emerging from previous observations. This SN was found in an ultralong GRB with a duration of $>10{,}000$ s (Gendre et al. 2013; Levan et al. 2014a), and in addition to being more luminous than most GRB-SNe was also spectrally very different. In particular, while it also appears to belong to the SN Ic population, it was far bluer and more UV luminous than other GRB-SNe, and its spectrum bore a strong resemblance to the spectra of type I SLSNe, which also show weak absorption lines on an extremely blue continuum (e.g. Mazzali et al. 2016). Other ULGRBs do not show such strong apparent SNe, and so it remains unclear if this burst is a unique and unusual object, or if it in fact represents a broader range of SNe properties that should be considered, perhaps from cases where the central engine of the GRB begins to impact the supernova itself, as is likely the case for SLSNe (e.g. Cano et al. 2016).

2.5 Large Scale Environments

Additional constraints on the progenitors of long GRBs can be obtained from both their large scale and small scale environments. The metallicities of their host galaxies are of particular interest (in practice, the metallicitics of the small scale environments are also of interest, but are not readily accessible with current technology, since one arcsecond seeing typically corresponds to several kiloparsecs at the redshift of a typical long GRB). Such constraints can either come from direct spectroscopic observations, or via inferences using the well-known correlation between mass (or luminosity) and metallicity. GRBs can provide exceptional diagnostics of metallicity at high redshift via studies of UV absorption lines in concert with the direct detection of Lyα (e.g. Fynbo et al. 2009). However, this is predominantly for higher redshift bursts, since Lyα does not enter the optical window until $z > 2$, and a measurement of the hydrogen column density $N_{\rm H}$ along the line of sight is necessary to obtain the ratio of a given element to hydrogen (e.g. O/H, Fe/H). For this reason the metallicities of the host galaxies where the GRB-SNe connection has been determined are predominantly obtained either from emission line diagnostics such as, $R_{23} = ([\text{O}\,\textsc{ii}](3727\ \text{Å}) + [\text{O}\,\textsc{iii}](4959 + 5007\ \text{Å}))/\text{H}\beta$ or $\text{N2} = \log[\text{N}\,\textsc{ii}]/\text{H}\alpha$, or indeed more complex approaches to line ratios that attempt to remove degeneracies that exist is the more simple examples (e.g. Dopita et al. 2016). This provides some insight into the chemical state of the gas-phase in these galaxies, although several important caveats should be noted in the use of these metallicities as a direct insight into GRB progenitors. Firstly, they are not a measurement of the metallicity of the progenitor star, and indeed, even if they are an accurate measurement of the oxygen metallicity of the star, this is still some way from the [Fe/H] ratios that are more commonly used in distinguishing different stellar evolution pathways. Secondly, significant metallicity gradients can exist within galaxies, and so the use of global proxies essentially derives central metallicities (i.e. those of the brightest regions), which may, or may not be indicative of the regions hosting a given transient event. Thirdly, in some cases, the GRB may be associated with a satellite galaxy of the presumed host. For example, Kelly et al. (2013) show that GRB 130702A (at $z = 0.145$), lies in a satellite galaxy at a significant offset (18 kpc). However, in other cases it is likely that more proximate galaxy/satellite pairs are not adequately resolved, at least by ground-based imaging, leading to the misidentification of the host. In extremum, there can even be chance alignment in which a background galaxy aligns with a foreground system leading not only to the misidentification of the precise host galaxy but the incorrect determination of the redshift. This has recently been demonstrated for an extremely well studied optically dark (i.e. no optical afterglow) GRB 080219B, which, rather than lying at $z = 0.41$ is in fact in a background galaxy at $z = 1.96$ (Perley et al. 2016). In these cases, the use of metallicity measurements inferred from the "apparent" host is clearly incorrect, and potentially problematic. For example, the case of GRB 080219B has been used to argue for the presence of GRBs in high metallicity environments, but in practice provides no such evidence since this metallicity is simply of the foreground system.

The range of metallicities seen in long GRB host galaxies (as well as short GRBs and luminous SNe) is shown in Fig. 3. It is clear that GRBs favour a lower metallicity than is typical in the local Universe, although given the redshift range considered this is not surprising. Interestingly, the metallicities of long GRB hosts at large appear systematically higher

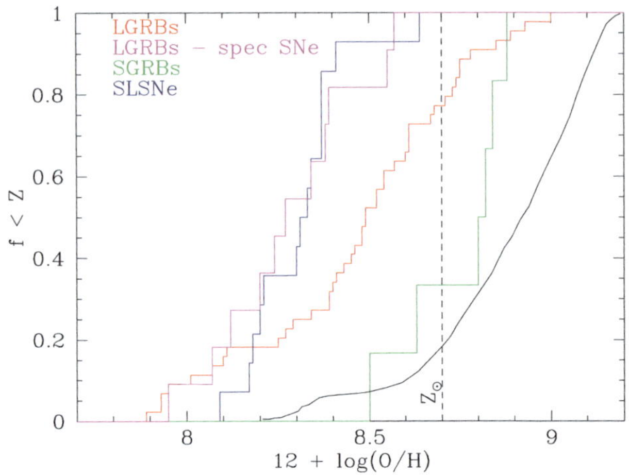

Fig. 3 Host galaxy metallicities for long GRBs from Krühler et al. (2015), short GRBs (from Berger 2009) and SLSNe (from Leloudas et al. 2015). It is striking that the metallicities of LGRBs appear significantly lower than for the SGRBs, or indeed the general galaxy population (Graham and Fruchter 2013). However, those of the GRBs with apparent SNe appear to be somewhat lower still, and are very similar to those observed in SLSNe host galaxies

than those of the examples in which a spectroscopic signature of the SNe has been seen, with the median metallicity of the GRB-SNe sample being 0.2–0.3 dex lower. This may well be due to a bias against dusty systems, since obscured star formation is typically more metal rich, and it is clear that dusty GRBs arise in more luminous, and likely metal-rich host galaxies. In any case, it also interesting to note, that when considering only the GRBs in which spectroscopic SNe signatures are seen there is an apparent similarity between their metallicity distribution and those of SLSNe. The degree to which these differences arise due to observational selection effects, or alternatively due to genuine astrophysical differences is of significant interest for further study. It may reflect genuine difference between populations, or perhaps that even in the relatively low redshift regime there may be significant selection effects (e.g. Japelj et al. 2016).

2.6 Small Scale Environments

The large scale environments can provide information about the chemical state of GRB hosts that is not currently available at smaller scales, due to both large luminosity distance, and the inability of large aperture telescopes to probe with sufficiently high resolution. However, star formation takes place on small spatial scales, from less than a parsec to tens of parsecs depending on the mode and intensity of star formation (e.g. Portegies Zwart et al. 2010). It is these small scale environments that in practice carry most information about the GRB progenitor itself, free from dilution from the light of the remaining galaxy. At a typical GRB redshift, the *HST* resolution (0.1 arcsec) still corresponds to physical scales of several hundred parsecs, and so while we are able to track the stellar populations in the regions of the GRBs, isolating the stellar population responsible for the GRB remains challenging. Two approaches can be considered when studying GRB locations. The first is simply to identify the GRB location relative to its host galaxy by measuring its offset, either relative to the galaxy host (e.g. Bloom et al. 2002), or perhaps to local regions of intense star formation, if such resolution is available (e.g. Hammer et al. 2006). The alternative is to try to study the population under the burst, despite the poor spatial resolution. This may be highly diagnostic for long GRBs, since the young massive stars are by far the most luminous in the host galaxy, especially at UV wavelengths (Fruchter et al. 2006). Indeed, at the typical redshifts of many GRBs rest frame UV light is redshifted into the optical window, making sensitive high-resolution imagery with *HST* an ideal route to characterising the immediate massive star populations surrounding GRBs. In addition, these hot massive stars also excite the gas phase of the interstellar medium, making narrow-band observations a sensitive probe of the young massive star population (James and Anderson 2006).

Such an approach has been taken by various groups studying both supernovae and GRBs. Fruchter et al. (2006) consider the total fraction of the galaxy light in pixels of lower surface brightness than the pixel containing the GRB or SNe, the so-called F_{light} parameter. They show that GRBs are highly concentrated on the light of their host galaxy, while core collapse SNe broadly trace the distribution of rest-frame UV-light. In other words, the probability of a SN occurring in a given pixel is approximately proportional to the brightness of that pixel, in contrast, the probability of a GRB occurring in a given pixel is proportional to the brightness of that pixel squared.[5] Interestingly, further studies of local SNe show that a similar effect can be seen in the distribution of hydrogen-rich SNe II and hydrogen/helium poor SNe Ic (James and Anderson 2006; Kelly et al. 2008; Anderson et al. 2012). This is likely understood because of the strong correlation between stellar mass and luminosity. In the low mass range (10–20 $\mathrm{M}_\odot$) $L \sim M^3$, while for higher mass stars (50–100 $\mathrm{M}_\odot$) $L \sim M^2$ (Yusof et al. 2013) such that the most massive stars dominate the UV budget. In this sense, if core collapse SNe arise from essentially every star with an initial mass $>8\ \mathrm{M}_\odot$, then GRBs must arise from stars significantly more massive than this. Studies based on the expected distribution of stars in galaxies (Raskin et al. 2008), and on young clusters in nearby galaxies (Larsson et al. 2007) provide a consistent picture in which GRB progenitors have initial masses of $>40\ \mathrm{M}_\odot$, representing one of the few ways in which the masses of GRB progenitors can be ascertained.

2.7 Summary: LGRB Progenitors

The emerging picture for long GRB progenitors appears clear. A long GRB progenitor is a massive star, typically born at sub-solar metallicity, in which the hydrogen envelope is lost/burned prior to its explosion as a broad-lined type Ic supernova. There is some observational evidence that suggests these stars are very massive, for example, the large Nickel yields required to achieve absolute magnitudes a factor 10 brighter than SNe Ic (e.g. Mazzali et al. 2007a), or the location of the bursts on the brightest regions of their hosts (Fruchter et al. 2006; Svensson et al. 2010) could both be interpreted as signatures of very massive stars, perhaps with an initial mass at the zero-age main sequence (ZAMS) of $M_{\mathrm{ZAMS}} > 40\ \mathrm{M}_\odot$. However, the locations of long GRBs are similar to those of the bulk SNe Ic population (Kelly et al. 2008), most of which can be explained by the explosion of initially far less massive stars (e.g. Mazzali et al. 2007b,c), and more complex models, involving the impact of chemically homogeneous evolution (e.g. Szécsi et al. 2015), or binaries (e.g. Stanway et al. 2016) on the observed environments may be necessary to provide tighter constraints on the progenitor mass, as well as on the progenitor rotation. These scenarios are not currently well constrained observationally, and theoretical progress is discussed in Sect. 5. None-the-less, the progress which has been made towards the progenitors of long GRBs has been remarkable, and while there remain many open questions about the details of long GRB progenitors, it is fair to say that barring a few special examples, we now know what forms the vast majority of the long GRB population.

[5] The squared exponent here is approximate, and does not carry any specific meaning.

3 Observational Constraints on Short GRB Progenitors

3.1 Defining Short GRBs

Short GRBs are traditionally defined as those with durations of $t_{90} < 2$ s. Observations with *Konus* and *BATSE* suggested they form a separate population from the long GRBs, and so may well also have different progenitors (Mazets et al. 1981, 1982; Kouveliotou et al. 1993). Following the identification of supernovae signatures in the afterglows of long GRBs, the favoured model for short bursts rapidly became the merger of two compact objects (neutron star—neutron star or neutron star—black hole), a model which had previously been popular for the long GRB population (e.g. Narayan et al. 1992). The reason for this was two-fold; both that the rapid time scale for the merger, combined with the clean environment (compared to e.g a collapsing massive star) meant that short durations were more naturally expected in merger scenarios. The second was the rather more speculative assertion that if mergers did not produce the long bursts, they must either produce the short bursts, or no high energy transient at all. Given this historical approach, it is unsurprising that the observational history of short GRBs has largely been benchmarked against this expectation, and hence the discussion of short GRBs has been largely framed as though their progenitors are known, even prior to the accrual of the significant observational data that now supports a merger origin.

Short bursts now represent approximately 10 % of the bursts observed by *Swift*, down somewhat from the 25 % observed in the BATSE observations. It seems likely that a significant element of this is due to the differing spectral responses of the different instruments, with BAT operating in a rather softer 15–150 keV band compared to BATSE (50–>300 keV). This has led to some discussion of whether the duration split between the two populations is correctly set at 2 seconds. Indeed, the duration distribution of *Swift* GRBs is far less clearly bimodal than that for BATSE, and so one cannot directly distinguish a "cut-off" based on these observations alone. Several alternative attempts in the early years of *Swift* addressed the potential problem by suggesting the use of additional diagnostics to distinguish between different progenitor types (e.g. Levan et al. 2007; Zhang et al. 2009). Indeed, Zhang et al. (2009) argued for a distinction between Type I (compact object merger) and Type II (collapsar) events based on a decision tree of multiple observational properties. These additional constraints often used rather more complex properties (for example the nature of the host galaxy, or the presence/absence of a supernova component) and so also risked significant confirmation bias, in finding bursts which met the expectations based on only a handful of systems and omitting other, potentially valuable systems. Using theoretical considerations, largely based on the break-out time for a collapsar, combined with the different spectral responses of different observatories Bromberg et al. (2013b) have suggested that very different values for T_{90} would be needed to identify the split between long and short GRBs for different missions. In particular, 2 s for BATSE and 0.7 s for *Swift*.[6] The essence of the argument is that hardness is a more important distinction than duration at durations where there is likely to be significant overlap between the two populations. This is shown graphically in Fig. 4, where the *Swift* short GRBs can be seen to be typically much softer than those detected by BATSE. Applying a cut of this nature would remove some (although by no means all) of the best studied "short GRBs" from consideration, since they would then be more likely to be collapsars.

[6] At 50 % probability. Since the standard assumption is to think of the duration distribution of long and short GRBs as Gaussians, there is inevitably overlap in the two populations.

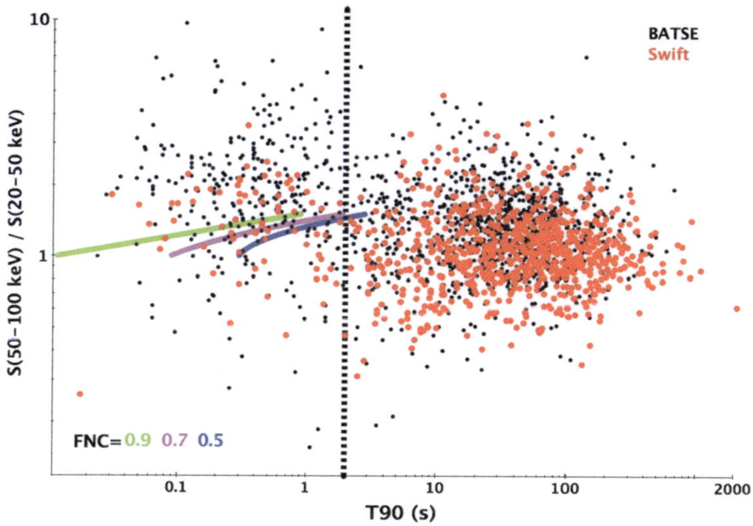

Fig. 4 The hardness-duration distribution of GRBs detected by *Swift* (*red*) and BATSE (*black*). The *x-axis* shows the duration over which 90 % of the total fluence is recorded (T_{90}, while the *y-axis* shows the γ-ray hardness, defined as the ratio of flux between two adjacent channels (the channels chosen are those used by BATSE). The *Swift* data have been placed on the same scale by extrapolating the prompt spectral fits (either as a power-law or cut-off power-law) to the appropriate bands. The *dotted vertical line* shows the notional distinction between short and long GRBs at two seconds, but there is clearly significant overlap. It is apparent from the BATSE distribution that short GRBs are *on average* harder than long GRBs. However, for *Swift* the two distributions look very similar. Given this the contribution of the long GRB population is larger for *Swift* bursts at durations of <2 s than for BATSE. The lines suggested by Bromberg et al. (2013a) as distinguishing long and short GRBs at 50, 70 and 90 % confidence are also shown (where FNC is the fraction of non-collapsars), bursts below these lines are more likely to be collapsars from the short duration tail of the long GRB population. For *Swift*, only bursts above the 90 % *green line* are highly likely to be genuine members of the short burst population

3.2 Precise Locations from Afterglows

While progress towards the origin of long GRBs proceeded at pace from the discovery of the first afterglows to the identification of broad-lined SNe-Ic, pinpointing the progenitors of the short GRBs remained much more challenging. Firstly, short GRBs are typically significantly fainter, or more precisely exhibit markedly lower fluence than long GRBs despite similar peak fluxes (a parameter which more accurately describes their detectability to many γ-ray observatories). Since broad correlations exist between the fluence of the prompt emission and the brightness of either the X-ray or optical afterglow (Gehrels et al. 2008; Nysewander et al. 2009b) it is perhaps unsurprising that the afterglows of short GRBs escaped detection while the afterglow revolution was transforming our knowledge of the long GRB phenomena. Indeed, while a handful of short GRBs were detected and reasonably localised during the long GRB afterglow revolution, no successful afterglow campaigns were made prior to the launch of *Swift*. Its ability to re-point its X-ray and UV-optical telescopes rapidly provided the first X-ray afterglow detection for GRB 050509B (Gehrels et al. 2005; Hjorth et al. 2005a; Bloom et al. 2006). Interestingly, the first optical afterglow detection actually came from a HETE-2 burst, GRB 050709 (Hjorth et al. 2005b; Fox et al. 2005), which with the detection of another HETE-2 burst in early 2006 (de Ugarte Postigo et al. 2006; Levan et al. 2006), perhaps suggested that previous searches had been rather unlucky, although doubtless the ability of *Swift* to rapidly repoint and detect their X-ray afterglows was also an important component in the discovery of their optical counterparts.

As with long GRBs, the discovery of afterglows to short GRBs has revolutionised their study. Perhaps the most important diagnostic enabled by an afterglow is a precise location on the sky, and hence the ability to study the galaxy population hosting short GRBs. The first short burst with an afterglow was GRB 050509B, and its location was striking, being offset ∼30 kpc from an extremely massive galaxy in a merging cluster system at $z = 0.225$ (Gehrels et al. 2005; Bloom et al. 2006). Deep searches for star formation in this host galaxy revealed no sign, and immediately suggest that some short GRBs arise from ancient populations. However, this burst was localised only by its X-ray afterglow, and the large resulting error box contained additional background galaxies, many of which had blue colours consistent with star formation. Hence, while the probability of chance alignment was low, the association did not clinch the origin of at least some short GRBs in ancient populations. The discovery of the optical afterglow of GRB 050724 in a clearly elliptical host galaxy further strengthened this argument (Berger et al. 2005), and while bursts in low star formation rate host galaxies are clearly in a minority, perhaps 10–20 % of short GRBs overall do arise from elliptical hosts (Fong et al. 2013). Equally, while the remaining 80–90 % of the host galaxies do show signs of significant star formation, they remain distinct from the hosts of the long bursts in terms of their stellar masses and star formation rates. While long GRBs show a strong preference for star-forming dwarfs, with low metallicities and high specific star formation rates (e.g. Fruchter et al. 2006; Savaglio et al. 2009; Svensson et al. 2010; Graham and Fruchter 2013), the short bursts are in rather more typical galaxies, with a range of metallicities and higher stellar masses, therefore they appear to sample the entirety of the galaxy population (Leibler and Berger 2010; Fong et al. 2013).

Beyond the nature of the galaxies themselves, there is significant information contained within the distribution of the short GRBs on their hosts. Again, the differences between long and short GRBs become apparent rapidly in this distribution. While long GRBs are highly concentrated on their host light (Bloom et al. 2002; Fruchter et al. 2006; Svensson et al. 2010, see Sect. 2.6) the short GRBs are frequently scattered on their host light, and at large projected radii, in some cases there is apparently no underlying host galaxy, despite relatively bright galaxies nearby in the field, suggesting that these progenitors have been kicked from their birth sites (Berger 2010; Tunnicliffe et al. 2014), although it remains plausible that some lie at much higher redshifts in galaxies too faint for current observations (Levan et al. 2006; Berger et al. 2007). Beyond this much broader distribution, those short GRBs that do lie on galaxies tend to appear on much fainter pixels than those in the long GRBs (Fong et al. 2010; Fong and Berger 2013), and in particular show essentially no association with the blue light of their host galaxies. Taken together these properties demonstrate that short GRB progenitors are frequently old, and more importantly are also kicked from their birthplaces. These are all consistent with the expectations of binary mergers, in which the neutron stars receive substantial space velocities (kicks) from a combination of natal kicks and binary mass loss (e.g. Arzoumanian et al. 2002).

3.3 Kilonovae and Macronovae

The large scale observations are clearly broadly in favour of a model in which short GRBs arise from compact object mergers. However, they are in themselves not a smoking gun of the merger itself. This situation is reminiscent of early studies of long GRBs in which the star-forming host galaxies offered evidence of a link to massive stars and supernovae, but did not conclusively prove it. Since afterglows are generated at large radii and are essentially a product of the interaction of an outgoing relativistic shock with an external medium, they themselves are not uniquely diagnostic of the progenitor. Hence, some additional signature that could in principle only arise in compact object mergers was necessary. The most likely scenario is to identify a faint radioactively powered transient, created by nucleosynthesis in the neutron-rich material available in a compact object merger containing a neutron star. This material may reside either in the accretion disc, or be ejected into tidal tails. None-the-less it was recognised that for typical properties of a merger such a transient would yield a rapidly evolving, faint transient, significantly faster and fainter than normal core-collapse SNe (e.g. Li and Paczyński 1998). Like a supernova, these events would start faint and rise to peak on timescales of hours to days, and so would be seen as photometric bumps, interrupting the otherwise smooth decay of the afterglow. These events have had a range of names over the years, but those which are commonly used today are either kilonova (referring in essence to something about 1000 times brighter than a nova), or macronova. The precise properties of these transients were uncertain, and while some populations of faint, fast transients have been uncovered by recent synoptic sky surveys (e.g. Kasliwal et al. 2012; Foley et al. 2013), none have been interpreted as NS–NS or NS–BH mergers (although some have been suggested to arise from NS-white dwarf mergers, through channels similar to those creating NS–NS binaries (Metzger 2012; Lyman et al. 2014b, 2016b). Hence, there is little *observational* evidence as to the signatures that observers should search for. Indeed, early, deep observations of several short GRBs failed to uncover any sign of the moderate to late time "bumps" expected in this scenario (e.g. Hjorth et al. 2005a) and ruled out the most optimistic scenarios for these transients (Metzger and Berger 2012).

However, an important revelation came from detailed calculations conducted in 2013 (Barnes and Kasen 2013). It had long been recognised that NS–NS mergers were promising sites for r-process nucleosynthesis, and that NS–NS mergers (depending on the rates and individual r-process outputs) could be important, if not dominant sites for the creation of the heaviest elements (Rosswog et al. 1999, 2003). Indeed, a range of recent studies from the abundances of radioactive elements on the sea floor (Wallner et al. 2015) to in depth analysis of metal poor stars in the Milky Way (e.g. Macias and Ramirez-Ruiz 2016) suggest that the r-process abundances are not in an equilibrium that would be expected if they were regularly replenished with small amounts of additional material, as might be expected for core collapse supernovae. Instead, these analyses prefer a scenario in which rare events, with significantly more mass per event are dominant, favouring a merger origin. This nucleosynthesis should naturally power a luminous transient, but because of the dominance of r-process elements its evolution might be quite different to those previously assumed based on our studies of supernovae. Indeed, the opacities that were assumed in the earlier predictions were predominantly those of iron group elements which are dominant in normal SNe. The impact of including new opacities for heavy elements, in particular, lanthanides, was profound. The heavy opacity extinguishes essentially all the optical light for external observers and means that earlier observations were, in essence, unconstraining of the nucleosynthesis taking place. These observations instead suggested that IR observations were the most promising route to the identification of a kilonova (Barnes and Kasen 2013).

This supposition was observationally tested with *HST* observations of the short GRB 130603B in June 2013. These observations took place in two bands, one in the optical and one in the IR, for a burst at $z = 0.35$ (de Ugarte Postigo et al. 2014). Although only two epochs were obtained they successfully showed a fading IR source, while nothing was seen in the optical (Berger et al. 2013; Tanvir et al. 2013). A comparison with the IR afterglow decay suggested a magnitude much brighter than expected, and hence a re-brightening from

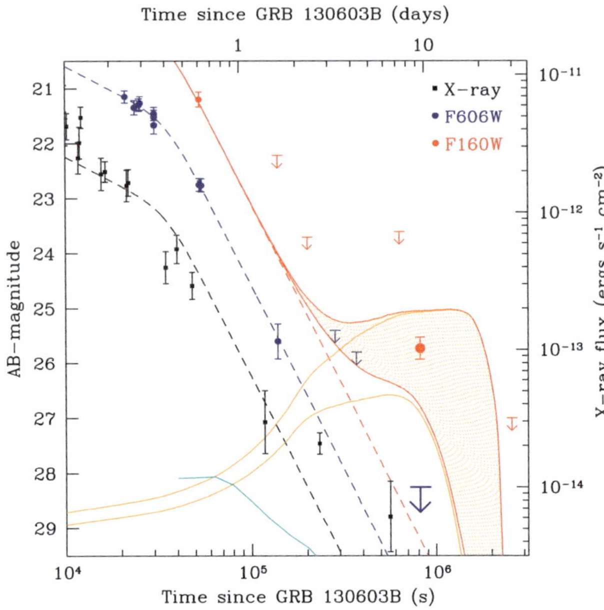

Fig. 5 The *light curve* of the kilonova seen in GRB 130603B (updated from Tanvir et al. 2013). The points represent the X-ray (*black*) optical (*blue*) and IR (*red*) photometry of the afterglow, along with their expected decay. The *shaded region* is the expected kilonova range for 10^{-1} and 10^{-2} $M_\odot$ of material from Barnes and Kasen (2013). While the *cyan line* shows the very faint expected optical emission. The re-brightening in the IR is strongly suggestive of a kilonova, although it is also relevant to note that late time X-ray observations by Fong et al. (2014) also imply that the X-ray lies above expectations

an associated kilonova (Fig. 5). The luminosity of the IR source was very similar to the prediction of kilonovae using the improved lanthanide opacities, and providing the first direct evidence for the origin of short GRBs in compact binary mergers.

Unfortunately, opportunities to further hone our understanding remain limited due to the rate of short GRBs at sufficiently low redshift, and so to date there are limited constraints on the true properties of this KNe. A detailed re-analysis of some archival short GRBs provides some evidence for similar components in their light curves (e.g. Jin et al. 2015), although other short GRBs show optical or X-ray bumps apparently not associated with the same physical process (Perley et al. 2009). Indeed, recent work, partly motivated by an apparent X-ray bump co-incident with the infrared kilonova in GRB 130603B (Fong et al. 2014) has focused on possible long-lived X-ray manifestations of mergers. In particular, in the form of scattered X-ray's from the central engine (either millisecond magnetar or black hole) (Kisaka et al. 2015), the absorption and re-radiation of X-rays to provide IR and X-ray signals (Kisaka et al. 2016), or direct emission of X-rays from the engine itself (Sun et al. 2016), perhaps creating isotropic X-ray emission that could be of some value in searches for mergers without a GRB trigger (see below). However, given the paucity of observations to date we have still to distinguish between various possible suggestions for the emission processes at play, including the role of X-ray power, the true nuclear yields and the balance between material ejected into the accretion disc or into tidal tails around the merger.

3.4 Gravitational Wave Detection

The mergers of compact objects were long expected to be the first observed gravitational wave signatures, and indeed NS–NS and NS–BH were long considered prime suspects for this. The recent discovery of a binary black hole merger with a particularly high total mass (>60 $M_\odot$) is therefore rather surprising at first sight (Abbott et al. 2016b), although it should be noted that the significantly higher masses here than normally considered result in a far larger horizon for massive BH–BH than for NS–NS systems (the measured strain scales approximately as M_1M_2), and so a significant astrophysical population of BH–BH binaries may make them preferentially detected (Abbott et al. 2016a,c). However, at the strain sensitivities now reached by the next generation gravitational wave detectors, and given the inferred rates of NS–NS and NS–BH mergers from both population synthesis and observed populations (see below and Abadie et al. (2010)) it is expected that mergers containing neutron stars could be found shortly.

The simultaneous detection of GW and GRB signals offers significant advantages. The GRB signal provides a precise time and location for a GW search. This in turn dramatically reduces the number of trials that must be run on the gravitational wave interferometer data stream, and means that the effective sensitivity increases substantially, perhaps by a factor $\sim$2 or more (e.g. Dietz et al. 2013). However, these advantages also extend to the study of short GRB progenitors. For example, the chirp mass determined by the GW detection would immediately identify a system as either a NS–NS or NS–BH merger. Indeed, the combination of GRB, KNe, and GW detection provides a series of different routes to probing the compact binary population over very different distance scales (see Fig. 6). In particular, SGRBs can be detected out to $z \sim 1$ or beyond (e.g. Graham et al. 2009; Thöne et al. 2011b), but likely only illuminate a small fraction of the sky due to their relativistic jets. This makes the probability of joint GRB–GW triggers relatively small, as the fraction of mergers in a volume limited sample where the GRB jet is aligned with the observer is low. Beaming is highly uncertain in short-GRBs, but beaming factors $>$10–100 seem likely (e.g. Chen and Holz 2013). This is somewhat offset since the relativistic jets are visible to observers face on to the merging binary, a geometry which also maximises the detectability of the gravitational waves (see Fig. 6 and Nissanke et al. 2011). The most promising counterparts are therefore systems that may emit isotropically. Kilonova are visible to sensitive IR searches over distances of several hundred Mpc, well matched to the sensitivities of current gravitational wave detectors. Similarly, isotropic X-rays with modest luminosity ($L_X \sim 10^{43}$ erg s^{-1}) should also be visible to *Swift* or other X-ray telescopes out to distances $>$100 Mpc. It is likely that both gravitational wave detections, and observations from the γ-ray to IR, and possibly beyond (for example magnetars should provide long-lived radio emission, e.g. Gompertz et al. 2015; Fong et al. 2016) will be necessary to fully constrain compact object mergers, their rate, and their role in r-process production.

4 Central Engines: Linking the Most Luminous Explosions

While progress towards the nature of the progenitor stars of long GRBs has been impressive there has been increasing focus on the nature of the compact object formed in the core of the collapsing star. These objects are the engines which drive the explosions and create the ultra-relativistic jets which ultimately pierce the star. Traditional models extract the GRB energy through some form of accretion onto a nascent black hole (Woosley 1993; Fryer et al. 1999). Accretion rates of up to several solar masses per second at typical efficiencies ($\sim$10 %) can create the necessary luminosities to explain both long and short GRBs (e.g. Oechslin and Janka 2006), provided that a sufficiently low density of baryons exists, which seems plausible along the rotation axis of rapidly rotating stars. Recently there has been a significant shift of focus to consider energy input from a millisecond-magnetar, a newly formed highly magnetic neutron star with $B > 10^{14}$ G, and spin periods of a few milliseconds (Metzger et al. 2011). Note that these, newly formed magnetars are quite different from the population of magnetars identified in the Milky Way as Soft Gamma Repeaters (SGRs)

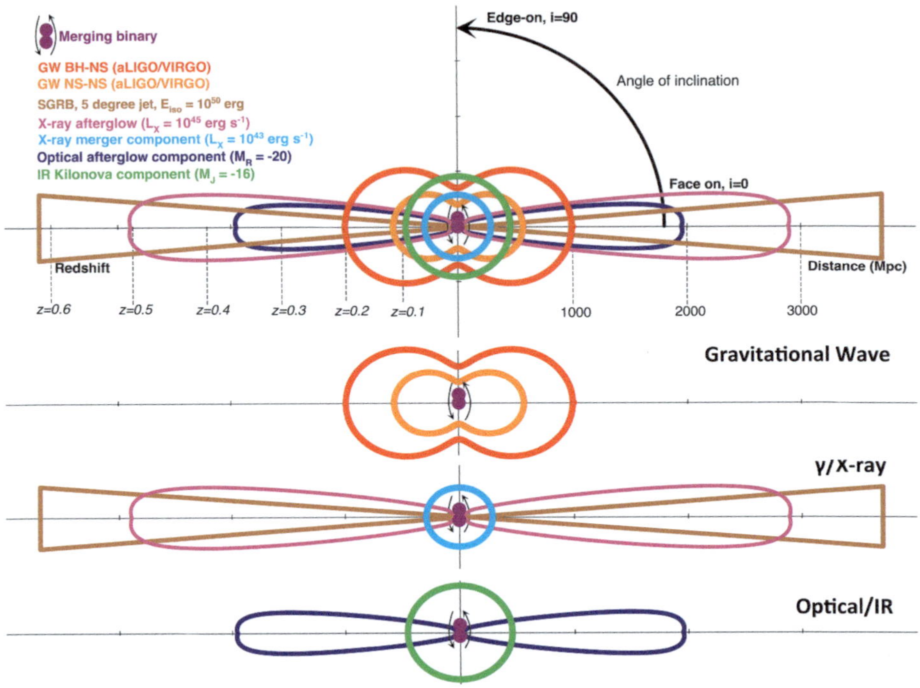

Fig. 6 The characteristic visibility horizons for the detection of different observational manifestations of compact object mergers. The *top panel* shows all the likely signatures overplotted for comparison, while they are broken down by detection route in the *lower panels*. Mergers can be detected by the Advanced generation of LIGO and VIRGO detectors out to hundreds of Mpc for both NS–NS and NS–BH mergers (Nissanke et al. 2011), although there is a strong bias towards face-on events, which can be observed to distances factors of >2 larger than for edge on systems. This increases the probability of observing them in coincidence with short GRBs, which are also likely to be visible only to face on observers. However, such overlapping events will likely remain in a minority. Other wavelengths of emission are likely to require triggered observations after the detection of a gravitational wave or short-GRB trigger. The horizons given are based on typical depths likely to be achieved fairly short exposures, several hours–days after the trigger with the *Swift* BAT ($S_\gamma = 1 \times 10^{-7}$ erg cm^{-2}, although the BAT in practice triggers on short GRBs based on their peak flux), *Swift* XRT (limiting $F_X \sim 1 \times 10^{-12}$ erg s^{-1} cm^{-2} in ~1000 s), ground based robotic optical telescopes (R = 21) and the VISTA survey telescope in the IR (J(AB) ~ 22). These horizons can obviously be scaled to different depths or time after trigger for a given assumption of temporal behaviour. In particular, observations of short-GRBs require much smaller areal coverage and so probe much deeper and farther. For example, *HST* can see kilonova signatures out to $z \sim 0.4$. Furthermore, uncertainty as to the detailed properties of the emission means that any such horizons should be viewed as indicative rather than precise. In particular, we properties of off-axis GRB emission which enables bursts to be viewed at larger angles than a direct GRB remains to be directly tested, and the properties of isotropic optical/IR (Barnes and Kasen 2013) or X-ray kilonova/macronova (Kisaka et al. 2015; Sun et al. 2016) that might be powered either directly by radioactivity, or by the action of a central engine such as a magnetar, remain uncertain. For SGRBs beamed signals have been the dominant source of information to date, however, it seems likely that isotropic signatures will be more promising in the majority of gravitational wave transients

and Anomalous X-ray Pulsars. The Galactic systems have typically lower fields, and importantly much slower rotation rates (several seconds) than those powering GRBs (see e.g., Olausen and Kaspi 2014). Indeed, even if the spin-down evolution of the Galactic magnetars is reversed, it seems unlikely that their early properties were consistent with those postulated as GRB central engines (Rea et al. 2015). The strong magnetic fields result in rapid extraction of energy from the dipole field, and the total energy extracted can approach 10^{52} erg, comparable to the typical isotropic energy releases of GRBs, and significantly above the beaming corrected values. Indeed an apparent upper limit on the kinetic energy releases of GRB-supernovae of around 10^{52} erg provides additional support for such a model (Mazzali et al. 2014). These millisecond-magnetars would eventually spin down to much longer periods, at which point they would either become dormant or, depending on their mass may collapse to black holes at the point at which they cease to be centrifugally supported.

In any case, it is clear that in the majority of GRBs this central engine inputs a large fraction of a solar rest mass on a timescale of tens to hundreds of seconds, perhaps extending to several thousand seconds in some cases. Indeed, evidence for prolonged energy injection existed before the launch of *Swift* (Vaughan et al. 2004; Watson et al. 2006), but was only clearly identified following the direct detection of X-ray flares in the afterglows bursts early in the *Swift* era (Burrows et al. 2005; Nousek et al. 2006), and was surprisingly found in both long bursts and short bursts (Berger et al. 2005). While in long bursts this late activity might naturally arise from processes involving fallback accretion, such emission was not naturally expected in the case of short GRBs should they arise from neutron star mergers, since the merger itself should be over very rapidly, with a relatively clean environment. Hence, if a common origin for the long and short GRBs was in accretion onto the nascent black hole we might expect the clean environment to provide a significant observational distinction between the two scenarios. While short GRB afterglows are markedly fainter, the presence of long-lived emission does not offer such easy solutions. This may be due to instabilities that build up in the accretion discs around both long and short GRBs and deliver similar flares (Perna et al. 2006a), or because some fraction of short GRBs arise from black hole-neutron star mergers, in which the neutron star is tidal shredded over several pericenter passages (e.g. Davies et al. 2005).

However, the millisecond-magnetar model may be particularly appealing here. Millisecond-magnetars are likely created from the collapse of some fraction of massive stars (see below) while they could also be formed via the merger of either white dwarfs or neutron stars, provided the final mass is below the maximum mass of a neutron star (Usov 1992; Levan et al. 2006). These neutron stars with extreme magnetic fields can have rapid spin-down times, or may even be unstable and centrifugally supported, such that as magnetic braking slows their rotation from $\sim$1 ms at birth they ultimately collapse to form black holes. Importantly, millisecond-magnetars provide a route to providing energy input into the GRB afterglow on timescales of minutes to hours (or potentially even longer) after the initial burst. While the models have numerous free parameters, millisecond-magnetar models can explain many long and short GRB light curves (e.g. Rowlinson et al. 2013). However, this injection of energy is not without its problems, since it should in many cases yield a strong radio afterglow. The absence of such afterglows may be difficult to remedy with rapidly spinning magnetar models (e.g. Fong et al. 2016).

Interestingly, central engine models have become increasingly popular in explaining not only GRBs but also the most luminous SNe. In this case, the energy extracted from the millisecond-magnetar re-energises the outgoing supernova shock and creates the additional luminosity, boosting the original luminosity by a factor of 100 or more (Kasen and Bildsten 2010). The difference between GRB magnetars and those postulated in the SLSNe lies in the duration of the energy input. Millisecond-magnetars driving SNe must do much of their energy injection at late times when the supernova is large, otherwise, the energy may do work on the ejecta (increasing the ejecta velocity) but not create luminosity. In contrast, in most GRBs, the spin-down times of the millisecond-magnetars are short, such that most of their work is done more quickly. Broadly speaking GRB millisecond magnetars require high fields ($B \sim 10^{16}$ G) and spin down on timescales of hundreds of seconds, those powering SLSNe have more modest fields ($B \sim 10^{14-15}$ G) but spin down scales of days to weeks (e.g. Metzger et al. 2015). The ultra-long GRBs (Levan et al. 2014a) offer an interesting intermediate population in which millisecond-magnetars may be active for hours to days, but not longer. In this case the detection of a luminous (although not super luminous) supernova, SN2011kl associated with GRB 111209A (Greiner et al. 2015a) is of particular interest in providing evidence of a direct link between the progenitors of GRBs and those of SLSNe.

While central engine models are now increasingly used to explain a variety of exotic explosions, it is important to note that the engines themselves are not the same. Indeed, both black hole accretion and the magnetic extraction of rotational energy are considered as routes. The stars that create either black holes or rapidly spinning neutron stars may, in fact, be quite different immediately prior to the collapse, and so distinguishing between the different models is important, not only in understanding the explosions but in characterising their progenitors. Such a task is not trivial since the central engine itself is hidden at the core of the explosion. While the difference between asymmetric supernovae and ringing down magnetars is potentially distinguishable via gravitational wave observations (see e.g., Fryer et al. 2001; Davies et al. 2002; Rowlinson et al. 2013), such work likely beyond the capability of the current generation of detectors, that can only see such signatures for very local (e.g. local group) supernovae. However, the rate of energy deposition and the total energy budget are different between the two models. For example, in the millisecond-magnetar case rotational energy is released following $\dot{E_{rot}} = I\Omega\dot{\Omega}$, where E_{rot} is the rotational energy, I the moment of inertia and Ω the spin frequency. In this case the total energy budget is $E_{rot} = \frac{1}{2}I\Omega^2$ (Lorimer and Kramer 2004). For a 1.4 $M_\odot$ neutron star, spinning at a period of a millisecond (approximately the maximum spin rate for most neutron star equations of state) the total energy is of order $E_{rot} \sim 10^{52}$ erg. Alternatively, in the black hole case, the late time accretion is likely to follow the fall-back rate of $t^{-5/3}$, and the total energy is governed simply by the total mass accreted $E_{acc} = \eta m_{acc}c^2$ or $\dot{E}_{acc} = \eta \dot{m}_{acc}c^2$. For an efficiency of 10 % and a massive star with $\sim$10 $M_\odot$ of material in-falling (either directly, or via fall-back) the total energy is a much larger $E_{acc} \sim 2 \times 10^{54}$ erg. Although the emission geometry can make it difficult to measure the true total energy of a given explosion, it is apparent that the most energetic GRBs and supernovae appear to exceed the limit for neutron star energy (e.g. Metzger et al. 2015), potentially posing a challenge to such models. However, there are some possibilities, such as very massive neutron stars (Metzger et al. 2015), that can provide a modest boost (factors of a few) to the total rotational energy, and so do enable such models to remain plausible, even at the high energy end of the distribution.

5 Massive Star Progenitors

5.1 The Role of Rotation

Observations make it clear that massive stars are now required in at least the vast majority, and probably all, long-duration GRBs. These massive stars must have somehow lost their hydrogen and helium envelope, create significant quantities of nickel, and have locations in galaxies consistent with the youngest and most massive stars. However, it is likely that not all stars with these conditions will launch a GRB since it would require very narrow beaming angles in GRBs for the GRB rate to match the massive star rate e.g. 1 degree for progenitors with $M_{\rm ZAMS} > 40\ M_\odot$ (Podsiadlowski et al. 2004). Indeed, the creation of a GRB is likely to require the specific conditions that can give rise to a central engine to power the burst. In the black hole engine model a centrifugally supported disc is formed outside a newly created black hole, the GRB jet can then be launched either through electrodynamic processes, or via neutrino—antineutrino annihilation off the disc (note that the T^2 dependence of the weak interaction cross section means that at the extreme temperatures of these discs the cross section for neutrino interactions is significant). Alternatively, in millisecond-magnetar models, the magnetar must be created, and have sufficient field and spin to energise the explosion. In both of these scenarios a crucial factor in the creation of the GRB itself arises from *rotation*.

In particular, the critical rotation is that of the core immediately prior to core collapse. In order to create a centrifugally supported accretion disc at the innermost stable orbit of a black hole a minimum specific angular momentum (j) is required (i.e. j is the angular momentum (L) per unit mass (M), so $j = L/M$);

$$j > \frac{\sqrt{6}GM}{c}. \tag{4}$$

For typical core parameters this corresponds to $j > 10^{16}$ cm^2 s^{-1}. A newly formed millisecond-magnetar (radius $\sim$10–20 km, and spin period 1 ms) has a rather similar specific angular momentum, and so the rotation properties of the core prior to core collapse are likely the same for the millisecond-magnetar or black hole scenarios.

In principle, such angular momentum should be easy to attain. As the core grows (and increases in density) during the main sequence its rotation should increase, potentially even to the point of breakup where the centrifugal force at the equator is equal to the gravitational force. However, in practice, angular momentum is effectively transported *outwards* in stars, both from the core to the envelope and subsequently from the envelope into the interstellar medium. The latter stage of this process is well understood, since mass loss from the star (at its surface) carries away angular momentum, resulting in a star with lower specific angular momentum. This process is strongly metallicity dependent, with the mass loss rate of iron group elements scaling broadly as metallicity, $Z^{0.7-0.8}$ (Vink et al. 2001; Vink and de Koter 2005). Angular momentum transport within the star is less well understood, although it seems likely that magnetic torques would be created by differential rotation of the core relative to the outer layers, and that these would ultimately create a rotation of the core that was tied to that of the envelope (Spruit 2002). Indeed, while such a model is far from universally accepted, its use does provide a reasonable match to the spins observed in neutron stars (Heger et al. 2005; Suijs et al. 2008). Since this mechanism couples the rotation of the core to that of the envelope, it therefore follows that the core is effectively braked by radiatively driven mass loss (Langer 1998), and so even at modest metallicity (e.g. somewhat less than solar) most massive stars will fall short of the critical specific angular momentum by an order of magnitude or more. Indeed, the apparent lack of angular momentum in most stars creates a signicant problem in understanding GRB creation. GRB-SNe are exclusively hydrogen poor events requiring no hydrogen envelope. However, standard pictures to create stars of this type remove the envelope via winds or binary interactions, through which significant angular momentum is also lost. At first sight, then the requirements of hydrogen deficiency and rapid rotation cannot both easily be met. However, there are solutions that may provide the necessary conditions for GRB creation.

5.2 Single Star Scenarios

We can consider solutions that involve both single and binary stars. In this sense we take single stars to mean stars that have either lived their entire lives (from zero age main sequence onwards) as single stars, or those in wide binaries in which there is no significant interaction of the two components.[7] "Normal" massive single stars evolve to have a strong central concentration, in which material in the core is used as the fuel for fusion of progressively heavier elements. On the main sequence, this core gradually builds from hydrogen to helium, and the outer layers of the star remain (chemically) as they were at the beginning of the main sequence. However, in rapidly rotating stars the mixing timescale can be shorter than the nuclear timescale, such that material synthesised in the core of the star is broadly mixed throughout the star, creating a star which undergoes so-called *chemically homogeneous evolution* (e.g. Yoon and Langer 2005; Woosley and Heger 2006; Yoon et al. 2012). This is possible because rotation creates hydrodynamic instabilities at the boundary between the convective core (where hydrogen fusion occurs) and the radiative envelope of the star. Of particular importance is the Eddington–Sweet circulation (Eddington 1926; Sweet 1950) which is driven by a thermal imbalance in rotating stars and may result in a short timescale of chemical mixing with a sufficiently high rotation rate. Because these stars evolve homogeneously they do not create the standard core-envelope structure, do not undergo giant branch phases, and do not experience the core-envelope breaking that dramatically slows rotation in the cores of slowly rotating massive stars. However, this phase must be initiated early in the life of the star before a significant chemical gradient at the boundary between the hydrogen-burning convective core and the radiative envelope is built up, as otherwise, rotationally-induced mixing becomes too inefficient to make the star undergo the chemically homogeneous evolution. This process of chemically homogeneous evolution further removes the hydrogen from the star (since it is burned) while not resulting in such significant mass loss, enabling relatively massive cores to be built from more modest initial masses (Yoon and Langer 2005).

However, chemically homogeneous evolution is not common. There is little evidence for it in massive stars in the Milky Way, although observations of massive stars in the Magellanic Clouds do show some that match the expectations of chemically homogeneous evolution (Martins et al. 2009). This suggests that there may well be some metallicity dependence on stars which undergo such an evolutionary pathway. Indeed, angular momentum loss due to stellar winds is important even early in the lives of stars, so that stars have a certain threshold rotational velocity which is a function of both their stellar mass and chemical composition. The overall impact of this is that single stars can only evolve homogeneously, and hence create GRBs if they are of low metallicity. Indeed, the predicted rates of GRBs from chemically homogeneous stars drop rapidly at metallicity around 0.2 $Z_\odot$ (see Fig. 7). Interestingly, this is very similar to the critical metallicity of 0.3 $Z_\odot$ inferred from detailed observations of some host galaxies by Graham and Fruchter (2015a), but somewhat lower than that inferred from larger samples (typically at higher redshift) either via spectroscopy (Krühler et al. 2015) or photometric proxies (Perley et al. 2015). However, at these low metallicities, the contribution of these effectively single stars could be very important.

5.3 Binary Star Scenarios

Binary stars also provide a natural route of retaining or gaining angular momentum. In particular, stars may be spun up either via the accretion of material with high specific angular momentum, via tidal locking in tight binaries, or possibly even by direct mergers. It is pertinent to understand binary *population* properties in general, since these populations, in turn,

[7] Indeed, it has been suggested that these wide binaries are actually the best test-beds for single stars, since when observing a single star it is extremely difficult to tell if it has formerly been in a binary (Sana et al. 2012).

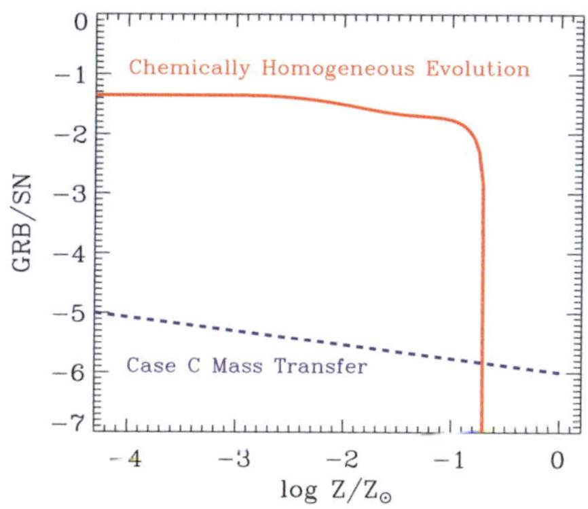

Fig. 7 The expected relative rates for the creation of rapidly rotating massive stellar cores through late binary mass transfer after the main sequence (so called case C, *blue line* Podsiadlowski et al. 2010), and through chemically homogeneous evolution (*red line* Yoon et al. 2006). Both are plotted as a function of the metallicity of the star. It is clear that at low metallicities the effectively single star rate is much larger than the Case C mass transfer rate, and so will dominate over it as a route to GRB creation. In practice massive rapidly rotating stellar cores can be created through routes other than case C mass transfer (mergers etc.), but none-the-less it appears that a significant contribution from homogeneously evolving stars is likely at low metallicities. Any high metallicity events would necessarily arise through binary routes

provide direct constraints on the rate of much rarer events, such as those giving rise to both long and short GRBs. It is now well-established, at least in the general field of our Milky Way galaxy and in low-density star-forming regions, that of the most massive stars at least 60 %–80 % occur in main sequence–main sequence binary systems (e.g Mason et al. 1998; Kouwenhoven et al. 2005; Raghavan et al. 2010; Sana et al. 2010, 2011, 2012; Kobulnicky et al. 2014).

The highest binary fractions involving massive, and hence young, stars are found in populous compact star clusters (e.g. Elson et al. 1998; Hu et al. 2010, and references therein), interestingly, the places where the youngest most massive stars are likely to reside. This is intuitively exemplified by the much higher fraction of low- and high-mass X-ray binaries found in both globular clusters and actively star-forming regions (e.g. Coleiro and Chaty 2013) than among the general field star population. Indeed, interactions in the dense cores of these clusters tend to leave the most massive stars in binaries through exchange interactions. These simultaneously harden tight binaries, at the expense of widening, and potentially unbinding the wider binaries. This so-called hard/soft divide means that tight (hard) binaries get tighter, with their individual components more likely to interact and wider (soft) binaries become progressively wider, and less likely to interact (e.g. Heggie 1975).

Given the apparent preference that GRBs have for low(er) metallicity environments, studies in the Magellanic Clouds may be particularly insightful. For example, studies of the population in the 15–30 Myr-old compact Large Magellanic Cloud (LMC) star clusters NGC 1818 (Hu et al. 2010; de Grijs et al. 2013; Li et al. 2013), show that for binaries with an F-star primary and mass ratios, $q \equiv m_2/m_1 > 0.4$ (where m_1 and m_2 are the masses of the primary and secondary components, respectively), the cluster's binary fraction is $\sim$0.35. This suggests a total binary fraction for F stars of 0.55 to unity, depending on assumptions about the form of the mass-ratio distribution at low q. A similarly high binary fraction (covering the clusters' full observable mass ranges) is obtained for the equivalently young, compact cluster NGC 1805 (Li et al. 2014), as well as for the intermediate-age ($\sim$ a few $\times 10^8$ yr-old), massive LMC clusters NGC 1831 and NGC 1868 (Li et al. 2014).

However, these binary fractions represent the entire population of binary systems. In a surprising development de Grijs et al. (2013), noticed that the binary fraction in the core region of NGC 1818 was significantly lower than that in the cluster's outer regions and even in the surrounding field. It seems likely that this is the effect of the dynamical disruption of soft binaries on short timescales in the dense inner core of the cluster where interaction times are low. Indeed, subsequent N-body simulations confirm this scenario (Geller et al. 2013).

For a range of initial conditions, from smooth virialized density distributions to highly substructured and collapsing configurations it is possible to explain the structures of the cluster, although with a slight preference for structured initial conditions. These models produce the observed radial trend in binary frequency through disruption of soft binaries (with semi-major axes, $a > 3000$ AU), on approximately a crossing time, preferentially in the cluster core. Mass segregation subsequently causes the remaining binaries to sink towards the core (Geller et al. 2015). Thus, both a radial binary fraction distribution that falls towards the core (as observed for NGC 1818) and one that rises towards the core (as for dynamically older star clusters) can arise naturally from the same evolutionary sequence owing to binary disruption and mass segregation in rich star clusters. Indeed, the radial distribution of the binary fraction in another very young LMC cluster NGC 1805, showed an enhanced binary fraction in its core. This cluster is dynamically much older than NGC 1818, however, so that the effects of dynamical mass segregation had already become dominant (Geller et al. 2015). Such complex evolution within clusters means that studies of locations that attempt to derive the binarity of GRB progenitors (e.g. due to kicks to the binaries during a previous supernova explosion) are challenging since their initial locations might be widely varied.

Once binaries are formed, their evolution is dictated both by the evolution of the individual components, but crucially, also by the interaction between the two stars, which can have a marked impact on their evolutionary pathways. The introduction of binary channels greatly increases the diversity of stellar evolution pathways and their associated remnants. Indeed, it is visible on galaxy-wide scales where, for example, the rejuvenation of older stars via binary interactions can make them look much younger than they (or the bulk of the stellar population) really are (Eldridge and Stanway 2009; Stanway et al. 2016).

Because of the increased diversity of stellar products that can be created through binary evolution (e.g. unusual composition, rotation, mass etc), binary channels are popular routes to create exotic (and rare) stellar transients. Through binary evolution it is straightforward to remove the stellar envelope (and provide a stripped envelope core collapse supernova), while tidal locking in tight binaries can in principle provide the necessary angular momentum in the core to enable disc or millisecond-magnetar formation. However, the impacts of binary evolution go beyond the straightforward changes on the outer layers of the star due to the impact of centrifugal force and rotationally-induced chemical mixing. Hence, detailed studies require the construction of complex models that can simultaneously model angular momentum transport within individual stars and the binary components, as well as tracking mass loss and nuclear burning. Such calculations have been attempted by several teams (e.g. Yoon et al. 2010).

As with single stars, the crucial parameter to track through binary routes is the specific angular momentum of the stellar core immediately prior to core collapse. Stars can be spun up via mass transfer from a companion star, but are spun down if they themselves lose mass. This creates similar problems to those seen in massive single stars, in which it is very difficult to both lose the hydrogen envelope and retain both sufficient mass and rotation to obtain the necessary conditions for GRB creation. Several authors have considered late time mass transfer during a late stage of core helium burning or after core helium exhaustion (so-called Case C mass transfer) as a possible route to obtaining GRB-like conditions. This late time mass transfer can enable the formation of a massive stellar core, even at high metallicity (Brown and Lee 2004). Alternatively, depending on the properties of the binary this mass transfer may result in the formation of a common envelope, which will unbind the envelope of the star at the cost of the orbital energy (and hence separation). The resulting binary (after removal of the envelope) is a tight binary containing a carbon-oxygen core. Tidal locking within this binary may match the rotation period of the core to the period of the binary (typically a few hours) in which case the requirements on specific angular momentum can be met (e.g. Lee et al. 2002). Indeed, tidal locking through a variety of routes including neutron stars, post main sequence stars and low mass main sequence stars has been considered as a route of creating GRBs (Izzard et al. 2004; van den Heuvel and Yoon 2007), although detailed calculations disfavour this scenario (Detmers et al. 2008). Alternatively, evolution in binaries with relatively extreme mass ratios may also create the necessary conditions. Here mass transfer occurs from a massive star evolving away from the main sequence, and onto a low mass star on the main sequence. In this scenario, a further mass transfer from the low mass companion can occur on the CO core of the primary star. This, in turn, leads to both the spin up of the core, and the explosive ejection of the common envelope (Podsiadlowski et al. 2010).

Finally, a route to obtaining the necessary angular momentum may be the direct merger of two stars. Such mergers do happen, for example, common envelope mergers have been used to explain unusual stellar outbursts of lower mass stars, such as V838 Mon, V1309 Sco or R136a1 (e.g. Tylenda and Soker 2006; Tylenda et al. 2011; Banerjee et al. 2012). More massive mergers may create the elusive Thorne–Zytkov objects. One of these has recently been claimed in the SMC (Levesque et al. 2014), although this is controversial, in particular, it seems unclear if it actually resides in the SMC (Worley et al. 2016), or is a foreground source (Maccarone and de Mink 2016). Indeed, at higher masses the merger of either two Helium cores, or perhaps a black hole (or neutron star) with a Helium core has been proposed as a GRB mechanism, with the latter suggested as a possible origin for the Christmas-day burst, GRB 101225A (e.g. Thöne et al. 2011a). In these mergers the orbital angular momentum eventually is combined within the single merged object. In the case of He-star—He-star mergers this dramatically increases the total mass, and there is little time between the merger and the supernova. For black hole–He star mergers, the black hole accretion can essentially create the GRB immediately.

It is clear that there are a wide variety of binary channels that can give rise to the necessary conditions for GRB production. All of these scenarios are likely to occur in nature. However, it is less clear if the most exotic events can create GRBs at the necessary rate to explain the observed populations of GRBs. In each of these scenarios, there is a limited parameter space over which the channel will work (in terms of mass, mass ratio, and initial separation), and so a given route is likely to provide only a modest rate of GRB-like events (see Fig. 7). Hence, while many stars are in binaries, the rates of creation of rapidly rotating massive cores through any one of these routes is significantly lower than the rate obtained through chemically homogeneous evolution for any star formation occurring at substantially sub-solar metallicity.

5.4 Predicted Long GRB Rates

Determining the rates of GRBs, and of their various progenitor channels is fraught with difficulty. Observationally, GRB detection is a sensitive function of both the properties of the detector (area, energy range etc.) and of the burst (hardness, peak flux, total fluence etc.). Furthermore, GRBs exhibit a broad range of luminosities, and so an extrapolation to the total rate of GRB-like events requires a significant correction factor. Indeed, it is apparent that there is a population of local, low-luminosity GRBs whose volumetric rate exceeds those of the more distant bursts by a factor of several hundred. If these are a separate population, or the faint end of a luminosity function remains uncertain (e.g. Chapman et al. 2007; Liang et al. 2007b), although the similarities in the supernovae in both the lowest and highest luminosity examples (e.g. Galama et al. 2000; Pian et al. 2006b; Xu et al. 2013; Levan et al. 2014b) suggests that similar physical mechanisms are at play. Finally, GRBs are relativistically beamed, and illuminate only a small fraction of the sky (Frail et al. 2001; Bloom et al. 2003). The beaming fraction for a given burst can be obtained from the so-called jet break when the relativistic jet expands laterally (Frail et al. 2001), but is a challenging and still highly uncertain parameter that significantly impacts the observed rates.

Similarly, when determining the likely rates of GRB production via the various routes considered above there are major uncertainties that enter. What is the relevant metallicity distribution for stars throughout the Universe, and how does this impact those that will undergo chemically homogeneous evolution (Langer and Norman 2006)? Are there any special environments in which the apparently universal top end of the initial mass function can become more top heavy (Bastian et al. 2010)? What is the binary fraction, and the range of initial separations? How is the mass ratio of binaries distributed, are these universal, and might this have any impact on the final products of stellar evolution (Li et al. 2013)? What is the efficiency of common envelope evolution (e.g. Nelemans and Tout 2005)? While these parameters can be studied from detailed simulations (or observations) (e.g. Yoon et al. 2006), or explored via rapid population synthesis (e.g. Izzard et al. 2004) there are still order of magnitude uncertainties in the true rate. Podsiadlowski et al. (2004) attempted to compare the rates of GRBs and massive stars by directly comparing the rates of GRBs (for some assumptions about beaming angles) with the rates of stars of a given mass within a typical galaxy. In Table 1 we provide an updated version of this table, including estimates of the low- and high- luminosity GRBs separately, as well as different types of SNe and massive stars. We also provide the rates of massive stars for a typical (i.e. Milky Way-like) galaxy, as well as the rates of massive stars below some threshold metallicity in the local Universe (extrapolated from Graham and Schady 2015b). We note that these scaled values assume a constant factor of 10^7 between the volume averaged rate and the galactic rate, and so does not take into account chemical differences between galaxies of different masses.

The rate of classical, high luminosity long GRBs, for a beaming angle of a few degrees is remarkably similar to the rate of low metallicity $>40\ M_{\odot}$ stars observed in the local Universe. Since both low metallicity (Graham and Fruchter 2015a; Perley et al. 2015) and high mass (Larsson et al. 2007; Raskin et al. 2008) have been observationally linked to GRBs this may suggest that a reasonable fraction of initially very massive stars create long GRBs. Although such interpretation is challenging because many high-luminosity long GRBs are at much higher redshift, where star formation rate densities are higher. None-the-less, this suggests that the rate of production of long GRBs is not much lower than the rate of production of massive ($>40\ M_{\odot}$), low metallicity ($<0.25\ Z_{\odot}$) stars. In turn, this implies that they are probably not created only from stars which undergo very rare and unusual interactions. For example, for a typical 5-degree beaming angle, the LGRB rate is only a factor of 3 lower than the low metallicity massive star rate. While alternative (wider) beaming angles could lower the rate of GRBs by an order of magnitude or more, it still seems unlikely that the rarest channels, involving $<1\ \%$ of stars are likely to be creating long GRBs. Indeed, if the

Table 1 Approximate rates of various engine driven transients, and of massive stars at all metallicities, and below a set metallicity. Various attempts have been made to determine these rates, and the numbers given are rounded and approximate, rather than representing the full range of possibilities (which in some cases are quite large). A typical galaxy is assumed to have a B-band luminosity of 10^{10} $L_\odot$, and we assume a local B-band luminosity density of 10^8 Mpc^{-3} (e.g. Calura and Matteucci 2003), although clearly, this value evolves significantly over cosmic history. We note that the numbers per galaxy have been directly scaled from the volume averaged values with a fixed scaling of 10^7. Since stars of a given chemical makeup are not the same "per galaxy" due to the mass-metallicity relation (Tremonti et al. 2004), the relative numbers of low metallicity stars should be viewed with caution since a given galaxy might contain stars entirely of low metallicity (if it was of low mass), or very few stars of low metallicity (if it was of high mass), i.e. if low metallicity is a requirement for GRB production then the rate per galaxy at low metallicity is boosted in low mass/metallicity galaxies, and depressed in high mass/metallicity galaxies. The massive star corrections to low metallicity (formally $12 + \log(O/H) < 8.4$ following Graham and Schady (2015b)) are at $z \sim 0$ where 10 % of the star formation lies below this metallicity. This increases rapidly, and the correction likely becomes less than a factor 2 by $z \sim 2$.

Object	Rate (galaxy^{-1} yr^{-1})	Rate (Gpc^{-3} yr^{-1})	References
Transients			
LGRB	8×10^{-8}	0.8	Sun et al. (2015)
$-1°$	6.6×10^{-4}	6600	
$-5°$	2.6×10^{-5}	260	
$-20°$	1.6×10^{-6}	16	
LLGRB	1.6×10^{-5}	160	Sun et al. (2015)
$-1°$	0.11	1.1×10^6	
$-5°$	4.2×10^{-3}	42 000	
$-20°$	2.6×10^{-4}	2600	
SGRB	2.0×10^{-7}	2.0	Sun et al. (2015)
$-1°$	1.3×10^{-3}	13 000	
$-5°$	5.3×10^{-5}	530	
$-20°$	3.0×10^{-6}	30	
ULGRB	1×10^{-8}	0.1	Gendre et al. (2013), Prajs et al. (2016)
$-1°$	6.6×10^{-5}	660	
$-5°$	2.5×10^{-6}	25	
$-20°$	2.0×10^{-7}	2.0	
rTDE	3.0×10^{-9}	0.03	Sun et al. (2015)
SLSNe*	3.0×10^{-6}	30	Quimby et al. (2013), Prajs et al. (2016)
Massive stars			
20 $M_\odot$ all Z	2×10^{-3}	20 000	Podsiadlowski et al. (2004)
20 $M_\odot$ $Z < 1/4\ Z_\odot$	2×10^{-4}	2000	Graham and Schady (2015b)
40 $M_\odot$ all Z	6×10^{-4}	6000	
40 $M_\odot$ $Z < 1/4\ Z_\odot$	6×10^{-5}	600	
80 $M_\odot$ all Z	2×10^{-4}	2000	
80 $M_\odot$ $Z < 1/4\ Z_\odot$	2×10^{-5}	200	
Compact binaries			
NS–NS	6×10^{-5}	600	Abadie et al. (2010)
NS–BH	2×10^{-7}	2	
BH–BH	5×10^{-7}	5	Abbott et al. (2016a)

low-luminosity GRBs are arising from a similar population, with similar progenitor masses and low metallicities then they would need to be born from a significant fraction of the very massive stars, even in the case of no beaming. Even with no beaming correction, the rate per galaxy of low-luminosity GRBs is $\sim 2 \times 10^{-5}$ galaxy^{-1} yr^{-1}, again only a factor of a few lower than the low metallicity 40 $M_{\odot}$ stars. Indeed, since these low-luminosity GRBs are at low redshift there is less concern about evolution over cosmic history, and it seems likely that the low-luminosity GRBs are either lower-mass stars or nearly isotropic in emission in order to avoid low luminosity GRB rates that approach, or even exceed the massive star formation rate. It is also relevant to compare the rates to those of other transients, for example, the newly uncovered population of ULGRBs (Levan et al. 2014a) that may be linked to the SLSNe (Greiner et al. 2015a). The beaming corrections in this case are particularly uncertain, but for plausible beaming rates, comparable to those of long GRBs the inferred rate of ULGRBs is lower, of the same order of magnitude as the SLSNe rate. However, given the substantial uncertainties in both rate calculations, it is possible that this similarity is purely coincidental.

6 Constraints from Local Stellar Populations

6.1 O Stars

It is apparent that long GRBs arise primarily from a subset of massive ($M_{\text{ZAMS}} \geq 40$ $M_{\odot}$), moderately metal-poor stars at cosmological redshifts. Although the nearest long GRBs lie at distances of tens to hundreds of megaparsec, we can study individual massive stars within appropriate environments in the Local Group.

The overwhelming majority of stars are believed to have their origins in dense star clusters, intermediate density OB associations or low-density star forming regions. If stars are randomly drawn from a Kroupa (2001) Initial Mass Function (IMF), cluster masses over $\sim$100 $M_{\odot}$ are required to produce one star capable of ending its life as a core-collapse supernova. However, the most massive stars are preferentially found in the more massive clusters. In particular, stochastic sampling of the initial mass function means that the probability of a low mass cluster forming a very massive star is low (i.e. the massive star content of say 10, 100 $M_{\odot}$ associations is less than that of a single 1000 $M_{\odot}$ cluster). Indeed, it has been suggested that the maximum mass of a star in a cluster ($M_{*,max}$) is related to the cluster total mass (M_C), such that $M_{*,max} \approx 0.39 M_C^{2/3}$ (Weidner et al. 2010). In this case, for a star forming region to include at least one star with an initial mass of at least $\sim$40 $M_{\odot}$, a total mass 10^3 $M_{\odot}$ is required, with 10 such stars hosted by 10^4 $M_{\odot}$ clusters.

The upper limit to star cluster mass is a sensitive function of how vigorously stars are being formed, so one would not expect long GRBs to occur in galaxies with low specific star formation rates. Therefore if progenitors of long GRBs arise from very massive stars, they will exclusively occur in galaxies with high specific star formation rates. Indeed, Kelly and Kirshner (2012) conclude that low z broad-lined (BL) SNe Ic arise exclusively from hosts with the highest star-forming intensities, with Kelly et al. (2008) having earlier established that long GRBs and low z SN Ic-BL originate from similar environments.

Since long GRBs strongly favour host galaxies with metallicities below 1/2 solar, the focus of our attention in the Local Group will be primarily the Large and Small Magellanic Clouds, with 1/2 and 1/5 solar metallicity, respectively. The most massive young star-forming region in the SMC, N66/NGC 346, hosts several tens of O stars, whereas many hundreds of O stars are known in the Tarantula Nebula region of the LMC.

The VLT FLAMES Tarantula Survey (VFTS Evans et al. 2011) has enabled properties of 800 OB stars in this region to be determined, revealing that 50 % of O stars are affected by binary evolution (Sana et al. 2013), with relatively low rotation rates, aside from a few very fast rotators presumably arising from binary interactions (Ramírez-Agudelo et al. 2015). Several star clusters are located within the Tarantula Nebula, most noticeably R136, the youngest (1.5 Myr) very high-mass cluster within the Local Group. Crowther et al. (2016) exploit HST/STIS spectroscopy to reveal that over three dozen stars more massive than 40 $M_{\odot}$ are located within the central parsec.

Nearly 30 very massive stars (VMS, $\geq$100 $M_{\odot}$) are located in the Tarantula Nebula, the majority within the R136 region, but others up to 100 parsecs away (Crowther et al. 2016). Some of the more remote VMS are plausible runaways from R136, but others likely were born in significantly low-density regions. In contrast, no VMS are located in SMC's N66 star-forming region, although it hosts a triple high mass system HD 5980.

In summary, large numbers of candidate long GRB progenitors exist in the LMC/SMC as far as their initial masses are concerned, yet there is no evidence that these are born with sufficiently high rotational rates for chemically homogeneous evolution. Rapid rotation ($v \sin i > 500$ km s^{-1}) is observed for small numbers of O stars in the Tarantula Nebula, presumably spun-up via binary evolution (de Mink et al. 2013). Similarly, no examples of rapid rotators are observed in N66 (Mokiem et al. 2006). Lamb et al. (2016) identify a large population of SMC field Oe stars, presumably arising from rapid rotation, although these too may result from close binary evolution. This in itself raises interesting questions, since if a significant population of sufficiently massive and low metallicity stars can be identified in the local Universe, but essentially none of these appear as viable GRB progenitors then the fraction of massive low metallicity stars creating GRBs must be small. However, this begins to create tension with the necessary observed rates, especially when lower luminosity GRBs are considered (see Sect. 5.4).

6.2 Wolf–Rayet Stars

The immediate progenitors of long GRBs are massive stars whose hydrogen-rich envelopes have been stripped away to reveal compact cores, i.e. classical Wolf–Rayet (WR) stars (Crowther 2007). These stars possess dense, fast outflows with atmospheric compositions characteristic of hydrogen-burning (WN-type) or helium-burning (WC-type). In common with their O type progenitors, the strength of WR winds scales with metallicity (Vink and de Koter 2005), so metal-poor WR stars possess, lower density, slower outflows than their Milky Way counterparts (Hainich et al. 2015b).

The observed association between long GRBs and SN Ic-BL suggests the presence of very little helium in the progenitor star, favouring WC stars. The Magellanic Clouds host approximately 150 WR stars, of which only 15 % belong to the carbon sequence or rare oxygen sequence (WO). The binary frequency of Magellanic Cloud WR stars is approximately 40 %, similar to that observed in the Milky Way, with a lower binary fraction amongst WC and WO stars (Bartzakos et al. 2001).

Analysis of single WC and WO stars in the LMC (Crowther et al. 2002; Tramper et al. 2015) reveals He core masses of 10–20 $M_{\odot}$. The sole SMC carbon/oxygen sequence WR star is in a close binary system, although Shenar et al. (2016) argue that the binary channel does not dominate the formation of WR stars at this metallicity.

The dense outflows from WR stars prevent direct rotational velocity measurements, but searches for non-spherical geometries can be done via linear spectropolarimetry. None of the four LMC WC stars observed by Vink (2007) exhibited the characteristic line depolarization with respect to the continuum, ruling out rapid rotation in these cases. Martins et al. (2009) have claimed that at least one of the hydrogen-rich WN stars in the SMC is the product of chemically homogeneous evolution. However, this single star might represent the product of spin-up arising from close binary evolution since such high initial rotation rates are not currently established amongst massive O stars in the SMC.

Finally, studies have been undertaken investigating the locations of WR stars relative to their host light, in principle directly comparable to the locations of transient phenomena (Leloudas et al. 2010; Bibby and Crowther 2012). These broadly show the locations of stripped envelope SNe match those of the WR population (Bibby and Crowther 2012), and that WC stars are more concentrated on their host light than WN system (Leloudas et al. 2010).

In summary, there are no robust long GRB progenitors amongst the Magellanic Cloud WR population at present, although at least one SMC WR star has been claimed to be the result of chemically homogeneous evolution, and their weak winds make measurements of their rotational rates via spectropolarimetry extremely challenging.

7 Summary and Open Questions

We are now approaching 20 years since the discovery of the first GRB afterglows, and to date over 1000 have been discovered. Intensive observations of the bursts themselves, their afterglows, associated supernovae and host galaxies have provided firm links to their progenitors in several cases, and in particular the link between long GRBs and type Ic supernovae appears secure. In turn, the link to young massive stars, coupled with their extreme luminosity makes GRBs a powerful probe of the distant Universe from the discovery of some of the most distant galaxies, to detailed work studying the buildup of stellar mass and metals across cosmic history. However, while this utility is clear, there remain many questions in addressing the nature of the progenitors themselves, and this, in turn, has a systematic impact on the cosmological utility of GRBs (e.g. in determining the correct factor to convert a GRB rate to a star formation rate at a given redshift/metallicity). In this review, we have highlighted much of the progress that has been made through a combination of intensive observation, large-scale numerical calculation and direct observation of local analogue populations. However, there are many questions that remain open, and will be the focus of research in the coming years, in particular;

- What fraction of massive O-stars are really required to create GRBs? What does this mean about routes to their progenitors?
- What is the true metallicity dependence for GRB production?
- How important is binary evolution in the creation of GRBs?
- Are any of the locally observed massive, rapidly rotating stars actually likely to create a GRB? Have any in the past that we can identify?
- What is the diversity of supernovae seen in long GRBs? Can they be standard candles? Is the presence of a luminous SN in one case a common or extremely rare occurrence?
- What is the range of kilonova/macronova properties seen in short GRBs? Can these create an important contribution to r-process enrichment, and explain the levels seen in the sea-floor on Earth?
- What is the role of NS–NS and NS–BH mergers in short GRB production? Can any BH–BH mergers make GRBs? What will simultaneous gravitational wave signatures tell us?

The answers to these questions are likely to come both from the application of the traditional techniques described above, with improved telescope aperture and response (e.g. JWST, E-ELT, GMRT etc.), or the every increasing computational power, and through new routes, such as direct multi-messenger observations (gravitational waves, neutrinos) that are now producing astrophysical detections. The answers to these questions are challenging, but it is likely that many will be uncovered with a degree of certainty in the coming decade.

Acknowledgements We thank the organisers and ISSI Beijing for a productive and well organised meeting in April 2015, and the referee for a very constructive and valuable report that has improved this manuscript. AJL acknowledges support from the Science and Technology Facilities Council under programme ST/L000733/1 and from the Philip Leverhulme Trust via a Leverhulme Prize. He is grateful for productive conversations with Nial Tanvir. RdG acknowledges financial support from the National Natural Science Foundation of China (grants 11373010, 11633005 and U1631102). He also acknowledges helpful insights into the evolution of binary systems provided by the participants of the 2014 conference on 'Binary systems, their evolution and environments' (particularly Philipp Podsiadlowski, Chris Belczyński and Dany Vanbeveren). SCY was supported by the Basic Science Research (2013R1A1A2061842) program through the National Research Foundation of Korea (NRF).

References

J. Abadie, B.P. Abbott, R. Abbott, M. Abernathy, T. Accadia, F. Acernese, C. Adams, R. Adhikari, P. Ajith, B. Allen et al., Topical review: predictions for the rates of compact binary coalescences observable by ground-based gravitational-wave detectors. Class. Quantum Gravity **27**(17), 173001 (2010). doi:10.1088/0264-9381/27/17/173001

B.P. Abbott, R. Abbott, T.D. Abbott, M.R. Abernathy, F. Acernese, K. Ackley, C. Adams, T. Adams, P. Addesso, R.X. Adhikari et al., Astrophysical implications of the binary black hole merger GW150914. Astrophys. J. Lett. **818**, 22 (2016a). doi:10.3847/2041-8205/818/2/L22

B.P. Abbott, R. Abbott, T.D. Abbott, M.R. Abernathy, F. Acernese, K. Ackley, C. Adams, T. Adams, P. Addesso, R.X. Adhikari et al., Observation of gravitational waves from a binary black hole merger. Phys. Rev. Lett. **116**(6), 061102 (2016b). doi:10.1103/PhysRevLett.116.061102

B.P. Abbott, R. Abbott, T.D. Abbott, M.R. Abernathy, F. Acernese, K. Ackley, C. Adams, T. Adams, P. Addesso, R.X. Adhikari et al., The Rate of Binary Black Hole Mergers Inferred from Advanced LIGO Observations Surrounding GW150914. ArXiv e-prints (2016c)

L. Amati, F. Frontera, C. Guidorzi, Extremely energetic Fermi gamma-ray bursts obey spectral energy correlations. Astron. Astrophys. **508**, 173–180 (2009). doi:10.1051/0004-6361/200912788

J.P. Anderson, S.M. Habergham, P.A. James, M. Hamuy, Progenitor mass constraints for core-collapse supernovae from correlations with host galaxy star formation. Mon. Not. R. Astron. Soc. **424**, 1372–1391 (2012). doi:10.1111/j.1365-2966.2012.21324.x

C.R. Angus, A.J. Levan, D.A. Perley, N.R. Tanvir, J.D. Lyman, E.R. Stanway, A.S. Fruchter, A Hubble Space Telescope survey of the host galaxies of superluminous supernovae. Mon. Not. R. Astron. Soc. **458**, 84–104 (2016). doi:10.1093/mnras/stw063

Z. Arzoumanian, D.F. Chernoff, J.M. Cordes, The velocity distribution of isolated radio pulsars. Astrophys. J. **568**, 289–301 (2002). doi:10.1086/338805

S. Banerjee, P. Kroupa, S. Oh, The emergence of super-canonical stars in R136-type starburst clusters. Mon. Not. R. Astron. Soc. **426**, 1416–1426 (2012). doi:10.1111/j.1365-2966.2012.21672.x

J. Barnes, D. Kasen, Effect of a high opacity on the light curves of radioactively powered transients from compact object mergers. Astrophys. J. **775**, 18 (2013). doi:10.1088/0004-637X/775/1/18

P. Bartzakos, A.F.J. Moffat, V.S. Niemela, Magellanic cloud WC/WO Wolf–Rayet stars—I. Binary frequency and Roche lobe overflow formation. Mon. Not. R. Astron. Soc. **324**, 18–32 (2001). doi:10.1046/j.1365-8711.2001.04126.x

N. Bastian, K.R. Covey, M.R. Meyer, A universal stellar initial mass function? A critical look at variations. Annu. Rev. Astron. Astrophys. **48**, 339–389 (2010). doi:10.1146/annurev-astro-082708-101642

E. Berger, The host galaxies of short-duration gamma-ray bursts: luminosities, metallicities, and star-formation rates. Astrophys. J. **690**, 231–237 (2009). doi:10.1088/0004-637X/690/1/231

E. Berger, A short gamma-ray burst "No-host" problem? Investigating large progenitor offsets for short GRBs with optical afterglows. Astrophys. J. **722**, 1946–1961 (2010). doi:10.1088/0004-637X/722/2/1946

E. Berger, W. Fong, R. Chornock, An r-process kilonova associated with the short-hard GRB 130603B. Astrophys. J. Lett. **774**, 23 (2013). doi:10.1088/2041-8205/774/2/L23

E. Berger, P.A. Price, S.B. Cenko, A. Gal-Yam, A.M. Soderberg, M. Kasliwal, D.C. Leonard, P.B. Cameron, D.A. Frail, S.R. Kulkarni, D.C. Murphy, W. Krzeminski, T. Piran, B.L. Lee, K.C. Roth, D.-S. Moon, D.B. Fox, F.A. Harrison, S.E. Persson, B.P. Schmidt, B.E. Penprase, J. Rich, B.A. Peterson, L.L. Cowie, The afterglow and elliptical host galaxy of the short γ-ray burst GRB 050724. Nature **438**, 988–990 (2005). doi:10.1038/nature04238

E. Berger, D.B. Fox, P.A. Price, E. Nakar, A. Gal-Yam, D.E. Holz, B.P. Schmidt, A. Cucchiara, S.B. Cenko, S.R. Kulkarni, A.M. Soderberg, D.A. Frail, B.E. Penprase, A. Rau, E. Ofek, S.J.B. Burnell, P.B. Cameron, L.L. Cowie, M.A. Dopita, I. Hook, B.A. Peterson, P. Podsiadlowski, K.C. Roth, R.E. Rutledge, S.S. Sheppard, A. Songaila, A new population of high-redshift short-duration gamma-ray bursts. Astrophys. J. **664**, 1000–1010 (2007). doi:10.1086/518762

D. Bersier, A.S. Fruchter, L.-G. Strolger, J. Gorosabel, A. Levan, I. Burud, J.E. Rhoads, A.C. Becker, A. Cassan, R. Chornock, S. Covino, R.S. de Jong, D. Dominis, A.V. Filippenko, J. Hjorth, J. Holmberg, D. Malesani, B. Mobasher, K.A.G. Olsen, M. Stefanon, J.M. Castro Cerón, J.P.U. Fynbo, S.T. Holland, C. Kouveliotou, H. Pedersen, N.R. Tanvir, S.E. Woosley, Evidence for a supernova associated with the X-ray flash 020903. Astrophys. J. **643**, 284–291 (2006). doi:10.1086/502640

J.L. Bibby, P.A. Crowther, The Wolf–Rayet population of the nearby barred spiral galaxy NGC 5068 uncovered by the very large telescope and gemini. Mon. Not. R. Astron. Soc. **420**, 3091–3107 (2012). doi:10.1111/j.1365-2966.2011.20220.x

J.S. Bloom, D.A. Frail, S.R. Kulkarni, Gamma-ray burst energetics and the gamma-ray burst Hubble diagram: promises and limitations. Astrophys. J. **594**, 674–683 (2003). doi:10.1086/377125

J.S. Bloom, S.R. Kulkarni, S.G. Djorgovski, The observed offset distribution of gamma-ray bursts from their host galaxies: a robust clue to the nature of the progenitors. Astron. J. **123**, 1111–1148 (2002). doi:10.1086/338893

J.S. Bloom, S.R. Kulkarni, S.G. Djorgovski, A.C. Eichelberger, P. Côté, J.P. Blakeslee, S.C. Odewahn, F.A. Harrison, D.A. Frail, A.V. Filippenko, D.C. Leonard, A.G. Riess, H. Spinrad, D. Stern, A. Bunker, A. Dey, B. Grossan, S. Perlmutter, R.A. Knop, I.M. Hook, M. Feroci, The unusual afterglow of the γ-ray burst of 26 March 1998 as evidence for a supernova connection. Nature **401**, 453–456 (1999). doi:10.1038/46744

J.S. Bloom, J.X. Prochaska, D. Pooley, C.H. Blake, R.J. Foley, S. Jha, E. Ramirez-Ruiz, J. Granot, A.V. Filippenko, S. Sigurdsson, A.J. Barth, H.-W. Chen, M.C. Cooper, E.E. Falco, R.R. Gal, B.F. Gerke, M.D. Gladders, J.E. Greene, J. Hennanwi, L.C. Ho, K. Hurley, B.P. Koester, W. Li, L. Lubin, J. Newman, D.A. Perley, G.K. Squires, W.M. Wood-Vasey, Closing in on a short-hard burst progenitor: constraints from early-time optical imaging and spectroscopy of a possible host galaxy of GRB 050509b. Astrophys. J. **638**, 354–368 (2006). doi:10.1086/498107

O. Bromberg, E. Nakar, T. Piran, R. Sari, Short versus long and collapsars versus non-collapsars: a quantitative classification of gamma-ray bursts. Astrophys. J. **764**, 179 (2013a). doi:10.1088/0004-637X/764/2/179

O. Bromberg, E. Nakar, T. Piran, R. Sari, Short versus long and collapsars versus non-collapsars: a quantitative classification of gamma-ray bursts. Astrophys. J. **764**, 179 (2013b). doi:10.1088/0004-637X/764/2/179

G.E. Brown, C.-H. Lee, The case for case C mass transfer in the galactic evolution of black hole binaries. New Astron. **9**, 225–237 (2004). doi:10.1016/j.newast.2003.10.005

D.N. Burrows, P. Romano, A. Falcone, S. Kobayashi, B. Zhang, A. Moretti, P.T. O'Brien, M.R. Goad, S. Campana, K.L. Page, L. Angelini, S. Barthelmy, A.P. Beardmore, M. Capalbi, G. Chincarini, J. Cummings, G. Cusumano, D. Fox, P. Giommi, J.E. Hill, J.A. Kennea, H. Krimm, V. Mangano, F. Marshall, P. Mészáros, D.C. Morris, J.A. Nousek, J.P. Osborne, C. Pagani, M. Perri, G. Tagliaferri, A.A. Wells, S. Woosley, N. Gehrels, Bright X-ray flares in gamma-ray burst afterglows. Science **309**, 1833–1835 (2005). doi:10.1126/science.1116168

N.R. Butler, D. Kocevski, J.S. Bloom, J.L. Curtis, A complete catalog of Swift gamma-ray burst spectra and durations: demise of a physical origin for pre-Swift high-energy correlations. Astrophys. J. **671**, 656–677 (2007). doi:10.1086/522492

F. Calura, F. Matteucci, The cosmic evolution of the galaxy luminosity density. Astrophys. J. **596**, 734–747 (2003). doi:10.1086/378195

S. Campana, V. Mangano, A.J. Blustin, P. Brown, D.N. Burrows, G. Chincarini, J.R. Cummings, G. Cusumano, M. Della Valle, D. Malesani, P. Mészáros, J.A. Nousek, M. Page, T. Sakamoto, E. Waxman, B. Zhang, Z.G. Dai, N. Gehrels, S. Immler, F.E. Marshall, K.O. Mason, A. Moretti, P.T. O'Brien, J.P. Osborne, K.L. Page, P. Romano, P.W.A. Roming, G. Tagliaferri, L.R. Cominsky, P. Giommi, O. Godet, J.A. Kennea, H. Krimm, L. Angelini, S.D. Barthelmy, P.T. Boyd, D.M. Palmer, A.A. Wells, N.E. White, The association of GRB 060218 with a supernova and the evolution of the shock wave. Nature **442**, 1008–1010 (2006). doi:10.1038/nature04892

Z. Cano, A new method for estimating the bolometric properties of Ibc supernovae. Mon. Not. R. Astron. Soc. **434**, 1098–1116 (2013). doi:10.1093/mnras/stt1048

Z. Cano, K.G. Johansson Andreas, K. Maeda, A self-consistent analytical magnetar model: the luminosity of γ-ray burst supernovae is powered by radioactivity. Mon. Not. R. Astron. Soc. **457**, 2761–2772 (2016). doi:10.1093/mnras/stw122

Z. Cano, A. de Ugarte Postigo, A. Pozanenko, N. Butler, C.C. Thöne, C. Guidorzi, T. Krühler, J. Gorosabel, P. Jakobsson, G. Leloudas, D. Malesani, J. Hjorth, A. Melandri, C. Mundell, K. Wiersema, P. D'Avanzo, S. Schulze, A. Gomboc, A. Johansson, W. Zheng, D.A. Kann, F. Knust, K. Varela, C.W. Akerlof, J. Bloom, O. Burkhonov, E. Cooke, J.A. de Diego, G. Dhungana, C. Farina, F.V. Ferrante, H.A. Flewelling, O.D. Fox, J. Fynbo, N. Gehrels, L. Georgiev, J.J. González, J. Greiner, T. Güver, O. Hartoog, N. Hatch, M. Jelinek, R. Kehoe, S. Klose, E. Klunko, D. Kopač, A. Kutyrev, Y. Krugly, W.H. Lee, A. Levan, V. Linkov, A. Matkin, N. Minikulov, I. Molotov, J.X. Prochaska, M.G. Richer, C.G. Román-Zúñiga, V. Rumyantsev, R. Sánchez-Ramírez, I. Steele, N.R. Tanvir, A. Volnova, A.M. Watson, D. Xu, F. Yuan, A trio of gamma-ray burst supernovae: GRB 120729A, GRB 130215A/SN 2013ez, and GRB 130831A/SN 2013fu. Astron. Astrophys. **568**, 19 (2014). doi:10.1051/0004-6361/201423920

G. Cavallo, M.J. Rees, A qualitative study of cosmic fireballs and gamma-ray bursts. Mon. Not. R. Astron. Soc. **183**, 359–365 (1978). doi:10.1093/mnras/183.3.359

R. Chapman, N.R. Tanvir, R.S. Priddey, A.J. Levan, How common are long gamma-ray bursts in the local Universe? Mon. Not. R. Astron. Soc. **382**, 21–25 (2007). doi:10.1111/j.1745-3933.2007.00381.x

H.-Y. Chen, D.E. Holz, Gamma-ray-burst beaming and gravitational-wave observations. Phys. Rev. Lett. **111**(18), 181101 (2013). doi:10.1103/PhysRevLett.111.181101

A. Coleiro, S. Chaty, Distribution of high-mass X-ray binaries in the Milky Way. Astrophys. J. **764**, 185 (2013). doi:10.1088/0004-637X/764/2/185

E. Costa, F. Frontera, J. Heise, M. Feroci, J. in't Zand, F. Fiore, M.N. Cinti, D. Dal Fiume, L. Nicastro, M. Orlandini, E. Palazzi, M. Rapisarda#, G. Zavattini, R. Jager, A. Parmar, A. Owens, S. Molendi, G. Cusumano, M.C. Maccarone, S. Giarrusso, A. Coletta, L.A. Antonelli, P. Giommi, J.M. Muller, L. Piro, R.C. Butler, Discovery of an X-ray afterglow associated with the γ-ray burst of 28 February 1997. Nature **387**, 783–785 (1997). doi:10.1038/42885

P.A. Crowther, Physical properties of Wolf–Rayet stars. Annu. Rev. Astron. Astrophys. **45**, 177–219 (2007). doi:10.1146/annurev.astro.45.051806.110615

P.A. Crowther, L. Dessart, D.J. Hillier, J.B. Abbott, A.W. Fullerton, Stellar and wind properties of LMC WC4 stars. A metallicity dependence for Wolf–Rayet mass-loss rates. Astron. Astrophys. **392**, 653–669 (2002). doi:10.1051/0004-6361:20020941

P.A. Crowther, S.M. Caballero-Nieves, K.A. Bostroem, J. Maíz Apellániz, F.R.N. Schneider, N.R. Walborn, C.R. Angus, I. Brott, A. Bonanos, A. de Koter, S.E. de Mink, C.J. Evans, G. Gräfener, A. Herrero, I.D. Howarth, N. Langer, D.J. Lennon, J. Puls, H. Sana, J.S. Vink, The R136 star cluster dissected with Hubble Space Telescope/STIS. I. Far-ultraviolet spectroscopic census and the origin of He II λ1640 in young star clusters. Mon. Not. R. Astron. Soc. **458**, 624–659 (2016). doi:10.1093/mnras/stw273

A. Cucchiara, A.J. Levan, D.B. Fox, N.R. Tanvir, T.N. Ukwatta, E. Berger, T. Krühler, A. Küpcü Yoldaş, X.F. Wu, K. Toma, J. Greiner, F.E. Olivares, A. Rowlinson, L. Amati, T. Sakamoto, K. Roth, A. Stephens, A. Fritz, J.P.U. Fynbo, J. Hjorth, D. Malesani, P. Jakobsson, K. Wiersema, P.T. O'Brien, A.M. Soderberg, R.J. Foley, A.S. Fruchter, J. Rhoads, R.E. Rutledge, B.P. Schmidt, M.A. Dopita, P. Podsiadlowski, R. Willingale, C. Wolf, S.R. Kulkarni, P. D'Avanzo, A photometric redshift of $z \sim 9.4$ for GRB 090429B. Astrophys. J. **736**, 7 (2011). doi:10.1088/0004-637X/736/1/7

M.B. Davies, A.J. Levan, A.R. King, The ultimate outcome of black hole-neutron star mergers. Mon. Not. R. Astron. Soc. **356**, 54–58 (2005). doi:10.1111/j.1365-2966.2004.08423.x

M.B. Davies, A. King, S. Rosswog, G. Wynn, Gamma-ray bursts, supernova kicks, and gravitational radiation. Astrophys. J. Lett. **579**, 63–66 (2002). doi:10.1086/345288

R. de Grijs, C. Li, Y. Zheng, L. Deng, Y. Hu, M.B.N. Kouwenhoven, Gravitational conundrum? Dynamical mass segregation versus disruption of binary stars in dense stellar systems. Astrophys. J. **765**, 4 (2013). doi:10.1088/0004-637X/765/1/4

S.E. de Mink, N. Langer, R.G. Izzard, H. Sana, A. de Koter, The rotation rates of massive stars: the role of binary interaction through tides, mass transfer, and mergers. Astrophys. J. **764**, 166 (2013). doi:10.1088/0004-637X/764/2/166

A. de Ugarte Postigo, A.J. Castro-Tirado, S. Guziy, J. Gorosabel, G. Jóhannesson, M.A. Aloy, S. McBreen, D.Q. Lamb, N. Benitez, M. Jelínek, S.B. Pandey, D. Coe, M.D. Pérez-Ramírez, F.J. Aceituno, M. Alises, J.A. Acosta-Pulido, G. Gómez, R. López, T.Q. Donaghy, Y.E. Nakagawa, T. Sakamoto, G.R. Ricker, F.R. Hearty, M. Bayliss, G. Gyuk, GRB 060121: implications of a short-/intermediate-duration γ-ray burst at high redshift. Astrophys. J. Lett. **648**, 83–87 (2006). doi:10.1086/507868

A. de Ugarte Postigo, I. Horváth, P. Veres, Z. Bagoly, D.A. Kann, C.C. Thöne, L.G. Balazs, P. D'Avanzo, M.A. Aloy, S. Foley, S. Campana, J. Mao, P. Jakobsson, S. Covino, J.P.U. Fynbo, J. Gorosabel, A.J. Castro-Tirado, L. Amati, M. Nardini, Searching for differences in Swift's intermediate GRBs. Astron. Astrophys. **525**, 109 (2011). doi:10.1051/0004-6361/201015261

A. de Ugarte Postigo, C.C. Thöne, A. Rowlinson, R. García-Benito, A.J. Levan, J. Gorosabel, P. Goldoni, S. Schulze, T. Zafar, K. Wiersema, R. Sánchez-Ramírez, A. Melandri, P. D'Avanzo, S. Oates, V. D'Elia, M. De Pasquale, T. Krühler, A.J. van der Horst, D. Xu, D. Watson, S. Piranomonte, S.D. Vergani, B. Milvang-Jensen, L. Kaper, D. Malesani, J.P.U. Fynbo, Z. Cano, S. Covino, H. Flores, S. Greiss, F. Hammer, O.E. Hartoog, S. Hellmich, C. Heuser, J. Hjorth, P. Jakobsson, S. Mottola, M. Sparre, J. Sollerman, G. Tagliaferri, N.R. Tanvir, M. Vestergaard, R.A.M.J. Wijers, Spectroscopy of the short-hard GRB 130603B. The host galaxy and environment of a compact object merger. Astron. Astrophys. **563**, 62 (2014). doi:10.1051/0004-6361/201322985

M. Della Valle, G. Chincarini, N. Panagia, G. Tagliaferri, D. Malesani, V. Testa, D. Fugazza, S. Campana, S. Covino, V. Mangano, L.A. Antonelli, P. D'Avanzo, K. Hurley, I.F. Mirabel, L.J. Pellizza, S. Piranomonte, L. Stella, An enigmatic long-lasting γ-ray burst not accompanied by a bright supernova. Nature **444**, 1050–1052 (2006). doi:10.1038/nature05374

J. Deng, N. Tominaga, P.A. Mazzali, K. Maeda, K. Nomoto, On the light curve and spectrum of SN 2003dh separated from the optical afterglow of GRB 030329. Astrophys. J. **624**, 898–905 (2005). doi:10.1086/429284

R.G. Detmers, N. Langer, P. Podsiadlowski, R.G. Izzard, Gamma-ray bursts from tidally spun-up Wolf–Rayet stars? Astron. Astrophys. **484**, 831–839 (2008). doi:10.1051/0004-6361:200809371

A. Dietz, N. Fotopoulos, L. Singer, C. Cutler, Outlook for detection of GW inspirals by GRB-triggered searches in the advanced detector era. Phys. Rev. D **87**(6), 064033 (2013). doi:10.1103/PhysRevD.87.064033

M.A. Dopita, L.J. Kewley, R.S. Sutherland, D.C. Nicholls, Chemical abundances in high-redshift galaxies: a powerful new emission line diagnostic. Astrophys. Space Sci. **361**, 61 (2016). doi:10.1007/s10509-016-2657-8

A.S. Eddington, *The Internal Constitution of the Stars* (1926)

J.J. Eldridge, E.R. Stanway, Spectral population synthesis including massive binaries. Mon. Not. R. Astron. Soc. **400**, 1019–1028 (2009). doi:10.1111/j.1365-2966.2009.15514.x

R.A.W. Elson, S. Sigurdsson, M. Davies, J. Hurley, G. Gilmore, The binary star population of the young cluster NGC 1818 in the Large Magellanic Cloud. Mon. Not. R. Astron. Soc. **300**, 857–862 (1998). doi:10.1046/j.1365-8711.1998.01948.x

C.J. Evans, W.D. Taylor, V. Hénault-Brunet, H. Sana, A. de Koter, S. Simón-Díaz, G. Carraro, T. Bagnoli, N. Bastian, J.M. Bestenlehner, A.Z. Bonanos, E. Bressert, I. Brott, M.A. Campbell, M. Cantiello, J.S. Clark, E. Costa, P.A. Crowther, S.E. de Mink, E. Doran, P.L. Dufton, P.R. Dunstall, K. Friedrich, M. Garcia, M. Gieles, G. Gräfener, A. Herrero, I.D. Howarth, R.G. Izzard, N. Langer, D.J. Lennon, J. Maíz Apellániz, N. Markova, F. Najarro, J. Puls, O.H. Ramirez, C. Sabín-Sanjulián, S.J. Smartt, V.E. Stroud, J.T. van Loon, J.S. Vink, N.R. Walborn, The VLT-FLAMES Tarantula Survey. I. Introduction and observational overview. Astron. Astrophys. **530**, 108 (2011). doi:10.1051/0004-6361/201116782

R.J. Foley, P.J. Challis, R. Chornock, M. Ganeshalingam, W. Li, G.H. Marion, N.I. Morrell, G. Pignata, M.D. Stritzinger, J.M. Silverman, X. Wang, J.P. Anderson, A.V. Filippenko, W.L. Freedman, M. Hamuy, S.W. Jha, R.P. Kirshner, C. McCully, S.E. Persson, M.M. Phillips, D.E. Reichart, A.M. Soderberg, Type Iax supernovae: a new class of stellar explosion. Astrophys. J. **767**, 57 (2013). doi:10.1088/0004-637X/767/1/57

W. Fong, E. Berger, The locations of short gamma-ray bursts as evidence for compact object binary progenitors. Astrophys. J. **776**, 18 (2013). doi:10.1088/0004-637X/776/1/18

W. Fong, E. Berger, D.B. Fox, Hubble space telescope observations of short gamma-ray burst host galaxies: morphologies, offsets, and local environments. Astrophys. J. **708**, 9–25 (2010). doi:10.1088/0004-637X/708/1/9

W-f. Fong, B.D. Metzger, E. Berger, F. Ozel, Radio Constraints on Long-Lived Magnetar Remnants in Short Gamma-Ray Bursts. ArXiv e-prints (2016)

W. Fong, E. Berger, R. Chornock, R. Margutti, A.J. Levan, N.R. Tanvir, R.L. Tunnicliffe, I. Czekala, D.B. Fox, D.A. Perley, S.B. Cenko, B.A. Zauderer, T. Laskar, S.E. Persson, A.J. Monson, D.D. Kelson, C. Birk,

D. Murphy, M. Servillat, G. Anglada, Demographics of the galaxies hosting short-duration gamma-ray bursts. Astrophys. J. **769**, 56 (2013). doi:10.1088/0004-637X/769/1/56

W. Fong, E. Berger, B.D. Metzger, R. Margutti, R. Chornock, G. Migliori, R.J. Foley, B.A. Zauderer, R. Lunnan, T. Laskar, S.J. Desch, K.J. Meech, S. Sonnett, C. Dickey, A. Hedlund, P. Harding, Short GRB 130603B: discovery of a jet break in the optical and radio afterglows, and a mysterious late-time X-ray excess. Astrophys. J. **780**, 118 (2014). doi:10.1088/0004-637X/780/2/118

D.B. Fox, D.A. Frail, P.A. Price, S.R. Kulkarni, E. Berger, T. Piran, A.M. Soderberg, S.B. Cenko, P.B. Cameron, A. Gal-Yam, M.M. Kasliwal, D.-S. Moon, F.A. Harrison, E. Nakar, B.P. Schmidt, B. Penprase, R.A. Chevalier, P. Kumar, K. Roth, D. Watson, B.L. Lee, S. Shectman, M.M. Phillips, M. Roth, P.J. McCarthy, M. Rauch, L. Cowie, B.A. Peterson, J. Rich, N. Kawai, K. Aoki, G. Kosugi, T. Totani, H.-S. Park, A. MacFadyen, K.C. Hurley, The afterglow of GRB 050709 and the nature of the short-hard γ-ray bursts. Nature **437**, 845–850 (2005). doi:10.1038/nature04189

D.A. Frail, S.R. Kulkarni, L. Nicastro, M. Feroci, G.B. Taylor, The radio afterglow from the γ-ray burst of 8 May 1997. Nature **389**, 261–263 (1997). doi:10.1038/38451

D.A. Frail, S.R. Kulkarni, R. Sari, S.G. Djorgovski, J.S. Bloom, T.J. Galama, D.E. Reichart, E. Berger, F.A. Harrison, P.A. Price, S.A. Yost, A. Diercks, R.W. Goodrich, F. Chaffee, Beaming in gamma-ray bursts: evidence for a standard energy reservoir. Astrophys. J. Lett. **562**, 55–58 (2001). doi:10.1086/338119

A.S. Fruchter, A.J. Levan, L. Strolger, P.M. Vreeswijk, S.E. Thorsett, D. Bersier, I. Burud, J.M. Castro Cerón, A.J. Castro-Tirado, C. Conselice, T. Dahlen, H.C. Ferguson, J.P.U. Fynbo, P.M. Garnavich, R.A. Gibbons, J. Gorosabel, T.R. Gull, J. Hjorth, S.T. Holland, C. Kouveliotou, Z. Levay, M. Livio, M.R. Metzger, P.E. Nugent, L. Petro, E. Pian, J.E. Rhoads, A.G. Riess, K.C. Sahu, A. Smette, N.R. Tanvir, R.A.M.J. Wijers, S.E. Woosley, Long γ-ray bursts and core-collapse supernovae have different environments. Nature **441**, 463–468 (2006). doi:10.1038/nature04787

C.L. Fryer, A.L. Hungerford, P.A. Young, Light-curve calculations of supernovae from fallback gamma-ray bursts. Astrophys. J. Lett. **662**, 55–58 (2007). doi:10.1086/519523

C.L. Fryer, S.E. Woosley, D.H. Hartmann, Formation rates of black hole accretion disk gamma-ray bursts. Astrophys. J. **526**, 152–177 (1999). doi:10.1086/307992

C.L. Fryer, S.E. Woosley, A. Heger, Pair-instability supernovae, gravity waves, and gamma-ray transients. Astrophys. J. **550**, 372–382 (2001). doi:10.1086/319719

J.P.U. Fynbo, D. Watson, C.C. Thöne, J. Sollerman, J.S. Bloom, T.M. Davis, J. Hjorth, P. Jakobsson, U.G. Jørgensen, J.F. Graham, A.S. Fruchter, D. Bersier, L. Kewley, A. Cassan, J.M. Castro Cerón, S. Foley, J. Gorosabel, T.C. Hinse, K.D. Horne, B.L. Jensen, S. Klose, D. Kocevski, J.-B. Marquette, D. Perley, E. Ramirez-Ruiz, M.D. Stritzinger, P.M. Vreeswijk, R.A.M. Wijers, K.G. Woller, D. Xu, M. Zub, No supernovae associated with two long-duration γ-ray bursts. Nature **444**, 1047–1049 (2006). doi:10.1038/nature05375

J.P.U. Fynbo, P. Jakobsson, J.X. Prochaska, D. Malesani, C. Ledoux, A. de Ugarte Postigo, M. Nardini, P.M. Vreeswijk, K. Wiersema, J. Hjorth, J. Sollerman, H.-W. Chen, C.C. Thöne, G. Björnsson, J.S. Bloom, A.J. Castro-Tirado, L. Christensen, A. De Cia, A.S. Fruchter, J. Gorosabel, J.F. Graham, A.O. Jaunsen, B.L. Jensen, D.A. Kann, C. Kouveliotou, A.J. Levan, J. Maund, N. Masetti, B. Milvang-Jensen, E. Palazzi, D.A. Perley, E. Pian, E. Rol, P. Schady, R.L.C. Starling, N.R. Tanvir, D.J. Watson, D. Xu, T. Augusteijn, F. Grundahl, J. Telting, P.-O. Quirion, Low-resolution spectroscopy of gamma-ray burst optical afterglows: biases in the Swift sample and characterization of the absorbers. Astrophys. J. Suppl. Ser. **185**, 526–573 (2009). doi:10.1088/0067-0049/185/2/526

A. Gal-Yam, Luminous supernovae. Science **337**, 927 (2012). doi:10.1126/science.1203601

A. Gal-Yam, D.B. Fox, P.A. Price, E.O. Ofek, M.R. Davis, D.C. Leonard, A.M. Soderberg, B.P. Schmidt, K.M. Lewis, B.A. Peterson, S.R. Kulkarni, E. Berger, S.B. Cenko, R. Sari, K. Sharon, D. Frail, D.-S. Moon, P.J. Brown, A. Cucchiara, F. Harrison, T. Piran, S.E. Persson, P.J. McCarthy, B.E. Penprase, R.A. Chevalier, A.I. MacFadyen, A novel explosive process is required for the γ-ray burst GRB 060614. Nature **444**, 1053–1055 (2006). doi:10.1038/nature05373

A. Gal-Yam, P. Mazzali, E.O. Ofek, P.E. Nugent, S.R. Kulkarni, M.M. Kasliwal, R.M. Quimby, A.V. Filippenko, S.B. Cenko, R. Chornock, R. Waldman, D. Kasen, M. Sullivan, E.C. Beshore, A.J. Drake, R.C. Thomas, J.S. Bloom, D. Poznanski, A.A. Miller, R.J. Foley, J.M. Silverman, I. Arcavi, R.S. Ellis, J. Deng, Supernova 2007bi as a pair-instability explosion. Nature **462**, 624–627 (2009). doi:10.1038/nature08579

T.J. Galama, N. Tanvir, P.M. Vreeswijk, R.A.M.J. Wijers, P.J. Groot, E. Rol, J. van Paradijs, C. Kouveliotou, A.S. Fruchter, N. Masetti, H. Pedersen, B. Margon, E.W. Deutsch, M. Metzger, L. Armus, S. Klose, B. Stecklum, Evidence for a supernova in reanalyzed optical and near-infrared images of GRB 970228. Astrophys. J. **536**, 185–194 (2000). doi:10.1086/308909

P.M. Garnavich, K.Z. Stanek, L. Wyrzykowski, L. Infante, E. Bendek, D. Bersier, S.T. Holland, S. Jha, T. Matheson, R.P. Kirshner, K. Krisciunas, M.M. Phillips, R.G. Carlberg, Discovery of the low-redshift optical afterglow of GRB 011121 and its progenitor supernova SN 2001ke. Astrophys. J. **582**, 924–932 (2003). doi:10.1086/344785

N. Gehrels, C.L. Sarazin, P.T. O'Brien, B. Zhang, L. Barbier, S.D. Barthelmy, A. Blustin, D.N. Burrows, J. Cannizzo, J.R. Cummings, M. Goad, S.T. Holland, C.P. Hurkett, J.A. Kennea, A. Levan, C.B. Markwardt, K.O. Mason, P. Meszaros, M. Page, D.M. Palmer, E. Rol, T. Sakamoto, R. Willingale, L. Angelini, A. Beardmore, P.T. Boyd, A. Breeveld, S. Campana, M.M. Chester, G. Chincarini, L.R. Cominsky, G. Cusumano, M. de Pasquale, E.E. Fenimore, P. Giommi, C. Gronwall, D. Grupe, J.E. Hill, D. Hinshaw, J. Hjorth, D. Hullinger, K.C. Hurley, S. Klose, S. Kobayashi, C. Kouveliotou, H.A. Krimm, V. Mangano, F.E. Marshall, K. McGowan, A. Moretti, R.F. Mushotzky, K. Nakazawa, J.P. Norris, J.A. Nousek, J.P. Osborne, K. Page, A.M. Parsons, S. Patel, M. Perri, T. Poole, P. Romano, P.W.A. Roming, S. Rosen, G. Sato, P. Schady, A.P. Smale, J. Sollerman, R. Starling, M. Still, M. Suzuki, G. Tagliaferri, T. Takahashi, M. Tashiro, J. Tueller, A.A. Wells, N.E. White, R.A.M.J. Wijers, A short γ-ray burst apparently associated with an elliptical galaxy at redshift $z = 0.225$. Nature **437**, 851–854 (2005). doi:10.1038/nature04142

N. Gehrels, J.P. Norris, S.D. Barthelmy, J. Granot, Y. Kaneko, C. Kouveliotou, C.B. Markwardt, P. Mészáros, E. Nakar, J.A. Nousek, P.T. O'Brien, M. Page, D.M. Palmer, A.M. Parsons, P.W.A. Roming, T. Sakamoto, C.L. Sarazin, P. Schady, M. Stamatikos, S.E. Woosley, A new γ-ray burst classification scheme from GRB060614. Nature **444**, 1044–1046 (2006). doi:10.1038/nature05376

N. Gehrels, S.D. Barthelmy, D.N. Burrows, J.K. Cannizzo, G. Chincarini, E. Fenimore, C. Kouveliotou, P. O'Brien, D.M. Palmer, J. Racusin, P.W.A. Roming, T. Sakamoto, J. Tueller, R.A.M.J. Wijers, B. Zhang, Correlations of prompt and afterglow emission in Swift long and short gamma-ray bursts. Astrophys. J. **689**, 1161–1172 (2008). doi:10.1086/592766

A.M. Geller, R. de Grijs, C. Li, J.R. Hurley, Consequences of dynamical disruption and mass segregation for the binary frequencies of star clusters. Astrophys. J. **779**, 30 (2013). doi:10.1088/0004-637X/779/1/30

A.M. Geller, R. de Grijs, C. Li, J.R. Hurley, Different dynamical ages for the two young and coeval LMC star clusters, NGC 1805 and NGC 1818, imprinted on their binary populations. Astrophys. J. **805**, 11 (2015). doi:10.1088/0004-637X/805/1/11

B. Gendre, G. Stratta, J.L. Atteia, S. Basa, M. Boër, D.M. Coward, S. Cutini, V. D'Elia, E.J. Howell, A. Klotz, L. Piro, The ultra-long gamma-ray burst 111209A: the collapse of a blue supergiant? Astrophys. J. **766**, 30 (2013). doi:10.1088/0004-637X/766/1/30

B.P. Gompertz, A.J. van der Horst, P.T. O'Brien, G.A. Wynn, K. Wiersema, Broad-band modelling of short gamma-ray bursts with energy injection from magnetar spin-down and its implications for radio detectability. Mon. Not. R. Astron. Soc. **448**, 629–641 (2015). doi:10.1093/mnras/stu2752

J. Gorosabel, J.P.U. Fynbo, A. Fruchter, A. Levan, J. Hjorth, P. Nugent, A.J. Castro-Tirado, J.M. Castro Cerón, J. Rhoads, D. Bersier, I. Burud, A possible bright blue supernova in the afterglow of GRB 020305. Astron. Astrophys. **437**, 411–418 (2005). doi:10.1051/0004-6361:20052783

J.F. Graham, A.S. Fruchter, The metal aversion of long-duration gamma-ray bursts. Astrophys. J. **774**, 119 (2013). doi:10.1088/0004-637X/774/2/119

J.F. Graham, A.S. Fruchter, The Relative Rate of LGRB Formation as a Function of Metallicity. ArXiv e-prints (2015a)

J.F. Graham, P. Schady, The Absolute Rate of LGRB Formation. ArXiv e-prints (2015b)

J.F. Graham, A.S. Fruchter, A.J. Levan, A. Melandri, L.J. Kewley, E.M. Levesque, M. Nysewander, N.R. Tanvir, T. Dahlen, D. Bersier, K. Wiersema, D.G. Bonfield, A. Martinez-Sansigre, GRB 070714B: discovery of the highest spectroscopically confirmed short burst redshift. Astrophys. J. **698**, 1620–1629 (2009). doi:10.1088/0004-637X/698/2/1620

J. Greiner, P.A. Mazzali, D.A. Kann, T. Krühler, E. Pian, S. Prentice, F. Olivares E., A. Rossi, S. Klose, S. Taubenberger, F. Knust, P.M.J. Afonso, C. Ashall, J. Bolmer, C. Delvaux, R. Diehl, J. Elliott, R. Filgas, J.P.U. Fynbo, J.F. Graham, A.N. Guelbenzu, S. Kobayashi, G. Leloudas, S. Savaglio, P. Schady, S. Schmidl, T. Schweyer, V. Sudilovsky, M. Tanga, A.C. Updike, H. van Eerten, K. Varela, A very luminous magnetar-powered supernova associated with an ultra-long γ-ray burst. Nature **523**, 189–192 (2015a). doi:10.1038/nature14579

R. Hainich, D. Pasemann, H. Todt, T. Shenar, A. Sander, W.-R. Hamann, Wolf–Rayet stars in the Small Magellanic Cloud. I. Analysis of the single WN stars. Astron. Astrophys. **581**, 21 (2015b). doi:10.1051/0004-6361/201526241

F. Hammer, H. Flores, D. Schaerer, M. Dessauges-Zavadsky, E. Le Floc'h, M. Puech, Detection of Wolf–Rayet stars in host galaxies of gamma-ray bursts (GRBs): are GRBs produced by runaway massive stars ejected from high stellar density regions? Astron. Astrophys. **454**, 103–111 (2006). doi:10.1051/0004-6361:20064823

A. Heger, S.E. Woosley, H.C. Spruit, Presupernova evolution of differentially rotating massive stars including magnetic fields. Astrophys. J. **626**, 350–363 (2005). doi:10.1086/429868

D.C. Heggie, Binary evolution in stellar dynamics. Mon. Not. R. Astron. Soc. **173**, 729–787 (1975). doi:10.1093/mnras/173.3.729

J. Hjorth, J. Sollerman, P. Møller, J.P.U. Fynbo, S.E. Woosley, C. Kouveliotou, N.R. Tanvir, J. Greiner, M.I. Andersen, A.J. Castro-Tirado, J.M. Castro Cerón, A.S. Fruchter, J. Gorosabel, P. Jakobsson, L. Kaper, S. Klose, N. Masetti, H. Pedersen, K. Pedersen, E. Pian, E. Palazzi, J.E. Rhoads, E. Rol, E.P.J. van den Heuvel, P.M. Vreeswijk, D. Watson, R.A.M.J. Wijers, A very energetic supernova associated with the γ-ray burst of 29 March 2003. Nature **423**, 847–850 (2003). doi:10.1038/nature01750

J. Hjorth, J. Sollerman, J. Gorosabel, J. Granot, S. Klose, C. Kouveliotou, J. Melinder, E. Ramirez-Ruiz, R. Starling, B. Thomsen, M.I. Andersen, J.P.U. Fynbo, B.L. Jensen, P.M. Vreeswijk, J.M. Castro Cerón, P. Jakobsson, A. Levan, K. Pedersen, J.E. Rhoads, N.R. Tanvir, D. Watson, R.A.M.J. Wijers, GRB 050509B: constraints on short gamma-ray burst models. Astrophys. J. Lett. **630**, 117–120 (2005a). doi:10.1086/491733

J. Hjorth, D. Watson, J.P.U. Fynbo, P.A. Price, B.L. Jensen, U.G. Jørgensen, D. Kubas, J. Gorosabel, P. Jakobsson, J. Sollerman, K. Pedersen, C. Kouveliotou, The optical afterglow of the short γ-ray burst GRB 050709. Nature **437**, 859–861 (2005b). doi:10.1038/nature04174

P. Höflich, J.C. Wheeler, L. Wang, Aspherical explosion models for SN 1998BW/GRB-980425. Astrophys. J. **521**, 179–189 (1999). doi:10.1086/307521

Y. Hu, L. Deng, R. de Grijs, Q. Liu, S.P. Goodwin, The binary fraction of the young cluster NGC 1818 in the Large Magellanic Cloud. Astrophys. J. **724**, 649–656 (2010). doi:10.1088/0004-637X/724/1/649

K. Hurley, S.E. Boggs, D.M. Smith, R.C. Duncan, R. Lin, A. Zoglauer, S. Krucker, G. Hurford, H. Hudson, C. Wigger, W. Hajdas, C. Thompson, I. Mitrofanov, A. Sanin, W. Boynton, C. Fellows, A. von Kienlin, G. Lichti, A. Rau, T. Cline, An exceptionally bright flare from SGR 1806-20 and the origins of short-duration γ-ray bursts. Nature **434**, 1098–1103 (2005). doi:10.1038/nature03519

R.G. Izzard, E. Ramirez-Ruiz, C.A. Tout, Formation rates of core-collapse supernovae and gamma-ray bursts. Mon. Not. R. Astron. Soc. **348**, 1215–1228 (2004). doi:10.1111/j.1365-2966.2004.07436.x

P.A. James, J.P. Anderson, The Hα galaxy survey. III. Constraints on supernova progenitors from spatial correlations with Hα emission. Astron. Astrophys. **453**, 57–65 (2006). doi:10.1051/0004-6361:20054509

J. Japelj, S.D. Vergani, R. Salvaterra, L.K. Hunt, F. Mannucci, Taking stock of SLSN and LGRB host galaxy comparison using a complete sample of LGRBs. ArXiv e-prints (2016)

Z.-P. Jin, X. Li, Z. Cano, S. Covino, Y.-Z. Fan, D.-M. Wei, The light curve of the macronova associated with the long-short burst GRB 060614. Astrophys. J. Lett. **811**, 22 (2015). doi:10.1088/2041-8205/811/2/L22

D.A. Kann, S. Klose, B. Zhang, D. Malesani, E. Nakar, A. Pozanenko, A.C. Wilson, N.R. Butler, P. Jakobsson, S. Schulze, M. Andreev, L.A. Antonelli, I.F. Bikmaev, V. Biryukov, M. Böttcher, R.A. Burenin, J.M. Castro Cerón, A.J. Castro-Tirado, G. Chincarini, B.E. Cobb, S. Covino, P. D'Avanzo, V. D'Elia, M. Della Valle, A. de Ugarte Postigo, Y. Efimov, P. Ferrero, D. Fugazza, J.P.U. Fynbo, M. Gålfalk, F. Grundahl, J. Gorosabel, S. Gupta, S. Guziy, B. Hafizov, J. Hjorth, K. Holhjem, M. Ibrahimov, M. Im, G.L. Israel, M. Jelínek, B.L. Jensen, R. Karimov, I.M. Khamitov, Ü. Kiziloğlu, E. Klunko, P. Kubánek, A.S. Kutyrev, P. Laursen, A.J. Levan, F. Mannucci, C.M. Martin, A. Mescheryakov, N. Mirabal, J.P. Norris, J.-E. Ovaldsen, D. Paraficz, E. Pavlenko, S. Piranomonte, A. Rossi, V. Rumyantsev, R. Salinas, A. Sergeev, D. Sharapov, J. Sollerman, B. Stecklum, L. Stella, G. Tagliaferri, N.R. Tanvir, J. Telting, V. Testa, A.C. Updike, A. Volnova, D. Watson, K. Wiersema, D. Xu, The afterglows of Swift-era gamma-ray bursts. I. Comparing pre-Swift and Swift-era long/soft (type II) GRB optical afterglows. Astrophys. J. **720**, 1513–1558 (2010). doi:10.1088/0004-637X/720/2/1513

D.A. Kann, P. Schady, F. Olivares E, S. Klose, A. Rossi, D.A. Perley, T. Krühler, J. Greiner, A. Nicuesa Guelbenzu, J. Elliott, F. Knust, R. Filgas, E. Pian, P. Mazzali, J.P.U. Fynbo, G. Leloudas, P.M.J. Afonso, C. Delvaux, J.F. Graham, A. Rau, S. Schmidl, S. Schulze, M. Tanga, A.C. Updike, K. Varela, Highly Luminous Supernovae associated with Gamma-Ray Bursts I: GRB 111209A/SN 2011kl in the Context of Stripped-Envelope and Superluminous Supernovae. ArXiv e-prints (2016)

D. Kasen, L. Bildsten, Supernova light curves powered by young magnetars. Astrophys. J. **717**, 245–249 (2010). doi:10.1088/0004-637X/717/1/245

M.M. Kasliwal, S.R. Kulkarni, A. Gal-Yam, P.E. Nugent, M. Sullivan, L. Bildsten, O. Yaron, H.B. Perets, I. Arcavi, S. Ben-Ami, V.B. Bhalerao, J.S. Bloom, S.B. Cenko, A.V. Filippenko, D.A. Frail, M. Ganeshalingam, A. Horesh, D.A. Howell, N.M. Law, D.C. Leonard, W. Li, E.O. Ofek, D. Polishook, D. Poznanski, R.M. Quimby, J.M. Silverman, A. Sternberg, D. Xu, Calcium-rich gap transients in the remote outskirts of galaxies. Astrophys. J. **755**, 161 (2012). doi:10.1088/0004-637X/755/2/161

P.L. Kelly, R.P. Kirshner, Core-collapse supernovae and host galaxy stellar populations. Astrophys. J. **759**, 107 (2012). doi:10.1088/0004-637X/759/2/107

P.L. Kelly, R.P. Kirshner, M. Pahre, Long γ-ray bursts and type Ic core-collapse supernovae have similar locations in hosts. Astrophys. J. **687**, 1201–1207 (2008). doi:10.1086/591925

P.L. Kelly, A.V. Filippenko, O.D. Fox, W. Zheng, K.I. Clubb, Evidence that gamma-ray burst 130702A exploded in a dwarf satellite of a massive galaxy. Astrophys. J. Lett. **775**, 5 (2013). doi:10.1088/2041-8205/775/1/L5

S. Kisaka, K. Ioka, T. Nakamura, Isotropic detectable X-ray counterparts to gravitational waves from neutron star binary mergers. Astrophys. J. Lett. **809**, 8 (2015). doi:10.1088/2041-8205/809/1/L8

S. Kisaka, K. Ioka, E. Nakar, X-ray-powered macronovae. Astrophys. J. **818**, 104 (2016). doi:10.3847/0004-637X/818/2/104

R.W. Klebesadel, I.B. Strong, R.A. Olson, Observations of gamma-ray bursts of cosmic origin. Astrophys. J. Lett. **182**, 85 (1973). doi:10.1086/181225

H.A. Kobulnicky, D.C. Kiminki, M.J. Lundquist, J. Burke, J. Chapman, E. Keller, K. Lester, E.K. Rolen, E. Topel, A. Bhattacharjee, R.A. Smullen, C.A. Vargas Álvarez, J.C. Runnoe, D.A. Dale, M.M. Brotherton, Toward complete statistics of massive binary stars: penultimate results from the Cygnus OB2 radial velocity survey. Astrophys. J. Suppl. Ser. **213**, 34 (2014). doi:10.1088/0067-0049/213/2/34

C.S. Kochanek, Failed supernovae explain the compact remnant mass function. Astrophys. J. **785**, 28 (2014). doi:10.1088/0004-637X/785/1/28

C.S. Kochanek, J.F. Beacom, M.D. Kistler, J.L. Prieto, K.Z. Stanek, T.A. Thompson, H. Yüksel, A survey about nothing: monitoring a million supergiants for failed supernovae. Astrophys. J. **684**, 1336–1342 (2008). doi:10.1086/590053

C. Kouveliotou, C.A. Meegan, G.J. Fishman, N.P. Bhat, M.S. Briggs, T.M. Koshut, W.S. Paciesas, G.N. Pendleton, Identification of two classes of gamma-ray bursts. Astrophys. J. Lett. **413**, 101–104 (1993). doi:10.1086/186969

C. Kouveliotou, S.E. Woosley, S.K. Patel, A. Levan, R. Blandford, E. Ramirez-Ruiz, R.A.M.J. Wijers, M.C. Weisskopf, A. Tennant, E. Pian, P. Giommi, Chandra observations of the X-ray environs of SN 1998bw/GRB 980425. Astrophys. J. **608**, 872–882 (2004). doi:10.1086/420878

M.B.N. Kouwenhoven, A.G.A. Brown, H. Zinnecker, L. Kaper, S.F. Portegies Zwart, The primordial binary population. I. A near-infrared adaptive optics search for close visual companions to A star members of Scorpius OB2. Astron. Astrophys. **430**, 137–154 (2005). doi:10.1051/0004-6361:20048124

P. Kroupa, On the variation of the initial mass function. Mon. Not. R. Astron. Soc. **322**, 231–246 (2001). doi:10.1046/j.1365-8711.2001.04022.x

T. Krühler, D. Malesani, J.P.U. Fynbo, O.E. Hartoog, J. Hjorth, P. Jakobsson, D.A. Perley, A. Rossi, P. Schady, S. Schulze, N.R. Tanvir, S.D. Vergani, K. Wiersema, P.M.J. Afonso, J. Bolmer, Z. Cano, S. Covino, V. D'Elia, A. de Ugarte Postigo, R. Filgas, M. Friis, J.F. Graham, J. Greiner, P. Goldoni, A. Gomboc, F. Hammer, J. Japelj, D.A. Kann, L. Kaper, S. Klose, A.J. Levan, G. Leloudas, B. Milvang-Jensen, A. Nicuesa Guelbenzu, E. Palazzi, E. Pian, S. Piranomonte, R. Sánchez-Ramírez, S. Savaglio, J. Selsing, G. Tagliaferri, P.M. Vreeswijk, D.J. Watson, D. Xu, GRB hosts through cosmic time. VLT/X-Shooter emission-line spectroscopy of 96 γ-ray-burst-selected galaxies at $0.1 < z < 3.6$. Astron. Astrophys. **581**, 125 (2015). doi:10.1051/0004-6361/201425561

J.B. Lamb, M.S. Oey, D.M. Segura-Cox, A.S. Graus, D.C. Kiminki, J.B. Golden-Marx, J.W. Parker, The runaways and isolated O-type star spectroscopic survey of the SMC (RIOTS4). Astrophys. J. **817**, 113 (2016). doi:10.3847/0004-637X/817/2/113

N. Langer, Coupled mass and angular momentum loss of massive main sequence stars. Astron. Astrophys. **329**, 551–558 (1998)

N. Langer, C.A. Norman, On the collapsar model of long gamma-ray bursts: constraints from cosmic metallicity evolution. Astrophys. J. Lett. **638**, 63–66 (2006). doi:10.1086/500363

J. Larsson, A.J. Levan, M.B. Davies, A.S. Fruchter, A new constraint for gamma-ray burst progenitor mass. Mon. Not. R. Astron. Soc. **376**, 1285–1290 (2007). doi:10.1111/j.1365-2966.2007.11523.x

C.-H. Lee, G.E. Brown, R.A.M.J. Wijers, Discovery of a black hole mass-period correlation in soft X-ray transients and its implication for gamma-ray burst and hypernova mechanisms. Astrophys. J. **575**, 996–1006 (2002). doi:10.1086/341349

W.-H. Lei, B. Zhang, E.-W. Liang, Hyperaccreting black hole as gamma-ray burst central engine. I. Baryon loading in gamma-ray burst jets. Astrophys. J. **765**, 125 (2013). doi:10.1088/0004-637X/765/2/125

C.N. Leibler, E. Berger, The stellar ages and masses of short gamma-ray burst host galaxies: investigating the progenitor delay time distribution and the role of mass and star formation in the short gamma-ray burst rate. Astrophys. J. **725**, 1202–1214 (2010). doi:10.1088/0004-637X/725/1/1202

G. Leloudas, J. Sollerman, A.J. Levan, J.P.U. Fynbo, D. Malesani, J.R. Maund, Do Wolf–Rayet stars have similar locations in hosts as type Ib/c supernovae and long gamma-ray bursts? Astron. Astrophys. **518**, 29 (2010). doi:10.1051/0004-6361/200913753

G. Leloudas, S. Schulze, T. Krühler, J. Gorosabel, L. Christensen, A. Mehner, A. de Ugarte Postigo, R. Amorín, C.C. Thöne, J.P. Anderson, F.E. Bauer, A. Gallazzi, K.G. Hełminiak, J. Hjorth, E. Ibar, D. Malesani, N. Morell, J. Vinko, J.C. Wheeler, Spectroscopy of superluminous supernova host galaxies. A preference of hydrogen-poor events for extreme emission line galaxies. Mon. Not. R. Astron. Soc. **449**, 917–932 (2015). doi:10.1093/mnras/stv320

A.J. Levan, N.R. Tanvir, A.S. Fruchter, E. Rol, J.P.U. Fynbo, J. Hjorth, G. Williams, E. Bergeron, D. Bersier, M. Bremer, T. Grav, P. Jakobsson, K. Nilsson, E. Olszewski, R.S. Priddey, D. Rafferty, J. Rhoads, The faint afterglow and host galaxy of the short-hard GRB 060121. Astrophys. J. Lett. **648**, 9–12 (2006). doi:10.1086/507625

A.J. Levan, P. Jakobsson, C. Hurkett, N.R. Tanvir, J. Gorosabel, P. Vreeswijk, E. Rol, R. Chapman, N. Gehrels, P.T. O'Brien, J.P. Osborne, R.S. Priddey, C. Kouveliotou, R. Starling, D. vanden Berk, K. Wiersema, A case of mistaken identity? GRB060912A and the nature of the long-short GRB divide. Mon. Not. R. Astron. Soc. **378**, 1439–1446 (2007). doi:10.1111/j.1365-2966.2007.11879.x

A.J. Levan, N.R. Tanvir, P. Jakobsson, R. Chapman, J. Hjorth, R.S. Priddey, J.P.U. Fynbo, K. Hurley, B.L. Jensen, R. Johnson, J. Gorosabel, A.J. Castro-Tirado, M. Jarvis, D. Watson, K. Wiersema, On the nature of the short-duration GRB 050906. Mon. Not. R. Astron. Soc. **384**, 541–547 (2008). doi:10.1111/j.1365-2966.2007.11953.x

A.J. Levan, N.R. Tanvir, S.B. Cenko, D.A. Perley, K. Wiersema, J.S. Bloom, A.S. Fruchter, A.d.U. Postigo, P.T. O'Brien, N. Butler, A.J. van der Horst, G. Leloudas, A.N. Morgan, K. Misra, G.C. Bower, J. Farihi, R.L. Tunnicliffe, M. Modjaz, J.M. Silverman, J. Hjorth, C. Thöne, A. Cucchiara, J.M.C. Cerón, A.J. Castro-Tirado, J.A. Arnold, M. Bremer, J.P. Brodie, T. Carroll, M.C. Cooper, P.A. Curran, R.M. Cutri, J. Ehle, D. Forbes, J. Fynbo, J. Gorosabel, J. Graham, D.I. Hoffman, S. Guziy, P. Jakobsson, A. Kamble, T. Kerr, M.M. Kasliwal, C. Kouveliotou, D. Kocevski, N.M. Law, P.E. Nugent, E.O. Ofek, D. Poznanski, R.M. Quimby, E. Rol, A.J. Romanowsky, R. Sánchez-Ramírez, S. Schulze, N. Singh, L. van Spaandonk, R.L.C. Starling, R.G. Strom, J.C. Tello, O. Vaduvescu, P.J. Wheatley, R.A.M.J. Wijers, J.M. Winters, D. Xu, An extremely luminous panchromatic outburst from the nucleus of a distant galaxy. Science **333**, 199 (2011). doi:10.1126/science.1207143

A.J. Levan, N.R. Tanvir, R.L.C. Starling, K. Wiersema, K.L. Page, D.A. Perley, S. Schulze, G.A. Wynn, R. Chornock, J. Hjorth, S.B. Cenko, A.S. Fruchter, P.T. O'Brien, G.C. Brown, R.L. Tunnicliffe, D. Malesani, P. Jakobsson, D. Watson, E. Berger, D. Bersier, B.E. Cobb, S. Covino, A. Cucchiara, A. de Ugarte Postigo, D.B. Fox, A. Gal-Yam, P. Goldoni, J. Gorosabel, L. Kaper, T. Krühler, R. Karjalainen, J.P. Osborne, E. Pian, R. Sánchez-Ramírez, B. Schmidt, I. Skillen, G. Tagliaferri, C. Thöne, O. Vaduvescu, R.A.M.J. Wijers, B.A. Zauderer, A new population of ultra-long duration gamma-ray bursts. Astrophys. J. **781**, 13 (2014a). doi:10.1088/0004-637X/781/1/13

A.J. Levan, N.R. Tanvir, A.S. Fruchter, J. Hjorth, E. Pian, P. Mazzali, R.A. Hounsell, D.A. Perley, Z. Cano, J. Graham, S.B. Cenko, J.P.U. Fynbo, C. Kouveliotou, A. Pe'er, K. Misra, K. Wiersema, Hubble space telescope observations of the afterglow, supernova, and host galaxy associated with the extremely bright GRB 130427A. Astrophys. J. **792**, 115 (2014b). doi:10.1088/0004-637X/792/2/115

A. Levan, P. Nugent, A. Fruchter, I. Burud, D. Branch, J. Rhoads, A. Castro-Tirado, J. Gorosabel, J.M. Castro Cerón, S.E. Thorsett, C. Kouveliotou, S. Golenetskii, J. Fynbo, P. Garnavich, S. Holland, J. Hjorth, P. Møller, E. Pian, N. Tanvir, M. Ulanov, R. Wijers, S. Woosley, GRB 020410: a gamma-ray burst afterglow discovered by its supernova light. Astrophys. J. **624**, 880–888 (2005). doi:10.1086/428657

E.M. Levesque, P. Massey, A.N. Żytkow, N. Morrell, Discovery of a Thorne–Żytkow object candidate in the Small Magellanic Cloud. Mon. Not. R. Astron. Soc. **443**, 94–98 (2014). doi:10.1093/mnrasl/slu080

C. Li, R. de Grijs, L. Deng, The binary fractions in the massive young Large Magellanic Cloud star clusters NGC 1805 and NGC 1818. Mon. Not. R. Astron. Soc. **436**, 1497–1512 (2013). doi:10.1093/mnras/stt1669

C. Li, R. de Grijs, L. Deng, Not-so-simple stellar populations in the intermediate-age Large Magellanic Cloud star clusters NGC 1831 and NGC 1868. Astrophys. J. **784**, 157 (2014). doi:10.1088/0004-637X/784/2/157

C. Li, R. de Grijs, L. Deng, X. Liu, Blue straggler evolution caught in the act in the Large Magellanic Cloud globular cluster Hodge 11. Astrophys. J. Lett. **770**, 7 (2013). doi:10.1088/2041-8205/770/1/L7

L.-X. Li, B. Paczyński, Transient events from neutron star mergers. Astrophys. J. Lett. **507**, 59–62 (1998). doi:10.1086/311680

E. Liang, B. Zhang, F. Virgili, Z.G. Dai, Low-luminosity gamma-ray bursts as a unique population: luminosity function, local rate, and beaming factor. Astrophys. J. **662**, 1111–1118 (2007a). doi:10.1086/517959

E. Liang, B. Zhang, F. Virgili, Z.G. Dai, Low-luminosity gamma-ray bursts as a unique population: luminosity function, local rate, and beaming factor. Astrophys. J. **662**, 1111–1118 (2007b). doi:10.1086/517959

Y.M. Lipkin, E.O. Ofek, A. Gal-Yam, E.M. Leibowitz, D. Poznanski, S. Kaspi, D. Polishook, S.R. Kulkarni, D.W. Fox, E. Berger, N. Mirabal, J. Halpern, M. Bureau, K. Fathi, P.A. Price, B.A. Peterson, A. Frebel, B. Schmidt, J.A. Orosz, J.B. Fitzgerald, J.S. Bloom, P.G. van Dokkum, C.D. Bailyn, M.M. Buxton, M. Barsony, The detailed optical light curve of GRB 030329. Astrophys. J. **606**, 381–394 (2004). doi:10.1086/383000

D.R. Lorimer, M. Kramer, *Handbook of Pulsar Astronomy* (2004)

R. Lunnan, R. Chornock, E. Berger, T. Laskar, W. Fong, A. Rest, N.E. Sanders, P.M. Challis, M.R. Drout, R.J. Foley, M.E. Huber, R.P. Kirshner, C. Leibler, G.H. Marion, M. McCrum, D. Milisavljevic, G. Narayan, D. Scolnic, S.J. Smartt, K.W. Smith, A.M. Soderberg, J.L. Tonry, W.S. Burgett, K.C. Chambers, H. Flewelling, K.W. Hodapp, N. Kaiser, E.A. Magnier, P.A. Price, R.J. Wainscoat, Hydrogen-poor superluminous supernovae and long-duration gamma-ray bursts have similar host galaxies. Astrophys. J. **787**, 138 (2014). doi:10.1088/0004-637X/787/2/138

R. Lunnan, R. Chornock, E. Berger, A. Rest, W. Fong, D. Scolnic, D.O. Jones, A.M. Soderberg, P.M. Challis, M.R. Drout, R.J. Foley, M.E. Huber, R.P. Kirshner, C. Leibler, G.H. Marion, M. McCrum, D. Milisavljevic, G. Narayan, N.E. Sanders, S.J. Smartt, K.W. Smith, J.L. Tonry, W.S. Burgett, K.C. Chambers, H. Flewelling, R.-P. Kudritzki, R.J. Wainscoat, C. Waters, Zooming in on the progenitors of superluminous supernovae with the HST. Astrophys. J. **804**, 90 (2015). doi:10.1088/0004-637X/804/2/90

J.D. Lyman, D. Bersier, P.A. James, Bolometric corrections for optical light curves of core-collapse supernovae. Mon. Not. R. Astron. Soc. **437**, 3848–3862 (2014a). doi:10.1093/mnras/stt2187

J.D. Lyman, A.J. Levan, R.P. Church, M.B. Davies, N.R. Tanvir, The progenitors of calcium-rich transients are not formed in situ. Mon. Not. R. Astron. Soc. **444**, 2157–2166 (2014b). doi:10.1093/mnras/stu1574

J.D. Lyman, D. Bersier, P.A. James, P.A. Mazzali, J.J. Eldridge, M. Fraser, E. Pian, Bolometric light curves and explosion parameters of 38 stripped-envelope core-collapse supernovae. Mon. Not. R. Astron. Soc. **457**, 328–350 (2016a). doi:10.1093/mnras/stv2983

J.D. Lyman, A.J. Levan, P.A. James, C.R. Angus, R.P. Church, M.B. Davies, N.R. Tanvir, Hubble Space Telescope observations of the host galaxies and environments of calcium-rich supernovae. Mon. Not. R. Astron. Soc. **458**, 1768–1777 (2016b). doi:10.1093/mnras/stw477

T.J. Maccarone, S.E. de Mink, Large proper motion of the Thorne-Żytkow object candidate HV 2112 reveals its likely nature as foreground Galactic S-star. Mon. Not. R. Astron. Soc. **458**, 1–5 (2016). doi:10.1093/mnrasl/slw004

b.P. Macias, E. Ramirez-Ruiz, A Stringent Limit on the Mass Production Rate of r-Process Elements in the Milky Way. ArXiv e-prints (2016)

K. Maeda, T. Nakamura, K. Nomoto, P.A. Mazzali, F. Patat, I. Hachisu, Explosive nucleosynthesis in aspherical hypernova explosions and late-time spectra of SN 1998bw. Astrophys. J. **565**, 405–412 (2002). doi:10.1086/324487

F. Martins, D.J. Hillier, J.C. Bouret, E. Depagne, C. Foellmi, S. Marchenko, A.F. Moffat, Properties of WNh stars in the Small Magellanic Cloud: evidence for homogeneous evolution. Astron. Astrophys. **495**, 257–270 (2009). doi:10.1051/0004-6361:200811014

A. Maselli, A. Melandri, L. Nava, C.G. Mundell, N. Kawai, S. Campana, S. Covino, J.R. Cummings, G. Cusumano, P.A. Evans, G. Ghirlanda, G. Ghisellini, C. Guidorzi, S. Kobayashi, P. Kuin, V. La Parola, V. Mangano, S. Oates, T. Sakamoto, M. Serino, F. Virgili, B.-B. Zhang, S. Barthelmy, A. Beardmore, M.G. Bernardini, D. Bersier, D. Burrows, G. Calderone, M. Capalbi, J. Chiang, P. D'Avanzo, V. D'Elia, M. De Pasquale, D. Fugazza, N. Gehrels, A. Gomboc, R. Harrison, H. Hanayama, J. Japelj, J. Kennea, D. Kopac, C. Kouveliotou, D. Kuroda, A. Levan, D. Malesani, F. Marshall, J. Nousek, P. O'Brien, J.P. Osborne, C. Pagani, K.L. Page, M. Page, M. Perri, T. Pritchard, P. Romano, Y. Saito, B. Sbarufatti, R. Salvaterra, I. Steele, N. Tanvir, G. Vianello, B. Weigand, K. Wiersema, Y. Yatsu, T. Yoshii, G. Tagliaferri, GRB 130427A: a nearby ordinary monster. Science **343**, 48–51 (2014). doi:10.1126/science.1242279

N. Masetti, E. Palazzi, E. Pian, A. Simoncelli, L.K. Hunt, E. Maiorano, A. Levan, L. Christensen, E. Rol, S. Savaglio, R. Falomo, A.J. Castro-Tirado, J. Hjorth, A. Delsanti, M. Pannella, V. Mohan, S.B. Pandey, R. Sagar, L. Amati, I. Burud, J.M. Castro Cerón, F. Frontera, A.S. Fruchter, J.P.U. Fynbo, J. Gorosabel, L. Kaper, S. Klose, C. Kouveliotou, L. Nicastro, H. Pedersen, J. Rhoads, I. Salamanca, N. Tanvir, P.M. Vreeswijk, R.A.M.J. Wijers, E.P.J. van den Heuvel, Optical and near-infrared observations of the GRB020405 afterglow. Astron. Astrophys. **404**, 465–481 (2003). doi:10.1051/0004-6361:20030491

B.D. Mason, T.J. Henry, W.I. Hartkopf, T. ten Brummelaar, D.R. Soderblom, A multiplicity survey of chromospherically active and inactive stars. Astron. J. **116**, 2975–2983 (1998). doi:10.1086/300654

E.P. Mazets, S.V. Golenetskii, V.N. Ilinskii, V.N. Panov, R.L. Aptekar, I.A. Gurian, M.P. Proskura, I.A. Sokolov, Z.I. Sokolova, T.V. Kharitonova, Catalog of cosmic gamma-ray bursts from the KONUS experiment data. I. Astrophys. Space Sci. **80**, 3–83 (1981). doi:10.1007/BF00649140

E.P. Mazets, S.V. Golenetskii, I.A. Gurian, V.N. Ilinskii, The 5 March 1979 event and the distinct class of short gamma bursts are they of the same origin. Astrophys. Space Sci. **84**, 173–189 (1982). doi:10.1007/BF00713635

P.A. Mazzali, J. Deng, K. Nomoto, D.N. Sauer, E. Pian, N. Tominaga, M. Tanaka, K. Maeda, A.V. Filippenko, A neutron-star-driven X-ray flash associated with supernova SN 2006aj. Nature **442**, 1018–1020 (2006). doi:10.1038/nature05081

P.A. Mazzali, R.J. Foley, J. Deng, F. Patat, E. Pian, D. Baade, J.S. Bloom, A.V. Filippenko, D.A. Perley, S. Valenti, L. Wang, K. Kawabata, K. Maeda, K-i. Nomoto, Keck and European Southern Observatory

very large telescope view of the symmetry of the ejecta of the XRF/SN 2006aj. Astrophys. J. **661**, 892–898 (2007a). doi:10.1086/517912
P.A. Mazzali, K.S. Kawabata, K. Maeda, R.J. Foley, K. Nomoto, J. Deng, T. Suzuki, M. Iye, N. Kashikawa, Y. Ohyama, A.V. Filippenko, Y. Qiu, J. Wei, The aspherical properties of the energetic type Ic SN 2002ap as inferred from its nebular spectra. Astrophys. J. **670**, 592–599 (2007b). doi:10.1086/521873
P.A. Mazzali, K.S. Kawabata, K. Maeda, R.J. Foley, K. Nomoto, J. Deng, T. Suzuki, M. Iye, N. Kashikawa, Y. Ohyama, A.V. Filippenko, Y. Qiu, J. Wei, The aspherical properties of the energetic type Ic SN 2002ap as inferred from its nebular spectra. Astrophys. J. **670**, 592–599 (2007c). doi:10.1086/521873
P.A. Mazzali, A.I. McFadyen, S.E. Woosley, E. Pian, M. Tanaka, An upper limit to the energy of gamma-ray bursts indicates that GRBs/SNe are powered by magnetars. Mon. Not. R. Astron. Soc. **443**, 67–71 (2014). doi:10.1093/mnras/stu1124
P.A. Mazzali, M. Sullivan, E. Pian, J. Greiner, D.A. Kann, Spectrum formation in superluminous supernovae (Type I). Mon. Not. R. Astron. Soc. **458**, 3455–3465 (2016). doi:10.1093/mnras/stw512
A. Melandri, E. Pian, V. D'Elia, P. D'Avanzo, M. Della Valle, P.A. Mazzali, G. Tagliaferri, Z. Cano, A.J. Levan, Diversity of gamma-ray burst energetics vs. supernova homogeneity: SN 2013cq associated with GRB 130427A. Astron. Astrophys. **567**, 29 (2014). doi:10.1051/0004-6361/201423572
B.D. Metzger, Nuclear-dominated accretion and subluminous supernovae from the merger of a white dwarf with a neutron star or black hole. Mon. Not. R. Astron. Soc. **419**, 827–840 (2012). doi:10.1111/j.1365-2966.2011.19747.x
B.D. Metzger, E. Berger, What is the most promising electromagnetic counterpart of a neutron star binary merger? Astrophys. J. **746**, 48 (2012). doi:10.1088/0004-637X/746/1/48
B.D. Metzger, D. Giannios, T.A. Thompson, N. Bucciantini, E. Quataert, The protomagnetar model for gamma-ray bursts. Mon. Not. R. Astron. Soc. **413**, 2031–2056 (2011). doi:10.1111/j.1365-2966.2011.18280.x
B.D. Metzger, B. Margalit, D. Kasen, E. Quataert, The diversity of transients from magnetar birth in core collapse supernovae. Mon. Not. R. Astron. Soc. **454**, 3311–3316 (2015). doi:10.1093/mnras/stv2224
M.R. Mokiem, A. de Koter, C.J. Evans, J. Puls, S.J. Smartt, P.A. Crowther, A. Herrero, N. Langer, D.J. Lennon, F. Najarro, M.R. Villamariz, S.-C. Yoon, The VLT-FLAMES survey of massive stars: mass loss and rotation of early-type stars in the SMC. Astron. Astrophys. **456**, 1131–1151 (2006). doi:10.1051/0004-6361:20064995
S. Mukherjee, E.D. Feigelson, G. Jogesh Babu, F. Murtagh, C. Fraley, A. Raftery, Three types of gamma-ray bursts. Astrophys. J. **508**, 314–327 (1998). doi:10.1086/306386
R. Narayan, B. Paczynski, T. Piran, Gamma-ray bursts as the death throes of massive binary stars. Astrophys. J. Lett. **395**, 83–86 (1992). doi:10.1086/186493
G. Nelemans, C.A. Tout, Reconstructing the evolution of white dwarf binaries: further evidence for an alternative algorithm for the outcome of the common-envelope phase in close binaries. Mon. Not. R. Astron. Soc. **356**, 753–764 (2005). doi:10.1111/j.1365-2966.2004.08496.x
R.J. Nemiroff, A century of gamma ray burst models. Comments Astrophys. **17**, 189 (1994)
S. Nissanke, J. Sievers, N. Dalal, D. Holz, Localizing compact binary inspirals on the sky using ground-based gravitational wave interferometers. Astrophys. J. **739**, 99 (2011). doi:10.1088/0004-637X/739/2/99
J.A. Nousek, C. Kouveliotou, D. Grupe, K.L. Page, J. Granot, E. Ramirez-Ruiz, S.K. Patel, D.N. Burrows, V. Mangano, S. Barthelmy, A.P. Beardmore, S. Campana, M. Capalbi, G. Chincarini, G. Cusumano, A.D. Falcone, N. Gehrels, P. Giommi, M.R. Goad, O. Godet, C.P. Hurkett, J.A. Kennea, A. Moretti, P.T. O'Brien, J.P. Osborne, P. Romano, G. Tagliaferri, A.A. Wells, Evidence for a canonical gamma-ray burst afterglow light curve in the Swift XRT data. Astrophys. J. **642**, 389–400 (2006). doi:10.1086/500724
M. Nysewander, A.S. Fruchter, A. Pe'er, A comparison of the afterglows of short- and long-duration gamma-ray bursts. Astrophys. J. **701**, 824–836 (2009a). doi:10.1088/0004-637X/701/1/824
M. Nysewander, A.S. Fruchter, A. Pe'er, A comparison of the afterglows of short- and long-duration gamma-ray bursts. Astrophys. J. **701**, 824–836 (2009b). doi:10.1088/0004-637X/701/1/824
R. Oechslin, H.-T. Janka, Torus formation in neutron star mergers and well-localized short gamma-ray bursts. Mon. Not. R. Astron. Soc. **368**, 1489–1499 (2006). doi:10.1111/j.1365-2966.2006.10238.x
S.A. Olausen, V.M. Kaspi, The McGill magnetar catalog. Astrophys. J. Suppl. Ser. **212**, 6 (2014). doi:10.1088/0067-0049/212/1/6
D.M. Palmer, S. Barthelmy, N. Gehrels, R.M. Kippen, T. Cayton, C. Kouveliotou, D. Eichler, R.A.M.J. Wijers, P.M. Woods, J. Granot, Y.E. Lyubarsky, E. Ramirez-Ruiz, L. Barbier, M. Chester, J. Cummings, E.E. Fenimore, M.H. Finger, B.M. Gaensler, D. Hullinger, H. Krimm, C.B. Markwardt, J.A. Nousek, A. Parsons, S. Patel, T. Sakamoto, G. Sato, M. Suzuki, J. Tueller, A giant γ-ray flare from the magnetar SGR 1806-20. Nature **434**, 1107–1109 (2005). doi:10.1038/nature03525
D.A. Perley, B.D. Metzger, J. Granot, N.R. Butler, T. Sakamoto, E. Ramirez-Ruiz, A.J. Levan, J.S. Bloom, A.A. Miller, A. Bunker, H.-W. Chen, A.V. Filippenko, N. Gehrels, K. Glazebrook, P.B. Hall, K.C. Hur-

ley, D. Kocevski, W. Li, S. Lopez, J. Norris, A.L. Piro, D. Poznanski, J.X. Prochaska, E. Quataert, N. Tanvir, GRB 080503: implications of a naked short gamma-ray burst dominated by extended emission. Astrophys. J. **696**, 1871–1885 (2009). doi:10.1088/0004-637X/696/2/1871

D.A. Perley, S.B. Cenko, A. Corsi, N.R. Tanvir, A.J. Levan, D.A. Kann, E. Sonbas, K. Wiersema, W. Zheng, X.-H. Zhao, J.-M. Bai, M. Bremer, A.J. Castro-Tirado, L. Chang, K.I. Clubb, D. Frail, A. Fruchter, E. Göğüş, J. Greiner, T. Güver, A. Horesh, A.V. Filippenko, S. Klose, J. Mao, A.N. Morgan, A.S. Pozanenko, S. Schmidl, B. Stecklum, M. Tanga, A.A. Volnova, A.E. Volvach, J.-G. Wang, J.-M. Winters, Y.-X. Xin, The afterglow of GRB 130427A from 1 to 10^{16} GHz. Astrophys. J. **781**, 37 (2014). doi:10.1088/0004-637X/781/1/37

D.A. Perley, N.R. Tanvir, J. Hjorth, T. Laskar, E. Berger, R. Chary, A. de Ugarte Postigo, J.P.U. Fynbo, T. Krühler, A.J. Levan, M.J. Michałowski, S. Schulze, The Swift GRB Host Galaxy Legacy Survey—II. Rest-Frame NIR Luminosity Distribution and Evidence for a Near-Solar Metallicity Threshold. ArXiv e-prints (2015)

D.A. Perley, T. Krühler, P. Schady, M.J. Michałowski, C.C. Thöne, D. Petry, J.F. Graham, J. Greiner, S. Schulze, S. Kim, A Revised Host Galaxy Association for GRB 020819B: A High-Redshift Dusty Starburst, Not a Low-Redshift Gas-Poor Spiral. ArXiv e-prints (2016)

R. Perna, P.J. Armitage, B. Zhang, Flares in long and short gamma-ray bursts: a common origin in a hyper-accreting accretion disk. Astrophys. J. Lett. **636**, 29–32 (2006a). doi:10.1086/499775

E. Pian, P.A. Mazzali, N. Masetti, P. Ferrero, S. Klose, E. Palazzi, E. Ramirez-Ruiz, S.E. Woosley, C. Kouveliotou, J. Deng, A.V. Filippenko, R.J. Foley, J.P.U. Fynbo, D.A. Kann, W. Li, J. Hjorth, K. Nomoto, F. Patat, D.N. Sauer, J. Sollerman, P.M. Vreeswijk, E.W. Guenther, A. Levan, P. O'Brien, N.R. Tanvir, R.A.M.J. Wijers, C. Dumas, O. Hainaut, D.S. Wong, D. Baade, L. Wang, L. Amati, E. Cappellaro, A.J. Castro-Tirado, S. Ellison, F. Frontera, A.S. Fruchter, J. Greiner, K. Kawabata, C. Ledoux, K. Maeda, P. Møller, L. Nicastro, E. Rol, R. Starling, An optical supernova associated with the X-ray flash XRF 060218. Nature **442**, 1011–1013 (2006b). doi:10.1038/nature05082

P. Podsiadlowski, P.A. Mazzali, K. Nomoto, D. Lazzati, E. Cappellaro, The rates of hypernovae and gamma-ray bursts: implications for their progenitors. Astrophys. J. Lett. **607**, 17–20 (2004). doi:10.1086/421347

P. Podsiadlowski, N. Ivanova, S. Justham, S. Rappaport, Explosive common-envelope ejection: implications for gamma-ray bursts and low-mass black-hole binaries. Mon. Not. R. Astron. Soc. **406**, 840–847 (2010). doi:10.1111/j.1365-2966.2010.16751.x

S.F. Portegies Zwart, S.L.W. McMillan, M. Gieles, Young massive star clusters. Annu. Rev. Astron. Astrophys. **48**, 431–493 (2010). doi:10.1146/annurev-astro-081309-130834

S. Prajs, M. Sullivan, M. Smith, A. Levan, N.V. Karpenka, T.D.P. Edwards, C.R. Walker, W.M. Wolf, C. Balland, R. Carlberg, A. Howell, C. Lidman, R. Pain, C. Pritchet, V. Ruhlmann-Kleider, The Volumetric Rate of Superluminous Supernovae at $z \sim 1$. ArXiv e-prints (2016)

P.A. Price, E. Berger, D.E. Reichart, S.R. Kulkarni, S.A. Yost, R. Subrahmanyan, R.M. Wark, M.H. Wieringa, D.A. Frail, J. Bailey, B. Boyle, E. Corbett, K. Gunn, S.D. Ryder, N. Seymour, K. Koviak, P. McCarthy, M. Phillips, T.S. Axelrod, J.S. Bloom, S.G. Djorgovski, D.W. Fox, T.J. Galama, F.A. Harrison, K. Hurley, R. Sari, B.P. Schmidt, M.J.I. Brown, T. Cline, F. Frontera, C. Guidorzi, E. Montanari, GRB 011121: a massive star progenitor. Astrophys. J. Lett. **572**, 51–55 (2002). doi:10.1086/341552

P.A. Price, S.R. Kulkarni, E. Berger, D.W. Fox, J.S. Bloom, S.G. Djorgovski, D.A. Frail, T.J. Galama, F.A. Harrison, P. McCarthy, D.E. Reichart, R. Sari, S.A. Yost, H. Jerjen, K. Flint, A. Phillips, B.E. Warren, T.S. Axelrod, R.A. Chevalier, J. Holtzman, R.A. Kimble, B.P. Schmidt, J.C. Wheeler, F. Frontera, E. Costa, L. Piro, K. Hurley, T. Cline, C. Guidorzi, E. Montanari, E. Mazets, S. Golenetskii, I. Mitrofanov, D. Anfimov, A. Kozyrev, M. Litvak, A. Sanin, W. Boynton, C. Fellows, K. Harshman, C. Shinohara, A. Gal-Yam, E. Ofek, Y. Lipkin, Discovery of GRB 020405 and its late red bump. Astrophys. J. **589**, 838–843 (2003a). doi:10.1086/374730

P.A. Price, S.R. Kulkarni, B.P. Schmidt, T.J. Galama, J.S. Bloom, E. Berger, D.A. Frail, S.G. Djorgovski, D.W. Fox, A.A. Henden, S. Klose, F.A. Harrison, D.E. Reichart, R. Sari, S.A. Yost, T.S. Axelrod, P. McCarthy, J. Holtzman, J.P. Halpern, R.A. Kimble, J.C. Wheeler, R.A. Chevalier, K. Hurley, G.R. Ricker, E. Costa, F. Frontera, L. Piro, GRB 010921: strong limits on an underlying supernova from the Hubble Space Telescope. Astrophys. J. **584**, 931–936 (2003b). doi:10.1086/345734

P.A. Price, D.W. Fox, S.R. Kulkarni, B.A. Peterson, B.P. Schmidt, A.M. Soderberg, S.A. Yost, E. Berger, S.G. Djorgovski, D.A. Frail, F.A. Harrison, R. Sari, A.W. Blain, S.C. Chapman, The bright optical afterglow of the nearby γ-ray burst of 29 March 2003. Nature **423**, 844–847 (2003c). doi:10.1038/nature01734

R.M. Quimby, F. Yuan, C. Akerlof, J.C. Wheeler, Rates of superluminous supernovae at $z = 0.2$. Mon. Not. R. Astron. Soc. **431**, 912–922 (2013). doi:10.1093/mnras/stt213

D. Raghavan, H.A. McAlister, T.J. Henry, D.W. Latham, G.W. Marcy, B.D. Mason, D.R. Gies, R.J. White, T.A. ten Brummelaar, A survey of stellar families: multiplicity of solar-type stars. Astrophys. J. Suppl. Ser. **190**, 1–42 (2010). doi:10.1088/0067-0049/190/1/1

O.H. Ramírez-Agudelo, H. Sana, S.E. de Mink, V. Hénault-Brunet, A. de Koter, N. Langer, F. Tramper, G. Gräfener, C.J. Evans, J.S. Vink, P.L. Dufton, W.D. Taylor, The VLT-FLAMES Tarantula Survey. XXI. Stellar spin rates of O-type spectroscopic binaries. Astron. Astrophys. **580**, 92 (2015). doi:10.1051/0004-6361/201425424

C. Raskin, E. Scannapieco, J. Rhoads, M. Della Valle, Using spatial distributions to constrain progenitors of supernovae and gamma-ray bursts. Astrophys. J. **689**, 358–370 (2008). doi:10.1086/592495

N. Rea, M. Gullón, J.A. Pons, R. Perna, M.G. Dainotti, J.A. Miralles, D.F. Torres, Constraining the GRB-magnetar model by means of the galactic pulsar population. Astrophys. J. **813**, 92 (2015). doi:10.1088/0004-637X/813/2/92

T.M. Reynolds, M. Fraser, G. Gilmore, Gone without a bang: an archival HST survey for disappearing massive stars. Mon. Not. R. Astron. Soc. **453**(3), 2889–2900 (2015)

J.E. Rhoads, The dynamics and light curves of beamed gamma-ray burst afterglows. Astrophys. J. **525**, 737–749 (1999). doi:10.1086/307907

S. Rosswog, E. Ramirez-Ruiz, M.B. Davies, High-resolution calculations of merging neutron stars—III. Gamma-ray bursts. Mon. Not. R. Astron. Soc. **345**, 1077–1090 (2003). doi:10.1046/j.1365-2966.2003.07032.x

S. Rosswog, M. Liebendörfer, F.-K. Thielemann, M.B. Davies, W. Benz, T. Piran, Mass ejection in neutron star mergers. Astron. Astrophys. **341**, 499–526 (1999).

A. Rowlinson, P.T. O'Brien, B.D. Metzger, N.R. Tanvir, A.J. Levan, Signatures of magnetar central engines in short GRB light curves. Mon. Not. R. Astron. Soc. **430**, 1061–1087 (2013). doi:10.1093/mnras/sts683

H. Sana, G. James, E. Gosset, The massive star binary fraction in young open clusters—III. IC 2944 and the Cen OB2 association. Mon. Not. R. Astron. Soc. **416**, 817 (2011)

H. Sana, Y. Momany, M. Gieles, G. Carraro, Y. Beletsky, V.D. Ivanov, G. de Silva, G. James, A MAD view of Trumpler 14. Astron. Astrophys. **515**, 26 (2010). doi:10.1051/0004-6361/200913688

H. Sana, S.E. de Mink, A. de Koter, N. Langer, C.J. Evans, M. Gieles, E. Gosset, R.G. Izzard, J.-B. Le Bouquin, F.R.N. Schneider, Binary interaction dominates the evolution of massive stars. Science **337**, 444 (2012). doi:10.1126/science.1223344

H. Sana, A. de Koter, S.E. de Mink, P.R. Dunstall, C.J. Evans, V. Hénault-Brunet, J. Maíz Apellániz, O.H. Ramírez-Agudelo, W.D. Taylor, N.R. Walborn, J.S. Clark, P.A. Crowther, A. Herrero, M. Gieles, N. Langer, D.J. Lennon, J.S. Vink, The VLT-FLAMES Tarantula Survey. VIII. Multiplicity properties of the O-type star population. Astron. Astrophys. **550**, 107 (2013). doi:10.1051/0004-6361/201219621

R. Sari, T. Piran, R. Narayan, Spectra and light curves of gamma-ray burst afterglows. Astrophys. J. Lett. **497**, 17–20 (1998). doi:10.1086/311269

S. Savaglio, K. Glazebrook, D. Le Borgne, The galaxy population hosting gamma-ray bursts. Astrophys. J. **691**, 182–211 (2009). doi:10.1088/0004-637X/691/1/182

b.T. Shenar, R. Hainich, H. Todt, A. Sander, W.-R. Hamann, A.F.J. Moffat, J.J. Eldridge, H. Pablo, L.M. Oskinova, N.D. Richardson, Wolf-Rayet stars in the Small Magellanic Cloud: II. Analysis of the binaries. ArXiv e-prints (2016)

S.J. Smartt, J.J. Eldridge, R.M. Crockett, J.R. Maund, The death of massive stars—I. Observational constraints on the progenitors of Type II-P supernovae. Mon. Not. R. Astron. Soc. **395**, 1409–1437 (2009). doi:10.1111/j.1365-2966.2009.14506.x

A.M. Soderberg, S.R. Kulkarni, E. Berger, D.W. Fox, M. Sako, D.A. Frail, A. Gal-Yam, D.S. Moon, S.B. Cenko, S.A. Yost, M.M. Phillips, S.E. Persson, W.L. Freedman, P. Wyatt, R. Jayawardhana, D. Paulson, The sub-energetic γ-ray burst GRB 031203 as a cosmic analogue to the nearby GRB 980425. Nature **430**, 648–650 (2004). doi:10.1038/nature02757

A.M. Soderberg, S.R. Kulkarni, E. Nakar, E. Berger, P.B. Cameron, D.B. Fox, D. Frail, A. Gal-Yam, R. Sari, S.B. Cenko, M. Kasliwal, R.A. Chevalier, T. Piran, P.A. Price, B.P. Schmidt, G. Pooley, D.-S. Moon, B.E. Penprase, E. Ofek, A. Rau, N. Gehrels, J.A. Nousek, D.N. Burrows, S.E. Persson, P.J. McCarthy, Relativistic ejecta from X-ray flash XRF 060218 and the rate of cosmic explosions. Nature **442**, 1014–1017 (2006). doi:10.1038/nature05087

M. Sparre, R.L.C. Starling, A search for thermal X-ray signatures in gamma-ray bursts—II. The Swift sample. Mon. Not. R. Astron. Soc. **427**, 2965–2974 (2012). doi:10.1111/j.1365-2966.2012.21858.x

H.C. Spruit, Dynamo action by differential rotation in a stably stratified stellar interior. Astron. Astrophys. **381**, 923–932 (2002). doi:10.1051/0004-6361:20011465

K.Z. Stanek, T. Matheson, P.M. Garnavich, P. Martini, P. Berlind, N. Caldwell, P. Challis, W.R. Brown, R. Schild, K. Krisciunas, M.L. Calkins, J.C. Lee, N. Hathi, R.A. Jansen, R. Windhorst, L. Echevarria, D.J. Eisenstein, B. Pindor, E.W. Olszewski, P. Harding, S.T. Holland, D. Bersier, Spectroscopic discovery of the supernova 2003dh associated with GRB 030329. Astrophys. J. Lett. **591**, 17–20 (2003). doi:10.1086/376976

E.R. Stanway, J.J. Eldridge, G.D. Becker, Stellar population effects on the inferred photon density at reionization. Mon. Not. R. Astron. Soc. **456**, 485–499 (2016). doi:10.1093/mnras/stv2661

R.L.C. Starling, K. Wiersema, A.J. Levan, T. Sakamoto, D. Bersier, P. Goldoni, S.R. Oates, A. Rowlinson, S. Campana, J. Sollerman, N.R. Tanvir, D. Malesani, J.P.U. Fynbo, S. Covino, P. D'Avanzo, P.T. O'Brien, K.L. Page, J.P. Osborne, S.D. Vergani, S. Barthelmy, D.N. Burrows, Z. Cano, P.A. Curran, M. de Pasquale, V. D'Elia, P.A. Evans, H. Flores, A.S. Fruchter, P. Garnavich, N. Gehrels, J. Gorosabel, J. Hjorth, S.T. Holland, A.J. van der Horst, C.P. Hurkett, P. Jakobsson, A.P. Kamble, C. Kouveliotou, N.P.M. Kuin, L. Kaper, P.A. Mazzali, P.E. Nugent, E. Pian, M. Stamatikos, C.C. Thöne, S.E. Woosley, Discovery of the nearby long, soft GRB 100316D with an associated supernova. Mon. Not. R. Astron. Soc. **411**, 2792–2803 (2011). doi:10.1111/j.1365-2966.2010.17879.x

R.L.C. Starling, K.L. Page, A. Pe'Er, A.P. Beardmore, J.P. Osborne, A search for thermal X-ray signatures in gamma-ray bursts—I. Swift bursts with optical supernovae. Mon. Not. R. Astron. Soc. **427**, 2950–2964 (2012). doi:10.1111/j.1365-2966.2012.22116.x

M.P.L. Suijs, N. Langer, A.-J. Poelarends, S.-C. Yoon, A. Heger, F. Herwig, White dwarf spins from low-mass stellar evolution models. Astron. Astrophys. **481**, 87–90 (2008). doi:10.1051/0004-6361:20080941l

b.H. Sun, B. Zhang, H. Gao, X-ray counterpart of gravitational waves due to binary neutron star mergers: light curves, luminosity functions, and event rate densities. ArXiv e-prints (2016)

H. Sun, B. Zhang, Z. Li, Extragalactic high-energy transients: event rate densities and luminosity functions. Astrophys. J. **812**, 33 (2015). doi:10.1088/0004-637X/812/1/33

K.M. Svensson, A.J. Levan, N.R. Tanvir, A.S. Fruchter, L.-G. Strolger, The host galaxies of core-collapse supernovae and gamma-ray bursts. Mon. Not. R. Astron. Soc. **405**, 57–76 (2010). doi:10.1111/j.1365-2966.2010.16442.x

P.A. Sweet, The importance of rotation in stellar evolution. Mon. Not. R. Astron. Soc. **110**, 548 (1950). doi:10.1093/mnras/110.6.548

D. Szécsi, N. Langer, S.-C. Yoon, D. Sanyal, S. de Mink, C.J. Evans, T. Dermine, Low-metallicity massive single stars with rotation. Evolutionary models applicable to I Zwicky 18. Astron. Astrophys. **581**, 15 (2015). doi:10.1051/0004-6361/201526617

N.R. Tanvir, R. Chapman, A.J. Levan, R.S. Priddey, An origin in the local Universe for some short γ-ray bursts. Nature **438**, 991–993 (2005). doi:10.1038/nature04310

N.R. Tanvir, D.B. Fox, A.J. Levan, E. Berger, K. Wiersema, J.P.U. Fynbo, A. Cucchiara, T. Krühler, N. Gehrels, J.S. Bloom, J. Greiner, P.A. Evans, E. Rol, F. Olivares, J. Hjorth, P. Jakobsson, J. Farihi, R. Willingale, R.L.C. Starling, S.B. Cenko, D. Perley, J.R. Maund, J. Duke, R.A.M.J. Wijers, A.J. Adamson, A. Allan, M.N. Bremer, D.N. Burrows, A.J. Castro-Tirado, B. Cavanagh, A. de Ugarte Postigo, M.A. Dopita, T.A. Fatkhullin, A.S. Fruchter, R.J. Foley, J. Gorosabel, J. Kennea, T. Kerr, S. Klose, H.A. Krimm, V.N. Komarova, S.R. Kulkarni, A.S. Moskvitin, C.G. Mundell, T. Naylor, K. Page, B.E. Penprase, M. Perri, P. Podsiadlowski, K. Roth, R.E. Rutledge, T. Sakamoto, P. Schady, B.P. Schmidt, A.M. Soderberg, J. Sollerman, A.W. Stephens, G. Stratta, T.N. Ukwatta, D. Watson, E. Westra, T. Wold, C. Wolf, A γ-ray burst at a redshift of $z \sim 8.2$. Nature **461**, 1254–1257 (2009). doi:10.1038/nature08459

N.R. Tanvir, A.J. Levan, A.S. Fruchter, J. Hjorth, R.A. Hounsell, K. Wiersema, R.L. Tunnicliffe, A 'kilonova' associated with the short-duration γ-ray burst GRB 130603B. Nature **500**, 547–549 (2013). doi:10.1038/nature12505

C.C. Thöne, A. de Ugarte Postigo, C.L. Fryer, K.L. Page, J. Gorosabel, M.A. Aloy, D.A. Perley, C. Kouveliotou, H.T. Janka, P. Mimica, J.L. Racusin, H. Krimm, J. Cummings, S.R. Oates, S.T. Holland, M.H. Siegel, M. de Pasquale, E. Sonbas, M. Im, W.-K. Park, D.A. Kann, S. Guziy, L.H. García, A. Llorente, K. Bundy, C. Choi, H. Jeong, H. Korhonen, P. Kubànek, J. Lim, A. Moskvitin, T. Muñoz-Darias, S. Pak, I. Parrish, The unusual γ-ray burst GRB 101225A from a helium star/neutron star merger at redshift 0.33. Nature **480**, 72–74 (2011a). doi:10.1038/nature10611

C.C. Thöne, S. Campana, D. Lazzati, A. de Ugarte Postigo, J.P.U. Fynbo, L. Christensen, A.J. Levan, M.A. Aloy, J. Hjorth, P. Jakobsson, E.M. Levesque, D. Malesani, B. Milvang-Jensen, P.W.A. Roming, N.R. Tanvir, K. Wiersema, M. Gladders, E. Wuyts, H. Dahle, Variable Lyα sheds light on the environment surrounding GRB 090426. Mon. Not. R. Astron. Soc. **414**, 479–488 (2011b). doi:10.1111/j.1365-2966.2011.18408.x

F. Tramper, S.M. Straal, D. Sanyal, H. Sana, A. de Koter, G. Gräfener, N. Langer, J.S. Vink, S.E. de Mink, L. Kaper, Massive stars on the verge of exploding: the properties of oxygen sequence Wolf–Rayet stars. Astron. Astrophys. **581**, 110 (2015). doi:10.1051/0004-6361/201425390

C.A. Tremonti, T.M. Heckman, G. Kauffmann, J. Brinchmann, S. Charlot, S.D.M. White, M. Seibert, E.W. Peng, D.J. Schlegel, A. Uomoto, M. Fukugita, J. Brinkmann, The origin of the mass-metallicity relation: insights from 53,000 star-forming galaxies in the Sloan Digital Sky Survey. Astrophys. J. **613**, 898–913 (2004). doi:10.1086/423264

R.L. Tunnicliffe, A.J. Levan, N.R. Tanvir, A. Rowlinson, D.A. Perley, J.S. Bloom, S.B. Cenko, P.T. O'Brien, B.E. Cobb, K. Wiersema, D. Malesani, A. de Ugarte Postigo, J. Hjorth, J.P.U. Fynbo, P. Jakobsson, On the nature of the 'hostless' short GRBs. Mon. Not. R. Astron. Soc. **437**, 1495–1510 (2014). doi:10.1093/mnras/stt1975

R. Tylenda, N. Soker, Eruptions of the V838 Mon type: stellar merger versus nuclear outburst models. Astron. Astrophys. **451**, 223–236 (2006). doi:10.1051/0004-6361:20054201

R. Tylenda, M. Hajduk, T. Kamiński, A. Udalski, I. Soszyński, M.K. Szymański, M. Kubiak, G. Pietrzyński, R. Poleski, Ł. Wyrzykowski, K. Ulaczyk, V1309 Scorpii: merger of a contact binary. Astron. Astrophys. **528**, 114 (2011). doi:10.1051/0004-6361/201016221

V.V. Usov, Millisecond pulsars with extremely strong magnetic fields as a cosmological source of gamma-ray bursts. Nature **357**, 472–474 (1992). doi:10.1038/357472a0

E.P.J. van den Heuvel, S.-C. Yoon, Long gamma-ray burst progenitors: boundary conditions and binary models. Astrophys. Space Sci. **311**, 177–183 (2007). doi:10.1007/s10509-007-9583-8

J. van Paradijs, P.J. Groot, T. Galama, C. Kouveliotou, R.G. Strom, J. Telting, R.G.M. Rutten, G.J. Fishman, C.A. Meegan, M. Pettini, N. Tanvir, J. Bloom, H. Pedersen, H.U. Nørdgaard-Nielsen, M. Linden-Vørnle, J. Melnick, G. van der Steene, M. Bremer, R. Naber, J. Heise, J. in't Zand, E. Costa, M. Feroci, L. Piro, F. Frontera, G. Zavattini, L. Nicastro, E. Palazzi, K. Bennett, L. Hanlon, A. Parmar, Transient optical emission from the error box of the γ-ray burst of 28 February 1997. Nature **386**, 686–689 (1997). doi:10.1038/386686a0

S. Vaughan, R. Willingale, P.T. O'Brien, J.P. Osborne, J.N. Reeves, A.J. Levan, M.G. Watson, J.A. Tedds, D. Watson, M. Santos-Lleó, P.M. Rodríguez-Pascual, N. Schartel, The discovery of an evolving dust-scattered X-ray halo around GRB 031203. Astrophys. J. Lett. **603**, 5–8 (2004). doi:10.1086/382785

J.S. Vink, Constraining GRB progenitor models by probing Wolf–Rayet wind geometries in the Large Magellanic Cloud. Astron. Astrophys. **469**, 707–711 (2007). doi:10.1051/0004-6361:20077302

J.S. Vink, A. de Koter, On the metallicity dependence of Wolf–Rayet winds. Astron. Astrophys. **442**, 587–596 (2005). doi:10.1051/0004-6361:20052862

J.S. Vink, A. de Koter, H.J.G.L.M. Lamers, Mass-loss predictions for O and B stars as a function of metallicity. Astron. Astrophys. **369**, 574–588 (2001). doi:10.1051/0004-6361:20010127

A. Wallner, T. Faestermann, J. Feige, C. Feldstein, K. Knie, G. Korschinek, W. Kutschera, A. Ofan, M. Paul, F. Quinto, G. Rugel, P. Steier, Abundance of live ^{244}Pu in deep-sea reservoirs on Earth points to rarity of actinide nucleosynthesis. Nat. Commun. **6**, 5956 (2015). doi:10.1038/ncomms6956

D. Watson, S.A. Vaughan, R. Willingale, J. Hjorth, S. Foley, J.P.U. Fynbo, P. Jakobsson, A. Levan, P.T. O'Brien, J.P. Osborne, K. Pedersen, J.N. Reeves, J.A. Tedds, M.G. Watson, The soft X-ray blast in the apparently subluminous GRB 031203. Astrophys. J. **636**, 967–970 (2006). doi:10.1086/498089

C. Weidner, P. Kroupa, I.A.D. Bonnell, The relation between the most-massive star and its parental star cluster mass. Mon. Not. R. Astron. Soc. **401**, 275–293 (2010). doi:10.1111/j.1365-2966.2009.15633.x

R. Willingale, J.P. Osborne, P.T. O'Brien, M.J. Ward, A. Levan, K.L. Page, The X-ray, optical and radio evolution of the GRB 030329 afterglow and the associated SN2003dh. Mon. Not. R. Astron. Soc. **349**, 31–38 (2004). doi:10.1111/j.1365-2966.2004.07475.x

S.E. Woosley, Gamma-ray bursts from stellar mass accretion disks around black holes. Astrophys. J. **405**, 273–277 (1993). doi:10.1086/172359

S.E. Woosley, A. Heger, The progenitor stars of gamma-ray bursts. Astrophys. J. **637**, 914–921 (2006). doi:10.1086/498500

S.E. Woosley, R.G. Eastman, B.P. Schmidt, Gamma-ray bursts and Type IC supernova SN 1998BW. Astrophys. J. **516**, 788–796 (1999). doi:10.1086/307131

C.C. Worley, M.J. Irwin, C.A. Tout, A.N. Żytkow, M. Fraser, R.G. Izzard, The proper motion of HV2112: a TŻO candidate in the SMC. Mon. Not. R. Astron. Soc. **459**, 31–35 (2016). doi:10.1093/mnrasl/slw034

D. Xu, A. de Ugarte Postigo, G. Leloudas, T. Krühler, Z. Cano, J. Hjorth, D. Malesani, J.P.U. Fynbo, C.C. Thöne, R. Sánchez-Ramírez, S. Schulze, P. Jakobsson, L. Kaper, J. Sollerman, D.J. Watson, A. Cabrera-Lavers, C. Cao, S. Covino, H. Flores, S. Geier, J. Gorosabel, S.M. Hu, B. Milvang-Jensen, M. Sparre, L.P. Xin, T.M. Zhang, W.K. Zheng, Y.C. Zou, Discovery of the broad-lined Type Ic SN 2013cq associated with the very energetic GRB 130427A. Astrophys. J. **776**, 98 (2013). doi:10.1088/0004-637X/776/2/98

S.-C. Yoon, N. Langer, Evolution of rapidly rotating metal-poor massive stars towards gamma-ray bursts. Astron. Astrophys. **443**, 643–648 (2005). doi:10.1051/0004-6361:20054030

S.-C. Yoon, A. Dierks, N. Langer, Evolution of massive population III stars with rotation and magnetic fields. Astron. Astrophys. **542**, 113 (2012). doi:10.1051/0004-6361/201117769

S.-C. Yoon, N. Langer, C. Norman, Single star progenitors of long gamma-ray bursts. I. Model grids and redshift dependent GRB rate. Astron. Astrophys. **460**, 199–208 (2006). doi:10.1051/0004-6361:20065912

S.-C. Yoon, S.E. Woosley, N. Langer, Type Ib/c supernovae in binary systems. I. Evolution and properties of the progenitor stars. Astrophys. J. **725**, 940–954 (2010). doi:10.1088/0004-637X/725/1/940

N. Yusof, R. Hirschi, G. Meynet, P.A. Crowther, S. Ekström, U. Frischknecht, C. Georgy, H. Abu Kassim, O. Schnurr, Evolution and fate of very massive stars. Mon. Not. R. Astron. Soc. **433**, 1114–1132 (2013). doi:10.1093/mnras/stt794

A. Zeh, S. Klose, D.H. Hartmann, A systematic analysis of supernova light in gamma-ray burst afterglows. Astrophys. J. **609**, 952–961 (2004). doi:10.1086/421100

B. Zhang, B.-B. Zhang, F.J. Virgili, E.-W. Liang, D.A. Kann, X.-F. Wu, D. Proga, H.-J. Lv, K. Toma, P. Mészáros, D.N. Burrows, P.W.A. Roming, N. Gehrels, Discerning the physical origins of cosmological gamma-ray bursts based on multiple observational criteria: the cases of z = 6.7 GRB 080913, z = 8.2 GRB 090423, and some short/hard GRBs. Astrophys. J. **703**, 1696–1724 (2009). doi:10.1088/0004-637X/703/2/1696

Space Sci Rev (2016) 202:81–111
DOI 10.1007/s11214-016-0270-3

Galaxy Formation and Evolution

Kentaro Nagamine[1,2] · **Naveen Reddy**[3] ·
Emanuele Daddi[4] · **Mark T. Sargent**[5]

Received: 12 December 2015 / Accepted: 13 June 2016 / Published online: 25 July 2016

Abstract In this chapter, we discuss the current status of observational and computational studies on galaxy formation and evolution. In particular, a joint analysis of star-formation rates (SFRs), stellar masses, and metallicities of galaxies throughout cosmic time can shed light on the processes by which galaxies build up their stellar mass and enrich the environment with heavy elements. Comparison of such observations and the results of numerical simulations can give us insights on the physical importance of various feedback effects by supernovae and active galactic nuclei.

In Sect. 1, we first discuss the primary methods used to deduce the SFRs, stellar masses, and (primarily) gas-phase metallicities in high-redshift galaxies. Then, we show how these quantities are related to each other and evolve with time.

In Sect. 2, we further examine the distribution of SFRs in galaxies following the 'Main Sequence' paradigm. We show how the so-called 'starbursts' display higher specific SFRs and SF efficiencies by an order of magnitude. We use this to devise a simple description of the evolution of the star-forming galaxy population since $z \sim 3$ that can successfully

✉ K. Nagamine
kn@ess.sci.osaka-u.ac.jp

N. Reddy
naveenr@ucr.edu

E. Daddi
edaddi@cea.fr

1 Theoretical Astrophysics, Department of Earth and Space Science, Graduate School of Science, Osaka University, 1-1 Machikaneyama, Toyonaka, Osaka 560-0043, Japan

2 Department of Physics & Astronomy, University of Nevada, Las Vegas, 4505 S. Maryland Pkwy, Las Vegas, NV 89154-4002, USA

3 Department of Physics & Astronomy, University of California, Riverside, 900 University Avenue, Riverside, CA 92521, USA

4 CEA Saclay, DSM/Irfu/Sérvice d'Astrophysique, Orme des Merisiers, 91191 Gif-sur-Yvette Cedex, France

5 Astronomy Centre, Department of Physics and Astronomy, University of Sussex, Brighton BN1 9QH, UK

reproduce some of the observed statistics in the infrared (IR) wavelength. We also discuss the properties of molecular gas.

In Sect. 3, we highlight some of the recent studies of high-redshift galaxy formation using cosmological hydrodynamic simulations. We discuss the physical properties of simulated galaxies such as luminosity function and escape fraction of ionizing photons, which are important statistics for reionization of the Universe. In particular the escape fraction of ionizing photons has large uncertainties, and studying gamma-ray bursts (which is the main topic of this conference) can also set observational constraints on this uncertain physical parameter as well as cosmic star formation rate density.

Keywords Galaxies · Galaxy formation · Galaxy evolution · Star formation · Cosmology · Metallicity · Numerical simulations

1 Star-Forming Galaxies (SFGs): Linking Star-Formation Rate, Stellar Mass, Metallicity, and Their Evolution (Primary Author: N. Reddy)

A joint analysis of the star-formation rates (SFRs), stellar masses ($M_\star$), and metallicities of galaxies throughout cosmic time can shed light on the processes by which galaxies build up their stellar mass, and enrich the interstellar and intergalactic media with heavy elements. In this chapter, we first discuss the primary methods used to deduce the SFRs, stellar masses, and (primarily) gas-phase metallicities in high-redshift galaxies. Then, we present a summary of how SFRs, stellar masses, and metallicities are related to each other, and how these relationships evolve with time. We conclude this section with future prospects of measuring these properties and their evolution.

The SFR, $M_\star$, and metallicity (Z) are three of the fundamental characteristics of galaxies that we can use to discern how galaxies evolve with cosmic time. Partly spurred by the advent of large multi-wavelength datasets, there has been significant progress over the past couple of decades to infer the SFRs, stellar masses, and metallicities of galaxies across a large range in lookback time. This progress has, in turn, given us new insights into how galaxies form stars and build up their stellar mass, and how the heavy element content in the ISM of galaxies may vary with time depending on a number of factors. Nevertheless, a number of critical questions remain, some of which may be addressed with future astronomical facilities. In this section, we discuss the current state of our knowledge regarding the SFRs, stellar masses, and metallicities of high-redshift galaxies; how these properties relate to those we see in local galaxies; and we will highlight some of the challenges to fully incorporating these measurements into a coherent picture of galaxy formation.

1.1 Calculating SFR, $M_\star$, and Metallicity

1.1.1 SFRs

A number of different multi-wavelength indicators may be used to estimate the SFRs of galaxies; some of the most common measures are summarized in Kennicutt (1998). The most frequently used indicator of star formation in *typical* star-forming galaxies (SFGs) at high redshift is the ultraviolet (UV) luminosity, as it is easily measured for distant galaxies. As the UV luminosity is susceptible to obscuration by interstellar dust, a correction must be applied to the UV luminosity. Typically, such corrections take advantage of the correlation between dustiness—as quantified by the ratio of the IR to UV luminosity, or essentially

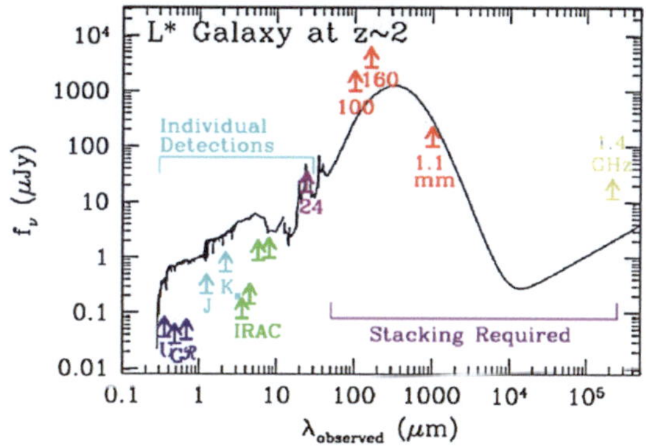

Fig. 1 Spectral energy distribution (SED) of a typical (L^*) galaxy at $z \sim 2$ (*black*), along with the detection limits of typical ground-based optical, near-IR, and IR imaging, and space-based mid-IR and far-IR imaging from *Spitzer* and *Herschel*, in the GOODS-North field. Also indicated is the detection limit of the VLA 1.4 GHz radio data in the GOODS-North field, as well as the detection limit for ALMA at 1.1 mm. Updated from Reddy et al. (2012a)

the dust-reprocessed light to that of the direct starlight—and the slope of the spectrum in the UV (Meurer et al. 1999). The wavelength dependence of dust obscuration implies that as the dustiness of a galaxy increases, the UV spectral slope becomes redder, as has been observed in local starburst galaxies (e.g., Meurer et al. 1999; Overzier et al. 2011), and subsequently the trends were shown for typical SFGs at $z \sim 2$ (e.g., Reddy et al. 2006; Daddi et al. 2007; Buat et al. 2012). The UV spectral slope is also sensitive to stellar population age, metallicity, and star-formation history (e.g., Kong et al. 2004; Seibert et al. 2005; Johnson et al. 2007; Dale et al. 2009; Reddy et al. 2010; Boquien et al. 2012; Wilkins et al. 2013), and so inferences of dustiness based on the UV slope must be viewed with caution, particularly for those galaxies where older stellar populations may contribute significantly to the near-UV flux of galaxies.

Ideally, one should obtain direct measurements of dust in high-redshift galaxies in order to more directly probe the amount of star formation in galaxies that may be obscured by dust. This may be accomplished by directly measuring dust emission in the mid- and far-infrared. While this technique can work well for the dustier and more luminous galaxies at high-redshift, current mid- and far-IR instrumentation (e.g., *Spitzer Space Telescope* and *Herschel Space Observatory*) are insufficiently sensitive to directly detect the dust emission from galaxies that are fainter than $\approx L^*_{\rm UV}$ (e.g., Reddy et al. 2006; Fig. 1). To remedy this difficulty, analyses of the *average* mid- and far-IR emission have been carried out by "stacking" images for many galaxies (Reddy et al. 2006; Daddi et al. 2007; Reddy et al. 2012a), but this comes at the expense of losing information on the scatter in the dust emission from galaxy-to-galaxy. Nonetheless, *Spitzer*, *Herschel*, and ALMA observations have made a transformative impact on our understanding of the dust contents of the most luminous galaxies at high redshift (see review by Casey et al. 2014).

The limitations of the UV slope and direct far-IR measurements in probing dusty star-formation may be mitigated by the use of other SFR indicators. For instance, the radio 21 cm emission from SFGs is dominated by the synchrotron emission of electrons accelerated by supernovae, and provides an extinction-free indicator of SFR. Radio imaging is still insufficiently sensitive to directly detect typical star-forming galaxies at $z \gtrsim 1$, and also may suffer from significant contamination if there is an active galactic nucleus (AGN). While X-ray emission may be used as another extinction-free measure of SFR (Ranalli et al. 2003) that can be measured for individual galaxies in deep Poisson-limited X-ray imaging, such emission may also vary with metallicity (Basu-Zych et al. 2013) and be affected by AGN.

The advent of multi-object near-infrared spectrographs (e.g., MOSFIRE, KMOS) have enabled measurements of the Balmer decrements ($H\alpha/H\beta$) of distant galaxies, as has been done locally for very large samples of galaxies (e.g., SDSS, Groves et al. 2012). For example, the recently commissioned MOSFIRE spectrograph on the Keck Telescope has been used to measure the Balmer decrements of a large sample of galaxies at $1.4 \leq z \leq 2.6$, enabling dust corrections to $H\alpha$ luminosities and dust-corrected estimates of the total SFR (Reddy et al. 2015). The Balmer lines provide the advantage of being less affected by SF history, as they arise only in the presence of hot, massive stars that can ionize hydrogen; and they can be measured now for individual galaxies without the need for spectral stacking (Reddy et al. 2015), as has been the case previously (e.g., Domínguez et al. 2013).

Finally, targeted observations with ALMA and other mm/sub-mm facilities, in combination with *Herschel* data, may be used to more accurately constrain dust temperature and mass and hence total infrared luminosities for more typical galaxies at high redshift. Similarly, gas masses measured with ALMA may also be used to estimate the amount of dust obscured star formation, albeit with large systematic uncertainties associated with the dust-to-gas ratio. More generally, resolved (kilo-parsec scale) imaging of the gas content (e.g., with IRAM or ALMA), combined with recombination line maps (e.g., made possible with the next generation of ground-based telescopes like TMT and ELT), will enable studies of the Kennicutt-Schmidt relation on resolved scales, thus clarifying the connection between gas surface density and SFR in high-redshift galaxies.

In the context of this conference, GRBs provide a unique avenue to constrain the shape of dust attenuation curve (e.g., Perley et al. 2013) which, in turn, will enable more accurate dust corrected SFRs. While each probe of SFR is subject to systematic uncertainties (e.g., such as the timescale for star formation, and the initial mass function), the use of multiple probes is generally indicated to provide the most robust measurement of SFRs of galaxies.

1.1.2 Stellar Masses

The primary method used to estimate stellar masses of distant galaxies is by fitting broad-band photometry with stellar population synthesis models for different SF histories, ages, and reddenings. Commonly used empirical and theoretical models include Bruzual and Charlot (2003), Maraston (2005), Fioc and Rocca-Volmerange (1997), and Leitherer et al. (1999). Typically a grid of such models is computed assuming a range of SF histories (e.g., exponentially-declining, rising, constant, or bursty/stochastic), with a range of ages from the age of the Universe to some minimum value (e.g., in some cases this may be set to the dynamical timescale expected for the galaxy; Reddy et al. 2012b), and some range of reddening assuming a dust attenuation curve (e.g., Calzetti et al. 2000). Generally, stellar masses derived in this manner tend to have less *random* uncertainty than other parameters derived from such fitting, though the systematic uncertainty may be quite large. Sources of systematic uncertainty include the choice of IMF; the degree to which the most current episode of star formation dominates the optical light, commonly referred to as the "out-shining" problem (Maraston 2005); and the contribution of thermally-pulsating AGBs to the rest-frame near-infrared light (e.g., Maraston 2005; Kriek et al. 2010).

The "SED-modeling" discussed above is usually performed using broadband photometry as typical SFGs at $z \gtrsim 1$ are too faint in the continuum for high signal-to-noise absorption spectroscopy to be feasible; the exception is for more massive galaxies at $z \sim 1$ where rest-optical continuum spectroscopy may be possible with longer integrations. Alternatively, one can deduce the stellar mass, or at least provide a rough upper limit, if dynamical mass estimates are available. At high-redshift, this dynamical masses have been computed using either resolved observations of some gas-sensitive tracer, such as the CO emission, or by converting an Hα luminosity to SFR, assuming a size, the Kennicutt-Schmidt relation, and a geometry for the galaxy (e.g., Erb et al. 2006b). Of course, while such measurements are subject to a number of systematic effects, they can in principle provide an independent handle on the stellar masses obtained from fitting the SEDs of galaxies.

1.1.3 Metallicities

Metallicities of high-redshift galaxies, specifically the *gas-phase* metallicities, are typically determined from strong-line abundance indicators that include some combination of the Balmer emission lines, and collisionally-excited forbidden oxygen and other weaker (e.g., [NII]) lines that arise from the ionized regions around massive stars. In rarer cases where a sufficiently high S/N and high spectral resolution spectrum has been obtained (e.g., for gravitationally lensed galaxies like cB58; Pettini et al. 2002), metallicities may be estimated from absorption lines either in the rest-UV (different sets of lines may be used to probe either the stellar metallicity, or the metal content of the ionized gas), or in the rest-optical (for stellar metallicities); see Shapley (2011) for a review of these metal indicators. Generally, direct electron temperature indicators which can be used to more robustly constrain the gas-phase metallicity are too weak to detect in individual high-redshift galaxy spectra, though access to these weaker features, like the [OIII]4363 line, will come with the next generation of large aperture telescopes (e.g., TMT, E-ELT, etc.).

There is a significant scatter and systematic offsets between different emission line indicators of metallicity (Kewley and Ellison 2008). Moreover, the application of locally-calibrated metallicity diagnostics to high-redshift galaxies may be inappropriate if the physical conditions (e.g., ionizing parameter, electron density) of the HII regions at high redshift differ from those found locally. Large rest-optical spectroscopic surveys of $z \gtrsim 1.4$ galaxies, made possible with multi-object near-IR spectrographs like MOSFIRE (McLean et al. 2012), have enabled the measurements of all the strong rest-optical emission lines for such galaxies. This, in turn, has enabled detailed assessments of the physical conditions in typical SFGs at high-redshift which underpin the metallicity measurements (e.g., Shapley et al. 2015).

1.2 The Relations Between SFR, $M_\star$, and Metallicity, and Their Evolution

1.2.1 SFR–$M_\star$ Relation

The evolution of the relationship between SFR and $M_\star$ may be used to deduce, at least on average, how galaxies are building their stellar mass with cosmic time. Trends between SFR and $M_\star$ have been found at essentially all redshifts where the relationship has been probed, including at $z \lesssim 1$ (Noeske et al. 2007), and $z \gtrsim 2$ (Reddy et al. 2006; Daddi et al. 2007; Rodighiero et al. 2011; Whitaker et al. 2014). Generally, there is a universal trend for SFGs where the SFR increases with stellar mass, but with an increasing normalization with redshift (Fig. 2).

Generally, many studies have advocated that the scatter in the SFR–$M_\star$ relation is relatively "tight" (i.e., $\lesssim 0.3$ dex), however there are a number of reasons why the measured scatter should be considered as a lower limit. First, sample selection (e.g., in the UV) may be biased against detecting the highest SFR galaxies (and $M_\star$) because they are very dusty, and

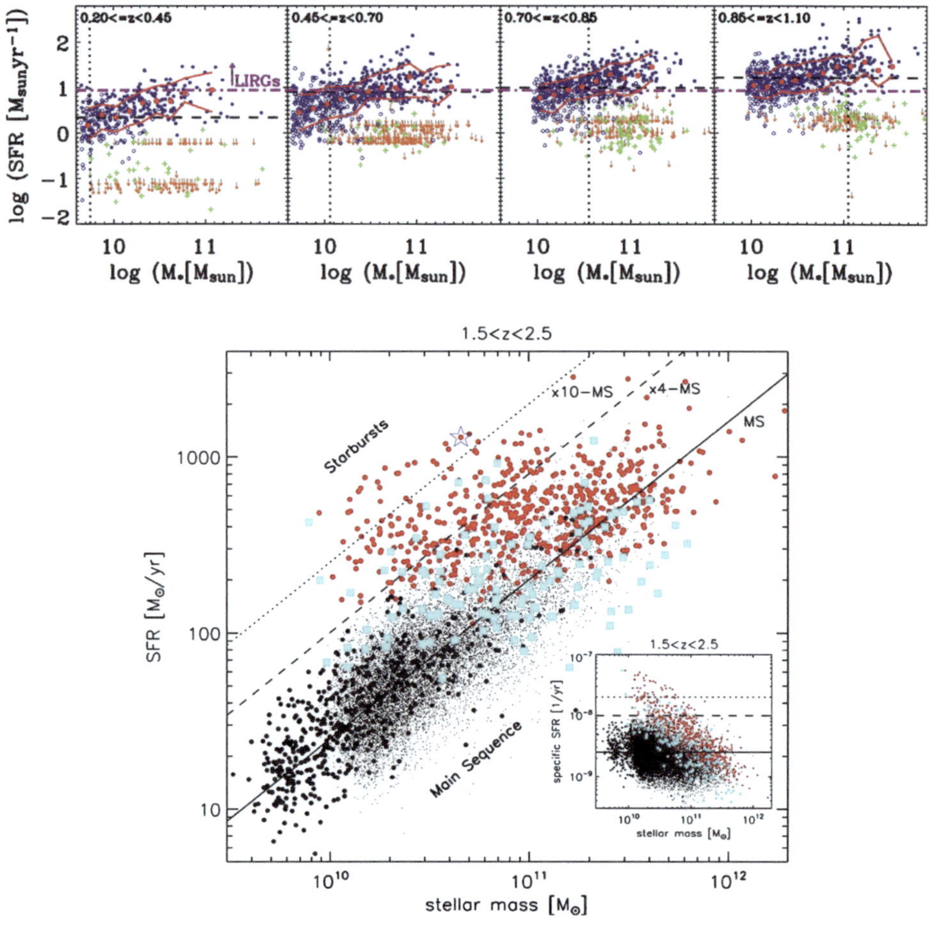

Fig. 2 *Top Row:* SFR vs. $M_\star$ for 2905 galaxies at $0.20 \lesssim z \lesssim 1.10$ from the AEGIS survey (from Noeske et al. 2007). *Bottom:* SFR vs. $M_\star$ for a sample of far-IR-detected (*red* and *cyan points*) and undetected (*black points*) galaxies at $z \sim 2$ (from Rodighiero et al. 2011)

thus one might underestimate the scatter in SFR at larger stellar masses (e.g., Daddi et al. 2007; Reddy et al. 2012b). On the other hand, while IR-selection may be more "complete" for the dustier and higher SFR galaxies, they will miss most of the typical SFGs. Further, when computing IR SFRs, one typically assumes a fixed conversion between IR luminosity and SFR. If this conversion changes from galaxy-to-galaxy, then the derived scatter in IR SFRs at a given $M_\star$ will be a lower limit to the true scatter. More generally, galaxy-to-galaxy variations in dust corrections, IMF, SF histories, etc., all imply that the scatter in the SFR–$M_\star$ relation, regardless of which SFR indicator is used, could be substantially larger than what has been claimed in the literature (Shivaei et al. 2015). Moreover, as many studies rely on photometric redshifts to derive stellar masses, and in some cases SFRs, the true scatter in SFR–$M_\star$ will be underestimated. In any case, the small scatter in the SFR–$M_\star$ relation has been interpreted as a reflection of the "smooth" manner in which galaxies build up their stellar mass (e.g, Davé 2008). However, given the uncertainties discussed above, we argue that the "scatter" derived for the SFR–$M_\star$ relation should be treated with caution.

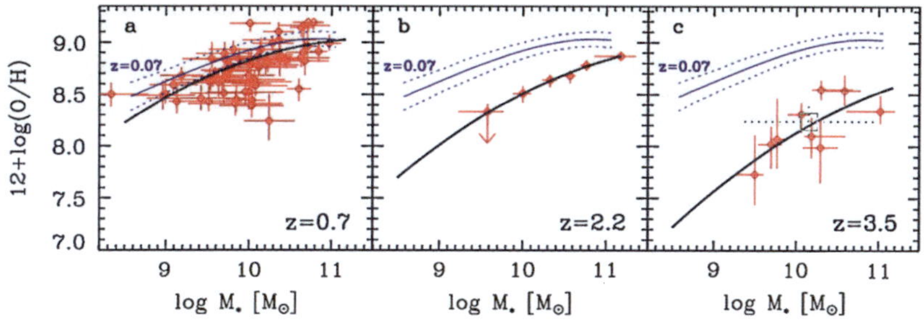

Fig. 3 Mass-metallicity relation at different redshifts, where the *blue solid* and *dotted lines* indicate the MZR at $z \sim 0.07$ from Kewley and Ellison (2008). Taken from Maiolino et al. (2008)

While the scatter may be difficult to pinpoint, the slope is generally more robust to the issues discussed above, except for that of sample selection and Malmquist bias where, for example, one is more prone to selecting galaxies with larger SFRs at a given low $M_\star$ for a flux- or line-selected (e.g, UV or Hα selected) sample (Reddy et al. 2012b). While rest-optical selection can mitigate some of this Malmquist bias, such samples can still be incomplete especially if they rely on photometric redshifts which are generally poorly determined for young, low-mass, and essentially flat-spectrum galaxies. Nonetheless, significant efforts are focused on constraining the slope of the SFR–$M_\star$ relation at low masses, with the expectation that such galaxies may build up their stellar mass in a different way than their brighter counterparts (e.g., Whitaker et al. 2014), perhaps because different feedback processes, or the same feedback process with a different efficiency, operate in different mass regimes. Resolved SED modeling (e.g., with *Hubble Space Telescope*), as well as resolved imaging of the gas and dust contents in these galaxies may help to further clarify the relationship between stellar populations, gas content, and feedback.

1.2.2 Mass–Metallicity Relation

Like the SFR–$M_\star$ relation, the mass–metallicity relation (MZR) has been found across a wide swath of redshift, at $z \sim 0$ (Tremonti et al. 2004), $z \sim 0.7$ (Savaglio et al. 2005), $z \sim 2.3$ (Erb et al. 2006a), and $z \gtrsim 3$ (Maiolino et al. 2008); see Fig. 3. These studies indicate that high-redshift galaxies are less metal-rich than local galaxies at a given stellar mass. To infer this evolution, however, previous studies had to rely on comparisons of the MZR that used different metallicity indicators at different redshifts (thus introducing some uncertainty in the evolution owing to systematic differences in metallicity indicators), and made the assumption that the local calibrations between strong-line emission indicators and the oxygen abundance apply at all redshifts.

What has been lacking previously is access to the same metallicity indicator at different redshifts, as well as an understanding of how such indicators scale with oxygen abundance at different redshifts. Recent advances in near-IR detector technology (e.g., MOSFIRE, KMOS) have begun to remedy this situation. For example, the MOSFIRE instrument enables access to the same strong rest-optical emission lines, including [OII], [OIII], Hβ and Hα, at high redshift ($1.4 \lesssim z \lesssim 2.6$) that are used to estimate metallicities in local galaxies. Further, the sensitivity of these observations is sufficient to obtain metallicity estimates for individual objects rather than for an ensemble (Fig. 4).

Surveys like MOSDEF, which is sensitive to a number of different strong rest-optical emission lines for $1.4 \lesssim z \lesssim 3.8$ galaxies, can enable detailed photoionization modeling to

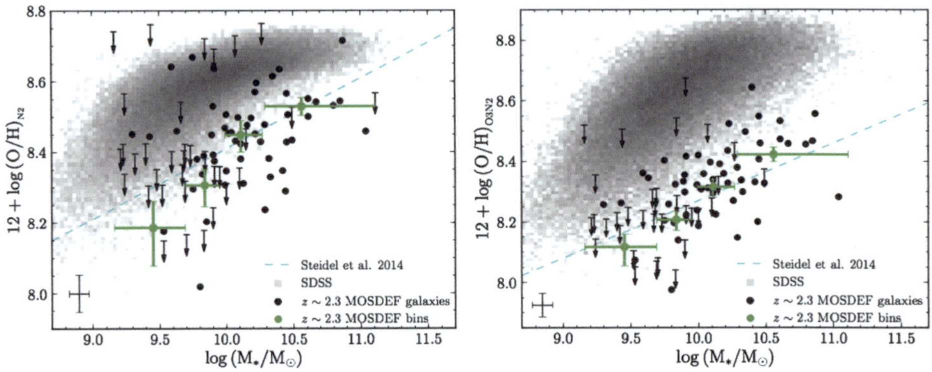

Fig. 4 Mass-metallicity relation at $z \sim 2.3$ based on data from the MOSFIRE Deep Evolution Field (MOSDEF) survey. Gas-phase metallicities are calculated using both the N2 (*left*) and O3N2 (*right*) indices, where upper limits are indicated by the *arrows*. Taken from Sanders et al. (2015)

jointly constrain the electron densities, ionizing parameters and hardness, and metallicities, of high-redshift galaxies. For example, preliminary studies suggest that for at a fixed ionization parameter, the high-redshift galaxies have the same metallicities as those of local galaxies (Shapley et al. 2015). Furthermore, these studies have suggested that high-redshift galaxies are more abundant in nitrogen at a given gas-phase metallicity relative to their local counterparts, perhaps due to the properties of massive stellar populations at high redshift (Masters et al. 2014; Shapley et al. 2015). Regardless, these larger samples will enable more detailed comparisons of the slope and scatter of the MZR relation to the predictions of galaxy simulations in order to deduce the effect of gas inflows and outflows on the metal content and SFRs of galaxies across cosmic time (e.g., Finlator and Davé 2008).

1.3 Connections Between SFR, $M_\star$, and Metallicity, and Future Directions

To first order, the stellar mass simply reflects the integral of the SFR with time (modulo the fraction of gas that is returned to the ISM). Similarly, the metallicity reflects the heavy element yield from this star formation. These connections motivated Mannucci et al. (2010) to propose a model whereby the SFRs, $M_\star$, and metallicities of galaxies lie on a "fundamental plane" (or "fundamental metallicity relation", FMR) that does not evolve, and that the observed redshift evolution of the MZR simply reflects that change in average SFRs probed at each redshift. However, more recent studies (e.g., Steidel et al. 2014; Troncoso et al. 2014; Sanders et al. 2015) suggest that high-redshift galaxies do not fall on the FMR defined by local galaxies. Current and forthcoming large surveys will enable detailed photo-ionization modeling and more robust estimates of the gas-phase metallicities over a large swath of cosmic time. This, in turn, should give us a better indication of the effects of gas inflows and outflows on the evolution of galaxies.

Aside from this impending improvement in our knowledge of the joint relationship between the SFRs, $M_\star$, and metallicities, future facilities and instruments will undoubtedly provide a more "resolved" picture of these relationships. Multi-object spectrographs on $\gtrsim 30$ m class telescopes will enable the measurement of weaker lines like [OIII]4363, giving us a direct metallicity indicator with which to calibrate all the strong line indicators. These facilities will also extend upon the currently known trends and scatter in the SFR–$M_\star$ and MZR relations to fainter galaxies that host the bulk of the star formation activity at high redshift (Reddy and Steidel 2009; Bouwens et al. 2015).

Combined with adaptive optics, the future ground-based facilities will enable a study of the connection between SFR, $M_\star$, and metallicity on kilo-parsec scales where the effects of galaxy feedback may be modeled more easily. Finally, we note that the *James Webb Space Telescope* will access the strong line indicators discussed above for galaxies at even higher redshifts (up to $z \sim 6$) where galaxies are evolving on more rapid timescales.

Probing faint galaxies at high redshift with gamma-ray bursts (GRBs), the topic of this conference, provides an alternative and complementary means of evaluating the connections between SFR, stellar mass, and metallicity. In particular, because of their stellar origin, GRBs can be found in very faint galaxies that would otherwise be "missed" in current surveys. Aside from opening up a new parameter space, GRBs also offer the opportunity to measure stellar metallicities using rest-UV lines (e.g., CIV; Berger et al. 2006; Sparre et al. 2014), or traditional metallicity diagnostics for DLAs that correspond to GRB host galaxies (Ledoux et al. 2009; Cucchiara et al. 2015). Understanding the context of GRB hosts in terms of the previously studied galaxy populations will require a more quantitative connection between the propensity for GRBs and host galaxy properties.

2 Global Properties of SFGs at $z < 4$ (Primary Author: E. Daddi & M.T. Sargent)

In this section, we further discuss modern understanding of the distribution of SFRs in galaxies following the Main Sequence paradigm. While the majority of SFGs are observed to be close to the relation between SFR and $M_\star$, a smaller subset of the population—so-called 'starbursts'—display specific SFRs and SF efficiencies that exceed those of normal (main-sequence) galaxies by up to an order of magnitude. This can be used to devise a simple description of the evolution of the star-forming galaxy population since $z \sim 3$ that can successfully reproduce the shape of the IR luminosity function of galaxies as a function of redshift, the IR number counts and Planck-Herschel power spectrum of the cosmic IR background, i.e., clustering of star formation. The evolution of star formation in galaxies is linked to the underlying properties of the molecular gas. Typical Main Sequence galaxies have warmer gas temperatures at high redshifts and display relatively stronger high-J CO emission, probably linked to their clumpier interstellar medium. Evolution at even higher redshifts is currently under investigation. A key driver is the expected decrease of metallicities towards early times.

At high and low redshift, the population of SFGs comprise a mix of (1) 'normal' SFGs following a narrow relation—the 'star-forming main sequence'—between SFR and $M_\star$ (e.g., Brinchmann et al. 2004; Salmi et al. 2012), and (2) 'starbursts' with a large excess specific SFR (sSFR) compared to typical objects on the SFR–$M_\star$ main sequence.

Investigations on the nature of the star-forming population have advanced our understanding by characterizing the systematic differences between starburst and normal galaxies while also revealing large homogeneity within each of these populations, at fixed redshift and also as a function of redshift: starburst galaxies tend to be more compact (e.g., Elbaz et al. 2011; Rujopakarn et al. 2011) than main-sequence galaxies. The latter show a stellar morphology that is well-fitted by exponential disks (e.g. Wuyts et al. 2011) (as opposed to high Sersic-n for the spheroids), they display IR line deficits (e.g., PAHs or in the far-IR the [CII]-line; Elbaz et al. 2011; Graciá-Carpio et al. 2011), both of which are consistent with a scenario in which very compact, more efficient star formation is taking place, in line with the up to tenfold higher SF efficiency (SFE) reported by molecular gas observations of distant SFGs (e.g. Daddi et al. 2010b; Genzel et al. 2010).

In the next sections we will show that a very simple characterization of the evolution of SFGs based on these two populations can account for a large variety of observables in galaxy formation and evolution, without the need of fine tuning or external parameters. Such a simple and self-consistent scheme (referred to as the 'Two Star-Formation Mode framework'; 2-SFM) allows one to predict basic properties of SFGs, like the IR luminosity function (LF), IR source counts, clustering, or molecular gas mass functions, starting from basic observables (e.g., the evolution of sSFR for main-sequence galaxies and stellar mass functions) and observed correlations (e.g., the main sequence or the integrated Kennicutt-Schmidt law connecting their gas content and SFR).

2.1 IR Properties of SFGs: Luminosity Functions, Source Counts and IR Clustering

Widely used, basic characterization of galaxy populations are the mass functions (MFs) and LFs. While the former is generally well reproduced by a Schechter function (e.g., Ilbert et al. 2010), the IR LF (i.e., the distribution of SFRs) is not, requiring for example double power law or exponentials (e.g., Le Floc'h et al. 2005; Magnelli et al. 2009). When accounting for the two SF modes among SFGs (bursty versus main sequence SF), this difference is naturally explained, suggesting that from the evolution of stellar mass and IR LFs, one can gather information on the interplay among the two SF modes at different redshifts. In the following we show how the contribution of main-sequence and starburst galaxies to IR LFs (hence to the SFR density) can be disentangled. We use this information to infer source counts from 24 to 1100 μm and predict the distribution of star formation among dark matter halos as a function of redshift.

2.1.1 IR LF at $z < 2.5$

The main sequence evolution with redshift out to $z \sim 2$ (e.g. Elbaz et al. 2011; Karim et al. 2011) and the evolution of galaxy stellar MFs to similar redshifts are well characterized, and can be jointly used to derive the IR LF evolution if the distribution of sSFR at fixed stellar mass is available. We use the results of Rodighiero et al. (2011) for SFGs at $z \sim 2$, decomposed into two log-normal components (see Fig. 5, left, corresponding to main-sequence and—likely interaction-induced—starburst-like SF activity, respectively; see Sargent et al. 2012 for more details and discussion of the underlying assumptions). The conversion of stellar MFs to IR LF is effectively a convolution of the MF and an evolutionary double-Gaussian kernel with (i) normalization fixed by the shape of the MF, and (ii) main-sequence peak position that—given the redshift—is uniquely determined by the position of the SF main sequence in the (s)SFR vs. $M_\star$ plane. As a result of this approach, we can distinguish the individual contribution to the IR LF of main sequence versus bursty SF (see Fig. 5, right).

Figure 1 shows that the stellar MF and the IR LF (i.e., the SFR distribution) of SFGs at $z \leq 2$ evolve consistently (see also Bell et al. 2007). The bright end of the IR LFs is entirely determined by starburst galaxies. These results imply that the contribution of starbursts to the cosmic SFR density is (8–14 %) perhaps marginally increasing with redshift at $z < 2$. We can account for the well-known observation that most ULIRGs at $z = 0$ are starbursts (e.g., Sanders and Mirabel 1996). Instead, the majority of ULIRGs at $z > 0.9$ live on the main-sequence. In fact, their high SFR ($> 100\ M_\odot$/yr) is not thought to be triggered by merging as in most local ULIRGs but is the result of large gas reservoirs in high-redshift disks (e.g., Daddi et al. 2010a; Tacconi et al. 2010; Geach et al. 2011). Local and distant ULIRGs are intrinsically different objects, in this respect, despite having similar SFRs by definition.

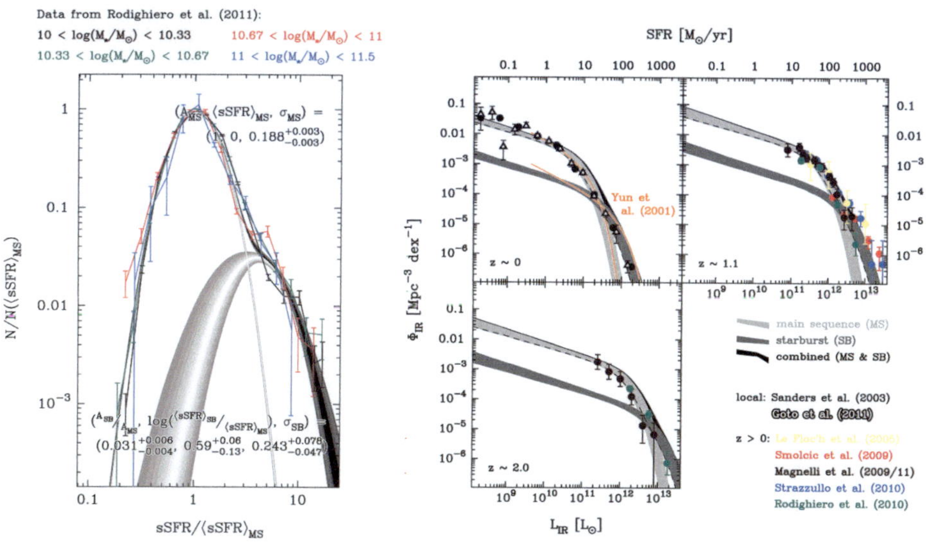

Fig. 5 *Left*: Distributions of sSFR are self-similar in mass bins (for $M_\star/M_\odot > 10^{10}$ from Rodighiero et al. 2011) and can be decomposed into two log-normal components, one accounting for the distribution of main-sequence galaxies, the other enhanced (s)SFRs in galaxies with starburstiness (likely due to interactions, affecting the galaxies at various degrees). *Right*: Contribution to the IR LFs (or equivalently, SFR distributions Kennicutt (1998)), resulting from normal (*light grey*) and burst-like (*dark grey*) galaxies. Literature measurements are displayed in *color*; see legend

2.1.2 IR Source Counts

An additional important observable on which evolutionary models of SFGs should be tested are IR number counts over a variety of multi-wavelength bands. Purely semi-analytical models (e.g., Lacey et al. 2010) have difficulties reproducing IR number counts, while phenomenological or hybrid models (e.g., Béthermin et al. 2011; Gruppioni et al. 2011) reach better results but are in general forced to adopt ad-hoc parameters to force agreement in the fit and thus imply evolution of the IR LFs which are not necessarily physically motivated. The 2-SFM framework introduced in Sect. 2.1.1 can be used to naturally predict IR galaxy counts.

In our framework we can assign SFRs (hence bolometric IR luminosities) to galaxies as a function of redshift. The main additional ingredient to predict their observed fluxes is determining their SED. We have been measuring the evolution of the average SED of galaxies through redshift (Magdis et al. 2012b) which motivate adopting a very simple SED library of templates: a single SED for both main sequence and starburst galaxies at fixed redshift but with both rising in dust temperature with rising redshift. The predictions of such a simple model agree nicely with measurements of differential galaxy counts from 24 μm to 1.4 GHz in Fig. 6. Overall, the observations of galaxy counts are well reproduced, showing the effectiveness of the key ingredient of the 2-SFM approach, which is the explicit distinction between normal and starbursting galaxies. When distinguishing main-sequence from bursty galaxies we can investigate which selection biases could be present toward normal or starburst galaxies in surveys defined at a variety of wavelengths and flux depth levels. Given the overall paucity of starbursts, it is main-sequence galaxies that dominate number counts, basically independently on flux densities and wavelengths. Nevertheless,

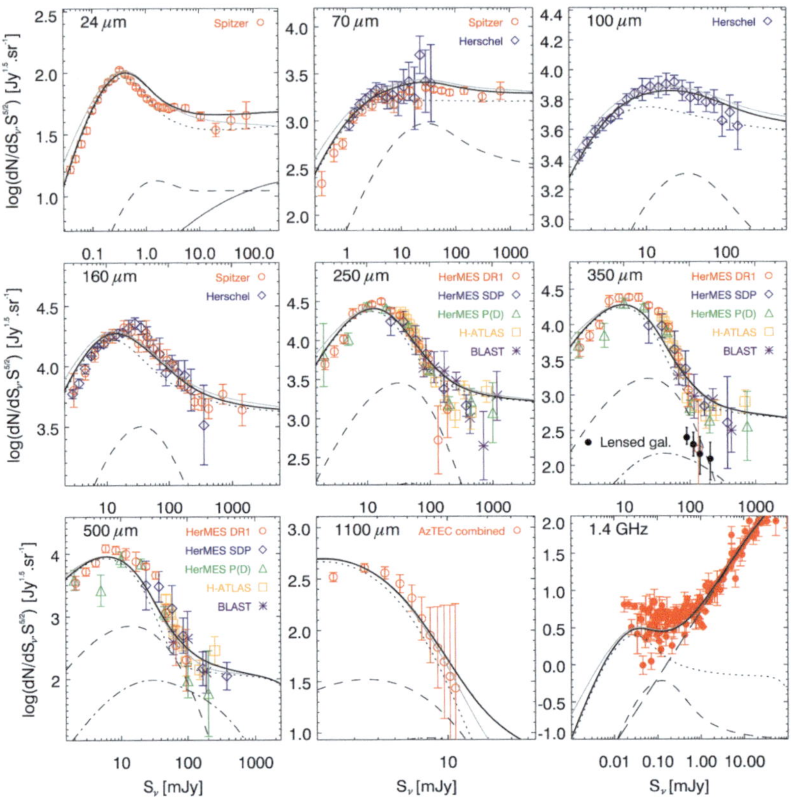

Fig. 6 24 μm to 1.4 GHz number counts (reproduced from Béthermin et al. 2012b). *Solid line*—total counts predicted by the 2-SFM framework, including a contribution for AGN-emission, mass-dependent dust attenuation and lensed sources; *gray line*—counts predicted when neglecting the three aforementioned refinements; *dotted line*—main-sequence contribution; *short-dashed line*—starburst contribution; *dot-dashed line*—lensed sources; *triple-dot-dashed line*—difference between counts with and without AGN contribution. At 1.4 GHz the model for SFGs was combined with the model of AGN-driven radio sources by Massardi et al. (2010; *long-dashed line*). Despite some deviance, e.g. at the faint end in some cases, this is a high successful reproduction of the observations given precedent literature and the lack of adjustable parameters

their overall importance can vary substantially and becomes quite relevant (∼ 30 %) around 30 mJy at 70 μm and 50 mJy at 350 and 500 μm.

Despite its simplicity and the lack of adjustable parameters, these 2-SFM predictions result in one of the best reproduction achieved so far to IR source counts, including counts per redshift slice at SPIRE wavelengths, which were poorly reproduced in earlier models.

2.1.3 Star Formation and Dark Matter Halos

This simple framework can been extended further, using observations of how stellar masses and dark matter masses are connected via measurements of clustering (e.g., Béthermin et al. 2014) and/or abundance matching techniques (e.g., Behroozi et al. 2013). This allows one in turn to distribute star formation among dark matter halos, without the need of further external parameter or tunings. Béthermin et al. (2013) showed that this approach successfully predicts the (cross-)power spectra of the cosmic infrared background (CIB), the cross-

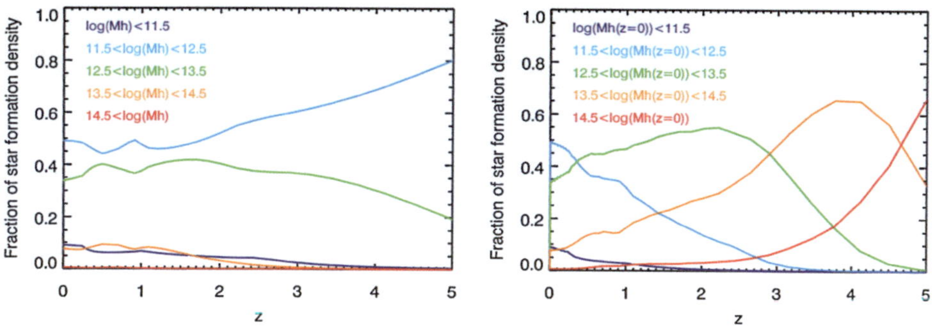

Fig. 7 *Both panels* show the fraction of SFR density as distributed among dark matter halos as a function of redshift, but considering the instantaneous dark matter mass at each redshift (*left*) or the halo mass where the activity will end up at $z = 0$ (*right*) (from Béthermin et al. 2013)

correlation between CIB and cosmic microwave background (CMB) lensing, and the correlation functions of bright, resolved infrared galaxies measured by Herschel, Planck, ACT, and SPT. This is an important result as it shows that we have a good basic understanding on how to associate activity to hosting dark matter halos, at least for typical structures (most activity in all redshifts is happening inside $11.5 < \log M_{\rm DM} < 13.5$ halos; Fig. 7). At the same time, given the well understood evolutionary behavior of dark matter halos with time, this approach can be used to investigate in which kind of dark matter halos at $z = 0$ will eventually end-up the SFR activity observed in the distant Universe (Fig. 7): this is typically borderline clusters with $10^{14} M_\odot$ $z = 0$ halos for $3 < z < 4.5$ activity, and $10^{13} M_\odot$ $z = 0$ small groups at $1 < z < 3$.

2.2 Molecular Gas in SFGs Near and Far

The evolution of molecular gas reservoirs in galaxies, including normal and starburst cases, can also be described in the powerful framework we have been sketching so far, based on the well determined evolution of the stellar MF at $z < 2.5$ and on the distribution of sSFR with naturally decomposes into normal and burst like modes (at fixed stellar mass, and invariantly to first order with redshifts and stellar mass). Similarly to what done for IR source counts in Sect. 2.1.2, we can attribute gas masses to galaxies based on their IR luminosity once they are classified into main sequence and starburst objects. The choice of the SFE is based on the different observational behavior of normal versus starburst galaxies (Sect. 2.2.1). We apply these relations to galaxy populations in the 2-SFM framework to show that the coherent increase of gas fraction ($f_{\rm gas}$) in normal galaxies to high redshift is driving the evolution of the main sequence to $z \sim 3$.

2.2.1 Gas Reservoirs in SFGs: Basic Scaling Relations

Roughly 50–100 galaxies at $z > 0.4$ selected to be on the main sequence have been followed-up in CO in recent years (e.g. Daddi et al. 2010a; Genzel et al. 2010; Tacconi et al. 2010; Geach et al. 2011; Tacconi et al. 2013), providing information on their molecular gas properties. We combine this high-z literature data with a roughly equally sized local sample—including massive galaxies from the HERACLES survey (Leroy et al. 2009) and AKARI-detected disks from the COLD GASS survey (Saintonge et al. 2011)—and plot their $L_{\rm IR}$ and CO-luminosity $L'_{\rm CO}$ in Fig. 8(left). The well-defined correlation between these observational

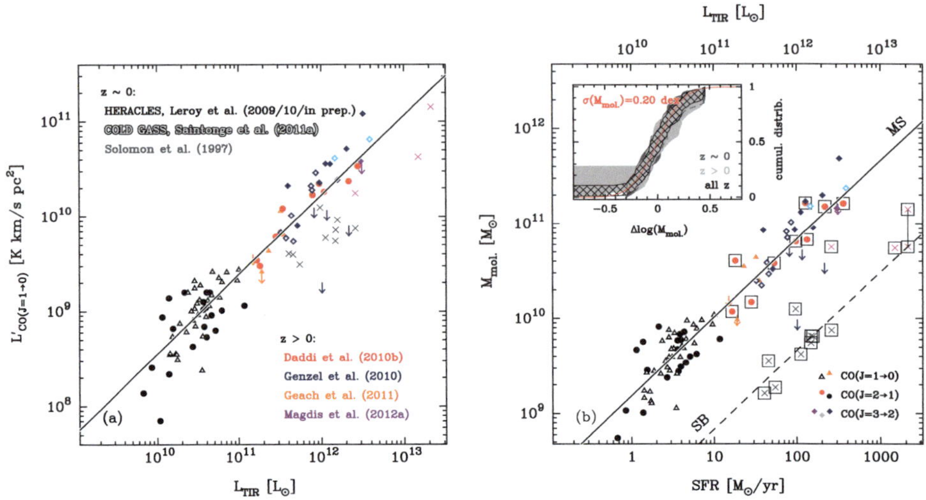

Fig. 8 Correlation between measures of SFR and gas content for massive ($M_\star > 10^{10}\ M_\odot$), main-sequence galaxies in the nearby universe and at $z > 0.4$ (see legends in panel (**a**) for data source). Selected starburst galaxies with measured conversion factors $\alpha_{\rm CO}$ from Solomon et al. (1997) and Magdis et al. (2012b) are plotted with *crosses*. (**a**): Correlation between infrared luminosity ($L_{\rm IR}$) and CO(1 → 0) luminosity ($L'_{\rm CO(1\rightarrow 0)}$; standard excitations corrections—e.g. Dannerbauer et al. (2009), Leroy et al. (2009)—were applied to $J > 1$ transitions, cf. legend in panel (**b**)) with best-fitting relation derived for the main-sequence galaxy sample overplotted in *black*. (**b**): Correlation between SFR and molecular gas mass ($M_{\rm mol.}$), the latter having been derived based on either (**a**) observational determinations of $\alpha_{\rm CO}$ (available for sources with *boxed symbols*) or (**b**) using a metallicity-dependent conversion factor (see text). The dispersion about the best–fit linear trend (*solid black line*) for main-sequence galaxies is approximately Gaussian with a dispersion $\sigma(M_{\rm mol.}) \sim 0.2$ dex (*red curve in inset*). *Dashed line*—offset locus with approx. 15-fold higher SFE for strong starburst galaxies

proxies of SFR and molecular gas mass suggests that there is a high degree of homogeneity between low- and high-redshift main-sequence galaxies in terms of their molecular gas properties. We measure a dispersion of 0.2 dex in the correlation between molecular gas mass $M_{\rm mol.}$ and SFR (see Fig. 8, right), where we have applied either (a) observational determinations of $\alpha_{\rm CO}$ (available for sources with boxed symbols) or (b) a metallicity-dependent conversion factor[1] to convert CO-luminosities to molecular gas masses. The best-fit, inverse integrated Kennicutt-Schmidt (K-S) relation plotted in Fig. 8 is slightly sublinear ($M_{\rm mol.} \propto {\rm SFR}^{\beta_{\rm K\text{-}S}}$, with $\beta_{\rm K\text{-}S} = 0.83$), but due to the small overlap in luminosity-space between local and $z > 0$ main-sequence galaxies a redshift-dependent integrated K-S relation with different slope and changing normalization currently cannot be excluded. Note, however, that by considering stacked samples of $z \sim 0$ and 2 main-sequence galaxies from Magdis et al. (2012b)—who inferred molecular gas masses indirectly by measuring of average dust masses—the correlation is found to extend over at least two orders of (partially) overlapping magnitude in SFR at both redshifts, suggesting that the integrated K-S relation

[1] Recent observational studies, as well as a numerical predictions on a metallicity-dependence of the conversion factor parametrized as $\alpha_{\rm CO} \propto Z^{\gamma}$ find a broad range of power-law exponents γ ranging from -0.5 to -2 (e.g. Feldmann et al. 2012; Genzel et al. 2012; Schruba et al. 2012). Here we adopt an intermediate value of the exponent, $\gamma \simeq -0.9$, which results in the best fit to the $z \sim 0$ CO LF by Keres et al. (2003) (see Fig. 10) within the 2-SFM framework and its underlying assumptions (see Sargent et al. 2014).

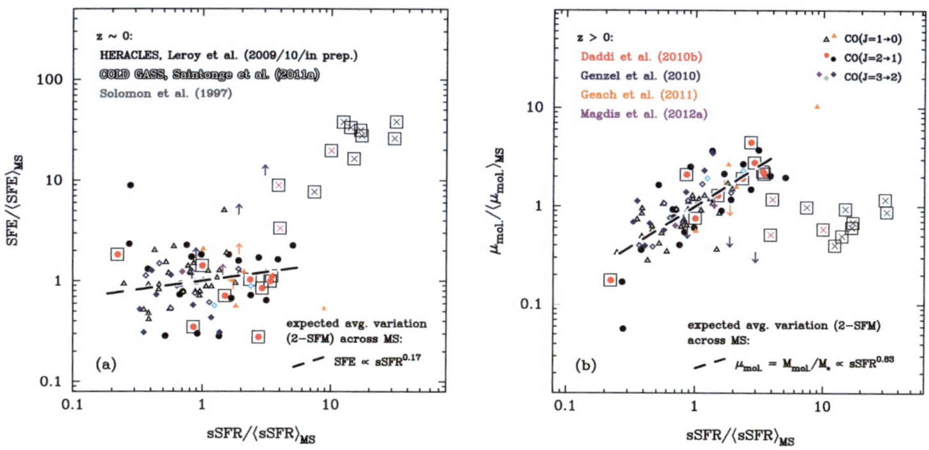

Fig. 9 Star-formation efficiency (*left*), and molecular gas mass fraction, $\mu_{\rm mol} \equiv M_{\rm mol}/M_\star$ (*right*) vs. sSFR for selected main-sequence galaxies and starbursts at $z < 3$ (all measurements normalized to the $M_\star$- and redshift-dependent average of the main-sequence and all symbols and data as in Fig. 8). When normalized to the characteristic main-sequence value, the homogeneous behavior of main-sequence galaxies at all redshifts becomes visible: SFE varies very little within the main sequence while $f_{\rm gas}$ steadily rises. Starbursting sources, on the other hand, have a gas content that is similar to the average main-sequence galaxy but display enhanced SFEs that lead to their excess (s)SFR

to first order is a universal function of redshift for massive, normal SFGs ($M_\star/M_\odot > 10^{10}$) at $z < 3$.

To further highlight this uniform behavior across a broad range of redshift, in Fig. 9 we plot SFE and gas fractions that have been normalized to the (mass- and redshift-dependent) average values of a galaxy located directly on the main-sequence and compare these to the analogously normalized sSFR (i.e. the offset from the main sequence). This representation of the data reveals that: (1) at all redshifts, the SFE in main-sequence galaxies in our "calibration sample" is almost independent of their position within the main sequence, while their SFR rises in lockstep with $f_{\rm gas}$; (2) starburst galaxies (here we consider a small sample of local and high-z starbursts with measured $\alpha_{\rm CO}$) have similarly-sized gas reservoirs as normal galaxies but are characterized by a strongly enhanced SFE. The dashed line in Fig. 9 illustrates the expected variation of SFE ($\propto$ sSFR$^{-0.17}$) and $f_{\rm gas}$ ($\propto$ sSFR$^{0.83}$) within the main sequence as it follows from the best-fit integrated K-S relation shown in Fig. 8(b), in agreement with the dust mass-based findings of Magdis et al. (2012b).

2.2.2 Redshift Evolution of Gas Fractions

Given the tight correlations between $M_\star$ and SFR, and between SFR and H_2-mass, it is straightforward to model the evolution of $f_{\rm gas}$ in main sequence galaxies where $M_{\rm gas}/M_\star \equiv$ const. $\times$ SFR$(M_\star)^{\beta_{\rm K\text{-}S}}/M_\star$. This results in predictions for gas fractions in main sequence galaxies at $z < 3$ in good agreement with observations (Magdis et al. 2012a) confirming that the gas content of secularly-evolving SFGs is an important driver of the cosmic sSFR-evolution. This is consistent with the observation in Fig. 9 that, also at fixed redshift, variations of sSFR across the main sequence are driven by varying $f_{\rm gas}$. Of course, all this agreement is also, in part, by construction. It would considerably worsen if the various means by which total molecular masses are estimated were to be incorrect, which would be surprising given the overall consistency from CO, dust and dynamical estimations.

2.2.3 Molecular Gas Mass Functions and the Molecular Gas History of the Universe

To estimate the redshift evolution of the distribution of molecular gas masses in normal galaxies we convert the main-sequence component of the SFR-distribution depicted in Fig. 5 to a gas mass distribution by means of the best-fitting integrated K-S relation of Fig. 8(b). For the starbursting population we assume that the SFE increase (with respect to the value of the main-sequence state prior to the onset of burst activity) in these systems scales with the boost in star-formation rate engendered by the burst (see Sargent et al. 2012 for details). This kind of behavior is suggested by the roughly similar excess in sSFR and SFE observed in Fig. 9(a) and also seen in numerical simulations of galaxy mergers (e.g. Di Matteo et al. 2007). The total molecular gas MF (i.e. including both normal and starburst galaxies) inferred using this approach for the SFG population at $z \sim 0$, 2 & 5 is shown in Fig. 10. Note that the H_2-MF is only observationally constrained in the local universe (Keres et al. 2003; Obreschkow and Rawlings 2009a) where the measurements are entirely consistent with the 2-SFM predictions. (This is also true of the predicted CO LFs—cf. Fig. 10—which follow from the H_2-MFs after (a) assigning to main-sequence galaxies α_{CO}-values based on statistically-derived metallicities, computed using the fundamental metallicity relation of Mannucci et al. (2010) given the known SFR and $M_\star$, and (b) for starbursts assumes a conversion factor that scales inversely with the SFE-enhancement. (See Sargent et al. 2014 for additional details and underlying assumptions/simplifications.) The dominant contribution to the predicted H_2-MF stems from main-sequence galaxies which hence also are the main contributors to the evolution of the cosmic abundance of molecular gas which is obtained by integration of the molecular gas mass functions and which is shown in Fig. 11. This indirect measurement of the comoving H_2-density reveals a roughly 5-fold increase of the molecular content of the Universe out to $z \sim 1.5$ where it roughly equals the stellar mass density and the cosmic abundance of atomic hydrogen. The increase traces the rise of the cosmic SFRD but is roughly five times smaller due to the sublinear relation between SFR and H_2-mass found in Fig. 8 which implies slowly rising SFEs for the typical SFGs (of stellar mass $\sim 5 \times 10^{10}\ M_\odot$, cf. Karim et al. 2011) that contribute most to the cosmic SFRD density over this time. Overall, there is a large range of predictions from alternative models at high redshifts (Fig. 10, emphasizing the need for observations to confirm our results based on scaling relations).

2.2.4 Evolution of the Interstellar Medium

While there are no convincing evidence yet that star formation laws are changing with redshift, the strong rise of the specific SFR of galaxies appears to be accompanied by substantial changes in the properties of the interstellar medium of Main Sequence galaxies, other than a simple augmentation of the gas reservoirs. This is manifested already in evolution in optical emission line ratios (e.g., Steidel et al. 2014), which requires variations in the radiation field that the gas is witnessing. From the point of view of the ISM, several studies now have converged to show that this evolution is paralleled for Main Sequence galaxies to a rise in the intensity of the radiation field which is heating dust (the parameter $\langle U \rangle$ scaling like $(1+z)^{1.2\text{–}1.8}$; Magdis et al. 2012b; Béthermin et al. 2015; Genzel et al. 2015). This is understood to be driven by a combination of rising star formation efficiencies and declining metallicities towards high-z (at fixed stellar mass; Magdis et al. 2012a, 2012b). On a similar manner, observations of CO excitation for Main Sequence galaxies also show evolution from low redshift disk-like SFGs. In particular, high-z galaxies have much stronger CO[5-4] transitions, which points to a higher relevance of denser and/or warmer gas (Daddi et al.

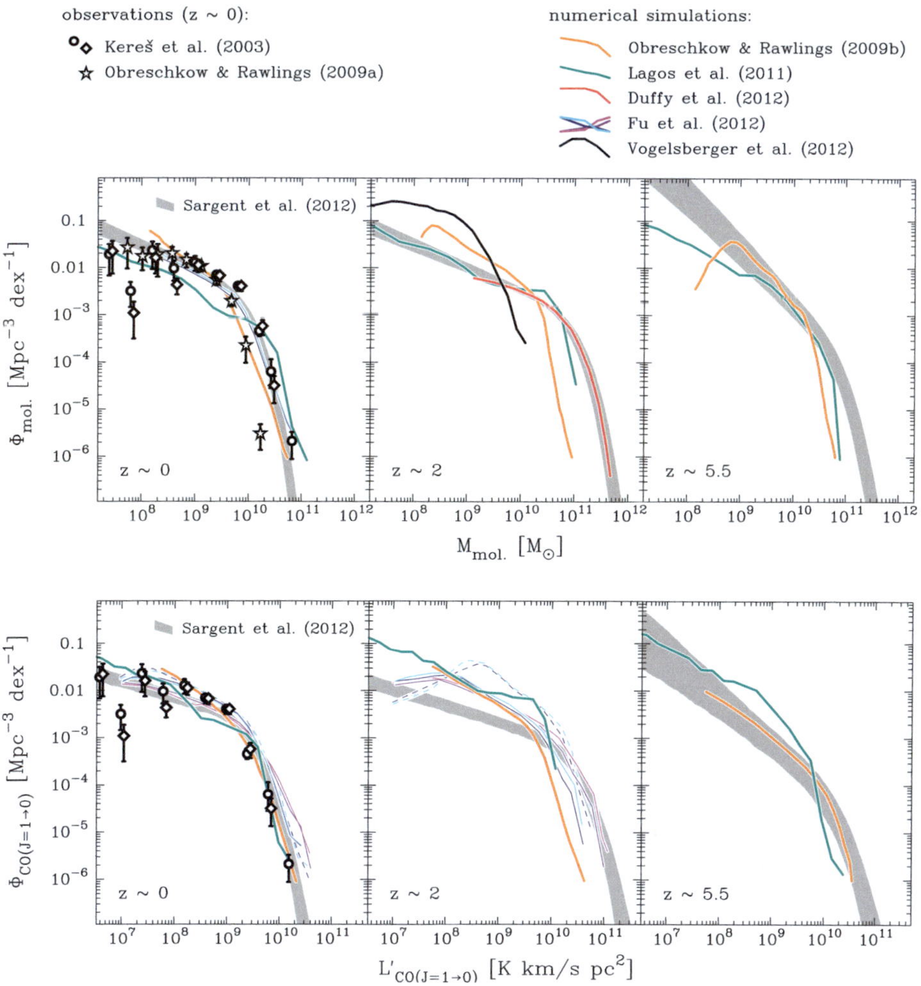

Fig. 10 Molecular gas MFs (*upper row*; *grey shading*) and CO(1-0) LFs (*lower row*) based on (1) the evolution of the stellar MF of SFGs, (2) the redshift evolution of the sSFR of main-sequence galaxies, (3) the distribution of main-sequence and starbursting galaxies in the SFR-$M_\star$-plane (see Fig. 5, left), and (4) a metallicity-dependent conversion factor α_{CO}. MFs/LFs include contributions both from 'secular' and burst-like star formation (the latter being characterized by increased SFEs with respect to the secular mode which is assumed to be typical of galaxies residing on the main sequence). *Colored lines*—predictions of recent semi-analytical models (see *legend* above figure) covering a similar redshift range as the empirically-motivated, predictive analysis of Sargent et al. (2014). At $z \sim 0$ we show observational constraints on the local MF/LF reported in Keres et al. (2003) and Obreschkow and Rawlings (2009b)

2015; Fig. 12). Empirically the high-J to low-J CO ratios among various populations is well correlated to their radiation field intensity $\langle U \rangle$ (hence dust temperature) as well as to star formation surface brightness (hence gas density). Simulations suggest that such a dense/warm gas is typically associated to clumpy star-forming regions (Bournaud et al. 2015). This is supported by observations but not fully proven yet, and there is debate on the real mass and relevance of clumpy star formation in the distant Universe. However, a massive clump formation event has recently been observed (Zanella et al. 2015).

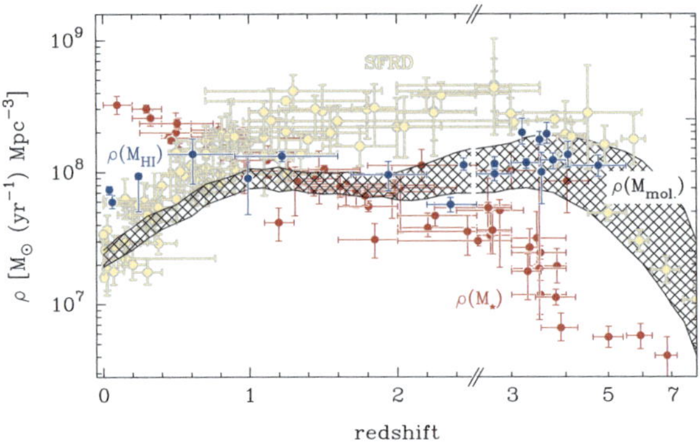

Fig. 11 Redshift-evolution of the comoving mass density (cf. *scale on left*) of molecular gas, $\rho(M_{\mathrm{mol.}})$. The *hatched region* shows the evolution inferred from the integration of the predicted molecular gas MFs in Fig. 10 (extended to $z > 2.5$ with a set of assumptions that successfully match high-z stellar MFs and UV-LFs, e.g. González et al. 2010; Bouwens et al. 2011). The evolution of $\rho(M_{\mathrm{mol.}})$ is compared to compilations of the cosmic evolution of the atomic hydrogen abundance ($\rho(M_{\mathrm{HI}})$; e.g. Bauermeister et al. 2010, *blue symbols*) and of $\rho(M_\star)$, the stellar mass density (*red symbols*; see, e.g., compilation in Marchesini et al. 2008). *Yellow symbols*—SFRD-evolution (cf. *scale on right*), as constrained by literature data

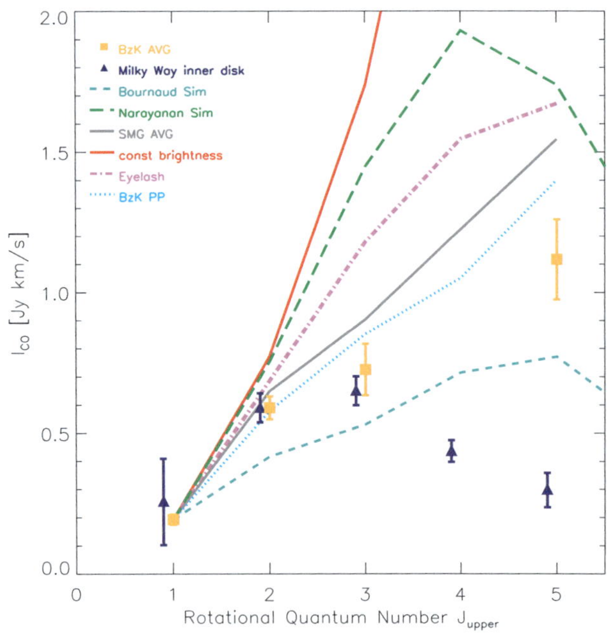

Fig. 12 Comparison of the average SLED of BzK galaxies to the MW SLED, the average of SMGs and the average (U)LIRGs SLED. All SLEDs are normalized to the CO[1-0] of the average BzK galaxy SLED, except the MW which is normalized using CO[2-1]. The results models and simulations are also shown. See Daddi et al. (2015) for references. We notice that the average SLED of lensed SPT sources (Spilker et al. 2014) is similar to that of SMGs

2.2.5 *Going to Earlier Epochs*

Star formation in the high-z Universe beyond $z \sim 3$ has been so far very well studied in the UV, but detailed IR investigations (required to trace dusty star formation and study the gas content) have become potentially feasible only since the advent of ALMA, as they require exquisite sensitivity. Initial results are coming in, and many more will come in the near future. In the meanwhile, a general worry to keep in mind, which could adversely af-

Fig. 13 The redshift evolution of dust to stellar mass ratios, for Main Sequence (*circles*) and Starburst (*stars*) galaxies, see Tan et al. (2014) for details and references. The (uncertain) evolution in the metallicity of massive galaxies at high redshift could imply a rapid drop of dust masses at $z > 4$

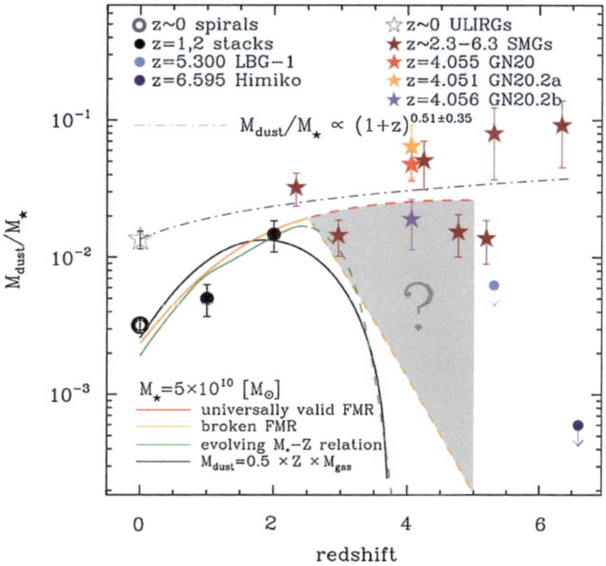

fect IR observations in the distant Universe is the possible general decline of metallicities, suggested to increase its pace earlier than $z \sim 3$ (e.g., Troncoso et al. 2014; Onodera et al., 2016). In turn, lowering the metallicity (Z) also dust masses are decreased, as one can write approximately $M_{dust} \sim 0.5 \times Z \times M_{gas}$ (roughly half of the metals being caught in dust). The possible evolution of dust to stellar mass ratios is depicted in Fig. 13. This could make IR fluxes of galaxies substantially fainter, especially in the Rayleigh-Jeans tail which is more directly proportional to the dust mass present, while the overall IR luminosity might be kept high by an increase of dust temperature. Early confirmation of these effects to $z \sim 4$ were provided by Béthermin et al. (2015). In turn, CO emission could be adversely affected as well as a roughly constant $M_{dust}/L'CO$ ratio is expected and observed at $z < 3$ for normal galaxies (Tan et al. 2013). Observations of distant GRBs could help shedding light on this issue (e.g., see Laskar et al. 2011).

2.3 Summary of Section 2

Motivated by the homogeneity of the bulk of SFGs (in particular those occupying the so-called "main sequence" of SFGs) across a broad range of redshift we have presented a simple framework (2-SFM) for the prediction of basic properties of the star-forming population. By explicitly distinguishing between (1) the population of 'normal' galaxies residing on the star-forming main sequence, and (2) starbursting galaxies for which we assume different characteristics (e.g. IR SEDs or a burst-dependent continuum of SFE enhancements) based our currently best observational understanding, we show that this simple description is capable of reproducing the evolution of the IR LF out to $z \sim 2.5$ and IR/radio source counts at 24 to 1100 μm and 1.4 GHz, and the distribution of IR light among dark matter halos. We use the 2-SFM framework to predict the molecular gas mass function (and $CO(1 \rightarrow 0)$ luminosity function of SFGs, a fundamental observable which so far has only been measured at $z = 0$ and the extension of which to higher redshift is a major goal of the ALMA era. The strong evolution of the inferred H_2-mass function to higher masses with increasing redshift suggest that the cosmic H_2-abundance was ~ 5 times larger at the peak of the cosmic SFH than in the local universe. Furthermore, we provide strong evidence that the higher gas fractions in distant galaxies are directly reflected in the higher sSFR of these systems. We discuss evidences for the evolution of ISM properties of galaxies to $z = 2$–4, manifested in higher dust temperatures and higher CO excitations. Further investigations at higher redshifts will be soon possible thanks to ALMA, barring the possible adverse effect of declining metallicities at high redshifts.

3 Galaxy Formation at High Redshift using Cosmological Hydrodynamic Simulations

Over the past quarter of a century, computational astrophysicists have been making tremendous advances in simulating structure formation in the Universe from shortly after the last scattering surface to the present time. In this section, we summarize some of the recent highlights of the study of high-z galaxy formation using cosmological hydrodynamic simulations.

Astronomers now have a standard cosmological model, the Λ cold dark matter (CDM) model, which can be described well with only six cosmological parameters: $(\Omega_M, \Omega_\Lambda, \Omega_b, h, \sigma_8, n_s) \approx (0.3, 0.7, 0.04, 0.7, 0.8, 0.96)$ (Komatsu et al. 2011; Hinshaw et al. 2013; Planck Collaboration et al. 2015). This simple fact itself is a remarkable achievement, and it is one of the highlights of cosmological studies obtained from the temperature anisotropy of cosmic microwave background (CMB) radiation. The predicted power spectrum of matter density fluctuations agrees well with various observational estimates, including CMB, galaxy clustering, number density of galaxy clusters, weak lensing, and Lyman-α forest (Tegmark et al. 2004). The collisionless dark matter (DM) in the ΛCDM model provides the back-bone of structure in our Universe, and it describes the matter density distribution very well on scales greater than ~ 1 Mpc. For example, nowadays one can easily run a simple N-body simulation with $\sim 500^3$ particles, which can reproduce the network of filamentary structures harboring dark matter halos and galaxies. The big question is, "Can we understand galaxy formation in the cosmological context of ΛCDM model?"

Based on the ΛCDM model, we now understand the cosmic time-line relatively well. Starting from the Big Bang at 13.8 billion years ago, the cosmic gas in our Universe experienced a major transition from being optically thick to thin, when the electrons combined with protons, and the Universe became transparent to the photons at $z \sim 1100$ (i.e., 'reionization' of the Universe). We now observe these photons as the CMB radiation that have traveled from the 'last scattering surface' to the present time without being scattered randomly.

From $z \sim 1100$ to $z \sim 20$, the 'dark ages' continued, during which the intergalactic medium (IGM) was fully neutral and there were no light sources in the Universe. Then the 'first stars' started to form in DM halos with masses $M_h \sim 10^6\ M_\odot$, and the bubbles of ionized region (i.e., the Stromgren spheres) expanded and percolated gradually (e.g., Iliev et al. 2006). Shortly after, the 'first galaxies' started to form in DM halos in the atomic-cooling halos with virial temperatures greater than $T \sim 10^4$ K and $M_h \sim 10^8\ M_\odot$. It is widely believed that these first stars and first galaxies are the sources which provided the ionizing photons necessary to reionize the Universe (e.g., Iliev et al. 2006; Yajima et al. 2011). From the measurements of electron scattering optical depth by CMB, it is estimated that the reionization took place at around $z \sim 6$–11 (Hinshaw et al. 2013; Planck Collaboration et al. 2015), and the Gunn-Peterson trough in the quasar spectrum tells us that the reionization was completed by $z \sim 6$ (Fan et al. 2006).

The first stars and first galaxies also enriched the IGM for the first time, and the metals change the cooling rate of gas in later galaxy formation processes via metal-line cooling. Therefore understanding the formation of first galaxies and evaluating their number density is important for both cosmology and galaxy formation theory. One of the key physical parameters that we need to determine is the escape fraction of ionizing photons, f_{esc}, from first galaxies as we will discuss further in the following subsections.

Galaxies have complex internal structures, e.g., bulge, disk, halo. The disk exhibits spiral arms, and the bulge size varies from galaxies to galaxies. In order to estimate f_{esc}, we need to understand the structures of galaxies better, and how the stars and gas are distributed within in them. This means that we want to understand the disappearance of the Hubble Sequence towards high-redshift, and how the SFR changes over time in different galaxies. One method to characterize the star formation in galaxies is to examine the Kennicutt-Schmidt relation, which has been performed extensively at low redshift as described in earlier sections. The same can be done at higher redshifts with ALMA, and computer simulations can provide useful information on galactic structure evolution at high-redshift where observational data are still scarce.

3.1 Computational Cosmology

To understand the history of structure formation in the Universe, computational cosmology (or numerical cosmology) provides a useful method, where we try to follow the formation and evolution of structures in our Universe self-consistently from first principles as much as possible. We start from the initial condition that is consistent with the fluctuations observed in the temperature fluctuations of CMB, then prepare a universe in a supercomputer assuming a set of cosmological parameters; namely, dark energy, dark matter, and baryons in an expanding universe. And then we simulate the structure formation as a function of time using the laws of gravity and fluid dynamics. The hope is to reproduce the Universe as we observe today at $z = 0$, after the simulation is evolved from $z \sim 200$ to $z = 0$.

To form galaxies in cosmological simulations, one has to include additional physics, such as radiative cooling and heating, star formation (SF), and feedback. For example, we use the modified version of GADGET-3 smoothed particle hydrodynamics (SPH) code (Springel 2005), which includes radiative heating by UV background radiation, cooling by H, He and metals (Choi and Nagamine 2009), multi-component variable velocity (MVV) wind model for supernova (SN) feedback (Choi and Nagamine 2011), self-shielding effect of UV background (Nagamine et al. 2010). The treatment of self-shielding is important for these cosmological simulations, and follow-up work have shown that our treatment works well to simulate HI absorbers (Bird et al. 2013).

The resolution of these cosmological simulations depend on the box size and particle counts. For example, our simulations have box sizes of 10–100 h^{-1} Mpc, the particle count ranges from 2×144^3 to 2×600^3, comoving gravitational softening length of a comoving few kpc to 700 h^{-1} pc. A simulation with 2×600^3 particles in a comoving 10 h^{-1} Mpc has a dark matter particle mass of $2.78 \times 10^5\ h^{-1}\ M_\odot$, gas particle mass of $5.65 \times 10^4\ h^{-1}\ M_\odot$, and gravitational softening length of comoving 670 h^{-1} pc. Usually it is difficult to cover a wide range of dark matter halo masses in one simulation box, therefore, we use different comoving box sizes (e.g., 10, 34, 100 h^{-1} Mpc) to cover a wide range of $M_h = 10^8$–$10^{13}\ h^{-1}\ M_\odot$. This composite method allows us to also examine galaxies with stellar masses $M_\star = 10^7$–$10^{11}\ h^{-1}\ M_\odot$ (Jaacks et al. 2012), from dwarfs to moderately massive galaxies.

In our simulations, the star formation is treated either by the fiducial pressure-based SF model (Schaye and Dalla Vecchia 2008; Choi and Nagamine 2010) or the H_2-based

SF model (Thompson et al. 2014). With the mass and spatial resolution given above for cosmological hydrodynamic simulations, it is still not possible to resolve the formation and destruction of giant molecular clouds in detail. Therefore many research groups resort to various subgrid multiphase interstellar medium (ISM) models, where each SPH particle is pictured as a multiphase hybrid gas (Yepes et al. 1997; Springel and Hernquist 2003). Only the thermodynamics of the hot phase is followed here, while the cold phase is assumed to stay cold with $T \sim 10^4$ K. The exchange of energy and mass between the hot and cold phase is assumed to be in equilibrium, as the exchanges in high-density regions take place on a much shorter time-scale than the global simulation time-steps.

The star formation is typically computed with an equation similar to $\dot{\rho}_\star = (1-\beta)\rho_c/t_\star$, where β is the instantaneous recycling fraction of gas after supernova explosions, and ρ_c is the cold phase gas density. The SF time-scale, $t_\star$, is usually taken to be proportional to the dynamical time of the gas, i.e., $t_\star = t_\star^0(\rho_g/\rho_{\rm th})^{-0.5}$, where the normalization $t_\star^0 \sim 2$ Gyr is needed to reproduce the observed normalization of Kennicutt-Schmidt law, and $\rho_{\rm th}$ is the SF threshold density. One of the problems of such simple SF law is that it does not have an explicit dependence on metallicity. Therefore we have revised our model and adopted the H_2-based SF law (Thompson et al. 2014), where the H_2 mass fraction is computed for each SPH particle and the H_2 mass density is used for the SF law instead of total cold gas density. Since the H_2 fraction is dependent on the gas metallicity Krumholz et al. (2009), this introduces the metallicity dependence of SFR naturally through the abundance of H_2 molecules. The H_2 molecules form on dust, and the model of Krumholz et al. (2009) provided simple approximate formulae to estimate H_2 fraction as a function of gas surface density and metallicity.

The effect of supernova feedback is considered, and a part of SN energy is given to nearby SPH particles as a kinetic energy. There are various ways of distributing SN energy, but some form of efficient kinetic feedback is necessary to distribute the metals and energy into the IGM. In the case of the MVV wind model by (Choi and Nagamine 2011), we compute the galactic properties such as stellar mass and SFR by grouping the star/gas/DM particles on the fly as the simulation runs, and compute the wind velocity based on the SFR of individual galaxies. Such a scaling between galactic wind velocity and galaxy SFR (or stellar mass) has been suggested by some observations of high-z galaxies (Weiner et al. 2009).

Once the star particles are created based on the above SFR estimate, each star particle carries physical tags such as the stellar mass, formation time, and metallicity. Based on these quantities, we can apply population synthesis models (e.g., Bruzual and Charlot 2003) to each star particle and compute the spectral output of simulated galaxies. Of course the simulation examples that we describe in this section is a limited example, and there have been other extensive efforts by many groups. For example, the Illustris project (e.g., Genel et al. 2014; Vogelsberger et al. 2014) and EAGLE simulations (Schaye et al. 2015) have performed even larger cosmological hydrodynamic simulations than described in this article and provided large galaxy catalogues to the astronomical community.

3.2 Evolution of Galaxy Stellar and Luminosity Functions at High-z

One of the key statistics for galaxy formation and evolution is the galaxy stellar mass function (GSMF) and luminosity function (LF). The shapes of these functions at $z = 0$ have been constrained well (Blanton et al. 2001; Cole et al. 2001), and they can be fit well with the Schechter function. However, the situation at higher redshift is more uncertain, in particular the faint-end slope, α, is not constrained well yet. At $z = 0$, the slope is $\alpha \sim -1.2$, and it becomes steeper to $\alpha \sim -1.6$ at $z \sim 3$ (Reddy and Steidel 2009).

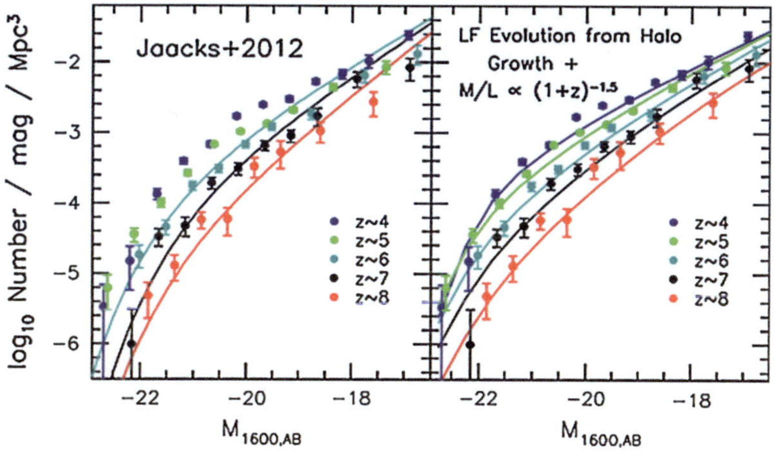

Fig. 14 Comparison of galaxy luminosity functions at $z = 4$–8 between cosmological hydrodynamic simulation (Jaacks et al. 2012, left panel) and observations (Fig. 19 of Bouwens et al. 2015). In *both panels*, the observational data is shown as data points, and either simulation or a simple model results are shown with Schechter fits. *Both panels* show the steepening of the faint-end slope α with increasing redshift at $z > 4$, e.g., from $\alpha = -1.64$ ($z = 4$) to -2.06 ($z = 6$)

Using the results of cosmological hydrodynamic simulations, several groups have shown that the GSMF and LF become steeper at the faint-end with increasing redshift, from $\alpha = -1.6$ at $z \sim 3$ to even $\alpha \sim -2$ at $z \sim 6$ (e.g., Nagamine et al. 2004; Night et al. 2006; Finlator et al. 2006; Lo Faro et al. 2009). More recent work has extended the comparison to $z > 6$, and some simulations indicated that the slope could become even steeper with $\alpha < -2.0$ (Jaacks et al. 2012; Bouwens et al. 2015), as shown in Fig. 14. However, the initial work by Jaacks et al. (2012) used an SF model which does not have any dependence on metallicity, and this may have lead to a slight overestimation of SF efficiency at high-z.

Jaacks et al. (2013) have examined the impact of lower metallicity on GSMF at high-z using the H_2-based SF model of Thompson et al. (2014). In this SF model, the mass fraction of molecular hydrogen, f_{H_2}, is computed within the framework of multiphase ISM model of Springel and Hernquist (2003), and f_{H_2} becomes lower with decreasing metallicity because H_2 molecules form on dust. Figure 15 shows the star formation rate (SFR) function at $z = 6$–8, in comparison with the observational data by Smit et al. (2012). The SFR function provides a more direct comparison between simulations and observations, as SFR is a more direct output from the simulation rather than spectral output that has to be computed by a population synthesis code with an assumption of a stellar initial mass function (IMF). The simulation with the H_2-based SF model yielded a flatter faint-end slope at SFR $< 1\ M_\odot\,\mathrm{yr}^{-1}$, which corresponds to a limiting rest-frame UV magnitude of $M_{\mathrm{uv}} \sim -16$. Within the observed range of magnitudes, the simulation result agrees very well with the observed data (red crosses). The flatter faint-end slope and the break at SFR $\sim 1\ M_\odot\,\mathrm{yr}^{-1}$ is caused by the inefficient star formation at low metallicities in faint galaxies at high-z in our simulations.

The faint-end slope of GSMF and LF is quite important for the ionizing photon budget, as these faint galaxies are the ones that provided the photons necessary to reionize the Universe. Jaacks et al. (2013) have shown that, even with the reduced number of faint galaxies with the H_2-based SF model, these galaxies can provide a sufficient number of ionizing photons for reionization by $z \sim 6$, and the contribution is in fact dominated by the lower mass and fainter galaxies with $M_\star \leq 10^8 M_\odot$.

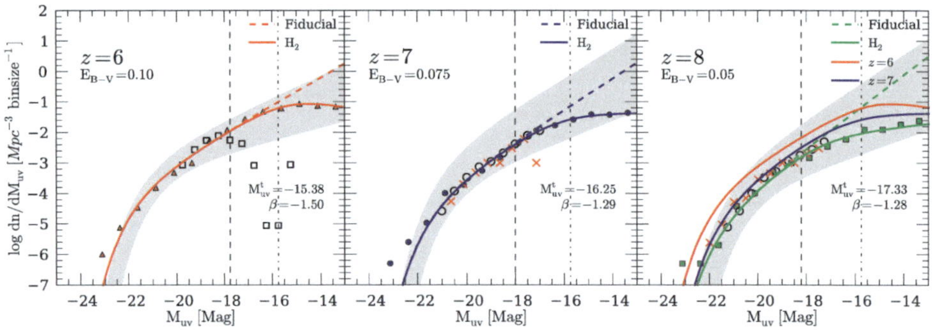

Fig. 15 SFR function at $z = 6, 7, 8$ in cosmological hydrodynamic simulations with H_2-based SF model (adopted from Jaacks et al. 2013). The data points are mostly simulation data, and the *red crosses* are the observed data by Smit et al. (2012). The fits shown by the *solid lines* are the modified Schechter function with an additional break at the faint-end, and the fits for $z = 6, 7, 8$ are compared in the *right-most panel*, where we can see that the galaxies become brighter and more numerous from $z = 8$ to $z = 6$. The *grey shade* is the range of Schechter fits to the observed data by Bouwens et al. (2011)

Integrating the observed galaxy UV luminosity function can give the estimates of UV luminosity density, which can be converted to the cosmic SFR density with an assumption of an IMF. However, since current galaxy observations cannot probe the faintest galaxies that might exist at high-redshift at $M_{UV} > -15$, it is difficult to constrain the SFR density accurately at high-z. Using GRBs can be useful in this context, as it is very bright and could be detected as high redshift as $z > 8$ (Kistler et al. 2009; Tanvir et al. 2009). These initial estimates, albeit from a very small sample, gives higher value of cosmic SFR density at $z > 6$ than the estimates based on UV luminosity function of galaxies.

A more proper treatment of ionizing photon budget requires a radiative transfer calculation to estimate the escape fraction of ionizing photons from galaxies, and there have been a number of interesting works that gave varying results on the values of f_{esc} based on radiative transfer calculations (e.g., Gnedin et al. 2008; Wise and Cen 2009; Razoumov and Sommer-Larsen 2010; Paardekooper et al. 2011). Yajima et al. (2011, 2015) have shown that f_{esc} is a decreasing function of halo mass at high-z, although with a large dispersion particularly for lower mass galaxies (Fig. 16). Significant amounts of gas around star-forming regions prevent the escape of ionizing photons in massive galaxies, whereas the gas distribution is somewhat more irregular in low-mass galaxies, enabling the photons to escape in a preferential direction. Folding in the estimated values of f_{esc} of each galaxies into the calculation of ionizing photon budget, they estimated that the faint galaxies at $z \geq 6$ can provide sufficient number of ionizing photons thanks to their large f_{esc}.

GRBs can also help to set constraints on f_{esc} of high-z galaxies. For example, Chen et al. (2007) showed that one can estimate f_{esc} of ionizing photons using the afterglows of long GRBs based on the optical depth of neutral hydrogen column density in the host ISM, free from the background subtraction uncertainties. This method is potentially very useful, because it can provide estimates of f_{esc} of sub-L^* galaxies where most GRBs occur and otherwise difficult to observe directly. These faint galaxies at the fainter end of the luminosity functions contributed the most to the ionizing photon budget, therefore having observational estimates of f_{esc} for these low-mass galaxies is quite important for the study of reionization of the Universe. Both Chen et al. (2007) and Fynbo et al. (2009) found small mean values of $\langle f_{esc} \rangle \sim 0.02$, however larger values have also been suggested by observations (e.g., Inoue et al. 2006; Smith et al. 2016) and from models based on GRB and galaxy observations (Wyithe et al. 2010; Dijkstra et al. 2014).

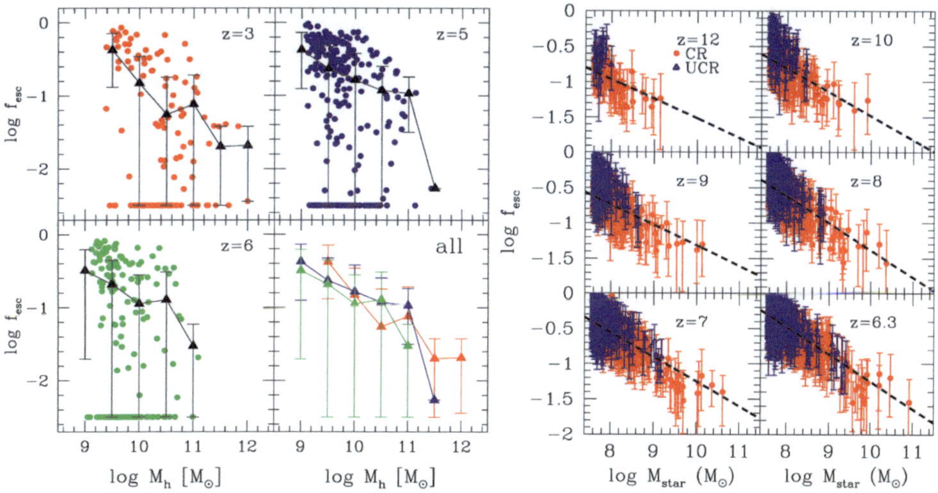

Fig. 16 Escape fraction of ionizing photons at $z = 3$–6 in cosmological SPH simulations (panel (a), Yajima et al. 2011), and that of UV photons from $z = 12$ to $z = 6$ (panel (b), Yajima et al. 2015). In general, it is a declining function with increasing M_h, but with a large scatter for lower mass galaxies in particular. Also there doesn't seem to be strong evolution in the relation from $z = 6$ to $z = 3$

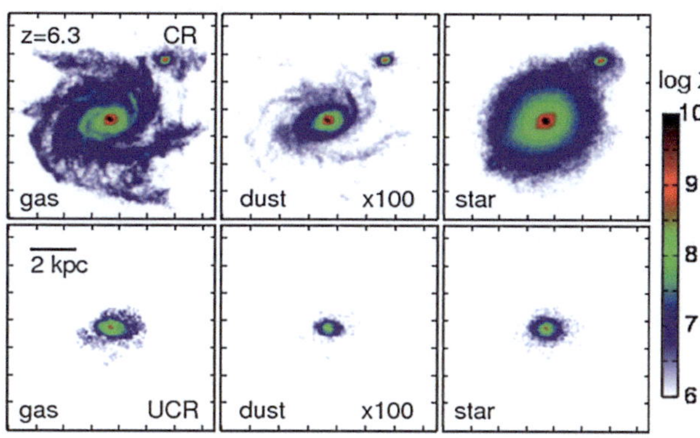

Fig. 17 Massive disk galaxy at $z \sim 6$ found in a cosmological SPH simulation (Yajima et al. 2015). The *upper row* represents the most massive galaxy in the constrained realization of a high-density region ($M_{\rm star} \sim 8.4 \times 10^{10} M_{\odot}$ and $M_{\rm gas} \sim 4.8 \times 10^{10} M_{\odot}$). The *lower row* shows the most massive galaxy in the less dense, unconstrained region with $M_{\rm star} \sim 1.5 \times 10^{9} M_{\odot}$ and $M_{\rm gas} \sim 3.1 \times 10^{9} M_{\odot}$

Romano-Díaz et al. (2011) and Yajima et al. (2015) found the existence of massive spiral disk galaxies at high redshift ($z \sim 10$ and 6, respectively) using zoom-in cosmological hydrodynamic simulations of a high-density region, as shown in Fig. 17. These massive galaxies contain a fair amount of dust, and therefore might be good targets for ALMA observations. In fact bright quasars have been already detected by ALMA via [CII] line (e.g., Maiolino et al. 2005), and perhaps ALMA can extend these detections down to lower mass-scale objects in the near future.

As these recent works have shown, it seems now possible to form massive disk galaxies *for a certain feedback models* at high-redshift in zoom-in cosmological hydrodynamic with sub-kpc resolution. Although Yajima et al. (2015) found a nice disk geometry for a massive galaxy in a high-density region at high-redshift, some of the more recent high-resolution zoom-in simulations employ more direct momentum injection rather than relying on sub-resolution kinetic feedback model (e.g., Hopkins et al. 2013; Muratov et al. 2015), and these more direct feedback models seem to be more efficient in ejecting gas from high-z star-forming galaxies, often disrupting the disk galaxy geometry. More future studies of galaxy formation will be performed using cosmological hydrodynamic simulations, and the impact of both supernova and AGN feedback will be assessed in more detail, giving insight into the mechanism of forming the main sequence of star formation and downsizing of galaxies.

Acknowledgements The authors would like to thank the organizers of the workshop "Gamma-Ray Bursts: A Tool to Explore the Young Universe", and in particular Emeric Le Floc'h, for inviting us to discuss the subject in this chapter. K. Nagamine acknowledges the partial support by JSPS KAKENHI Grant Number 26247022, the HST grant AR-12143.01-A provided by NASA through a grant from the Space Telescope Science Institute, which is operated by the Association of Universities for Research in Astronomy, Incorporated, under NASA contract NAS5-26555.

References

A.R. Basu-Zych, B.D. Lehmer, A.E. Hornschemeier, T.S. Gonçalves, T. Fragos, T.M. Heckman, R.A. Overzier, A.F. Ptak, D. Schiminovich, Astrophys. J. **774**, 152 (2013)

A. Bauermeister, L. Blitz, C.-P. Ma, Astrophys. J. **717**, 323 (2010)

P.S. Behroozi, R.H. Wechsler, C. Conroy, Astrophys. J. **770**, 57 (2013)

E.F. Bell et al., Astrophys. J. **663**, 834 (2007)

E. Berger, B.E. Penprase, S.B. Cenko, S.R. Kulkarni, D.B. Fox, C.C. Steidel, N.A. Reddy, Spectroscopy of GRB 050505 at $z = 4.275$: a $\log N(\mathrm{HI}) = 22.1$ DLA host galaxy and the nature of the progenitor. Astrophys. J. **642**, 979–988 (2006). doi:10.1086/501162

M. Béthermin, H. Dole, G. Lagache, D. Le Borgne, A. Penin, Astron. Astrophys. **529**, A4 (2011)

M. Béthermin, E. Daddi, G. Magdis et al., Astrophys. J. Lett. **757**, L23 (2012b)

M. Béthermin et al., Astron. Astrophys. **557**, A66 (2013)

M. Béthermin et al., Astron. Astrophys. **567**, 103 (2014)

M. Béthermin et al., Astron. Astrophys. **573**, 113 (2015)

S. Bird, M. Vogelsberger, D. Sijacki, M. Zaldarriaga, V. Springel, L. Hernquist, Moving-mesh cosmology: properties of neutral hydrogen in absorption. Mon. Not. R. Astron. Soc. **429**, 3341–3352 (2013). doi:10.1093/mnras/sts590

M.R. Blanton, J. Dalcanton, D. Eisenstein, J. Loveday, M.A. Strauss, M. SubbaRao, D.H. Weinberg, J.E.J. Anderson, et al., The luminosity function of galaxies in sdss commissioning data. Astron. J. **121**, 2358 (2001)

M. Boquien, V. Buat, A. Boselli, M. Baes, G.J. Bendo, L. Ciesla, A. Cooray, L. Cortese, S. Eales, G. Gavazzi, H.L. Gomez, V. Lebouteiller, C. Pappalardo, M. Pohlen, M.W.L. Smith, L. Spinoglio, Astron. Astrophys. **539**, A145 (2012)

F. Bournaud, E. Daddi, A. Weiß, F. Renaud, C. Mastropietro, R. Teyssier, Astron. Astrophys. **575**, A56 (2015)

R.J. Bouwens, G.D. Illingworth, P.A. Oesch, I. Labbé, M. Trenti, P. van Dokkum, M. Franx, M. Stiavelli, C.M. Carollo, D. Magee, V. Gonzalez, Ultraviolet luminosity functions from 132 $z \sim 7$ and $z \sim 8$ Lyman-break galaxies in the ultra-deep HUDF09 and wide-area early release science WFC3/IR observations. Astrophys. J. **737**, 90 (2011). doi:10.1088/0004-637X/737/2/90

R.J. Bouwens, G.D. Illingworth, P.A. Oesch, M. Trenti, I. Labbé, L. Bradley, M. Carollo, P.G. van Dokkum, V. Gonzalez, B. Holwerda, M. Franx, L. Spitler, R. Smit, D. Magee, UV luminosity functions at redshifts $z \sim 4$ to $z \sim 10$: 10000 galaxies from HST legacy fields. Astrophys. J. **803**, 34 (2015). doi:10.1088/0004-637X/803/1/34

J. Brinchmann, S. Charlot, S.D.M. White et al., Mon. Not. R. Astron. Soc. **351**, 1151 (2004)

G. Bruzual, S. Charlot, Stellar population synthesis at the resolution of 2003. Mon. Not. R. Astron. Soc. **344**, 1000 (2003)

V. Buat, S. Noll, D. Burgarella, E. Giovannoli, V. Charmandaris, M. Pannella, H.S. Hwang, D. Elbaz, M. Dickinson, G. Magdis, N. Reddy, E.J. Murphy, Astron. Astrophys. **545**, A141 (2012)

D. Calzetti, L. Armus, R.C. Bohlin, A.L. Kinney, J. Koornneef, T. Storchi-Bergmann, The dust content and opacity of actively star-forming galaxies. Astrophys. J. **533**, 682 (2000)

C.M. Casey, D. Narayanan, A. Cooray, Phys. Rep. **541**, 45 (2014)

H.-W. Chen, J.X. Prochaska, N.Y. Gnedin, A new constraint on the escape fraction in distant galaxies using γ-ray burst afterglow spectroscopy. Astrophys. J. Lett. **667**, 125–128 (2007). doi:10.1086/522306

J.-H. Choi, K. Nagamine, Effects of metal enrichment and metal cooling in galaxy growth and cosmic star formation history. Mon. Not. R. Astron. Soc. **393**, 1595–1607 (2009). doi:10.1111/j.1365-2966.2008.14297.x

J.-H. Choi, K. Nagamine, Effects of cosmological parameters and star formation models on the cosmic star formation history in ΛCDM cosmological simulations. Mon. Not. R. Astron. Soc. **407**, 1464–1476 (2010). doi:10.1111/j.1365-2966.2010.17029.x

J.-H. Choi, K. Nagamine, Multicomponent and variable velocity galactic outflow in cosmological hydrodynamic simulations. Mon. Not. R. Astron. Soc. **410**, 2579–2592 (2011). doi:10.1111/j.1365-2966.2010.17632.x

S. Cole, P. Norberg, C.M. Baugh, C.S. Frenk, J. Bland-Hawthorn, T. Bridges, R. Cannon, M. Colless, C. Collins, W. Couch, N. Cross, G. Dalton, R. De Propris, S.P. Driver, G. Efstathiou, R.S. Ellis, K. Glazebrook, C. Jackson, O. Lahav, I. Lewis, S. Lumsden, S. Maddox, D. Madgwick, J.A. Peacock, B.A. Peterson, W. Sutherland, K. Taylor, The 2dF galaxy redshift survey: near-infrared galaxy luminosity functions. Mon. Not. R. Astron. Soc. **326**, 255–273 (2001). doi:10.1046/j.1365-8711.2001.04591.x

A. Cucchiara, M. Fumagalli, M. Rafelski, D. Kocevski, J.X. Prochaska, R.J. Cooke, G.D. Becker, Astrophys. J. **804**, 51 (2015)

E. Daddi, M. Dickinson, G. Morrison, R. Chary, A. Cimatti, D. Elbaz, D. Frayer, A. Renzini, A. Pope, D.M. Alexander, F.E. Bauer, M. Giavalisco, M. Huynh, J. Kurk, M. Mignoli, Astrophys. J. **670**, 156 (2007)

E. Daddi et al., Astrophys. J. **713**, 686 (2010a)

E. Daddi et al., Astrophys. J. Lett. **714**, L118 (2010b)

E. Daddi et al., Astron. Astrophys. **577**, A46 (2015)

D.A. Dale, S.A. Cohen, L.C. Johnson, M.D. Schuster, D. Calzetti, C.W. Engelbracht, A. Gil de Paz, R.C. Kennicutt, J.C. Lee, A. Begum, M. Block, J.J. Dalcanton, J.G. Funes, K.D. Gordon, B.D. Johnson, A.R. Marble, S. Sakai, E.D. Skillman, L. van Zee, F. Walter, D.R. Weisz, B. Williams, S.-Y. Wu, Y. Wu, Astrophys. J. **703**, 517 (2009)

H. Dannerbauer, E. Daddi, D.A. Riechers et al., Astrophys. J. Lett. **698**, L178 (2009)

R. Davé, Mon. Not. R. Astron. Soc. **385**, 147 (2008)

P. Di Matteo, F. Combes, A.-L. Melchior, B. Semelin, Astron. Astrophys. **468**, 61 (2007)

M. Dijkstra, S. Wyithe, Z. Haiman, A. Mesinger, L. Pentericci, Evolution in the escape fraction of ionizing photons and the decline in strong Lyα emission from $z > 6$ galaxies. Mon. Not. R. Astron. Soc. **440**, 3309–3316 (2014). doi:10.1093/mnras/stu531

A. Domínguez, B. Siana, A.L. Henry, C. Scarlata, A.G. Bedregal, M. Malkan, H. Atek, N.R. Ross, J.W. Colbert, H.I. Teplitz, M. Rafelski, P. McCarthy, A. Bunker, N.P. Hathi, A. Dressler, C.L. Martin, D. Masters, Astrophys. J. **763**, 145 (2013)

D. Elbaz, M. Dickinson, H.S. Hwang et al., Astron. Astrophys. **533**, A119 (2011)

D.K. Erb, A.E. Shapley, M. Pettini, C.C. Steidel, N.A. Reddy, K.L. Adelberger, Astrophys. J. **644**, 813 (2006a)

D.K. Erb, C.C. Steidel, A.E. Shapley, M. Pettini, N.A. Reddy, K.L. Adelberger, Astrophys. J. **646**, 107 (2006b)

X. Fan, C.L. Carilli, B. Keating, Observational constraints on cosmic reionization. Annu. Rev. Astron. Astrophys. **44**, 415–462 (2006). doi:10.1146/annurev.astro.44.051905.092514

R. Feldmann, N.Y. Gnedin, A.V. Kravtsov, Astrophys. J. **747**, 124 (2012)

K. Finlator, R. Davé, Mon. Not. R. Astron. Soc. **385**, 2181 (2008)

K. Finlator, R. Davé, C. Papovich, L. Hernquist, The physical and photometric properties of high-redshift galaxies in cosmological hydrodynamic simulations. Astrophys. J. **639**, 672–694 (2006). doi:10.1086/499349

M. Fioc, B. Rocca-Volmerange, Astron. Astrophys. **326**, 950 (1997)

J.P.U. Fynbo, P. Jakobsson, J.X. Prochaska, D. Malesani, C. Ledoux, A. de Ugarte Postigo, M. Nardini, P.M. Vreeswijk, K. Wiersema, J. Hjorth, J. Sollerman, H.-W. Chen, C.C. Thöne, G. Björnsson, J.S. Bloom, A.J. Castro-Tirado, L. Christensen, A. De Cia, A.S. Fruchter, J. Gorosabel, J.F. Graham, A.O. Jaunsen, B.L. Jensen, D.A. Kann, C. Kouveliotou, A.J. Levan, J. Maund, N. Masetti, B. Milvang-Jensen, E. Palazzi, D.A. Perley, E. Pian, E. Rol, P. Schady, R.L.C. Starling, N.R. Tanvir, D.J. Watson, D. Xu, T. Augusteijn, F. Grundahl, J. Telting, P.-O. Quirion, Low-resolution spectroscopy of gamma-ray burst optical afterglows: biases in the swift sample and characterization of the absorbers. Astrophys. J. Suppl. Ser. **185**, 526–573 (2009). doi:10.1088/0067-0049/185/2/526

J.E. Geach, I. Smail, S.M. Moran, L.A. MacArthur, C.d.P Lagos, A.C. Edge, Astrophys. J. Lett. **730**, L19 (2011)

S. Genel, M. Vogelsberger, V. Springel, D. Sijacki, D. Nelson, G. Snyder, V. Rodriguez-Gomez, P. Torrey, L. Hernquist, Introducing the illustris project: the evolution of galaxy populations across cosmic time. Mon. Not. R. Astron. Soc. **445**, 175–200 (2014). doi:10.1093/mnras/stu1654

R. Genzel et al., Mon. Not. R. Astron. Soc. **407**, 2091 (2010)

R. Genzel, L.J. Tacconi, F. Combes et al., Astrophys. J. **746**, 69 (2012)

R. Genzel et al., Astrophys. J. **800**, 20 (2015)

N.Y. Gnedin, A.V. Kravtsov, H. Chen, Escape of ionizing radiation from high-redshift galaxies. Astrophys. J. **672**, 765–775 (2008). doi:10.1086/524007
V. González, I. Labbé, R.J. Bouwens et al., Astrophys. J. **713**, 115 (2010)
J. Graciá-Carpio, E. Sturm, S. Hailey-Dunsheath et al., Astrophys. J. Lett. **728**, 7 (2011)
B. Groves, J. Brinchmann, C.J. Walcher, Mon. Not. R. Astron. Soc. **419**, 1402 (2012)
C. Gruppioni, F. Pozzi, G. Zamorani, C. Vignali, Mon. Not. R. Astron. Soc. **416**, 70 (2011)
G. Hinshaw, D. Larson, E. Komatsu, D.N. Spergel, C.L. Bennett, J. Dunkley, M.R. Nolta, M. Halpern, R.S. Hill, N. Odegard, L. Page, K.M. Smith, J.L. Weiland, B. Gold, N. Jarosik, A. Kogut, M. Limon, S.S. Meyer, G.S. Tucker, E. Wollack, E.L. Wright, Nine-year Wilkinson Microwave Anisotropy Probe (WMAP) observations: cosmological parameter results. Astrophys. J. Suppl. Ser. **208**, 19 (2013). doi:10.1088/0067-0049/208/2/19
P.F. Hopkins, T.J. Cox, L. Hernquist, D. Narayanan, C.C. Hayward, N. Murray, Star formation in galaxy mergers with realistic models of stellar feedback and the interstellar medium. Mon. Not. R. Astron. Soc. **430**, 1901–1927 (2013). doi:10.1093/mnras/stt017
O. Ilbert, M. Salvato, E. Le Floc'h et al., Astrophys. J. **709**, 644 (2010)
I.T. Iliev, B. Ciardi, M.A. Alvarez, A. Maselli, A. Ferrara, N.Y. Gnedin, G. Mellema, T. Nakamoto, M.L. Norman, A.O. Razoumov, E.-J. Rijkhorst, J. Ritzerveld, P.R. Shapiro, H. Susa, M. Umemura, D.J. Whalen, Cosmological radiative transfer codes comparison project. I. The static density field tests. Mon. Not. R. Astron. Soc. **371**, 1057–1086 (2006). doi:10.1111/j.1365-2966.2006.10775.x
A.K. Inoue, I. Iwata, J.-M. Deharveng, The escape fraction of ionizing photons from galaxies at $z = 0$–6. Mon. Not. R. Astron. Soc. **371**, 1–5 (2006). doi:10.1111/j.1745-3933.2006.00195.x
J. Jaacks, J.-H. Choi, K. Nagamine, R. Thompson, S. Varghese, Steep faint-end slopes of galaxy mass and luminosity functions at $z \geq 6$ and the implications for reionization. Mon. Not. R. Astron. Soc. **420**, 1606–1620 (2012). doi:10.1111/j.1365-2966.2011.20150.x
J. Jaacks, R. Thompson, K. Nagamine, Impact of H_2-based star formation model on the $z \geq 6$ luminosity function and the ionizing photon budget for reionization. Astrophys. J. **766**, 94 (2013). doi:10.1088/0004-637X/766/2/94
B.D. Johnson, D. Schiminovich, M. Seibert, M. Treyer, D.C. Martin, T.A. Barlow, K. Forster, P.G. Friedman, P. Morrissey, S.G. Neff, T. Small, T.K. Wyder, L. Bianchi, J. Donas, T.M. Heckman, Y.-W. Lee, B.F. Madore, B. Milliard, R.M. Rich, A.S. Szalay, B.Y. Welsh, S.K. Yi, Astrophys. J. Suppl. Ser. **173**, 392 (2007)
A. Karim et al., Astrophys. J. **730**, 61 (2011)
R.C. Kennicutt, Annu. Rev. Astron. Astrophys. **36**, 189 (1998)
D. Keres, M.S. Yun, J.S. Young, Astrophys. J. **582**, 659 (2003)
L.J. Kewley, S.L. Ellison, Astrophys. J. **681**, 1183 (2008)
M.D. Kistler, H. Yüksel, J.F. Beacom, A.M. Hopkins, J.S.B. Wyithe, The star formation rate in the reionization era as indicated by gamma-ray bursts. Astrophys. J. Lett. **705**, 104–108 (2009). doi:10.1088/0004-637X/705/2/L104
E. Komatsu, K.M. Smith, J. Dunkley, C.L. Bennett, B. Gold, G. Hinshaw, N. Jarosik, D. Larson, et al., Seven-year Wilkinson Microwave Anisotropy Probe (WMAP) observations: cosmological interpretation. Astrophys. J. Suppl. Ser. **192**, 18 (2011). doi:10.1088/0067-0049/192/2/18
X. Kong, S. Charlot, J. Brinchmann, S.M. Fall, Mon. Not. R. Astron. Soc. **349**, 769 (2004)
M. Kriek, I. Labbé, C. Conroy, K.E. Whitaker et al., Astrophys. J. Lett. **722**, L64 (2010)
M.R. Krumholz, C.F. McKee, J. Tumlinson, The star formation law in atomic and molecular gas. Astrophys. J. **699**, 850–856 (2009). doi:10.1088/0004-637X/699/1/850
C.G. Lacey, C.M. Baugh, C.S. Frenk et al., Mon. Not. R. Astron. Soc. **405**, 2 (2010)
T. Laskar, E. Berger, R.-R. Chary, Astrophys. J. **739**, 1 (2011)
E. Le Floc'h et al., Astrophys. J. **632**, 169 (2005)
C. Ledoux, P.M. Vreeswijk, A. Smette, A.J. Fox, P. Petitjean, S.L. Ellison, J.P.U. Fynbo, S. Savaglio, Astron. Astrophys. **506**, 661 (2009)
C. Leitherer, D. Schaerer, J.D. Goldader, R.M.G. Delgado, C. Robert, D.F. Kune, D.F. de Mello, D. Devost, T.M. Heckman, Astrophys. J. Suppl. Ser. **123**, 3 (1999)
A.K. Leroy et al., Astron. J. **137**, 4670 (2009)
B. Lo Faro, P. Monaco, E. Vanzella, F. Fontanot, L. Silva, S. Cristiani, Faint Lyman-break galaxies as a crucial test for galaxy formation models. Mon. Not. R. Astron. Soc. **399**, 827–841 (2009). doi:10.1111/j.1365-2966.2009.15316.x
G.E. Magdis, E. Daddi, M. Sargent et al., Astrophys. J. Lett. **758**, L9 (2012a)
G.E. Magdis, E. Daddi, M. Béthermin et al., arXiv:1210.1035 (2012b)
B. Magnelli, D. Elbaz, R.R. Chary et al., Astron. Astrophys. **496**, 57 (2009)

R. Maiolino, P. Cox, P. Caselli, A. Beelen, F. Bertoldi, C.L. Carilli, M.J. Kaufman, K.M. Menten, et al., First detection of [CII]158 μm at high redshift: vigorous star formation in the early universe. Astron. Astrophys. **440**, 51–54 (2005). doi:10.1051/0004-6361:200500165

R. Maiolino, T. Nagao, A. Grazian, F. Cocchia, A. Marconi, F. Mannucci, A. Cimatti, A. Pipino, S. Ballero, F. Calura, C. Chiappini, A. Fontana, G.L. Granato, F. Matteucci, G. Pastorini, L. Pentericci, G. Risaliti, M. Salvati, L. Silva, Astron. Astrophys. **488**, 463 (2008)

F. Mannucci, G. Cresci, R. Maiolino, A. Marconi, A. Gnerucci, Mon. Not. R. Astron. Soc. **408**, 2115 (2010)

C. Maraston, Mon. Not. R. Astron. Soc. **362**, 799 (2005)

D. Marchesini, P.G. van Dokkum, N.M. Forster Schreiber, M. Franx, I. Labbe', S. Wuyts, The evolution of the stellar mass function of galaxies from $z = 4.0$ and the first comprehensive analysis of its uncertainties: evidence for mass-dependent evolution. ArXiv e-prints (2008)

M. Massardi, A. Bonaldi, M. Negrello et al., Mon. Not. R. Astron. Soc. **404**, 532 (2010)

D. Masters, P. McCarthy, B. Siana, M. Malkan, B. Mobasher, H. Atek, A. Henry, C.L. Martin, M. Rafelski, N.P. Hathi, C. Scarlata, N.R. Ross, A.J. Bunker, G. Blanc, A.G. Bedregal, A. Domínguez, J. Colbert, H. Teplitz, A. Dressler, Astrophys. J. **785**, 153 (2014)

I.S. McLean, C.C. Steidel, H.W. Epps, N. Konidaris, K.Y. Matthews, S. Adkins, T. Aliado, G. Brims, J.M. Canfield, J.L. Cromer, J. Fucik, K. Kulas, G. Mace, K. Magnone, H. Rodriguez, G. Rudie, R. Trainor, E. Wang, B. Weber, J. Weiss, in *Status of the Dark Energy Survey Camera (DECam) Project*. Society of Photo-Optical Instrumentation Engineers (SPIE) Conference Series, vol. 8446 (2012)

G.R. Meurer, T.M. Heckman, D. Calzetti, Astrophys. J. **521**, 64 (1999)

A.L. Muratov, D. Kereš, C.-A. Faucher-Giguère, P.F. Hopkins, E. Quataert, N. Murray, Gusty, gaseous flows of FIRE: galactic winds in cosmological simulations with explicit stellar feedback. Mon. Not. R. Astron. Soc. **454**, 2691–2713 (2015). doi:10.1093/mnras/stv2126

K. Nagamine, V. Springel, L. Hernquist, M. Machacek, Photometric properties of Lyman-break galaxies at $z = 3$ in cosmological SPH simulations. Mon. Not. R. Astron. Soc. **350**, 385–395 (2004). doi:10.1111/j.1365-2966.2004.07664.x

K. Nagamine, J. Choi, H. Yajima, Effects of ultraviolet background and local stellar radiation on the H I column density distribution. Astrophys. J. Lett. **725**, 219–222 (2010). doi:10.1088/2041-8205/725/2/L219

C. Night, K. Nagamine, V. Springel, L. Hernquist, Lyman break galaxies at $z = 4$–6 in cosmological smoothed particle hydrodynamics simulations. Mon. Not. R. Astron. Soc. **366**, 705–716 (2006). doi:10.1111/j.1365-2966.2005.09730.x

K.G. Noeske, B.J. Weiner, S.M. Faber, C. Papovich, D.C. Koo, R.S. Somerville, K. Bundy, C.J. Conselice, J.A. Newman, D. Schiminovich, E. Le Floc'h, A.L. Coil, G.H. Rieke, J.M. Lotz, J.R. Primack, P. Barmby, M.C. Cooper, M. Davis, R.S. Ellis, G.G. Fazio, P. Guhathakurta, J. Huang, S.A. Kassin, D.C. Martin, A.C. Phillips, R.M. Rich, T.A. Small, C.N.A. Willmer, G. Wilson, Astrophys. J. Lett. **660**, L43 (2007)

D. Obreschkow, S. Rawlings, Mon. Not. R. Astron. Soc. **394**, 1857 (2009a)

D. Obreschkow, S. Rawlings, Astrophys. J. Lett. **696**, L129 (2009b)

M. Onodera et al., Astrophys. J. **822**, 42 (2016)

R.A. Overzier, T.M. Heckman, J. Wang, L. Armus, V. Buat, J. Howell, G. Meurer, M. Seibert, B. Siana, A. Basu-Zych, S. Charlot, T.S. Gonçalves, D.C. Martin, J.D. Neill, R.M. Rich, S. Salim, D. Schiminovich, Astrophys. J. Lett. **726**, L7 (2011)

J.-P. Paardekooper, F.I. Pelupessy, G. Altay, C.J.H. Kruip, The escape of ionising radiation from high-redshift dwarf galaxies. Astron. Astrophys. **530**, 87 (2011). doi:10.1051/0004-6361/201116841

D.A. Perley, A.J. Levan, N.R. Tanvir, S.B. Cenko et al., Astrophys. J. **778**, 128 (2013)

M. Pettini, S.A. Rix, C.C. Steidel, K.L. Adelberger, M.P. Hunt, A.E. Shapley, Astrophys. J. **569**, 742 (2002)

Planck Collaboration, P.A.R. Ade, N. Aghanim, M. Arnaud, M. Ashdown, J. Aumont, C. Baccigalupi, A.J. Banday, R.B. Barreiro, J.G. Bartlett et al., Planck 2015 results. XIII. Cosmological parameters. ArXiv e-prints (2015)

P. Ranalli, A. Comastri, G. Setti, Astron. Astrophys. **399**, 39 (2003)

A.O. Razoumov, J. Sommer-Larsen, Ionizing radiation from $z = 4$–10 galaxies. Astrophys. J. **710**, 1239–1246 (2010). doi:10.1088/0004-637X/710/2/1239

N.A. Reddy, C.C. Steidel, A steep faint-end slope of the UV luminosity function at $z \sim 2$–3: implications for the global stellar mass density and star formation in low-mass halos. Astrophys. J. **692**, 778–803 (2009). doi:10.1088/0004-637X/692/1/778

N.A. Reddy, C.C. Steidel, D. Fadda, L. Yan, M. Pettini, A.E. Shapley, D.K. Erb, K.L. Adelberger, Astrophys. J. **644**, 792 (2006)

N.A. Reddy, D.K. Erb, M. Pettini, C.C. Steidel, A.E. Shapley, Astrophys. J. **712**, 1070 (2010)

N. Reddy, M. Dickinson, D. Elbaz, G. Morrison, M. Giavalisco, R. Ivison, C. Papovich, D. Scott, V. Buat, D. Burgarella, V. Charmandaris, E. Daddi, G. Magdis, E. Murphy, B. Altieri, H. Aussel, H. Dannerbauer, K. Dasyra, H.S. Hwang, J. Kartaltepe, R. Leiton, B. Magnelli, P. Popesso, Astrophys. J. **744**, 154 (2012a)

N.A. Reddy, M. Pettini, C.C. Steidel, A.E. Shapley, D.K. Erb, D.R. Law, Astrophys. J. **754**, 25 (2012b)
N.A. Reddy, M. Kriek, A.E. Shapley, W.R. Freeman, B. Siana, A.L. Coil, B. Mobasher, S.H. Price, R.L. Sanders, I. Shivaei ArXiv e-prints (2015)
G. Rodighiero, E. Daddi, I. Baronchelli, A. Cimatti, A. Renzini, H. Aussel, P. Popesso, D. Lutz, P. Andreani, S. Berta, A. Cava, D. Elbaz, A. Feltre, A. Fontana, N.M. Förster Schreiber, A. Franceschini, R. Genzel, A. Grazian, C. Gruppioni, O. Ilbert, E. Le Floch, G. Magdis, M. Magliocchetti, B. Magnelli, R. Maiolino, H. McCracken, R. Nordon, A. Poglitsch, P. Santini, F. Pozzi, L. Riguccini, L.J. Tacconi, S. Wuyts, G. Zamorani, Astrophys. J. Lett. **739**, L40 (2011)
E. Romano-Díaz, J.-H. Choi, I. Shlosman, M. Trenti, Astrophys. J. **738**, L19 (2011)
W. Rujopakarn, G.H. Rieke, D.J. Eisenstein, S. Juneau, Astrophys. J. **726**, 93 (2011)
A. Saintonge et al., Mon. Not. R. Astron. Soc. **415**, 32 (2011)
F. Salmi, E. Daddi, D. Elbaz et al., Astrophys. J. Lett. **754**, L14 (2012)
D.B. Sanders, I.F. Mirabel, Annu. Rev. Astron. Astrophys. **34**, 749 (1996)
R.L. Sanders, A.E. Shapley, M. Kriek, N.A. Reddy, W.R. Freeman, A.L. Coil, B. Siana, B. Mobasher, I. Shivaei, S.H. Price, L. de Groot, Astrophys. J. **799**, 138 (2015)
M.T. Sargent, M. Béthermin, E. Daddi, D. Elbaz, Astrophys. J. Lett. **747**, L31 (2012)
M.T. Sargent et al., Astrophys. J. **793**, 19 (2014)
S. Savaglio, K. Glazebrook, D. Le Borgne, S. Juneau, R.G. Abraham, H.-W. Chen, D. Crampton, P.J. McCarthy, R.G. Carlberg, R.O. Marzke, K. Roth, I. Jørgensen, R. Murowinski, Astrophys. J. **635**, 260 (2005)
J. Schaye, C. Dalla Vecchia, On the relation between the Schmidt and Kennicutt-Schmidt star formation laws and its implications for numerical simulations. Mon. Not. R. Astron. Soc. **383**, 1210–1222 (2008). doi:10.1111/j.1365-2966.2007.12639.x
J. Schaye, R.A. Crain, R.G. Bower, M. Furlong, M. Schaller, T. Theuns, C. Dalla Vecchia, C.S. Frenk, I.G. McCarthy, J.C. Helly, A. Jenkins, Y.M. Rosas-Guevara, S.D.M. White, M. Baes, C.M. Booth, P. Camps, J.F. Navarro, Y. Qu, A. Rahmati, T. Sawala, P.A. Thomas, J. Trayford, The EAGLE project: simulating the evolution and assembly of galaxies and their environments. Mon. Not. R. Astron. Soc. **446**, 521–554 (2015). doi:10.1093/mnras/stu2058
A. Schruba, A.K. Leroy, F. Walter et al., Astron. J. **143**, 138 (2012)
M. Seibert, D.C. Martin, T.M. Heckman, V. Buat, C. Hoopes, T. Barlow, L. Bianchi, Y.-I. Byun, J. Donas, K. Forster, P.G. Friedman, P. Jelinsky, Y.-W. Lee, B.F. Madore, R. Malina, B. Milliard, P. Morrissey, S. Neff, R.M. Rich, D. Schiminovich, O. Siegmund, T. Small, A.S. Szalay, B. Welsh, T.K. Wyder, Astrophys. J. Lett. **619**, L55 (2005)
A.E. Shapley, Annu. Rev. Astron. Astrophys. **49**, 525 (2011)
A.E. Shapley, N.A. Reddy, M. Kriek, W.R. Freeman, R.L. Sanders, B. Siana, A.L. Coil, B. Mobasher, I. Shivaei, S.H. Price, L. de Groot, Astrophys. J. **801**, 88 (2015)
I. Shivaei, N.A. Reddy, A.E. Shapley, M. Kriek, B. Siana, B. Mobasher, A.L. Coil, W.R. Freeman, R. Sanders, S.H. Price, L. de Groot, M. Azadi ArXiv e-prints (2015)
R. Smit, R.J. Bouwens, M. Franx, G.D. Illingworth, I. Labbé, P.A. Oesch, P.G. van Dokkum, The star formation rate function for redshift $z \sim 4$–7 galaxies: evidence for a uniform buildup of star-forming galaxies during the first 3 Gyr of cosmic time. Astrophys. J. **756**, 14 (2012). doi:10.1088/0004-637X/756/1/14
B.M. Smith, R.A. Windhorst, R.A. Jansen, S.H. Cohen, L. Jiang, M. Dijkstra, A.M. Koekemoer, R. Bielby, A.K. Inoue, J.W. MacKenty, R.W. O'Connell, J.I. Silk, Hubble Space Telescope Wide Field Camera 3 observations of escaping Lyman continuum radiation from galaxies and weak AGN at redshifts $z \sim$ 2.3–5. ArXiv e-prints (2016)
P.M. Solomon, D. Downes, S.J.E. Radford, J.W. Barrett, Astrophys. J. **478**, 144 (1997)
M. Sparre, O.E. Hartoog, T. Krühler, J.P.U. Fynbo, D.J. Watson, K. Wiersema, V. D'Elia, T. Zafar, P.M.J. Afonso, S. Covino, A. de Ugarte Postigo, H. Flores, P. Goldoni, J. Greiner, J. Hjorth, P. Jakobsson, L. Kaper, S. Klose, A.J. Levan, D. Malesani, B. Milvang-Jensen, M. Nardini, S. Piranomonte, J. Sollerman, R. Sánchez-Ramírez, S. Schulze, N.R. Tanvir, S.D. Vergani, R.A.M.J. Wijers, Astrophys. J. **785**, 150 (2014)
J.S. Spilker et al., Astrophys. J. **785**, 149 (2014)
V. Springel, The cosmological simulation code GADGET-2. Mon. Not. R. Astron. Soc. **364**, 1105–1134 (2005). doi:10.1111/j.1365-2966.2005.09655.x
V. Springel, L. Hernquist, Cosmological smoothed particle hydrodynamics simulations: a hybrid multiphase model for star formation. Mon. Not. R. Astron. Soc. **339**, 289–311 (2003). doi:10.1046/j.1365-8711.2003.06206.x
C.C. Steidel, G.C. Rudie, A.L. Strom, M. Pettini, N.A. Reddy, A.E. Shapley, R.F. Trainor, D.K. Erb, M.L. Turner, N.P. Konidaris, K.R. Kulas, G. Mace, K. Matthews, I.S. McLean ArXiv e-prints (2014)
L.J. Tacconi et al., Nature **463**, 781 (2010)
L.J. Tacconi et al., Astrophys. J. **768**, 74 (2013)

Q. Tan et al., Astrophys. J. **776**, L24 (2013)
Q. Tan et al., Astron. Astrophys. **569**, A98 (2014)
N.R. Tanvir, D.B. Fox, A.J. Levan, E. Berger, K. Wiersema, J.P.U. Fynbo, A. Cucchiara, T. Krühler, N. Gehrels, J.S. Bloom, J. Greiner, P.A. Evans, E. Rol, F. Olivares, J. Hjorth, P. Jakobsson, J. Farihi, R. Willingale, R.L.C. Starling, S.B. Cenko, D. Perley, J.R. Maund, J. Duke, R.A.M.J. Wijers, A.J. Adamson, A. Allan, M.N. Bremer, D.N. Burrows, A.J. Castro-Tirado, B. Cavanagh, A. de Ugarte Postigo, M.A. Dopita, T.A. Fatkhullin, A.S. Fruchter, R.J. Foley, J. Gorosabel, J. Kennea, T. Kerr, S. Klose, H.A. Krimm, V.N. Komarova, S.R. Kulkarni, A.S. Moskvitin, C.G. Mundell, T. Naylor, K. Page, B.E. Penprase, M. Perri, P. Podsiadlowski, K. Roth, R.E. Rutledge, T. Sakamoto, P. Schady, B.P. Schmidt, A.M. Soderberg, J. Sollerman, A.W. Stephens, G. Stratta, T.N. Ukwatta, D. Watson, E. Westra, T. Wold, C. Wolf, A γ-ray burst at a redshift of $z \sim 8.2$. Nature **461**, 1254–1257 (2009). doi:10.1038/nature08459
M. Tegmark, M.A. Strauss, M.R. Blanton, K. Abazajian, S. Dodelson, H. Sandvik, X. Wang, D.H. Weinberg, I. Zehavi, N.A. Bahcall, F. Hoyle, D. Schlegel, R. Scoccimarro, M.S. Vogeley, A. Berlind, T. Budavari, A. Connolly, D.J. Eisenstein, D. Finkbeiner, J.A. Frieman, J.E. Gunn, L. Hui, B. Jain, D. Johnston, S. Kent, H. Lin, R. Nakajima, R.C. Nichol, J.P. Ostriker, A. Pope, R. Scranton, U. Seljak, R.K. Sheth, A. Stebbins, A.S. Szalay, I. Szapudi, Y. Xu, J. Annis, J. Brinkmann, S. Burles, F.J. Castander, I. Csabai, J. Loveday, M. Doi, M. Fukugita, B. Gillespie, G. Hennessy, D.W. Hogg, Ž. Ivezić, G.R. Knapp, D.Q. Lamb, B.C. Lee, R.H. Lupton, T.A. McKay, P. Kunszt, J.A. Munn, L. O'Connell, J. Peoples, J.R. Pier, M. Richmond, C. Rockosi, D.P. Schneider, C. Stoughton, D.L. Tucker, D.E. vanden Berk, B. Yanny, D.G. York, Cosmological parameters from SDSS and WMAP. Phys. Rev. D **69**(10), 103501 (2004). doi:10.1103/PhysRevD.69.103501
R. Thompson, K. Nagamine, J. Jaacks, J.-H. Choi, Molecular hydrogen regulated star formation in cosmological smoothed particle hydrodynamics simulations. Astrophys. J. **780**, 145 (2014). doi:10.1088/0004-637X/780/2/145
C.A. Tremonti, T.M. Heckman, G. Kauffmann, J. Brinchmann, S. Charlot, S.D.M. White, M. Seibert, E.W. Peng, D.J. Schlegel, A. Uomoto, M. Fukugita, J. Brinkmann, Astrophys. J. **613**, 898 (2004)
P. Troncoso, R. Maiolino, V. Sommariva, G. Cresci, F. Mannucci, A. Marconi, M. Meneghetti, A. Grazian, A. Cimatti, A. Fontana, T. Nagao, L. Pentericci, Astron. Astrophys. **563**, A58 (2014)
M. Vogelsberger, S. Genel, V. Springel, P. Torrey, D. Sijacki, D. Xu, G. Snyder, D. Nelson, L. Hernquist, Introducing the Illustris Project: simulating the coevolution of dark and visible matter in the Universe. Mon. Not. R. Astron. Soc. **444**, 1518–1547 (2014). doi:10.1093/mnras/stu1536
B.J. Weiner, A.L. Coil, J.X. Prochaska, J.A. Newman, M.C. Cooper, K. Bundy, C.J. Conselice, A.A. Dutton, et al., Ubiquitous Outflows in DEEP2 spectra of star-forming galaxies at $z = 1.4$. Astrophys. J. **692**, 187–211 (2009). doi:10.1088/0004-637X/692/1/187
K.E. Whitaker, M. Franx, J. Leja, P.G. van Dokkum, A. Henry, R.E. Skelton, M. Fumagalli, I.G. Momcheva, G.B. Brammer, I. Labbé, E.J. Nelson, J.R. Rigby, Astrophys. J. **795**, 104 (2014)
S.M. Wilkins, A. Bunker, W. Coulton, R. Croft, T.D. Matteo, N. Khandai, Y. Feng, Mon. Not. R. Astron. Soc. **430**, 2885 (2013)
J.H. Wise, R. Cen, Ionizing photon escape fractions from high-redshift dwarf galaxies. Astrophys. J. **693**, 984–999 (2009). doi:10.1088/0004-637X/693/1/984
S. Wuyts, N.M. Förster Schreiber, A. van der Wel et al., Astrophys. J. **742**, 96 (2011)
J.S.B. Wyithe, A.M. Hopkins, M.D. Kistler, H. Yüksel, J.F. Beacom, Determining the escape fraction of ionizing photons during reionization with the GRB-derived star formation rate. Mon. Not. R. Astron. Soc. **401**, 2561–2571 (2010). doi:10.1111/j.1365-2966.2009.15834.x
H. Yajima, J.-H. Choi, K. Nagamine, Escape fraction of ionizing photons from high-redshift galaxies in cosmological SPH simulations. Mon. Not. R. Astron. Soc. **412**, 411–422 (2011). doi:10.1111/j.1365-2966.2010.17920.x
H. Yajima, I. Shlosman, E. Romano-Díaz, K. Nagamine, Observational properties of simulated galaxies in overdense and average regions at redshifts $z \sim 6$–12. Mon. Not. R. Astron. Soc. **451**, 418–432 (2015). doi:10.1093/mnras/stv974
G. Yepes, R. Kates, A. Khokhlov, A. Klypin, Hydrodynamical simulations of galaxy formation: effects of supernova feedback. Mon. Not. R. Astron. Soc. **284**, 235–256 (1997)
A. Zanella et al., Nature **521**, 54 (2015)

Space Sci Rev (2016) 202:113–144
DOI 10.1007/s11214-016-0237-4

Long-Duration Gamma-Ray Burst Host Galaxies in Emission and Absorption

Daniel A. Perley[1] · Yuu Niino[2] · Nial R. Tanvir[3] · Susanna D. Vergani[4] · Johan P.U. Fynbo[1]

Received: 1 November 2015 / Accepted: 20 January 2016 / Published online: 10 March 2016

Abstract The galaxy population hosting long-duration GRBs provides a means to constrain the progenitor and an opportunity to use these violent explosions to characterize the nature of the high-redshift universe. Studies of GRB host galaxies in emission reveal a population of star-forming galaxies with great diversity, spanning a wide range of masses, metallicities, and redshifts. However, as a population GRB hosts are significantly less massive and poorer in metals than the hosts of other core-collapse transients, suggesting that GRB production is only efficient at metallicities significantly below Solar. GRBs may also prefer compact galaxies, and dense and/or central regions of galaxies, more than other types of core-collapse explosion. Meanwhile, studies of hosts in absorption against the luminous GRB optical afterglow provide a unique means of unveiling properties of the ISM in even the faintest and most distant galaxies; these observations are helping to constrain the chemical evolution of galaxies and the properties of interstellar dust out to very high redshifts. New ground- and

✉ D.A. Perley
dperley@dark-cosmology.dk

Y. Niino
yuu.niino@nao.ac.jp

N.R. Tanvir
nrt3@leicester.ac.uk

S.D. Vergani
susanna.vergani@obspm.fr

J.P.U. Fynbo
jfynbo@dark-cosmology.dk

1 Dark Cosmology Centre, Niels Bohr Institute, University of Copenhagen, Juliane Maries Vej 30, 2100 Copenhagen Ø, Denmark

2 Division of Optical & Infrared Astronomy, National Astronomical Observatory of Japan, 2-21-1 Osawa, Mitaka, Tokyo 181-8588, Japan

3 Department of Physics and Astronomy, University of Leicester, University Road, Leicester LE1 7RH, UK

4 GEPI, Observatoire de Paris, PSL Research University, CNRS, Univ. Paris Diderot, Sorbonne Paris Cité, Place Jules Janssen, 92195 Meudon, France

space-based instrumentation, and the accumulation of larger and more carefully-selected samples, are continually enhancing our view of the GRB host population.

Keywords Gamma-ray bursts · Interstellar medium · Dust · High-redshift galaxies

1 Introduction

The study of long-duration gamma-ray burst (GRB) host galaxies is, in essence, an attempt to use each of these two different astronomical objects (GRBs and galaxies) to understand the other. By studying the population of galaxies that produces GRBs and the locations of the GRBs inside their hosts, we hope to identify the GRB progenitor and how it is formed. The ubiquitously star-forming nature of the GRB host population is one of several lines of evidence linking GRBs to massive stars—but how does the progenitor star evolve to its final state and why does it explode?

The field is equally driven by interest in reversing this line of investigation to use GRBs to understand distant galaxies. Spectroscopy of a GRB optical afterglow provides rich detail on the properties of the absorbing system in a way that is not possible with other observational methods. In addition, since GRBs can be detected from galaxies of arbitrarily faint luminosities (even if their hosts cannot be detected in emission), they may provide a way to quantify and characterize the contributions to star formation of galaxy populations too faint, distant, or dusty to be easily studied by more traditional means.

In the following chapter we will discuss a number of ways in which GRB hosts are used to bridge the gap in understanding between GRBs and high-z galaxies on both sides. We will first present what is known about the GRB host population in emission via studies of their host galaxy population after the burst had faded, and compare the population of known GRB hosts to other distant galaxy populations. Afterwards, we will outline the properties of GRB hosts in absorption, illustrating some of the ways in which spectroscopy of GRB afterglows can uniquely reveal the properties of distant galaxies.

2 Long GRB Hosts in Emission

2.1 Background and Theoretical Predictions

Since nearly any observational technique or wavelength range that can be used to study other populations of distant galaxies can also be applied to GRB hosts, and because known GRBs span a vast redshift range ($0.008 < z < 9.4$; Galama et al. 1998; Cucchiara et al. 2011), the observational study of GRB hosts involves many different kinds of data and similar challenges as high-z galaxy astronomy generally. For most of its history these investigations have been limited to small samples of objects and subject to uncertain selection effects, but in the past five years the scene has changed dramatically as the large sample of positions provided by the Swift satellite—nearly 1000 GRBs localized within $2''$ (Gehrels et al. 2004; Butler 2007; Evans et al. 2009), and several hundred within $0.5''$ and with known redshift (Perley and Kemper 2008)—is systematically exploited with a range of modern observatories, including powerful new instruments such as WFC3 on the Hubble Space Telescope and X-shooter at the Very Large Telescope.

The link between GRBs and the explosions of massive stars is now well-established (Hjorth and Bloom 2012), so we expect GRBs to be found only in galaxies that are undergoing active star-formation (with the exception of short GRBs, which are not addressed in this review chapter). Of course, the diversity of star-forming galaxies is vast, and any attempt to establish the expectations for *which* star-forming galaxies GRBs should be found in at what frequency requires establishing assumptions about how the environment influences their rate.

It is helpful to start by imagining the simplest possible case: the assumption that the GRB rate is independent of all factors other than the overall rate of star-formation itself, such that a fixed fraction of all newly-formed stars explode as GRBs without regard to any of the chemical, physical, or other properties of the galaxy in which those stars formed. Observationally, this implies that GRBs should stochastically sample the locations of cosmic star-formation throughout the volume of the Universe in which they can be observed. The probability that any given galaxy will host a GRB during some period of time is proportional to its SFR.

Theoretically, there are various reasons to expect that reality may be more complicated than this. The most frequently-invoked possibility is a dependence of the GRB rate on metallicity. A fundamental challenge in producing a GRB is the need to eliminate the progenitor star's hydrogen envelope (which would smother the jet, and produce emission lines in the associated SN that are not observed) without spinning down the angular momentum that the central engine—a fast-rotating black hole or neutron star—needs to launch the jet in the first place. The (initial) metallicity of a star can affect this process in several ways: a higher metal abundance produces stronger stellar winds, greater mass loss, and less interior mixing (Crowther et al. 2002; Heger et al. 2003; Vink and de Koter 2005; Hirschi et al. 2008). Low metal abundance will both discourage mass (angular momentum) loss and encourage mixing of the hydrogen layer into the core, and single-star models (MacFadyen and Woosley 1999; Hirschi et al. 2005; Yoon and Langer 2005; Yoon et al. 2006; Langer and Norman 2006; Woosley and Heger 2006) robustly predict that GRBs should occur exclusively or preferentially in very metal-poor ($Z < 0.2$–$0.3 Z_\odot$) environments. Binary channels offer alternative means of exchanging mass and/or angular momentum (Izzard et al. 2004; Fryer and Heger 2005; Podsiadlowski et al. 2010), although these may also be metal-sensitive to a lesser extent.

Other environmental factors aside from metallicity may also be expected to impact the likelihood of producing a gamma-ray burst. Variations in the stellar IMF or close-binary fraction could lead to more massive stars and/or more close binaries, and a higher rate of GRBs, in some environments (Davé 2008; Wang and Dai 2011). Some GRB progenitor models appeal to dynamical effects between multiple stars (van den Heuvel and Portegies Zwart 2013); these would only be important in very dense environments and would cause GRBs to prefer the most intensely star-forming galaxies with many dense young star clusters.

In any case, from the observer's standpoint the general methodology to distinguish the possibilities above is straightforward: catalog the known population of GRB hosts as thoroughly as possible and measure its parameter distribution across as many metrics as possible—and then compare these to the distribution expected under the environment-independent null hypothesis (and to the predictions of specific theories, if possible) to identify what factors do and do not matter for GRB production.

Calculating this expected distribution requires detailed knowledge of galaxy evolution and cosmic star-formation at a variety of redshifts, since a detailed mapping of the star-formation history as a function of a wide variety of observables is required. This knowledge can come directly from observations, or from theory. Many studies simply use the observed distribution of total star-formation rate as a function of galaxy parameters in known surveys and apply empirical laws relating galaxy properties to each other (e.g. the mass-metallicity relation) to calculate the cosmic GRB rate as a function of various observable parameters (Wolf and Podsiadlowski 2007; Kocevski et al. 2009; Trenti et al. 2015) and determine the best-fit rate-dependence model. Theoretical techniques have also been employed to avoid the dependence on observations from field surveys, which can be incomplete (or lacking entirely) at high redshifts and low galaxy luminosities. In particular, semi-analytic models of galaxy formation (Lapi et al. 2008; Campisi et al. 2009; Chisari et al. 2010), and cosmological hydrodynamic simulations (Nuza et al. 2007; Niino et al. 2011; Salvaterra et al. 2013; Elliott et al. 2015) have been employed. However, these techniques are still limited by the uncertain physics surrounding star formation and feedback, as well as the difficulty in achieving sufficient resolution to compute internal structure within galaxies and sufficient volume to reproduce the cosmological population of galaxies at the same time.

During this process, great care must be taken to be sure that the observed GRB host sample being analyzed is chosen without selection bias. There are at least two obvious potential sources of bias when constructing samples. The first is dust extinction: significant star-formation occurs in heavily dust-obscured regions (Casey et al. 2014); if these stars produce GRBs, then their optical afterglows will be obscured and the afterglow position will need to be measured to $\sim$ arcsecond precision at another wavelength (radio or X-rays) to localize the host; these data are not always available. Furthermore, luminous GRB hosts are easier to detect and more likely to be reported (and scrutinized in the literature) than low-luminosity hosts, especially at high redshift. For these reasons, the means by which a sample was constructed should always be kept in mind when interpreting observations.

2.2 Photometric Properties

It was noticed early on that the GRB host population is relatively faint in comparison to other star-forming galaxies found at the same redshift (Le Floc'h et al. 2003) with a curious lack of massive, luminous galaxies with large populations of older stars among the pre-Swift (1997–2004) host population. Of course, the number of small galaxies in the Universe is much larger than the number of massive galaxies, and the manner in which GRBs select galaxies means that their hosts are frequently expected to be faint compared to galaxies found by flux-limited surveys, even if GRB formation is environmentally-independent. A more quantitative treatment was adopted by Wolf and Podsiadlowski (2007), who used archival data from a sample of hosts at $0.2 < z < 1.0$ to calculate the rest-frame optical luminosity distribution for GRB hosts and compare it to a model of the star-formation weighted galaxy luminosity distribution. Consistent with earlier reports, they found a deficiency of GRBs in the brightest galaxies: although the difference is not enormous, and if interpreted as a metallicity "cap" on GRB production they estimated that the limit would have to be approximately Solar ($12 + \log[\mathrm{O/H}]_{\mathrm{KK04}} < 8.7 \pm 0.30$). While these studies primarily examined nearby GRBs, optical-afterglow-selected high-redshift samples from Swift (Laskar et al. 2011) also showed very few luminous hosts.

The colors and SEDs of the population have also attracted interest; the same pre-Swift samples used in the above studies appear to exhibit relatively blue colors, due to low extinctions and star-formation rates that are quite high for galaxies of their mass (Chary et al. 2002; Castro Cerón et al. 2006, 2010; Savaglio et al. 2009). While not quantified in detail, the apparent absence (or at least rarity) of more mature or dusty systems again suggested a trend towards a galaxy population that is metal-poor.

The pre-Swift GRB sample was localized primarily via optical afterglow, and the results above were subject to the significant caveat that they omitted dust-obscured "dark" GRBs (see Sect. 3.7). Such events were already known to exist in significant numbers in the

Swift sample (Cenko et al. 2009; Perley et al. 2010), and a handful of individual cases suggested their hosts may indeed be quite different from host galaxies selected optically (e.g., Levan et al. 2006; Berger et al. 2007; Castro-Tirado et al. 2007; Rol et al. 2007; Hashimoto et al. 2010; Hunt et al. 2011; Svensson et al. 2012). Systematic studies of large numbers of optically-reddened or optically-undetected bursts confirmed this, establishing that most (albeit not all) dust-obscured GRBs are hosted within luminous, massive, reddened systems of exactly the type that GRBs were previously found to be deficient in (Krühler et al. 2011; Rossi et al. 2012; Perley et al. 2013). Even so, heavily dust-obscured GRBs are not large contributors to the GRB population as a whole: events with $A_V > 1$ mag represent perhaps 20 % of the population in total (Greiner et al. 2011). Perley et al. (2013) therefore argued that they were not enough to eliminate the apparent deficiency of GRBs in luminous, red hosts.

Given the clear impact of the selection method on the nature of the host galaxy population that is probed, any proper quantitative assessment of the host population requires selecting a single sample (of both ordinary and obscured bursts) in a uniform manner, free of afterglow-related selection biases. These and similar considerations motivated the construction of several large uniform host-galaxy samples known by various acronyms: TOUGH (Hjorth et al. 2012, containing 69 hosts), BAT6 (Salvaterra et al. 2012; 58 hosts), and SHOALS (Perley et al. 2016a; 119 hosts). In each case, a subset of GRBs from the full Swift catalog is targeted for extensive host follow-up independent of afterglow properties (using the Swift XRT to localize the host if ground-based observations were not acquired) to provide as complete and unbiased a sample as possible. The selection criteria employed in constructing the three samples are similar, usually combining a Sun-distance constraint, a foreground-extinction limit, and a requirement that Swift slewed to the position within a short timeframe in order to acquire XRT observations. BAT6 and SHOALS also require a minimum peak flux and fluence (respectively) to exclude faint bursts although TOUGH does not, while TOUGH and BAT6 (but not SHOALS) exclude bursts near the celestial poles. By eliminating bursts with poor-observability in this way and pursuing spectroscopic redshift measurements for any unknown-redshift GRBs which pass the cuts (which are typically dark GRBs), these techniques have been able to raise the redshift completeness from an initial value of ~ 30 % (for the unrestricted Swift sample) to 90 % or better.

Vergani et al. (2015) analyzed the K-band luminosity distribution of low-z GRBs from BAT6, and found that even in this unbiased sample a strong deficiency of massive galaxies remained. Since the NIR luminosity is driven primarily by stellar mass, which in turn is strongly correlated with metallicity, this can be interpreted as evidence in favor of a metallicity bias, although does not necessarily rule out other models. The role of metallicity specifically can be tested by extending this analysis to higher redshifts: thanks to the evolving mass-metallicity relation (Tremonti et al. 2004; Erb et al. 2006), higher-z galaxies are more metal-poor at a given mass than lower-z galaxies, and the mass- and luminosity trends should therefore evolve with redshift in a consistent way (Fynbo et al. 2006; Kocevski et al. 2009).

This behavior has recently been confirmed by Perley et al. (2016b), who measured the luminosity/mass "ceiling" for efficient GRB production within a galaxy and found a significant increase in this value with redshift, to a degree consistent with a simple model in which the GRB rate is uniform with respect to metallicity below a critical value of $\log[\mathrm{O/H}] = 8.94$ (under the Kobulnicky and Kewley 2004 system) but drops by about an order of magnitude in galaxies more metal rich than this (Fig. 1). This critical value is somewhat higher than, but consistent with, the estimate by Wolf and Podsiadlowski (2007) obtained many years previously. A redshift-dependent luminosity cutoff has also been observed using rest-frame

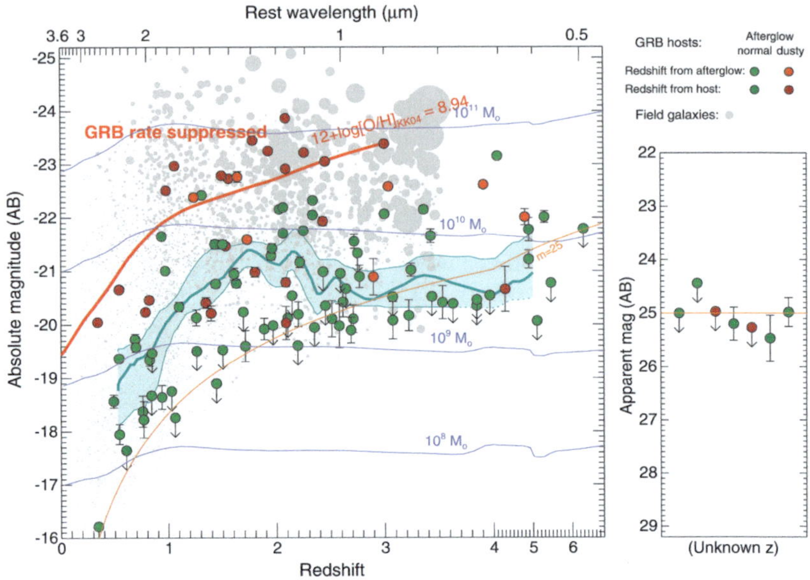

Fig. 1 Near-infrared luminosities of GRB hosts as a function of redshift for a large and unbiased sample of 119 GRB hosts from the SHOALS sample (Perley et al. 2016a,b), compared to star-forming galaxies (*gray*, from Kajisawa et al. 2011). The NIR luminosity can be used as a stellar mass proxy: the *horizontal blue curves* indicate equivalent stellar masses. GRBs sample galaxies of all masses and redshifts, but rarely occur in the most luminous galaxies, especially at low redshift ($z < 1.5$). This is probably because GRBs are strongly suppressed in metal-rich galaxies, leading to a soft "upper limit" on the host stellar mass that increases with z due to the evolving mass-metallicity relation: the *red curve* shows the luminosity of a galaxy at the metallicity threshold of $12 + \log[O/H] = 8.94$ (the value inferred by Perley et al. 2016b) as calculated using the mass-metallicity relation of Zahid et al. (2014). The strong correlation between host luminosity and the degree of attenuation of the afterglow can also be seen: nearly all GRBs with dusty or "dark" afterglows are hosted within galaxies at the upper end of the mass distribution

UV luminosities, which should also correlate with stellar mass and metallicity (albeit with significantly more scatter). Schulze et al. (2015) analyzed the distribution of UV luminosities for galaxies within TOUGH and found a strong preference towards faint galaxies at low redshift but not at high redshifts. Consistently, Greiner et al. (2015) found good agreement between the GRB host and LBG luminosity functions at $z \sim 3$. Curiously, however, at *very* high redshifts of $z > 5$, Schulze et al. (2015) again found a trend towards fainter galaxies: this is difficult to interpret and will need to be confirmed with larger samples.

2.3 Long-Wavelength Observations

Galaxies emit radiation beyond the UV/optical/near-infrared range as well: in particular, many of the most luminous galaxies contain copious amounts of dust that reprocesses nearly all of the UV starlight into FIR/submillimeter radiation (we will loosely refer to these types of galaxies as DSFGs, for dusty star-forming galaxies; Casey et al. 2014). The star-forming conditions in these systems are among the most extreme in the Universe, and the GRB fraction originating from these galaxies represents another way to investigate the impact of local conditions on the GRB efficiency.

The literature on this topic can be confusing: the importance of selection bias, the limited sensitivities of observations at these wavelengths and small samples, and the controversies over the exact contribution of luminous and dusty galaxies to cosmic star-formation in the first place often lead to contradictory conclusions (Berger et al. 2003; Tanvir et al. 2004; Le Floc'h et al. 2006; Michałowski et al. 2012). Studies using large samples (including dark bursts) and sensitive instruments (Perley et al. 2013, 2015; Hunt et al. 2014; Schady et al. 2014; Kohn et al. 2015) indicate that DSFG hosts are not uncommon, but still constitute a small minority ($\sim$ 10–30 %) of the entire GRB host population. This is similar to the fraction of cosmic star-formation found in these systems, so GRBs neither strongly prefer nor avoid DSFGs overall. On the other hand, GRB-selected DSFGs do not look like field-selected DSFGs: they are much lower in mass, but younger, higher in sSFR and in dust temperature (Michałowski et al. 2008; Perley et al. 2015). This could indicate that the apparent agreement in overall rate is in fact a coincidence: GRBs avoid metal-rich, dusty DSFGs but are strongly preferred in intensely star-forming environments within the less-massive and younger subset of the population.

More detailed investigation will be needed to confirm this trend. The Atacama Large Millimeter Array (ALMA) is expected to play major role in characterizing obscured star formation within GRB hosts in the future, since it is able to detect submillimeter emission from even "normal" ($\sim L_*$) galaxies out to very high redshifts. Indeed, ALMA observations have already detected submillimeter emission from a few GRB hosts, even out to $z \sim 3$ (Wang et al. 2012; Hatsukade et al. 2014). Observations of a larger sample will permit the full demographics of the population at these wavelengths to be studied in the near future.

Recently, several observers have also explored the possibility of studying GRB host galaxies in long-wavelength line emission, in particular in HI (Michałowski et al. 2015) and CO (Hatsukade et al. 2014). Samples of objects studied via these methods remain very small, but appear to be consistent with a general picture in which GRB hosts, and GRB sites within their hosts, are typically gas-rich but poor in metals and deficient in molecular gas.

2.4 Spectroscopic Properties

Spectroscopic measurements are more time-intensive than photometric ones and until recently have been limited to only the lowest-redshift GRBs, so the available catalog of spectroscopic host measurements is significantly smaller than what is known from photometric studies. Nevertheless, given the theoretical arguments presented earlier, the observed metallicity of the GRB host (which is estimated via emission-line ratios measured from optical spectra) is a key parameter of interest from the point of view of distinguishing GRB models, and so spectroscopy is critical to establish a firm understanding of GRB formation.

Some GRB hosts have been known to be very metal poor almost since the earliest studies. Stanek et al. (2006) analyzed a small sample and proposed a metallicity-dependent limit on the GRB luminosity (i.e., luminous GRBs can be produced only by very metal-poor galaxies, although less luminous GRBs can form at higher metallicity). This specific model has not been borne out by subsequent studies (Levesque et al. 2010b): nevertheless, the more general notion that the GRB rate is lower in metal-rich galaxies does appear to be supported. In particular, Modjaz et al. (2008) compared the metallicities of the hosts of GRBs to those of SNe Ic-BL without GRBs (GRBs are always or almost always associated with SN Ic-BL, but SN Ic-BL are observed without associated GRBs, even off-axis ones: Soderberg et al. 2006). They found that GRBs were hosted in significantly metal-poorer systems than the SNe Ic-BL: in fact, the two populations seemed to be divided at a critical value (of approximately $\log[O/H] = 8.5$ on the Kewley and Dopita 2002 scale) above which SNe Ic-BL were never associated with GRBs and below which they were always associated with GRBs. Both the GRB and SN samples were small, however, and follow-up of larger samples of GRBs including dark bursts has demonstrated that it is not quite this simple. In particular, several dark GRB hosts, as well as some more recent ordinary GRB hosts, have been found to be metal-rich (Levesque et al. 2010a; Elliott et al. 2013) and SN Ib/c without GRBs have been found in very metal-poor environments as well (Modjaz et al. 2011), so the most up-to-date GRB vs. star-formation and GRB vs. SN metallicity comparisons (Graham and Fruchter 2013) do show significant overlap. Nevertheless, a strong difference in metallicity between GRB hosts and all other classes of transient host remains—suggesting that metallicity plays a key role in encouraging the successful launch of the GRB jet.

Emission-line studies are currently undergoing a renaissance thanks largely to the prolific capabilities of a single new instrument: X-shooter at the VLT (Vernet et al. 2011). X-shooter has observed over 100 GRB hosts, and for the majority of these targets, multiple line detections permit chemical abundance analysis. The existing X-shooter sample is not unbiased, but it includes both obscured/dark and unobscured GRBs, and the derived properties present a picture that is quite similar to what is inferred from the photometric properties of GRB hosts (Krühler et al. 2015; Piranomonte et al. 2015). At every redshift, GRBs in very metal-rich (super-Solar) galaxies are uncommon but present, representing about 10 % of the sample. The median GRB host is moderately enriched ($\log[O/H] = 8.5$), and very metal-poor hosts are again uncommon. These properties further support the conclusion that the metallicity sensitivity of GRBs exhibits a sharp transition at around Solar metallicity that suppresses (but does not necessarily prevent) their formation in metal-rich (super-Solar) galaxies, but does not exert much impact on the rate in low versus moderate metallicity galaxies. It does not necessarily rule out the influence of other parameters on the GRB rate—however, the small number of luminous GRB hosts with measured metallicities tend to be metal-poorer than field-selected star-forming galaxies of the same mass (Levesque et al. 2010c; Graham and Fruchter 2013; Hashimoto et al. 2015), providing some evidence that metallicity is a dominant factor.

When interpreting these results, it should be kept in mind that individual galaxies also have significant internal chemical heterogeneity (e.g., Afflerbach et al. 1997; Smartt et al. 2001; Cioni 2009; Sanders et al. 2012), and the measured metallicity of a GRB host galaxy may therefore be different from the actual metallicity of the GRB progenitor. Niino (2011) predicted the expected metallicity distribution of long GRB host galaxies under the assumption in which long GRBs trace *only* low-metallicity stars using a model in which the internal variation of metallicity within galaxies is taken into account, assuming that a similar internal metallicity variation as seen in local galaxies such as the Milky Way also exists within GRB host galaxies. They found that both the existence of a few high metallicity hosts as well as the systematically low metallicities of more typical GRB hosts can be explained at the same time, even if the GRB progenitor is exclusively a moderately metal-poor ($< 0.3Z_\odot$) star (Fig. 2). While these results have not yet been updated to incorporate data from the large, unbiased samples mentioned earlier, they clearly suggest that the "true" metallicity threshold for GRB production is likely to be somewhat lower than the $\sim 1Z_\odot$ value preferred if all galaxies are assumed to be chemically homogeneous. Determining this value will require a careful accounting of the metallicity distribution within and between galaxies at a variety of masses and redshifts in the future. Spatially-resolved spectroscopy (which we will discuss in the next section) will also be informative.

Spectroscopic properties other than strong-line metallicity diagnostics have also been used to investigate the GRB progenitor and rate dependence, although so far the results of these analyses have been relatively inconclusive. For example, early samples suggested that strong Lyman-α emission might be much more common in GRB hosts than in star-forming galaxies generally (Fynbo et al. 2003; Jakobsson et al. 2005), although this was not confirmed by larger and more uniform samples (Milvang-Jensen et al. 2012). Other authors

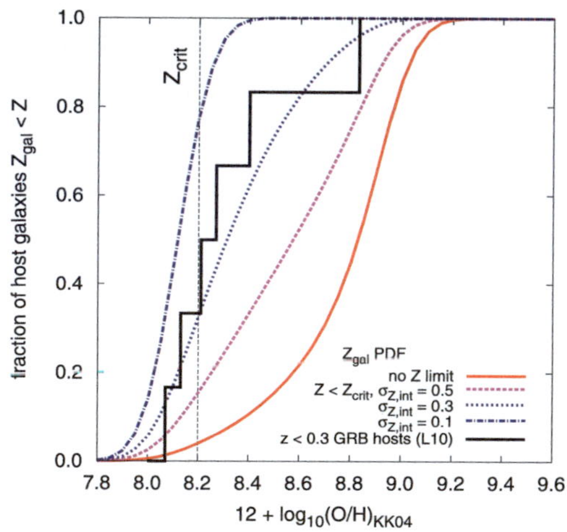

Fig. 2 The metallicity (expressed as the oxygen abundance, $12+\log(O/H)$) distribution of GRB host galaxies predicted in Niino (2011). The *red solid curve* represents the case in which long GRBs occur at the same rate regardless of metallicity. *Dashed*, *dotted*, and *dot-dashed curves* (*magenta*, *blue*, and *dark-blue*, respectively) represent cases in which only low metallicity stars below the assumed metallicity threshold of $12+\log(O/H) = 8.2$ (shown with vertical dashed line) produce GRBs when they collapse, and stars formed in each galaxy have variation of metallicity with $\sigma = 0.5$, 0.3, and 0.1 dex around the mean value of the galaxy, respectively. Due to the internal metallicity variation within each galaxy, GRB host galaxies frequently have higher metallicity than the assumed threshold. Metallicities of observed GRB host galaxies at $z < 0.3$ (from Levesque et al. 2010c, using the calibration of Kobulnicky and Kewley 2004, KK04) are also shown

have searched for Wolf-Rayet and other very short-lived starburst features in nearby hosts in an attempt to constrain the population age (Han et al. 2010) and found some indication that the progenitor is likely to be very young, but the small and possibly biased nature of the samples suitable for this analysis makes it difficult to make firm conclusions about the entire GRB population. Finally, detailed analysis of the rich data set provided by the large X-shooter host spectroscopic sample—which also provided information on specific star-formation rates, ionization states, and velocities–is likely to be a promising avenue for host investigations in the future (see also Krühler et al. 2015), although comparison of this type of spectroscopic data against field galaxy samples is not straightforward, since the samples of both GRB hosts and high-redshift field galaxies chosen to be targeted for deep spectroscopic follow-up are subject to complex biases.

2.5 Spatially Resolved Analysis

Most GRBs are sufficiently distant that their host galaxies appear only as marginally-extended point sources in typical ground-based seeing conditions and it is difficult to extract much information about their internal structure without spaceborne observations. However, a small number of GRB hosts are close enough that ground-based spatially-resolved analysis has been possible (Christensen et al. 2008; Thöne et al. 2008; Levesque et al. 2010a, 2011); Fig. 3 shows five of the nearest and best-resolved examples. Observations of these systems have given some support to the conclusions of Niino (2011), showing that in at least some instances the metallicity at the GRB site may be (slightly) lower than the metallicity of the host overall. However, even in these cases, the spatial resolution of these observations is limited to kpc scales. Significant metallicity variation within a galaxy is observed even on scales of < 1 kpc in the local universe (33 % of neighboring pairs of HII regions with

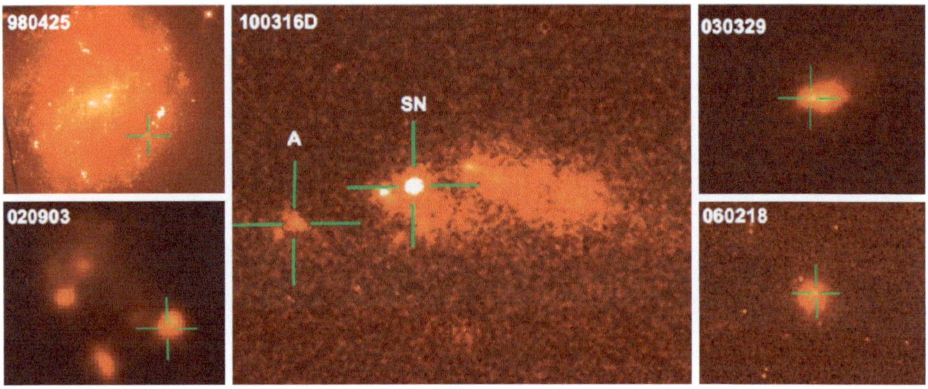

Fig. 3 Host galaxies of five of the nearest GRB host galaxies (from Starling et al. 2011). Most GRB hosts at $z \sim 0$ are irregular, compact, star-forming galaxies—quite unlike the host population of core-collapse supernovae, which is dominated by more massive spiral galaxies (Fruchter et al. 2006). Image reproduced with permission from Fig. 8 of Starling et al. (2011)

distance < 0.5 kpc in M31 have different oxygen abundance by more than 0.3 dex, Sanders et al. 2012). If such small scale variation of metallicity exists in GRB host galaxies, the metallicities observed on a few kpc scale may still be different from that of the direct environment (and the progenitor) of the long GRBs (Niino et al. 2015), a possibility that should be incorporated in future efforts.

For the vast majority of GRB hosts, only space-based observations (specifically, observations from the Hubble Space Telescope) provide a hope of spatially resolving the galaxy. While it is not yet practical to perform spatially resolved spectroscopy of GRB hosts from space, HST observations have nevertheless been instrumental in constraining the identity of the GRB progenitor in other ways—helping to constrain its age, and providing some evidence for influences beyond metallicity on the GRB rate.

Fruchter et al. (2006) and Wainwright et al. (2007) analyzed the morphologies of GRB hosts within a moderately large sample of pre-Swift observations with HST, and noted that GRB hosts are smaller than other galaxies and have a higher abundance of irregular and merging systems compared to the population expected for a "uniform" tracer. This behavior is expected for a metallicity-biased tracer, although morphologies are difficult to quantify, making it difficult to directly distinguish between progenitor models using this observation. More recently, Kelly et al. (2014) carried out a similar but more quantitative analysis, measuring the characteristic sizes (half-light radii) of galaxies in the GRB host sample in comparison to a variety of massive-stellar explosions (SN Ib/c and SN II). Consistent with earlier work, they found that the GRB hosts are significantly smaller. In particular, they found that they are smaller even at the same stellar mass. As the size of a galaxy at a fixed stellar mass does not correlate with metallicity significantly, this seems to provide strong evidence that the GRB progenitor must prefer dense stellar environments *in addition to* any preference towards low metallicity.

The location of a GRB within a resolved map of its host galaxy can also be used to constrain the progenitor, by studying whether or not it exhibits a preference or aversion for different types of environment on subgalactic scales. Such an analysis was first carried out by Bloom et al. (2002) using the nucleus-site offset distribution for a sample of 20 GRBs; they concluded that GRBs trace the galaxy light. Analysis based on the offset distribution alone is not sensitive to azimuthal structure of the source, however, and these data were subsequently re-analyzed by Fruchter et al. (2006) (and, later, by Svensson et al. 2010 and Blanchard et al. 2015 with expanded samples), who compared the UV surface brightness at the GRB host position compared to that of the galaxy as a whole, pixel-by-pixel. They found a significant preference for brighter regions (pixels) of the galaxy. Importantly, the same trend is not seen among ordinary core-collapse supernovae (but *is* seen in Type Ib/c supernovae; Kelly et al. 2008).

This result has been interpreted as suggestive of a very short progenitor lifetime (and, possibly, a large initial mass): on subgalactic scales, the sites of the most recent star-formation within a galaxy are expected to be brighter than locations for which a slightly longer time has passed since their formation episode, since the most massive stars have not yet exploded. If the GRB progenitor exploded faster than the "typical" star responsible for a galaxy's UV luminosity—as would be expected if the progenitor star was particularly massive—a concentration towards the brightest (youngest) concentrations of young stars is expected. In this model, the observational constraints suggest an initial mass in the range of > 20–$45 M_{\odot}$ (Larsson et al. 2007; Raskin et al. 2008).

However, it is also possible that is the actual density and not age-brightness effects that are responsible: regions of particularly intense star-formation may be more likely to produce very massive stars or close massive binaries, or interactions in dense stellar nurseries could also play a fundamental role. Indeed, the compact sizes of nearby GRB hosts noted by Kelly et al. (2014) is not well-explained by an age effect, but this result and the concentration result are naturally explained if GRBs preferentially occur in regions of vigorous star-formation.

2.6 GRB Hosts at High Redshift

Although studies of GRB hosts in emission are typically oriented towards using the galaxy properties to better understand the progenitor (e.g., measuring its metallicity tolerance), at the highest redshifts the direction of understanding reverses: instead, it is hoped that by observing the properties of the GRB host distribution we can better understand the properties of star-forming galaxies at that time. Indeed, a majority of the star-formation in the universe at $z \gtrsim 6$ is thought to occur in galaxies too faint to detect even in deep HST observations. However, GRBs from arbitrarily faint and distant galaxies are readily detectable, providing us a way to confirm and constrain these predictions independently. Of course, the GRB rate at high-z will also be affected to some degree by the progenitor's dependence on metallicity. But as high-z galaxies are thought to be safely well below the 0.3–1.0$Z_{\odot}$ threshold discussed above, there is reason to think that the impact of this limitation will be rather small in practice.

Until recently, searches for GRB hosts in emission had failed to find any above $z \sim 5$ (e.g., Stanway et al. 2011; Berger 2014). This is consistent with the bulk of star formation occurring in very faint galaxies in this era (Tanvir et al. 2012), indicative of a steep faint end slope of the galaxy luminosity function, as has also been found in successively deeper *HST* deep field campaigns (e.g., Bouwens et al. 2015). Since a steep luminosity function would allow a global star formation rate factors of several above that seen in galaxies detected to *Hubble* Ultra-Deep Field depths at $z \sim 8$, this reduces the tension between the far-ultraviolet radiation density required to maintain a reionized intergalactic medium, and the production of radiation from massive stars. It should be noted that addressing this problem via GRB hosts has the important advantage that no particular form of the galaxy luminosity function is required, since it ultimately depends only on the fraction of hosts found above the detection threshold (e.g., Trenti et al. 2012).

Recently, new, deeper *HST* observations (program GO13831, PI: Tanvir) have for the first time identified two high-redshift GRB hosts, specifically of GRBs 050904 and 140515A,

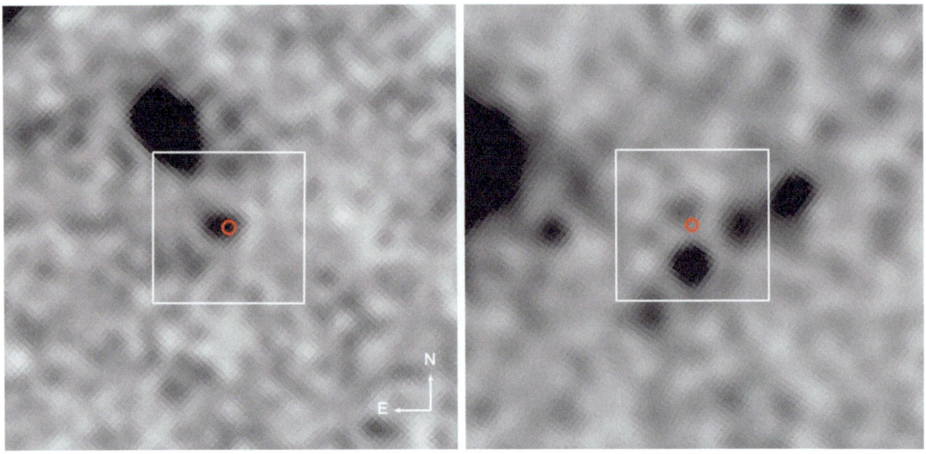

Fig. 4 Deep *HST* F140W filter images (lightly smoothed) of the locations of GRBs 050904 (*left*) and 140515A (*right*), from McGuire et al. (2015). Both bursts had redshifts of $z = 6.3$ and in each case the host is detected at AB magnitude ≈ 28, underlying the GRB positions (indicated by *red circles*). These are the first GRB host galaxies found in emission at $z > 5$. The *white boxes* are 2 arcsec on a side

at $z \approx 6.3$ (Fig. 4; McGuire et al. 2015). The luminosity of these hosts suggests they are consistent with being faint examples of the Lyman-break galaxies found in other deep *HST* surveys, roughly 20 % and 10 % of L_* at that redshift respectively (taking the luminosity function of Bouwens et al. 2015). In each case the afterglow spectroscopy had shown that the metallicities are < 10 % of Solar (Thöne et al. 2013; Hartoog et al. 2015; Chornock et al. 2013), confirming that galaxies at this epoch are well below the threshold on GRB production inferred by lower-redshift studies.

3 Long GRB Host Galaxies in Absorption

Thanks to their exceptional luminosities, GRB afterglows can also be used as probes of their host galaxies in *absorption*. Gas and dust in front of the GRB selectively absorb certain wavelengths of light, imprinting their signatures onto the optical and X-ray spectra of the afterglow where they can be revealed by spectroscopic observations. Other matter along the line of sight is also unveiled, including the inter-galactic medium (IGM) and any foreground galaxies along the sightline.

While this type of study can also be performed with other extragalactic continuum sources (in particular, QSOs), there are many factors that make GRBs particularly powerful absorption probes of galaxy evolution and the high redshift universe, even into the reionization epoch. First of all, GRBs and their afterglows can be extremely luminous (much more so than QSOs), making them accessible to current-day optical spectrographs even out to very high redshifts (Lamb and Reichart 2000; Bloom et al. 2009). Because GRBs are associated with the death of massive stars, they are indeed expected to exist at extremely high redshifts of $10 < z < 20$ (whereas QSOs, which require time for supermassive black holes to assemble, may not), so they uniquely probe chemical enrichment far into the epoch of reionization. But even at lower redshifts, GRBs have advantages. Their association with massive stars also means they are located within the disks of typical star-forming galaxies and provide a more effective means to select high-density gas columns (Fig. 5; Pontzen et al.

2010). Because the afterglow continuum is intrinsically expected to be a simple power-law, it can be used to constrain broad absorption features (such as DLA wings or, especially, extinction profiles) more reliably than QSOs, which have complex intrinsic spectra. Finally, because the afterglow emission fades away over time, the same GRB host galaxy characterized in absorption can also be studied in emission using the techniques outlined in the previous section. In particular, GRBs offer a unique opportunity to systematically investigate both the neutral gas using absorption lines in the afterglow spectra *and* the ionized gas using the emission lines in the host galaxy spectra.

The transient nature of GRBs is, on the other hand, also the main disadvantage of the use of these sources as cosmological probes. The optical/near-infrared afterglow observations necessary to provide a position of sufficient accuracy ($< 0.5''$) to point an optical spectrograph, as well as the spectroscopic observations themselves, must take place as early as possible after the GRB prompt emission detection. And unlike with QSOs, it is not possible to carry out observations over long integration times (more than a few hours) in order to increase the signal-to-noise (S/N) for faint objects. To date, less than 30 % of GRBs have a measured redshift, and less than ~10 % have a spectrum with a high S/N sufficient for detailed analysis of faint lines and uncommon chemical species.

To achieve the scientific goals mentioned above, medium/high resolution afterglow spectra with good S/N are necessary. Observations starting as soon after the GRB trigger as possible—when the afterglow is brightest—provide the best signal and are richer in information than observations taken later, so fast response to new GRBs is essential. For these reasons, target-of-opportunity afterglow follow-up programs exist at a variety of telescopes around the world, including Keck, Gemini, GTC, Subaru, and Magellan. However, it is the ESO Very Large Telescope (VLT) that has been the driving force in GRB afterglow spectroscopy for much of the past decade, for a variety of reasons. First, year-round target-of-opportunity observations (including dedicated rapid response-mode observations executed by the Observatory staff) are available. Even more importantly, X-shooter has been just as revolutionary in the study of GRB afterglows in absorption as it has been in studies of GRB hosts in emission: its wide wavelength coverage and ideal combination of resolution and sensitivity have permitted detailed studies of the environments of even faint and high-redshift afterglows.

3.1 Damped Lyman Alpha Systems

The most prominent absorption line seen in high-redshift GRB spectra is, unsurprisingly, the Lyman-α line of neutral hydrogen (HI) at 1217 Å, which is accessible to ground-based spectrographs at redshifts higher than approximately $z \gtrsim 1.9$. The strength of this line causes it to inevitably saturate, but the absorption column can be derived by fitting the wings of the profile using the same techniques derived for QSOs, producing a direct and accurate measurement of the neutral hydrogen column density N_{HI}. In addition to this Lyman-alpha absorption, the afterglow spectra also typically reveal many metal lines at the same redshift, present in both their low and high ionization state. It is also occasionally possible to detect the absorption line from molecules such as H_2, as well as fine structure lines associated with gas excited by the GRB afterglow itself. The combination of a neutral gas column and measurements of numerous metal lines can be used for a variety of purposes, described in the following sections.

3.2 Metallicity

The metallicity of the intervening gas towards a GRB can be directly measured using the ratio of the strengths of metal lines and HI in absorption, even in galaxies up to the highest

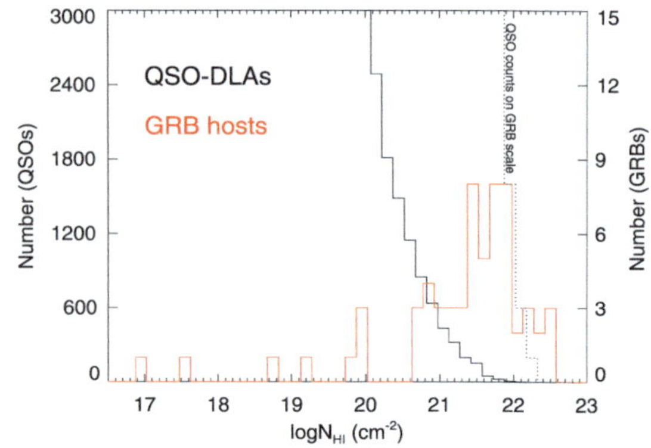

Fig. 5 HI column density measurements for more than 12000 QSO DLAs and for 52 GRB absorbers. The QSO histogram is plotted with a full-drawn line relative to the left-hand ordinate axis and with a *dotted line* relative to the right-hand ordinate axis. As seen, typical GRBs probe column densities that are extremely rare or absent among QSO-DLAs

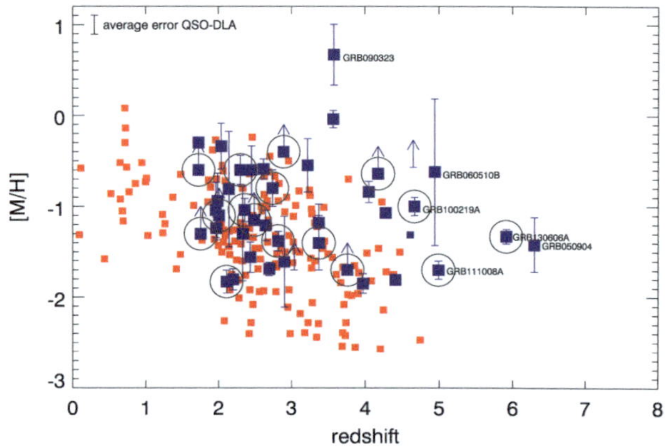

Fig. 6 Absorption-based metallicity measurements as a function of redshift for QSO intervening-system DLAs (*red symbols*) and GRB host DLAs (*blue symbols*). The two populations evolve differently with redshift, probably because they probe different locations within the galaxy: with GRB DLAs the metallicity of the inner regions of star-forming galaxies are probed, while QSO DLAs typically probe more outlying regions of foreground galaxies. Data from Thöne et al. (2013) and Sparre et al. (2014)

redshift. For example, X-shooter observations have succeeded in measuring the metallicity of GRB 130606A at $z = 5.9$ (Hartoog et al. 2015; see also Chornock et al. 2013 and Totani et al. 2014). By comparing systems at high redshift to others at more moderate redshifts, GRB DLAs allow us to probe the metallicity evolution of the ISM of star-forming galaxies. The data collected to date show that GRBs explode generally in sub-solar metallicity environments—but measured metallicities are not extremely low, even at high redshift, demonstrating that galaxies were already metal enriched (see Fig. 6).

3.3 Dust Depletion

Refractory elements, such as Fe, Ni, and Cr, can be heavily depleted into dust grains (e.g. Savage and Sembach 1996), and thus can be missing from the gas-phase abundances. To estimate the level of depletion in the ISM (and therefore the presence of dust), the relative abundance of heavily depleted species into dust grains, as Fe, towards those undergoing little depletion, as Zn, is used. Then, using different methods (e.g. Vladilo et al. 2006; Bohlin et al. 1978; Prochaska and Wolfe 2002; De Cia et al. 2013), it is possible to estimate the iron dust-phase column density, the dust-to-gas ratio, and the flux attenuation. The depletion patterns can be compared to the Galactic ones to estimate the origin of the gas and of dust, and/or to look for evolution of the dust-to-metal ratio (e.g. De Cia et al. 2013; Zafar and Watson 2013).

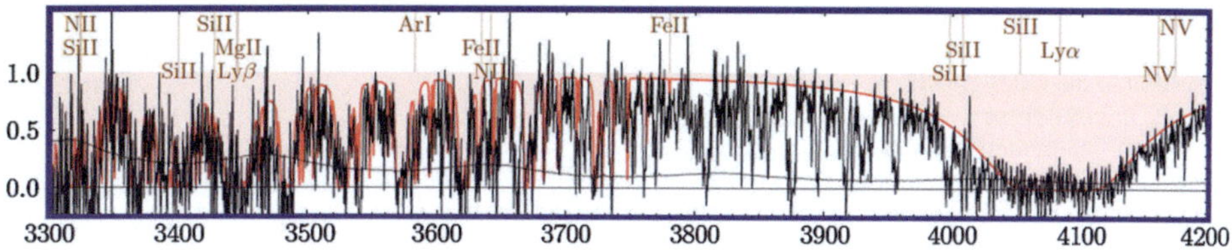

Fig. 7 X-shooter spectrum of GRB 120815A between 3300 and 4200 Å illustrating the presence of H_2 absorption. The *solid red line* denotes the synthetic H_2 model (including HI Lyman-alpha and Lyman-beta absorptions). Also marked are prominent metal absorption lines previously detected in GRB-DLAs and foreground absorbers. From Krühler et al. (2013, Fig. 8); reproduced with permission from Astronomy & Astrophysics, © ESO

GRB afterglows can also be used to study the shape of the dust extinction curve, as will be discussed in Sect. 3.7.

3.4 Molecules

As GRB DLAs should probe the gas associated with star-forming regions, a direct detection of absorption lines due to molecules is expected in their afterglow spectra. The detection of absorption lines from rotationally and vibrationally-excited H_2 transitions, however, has proven to be difficult: these H_2 transitions are typically blended with the Lyman-alpha forest, and disentangling the often-faint H_2 features from the forest is much easier in spectra with high spectral resolution and S/N. This favors bright afterglows. Physically, however, the H_2 column density is correlated with metallicity and dust depletion, and a large H_2 column is more likely for sightlines that are significantly obscured (and therefore much fainter in the rest-frame UV). As a result, afterglows that are both bright and heavily H_2-absorbed are uncommon. The combination of resolution and sensitivity available with the X-shooter spectrograph should make the detections of molecules more frequent than in the past. Indeed, three detections of H_2 and one of CH^+ in X-shooter afterglow spectra have been published to date (Fig. 7; Krühler et al. 2013; D'Elia et al. 2014; Friis et al. 2015; Fynbo et al. 2014), compared to only one robust detection of molecules in GRB DLAs before the advent of X-shooter (Prochaska et al. 2009).

In general, the fraction of H_2 found along GRB sightlines seems to be quite low compared to the HI content. The comparison of the content of molecules and neutral hydrogen may be useful to better understand the processes triggering star-formation, especially if extended to HI and CO detections (or limits) obtained at millimeter and radio wavelengths with ALMA and ATCA (e.g. Hatsukade et al. 2014; Michałowski et al. 2015).

3.5 Distance of the Gas from the GRB

The GRB afterglow radiation is intense enough to have an important impact on its environment at the time of the explosion. In particular, the UV radiation ionizes the neutral gas (e.g. Perna and Lazzati 2002) and destroys molecules and dust grains up to tens of parsecs away (Waxman and Draine 2000; Draine and Hao 2002). The metastable states of existing species (O I, Si II, Fe II) are populated by UV pumping followed by radiative cascade (Prochaska et al. 2006; Vreeswijk et al. 2008).

From the presence or absence of neutral, low and high ionization metal line absorptions, and thanks to the different fine-structure and metastable line transitions (due to UV pumping followed by radiative cascade) it is possible to establish the distances of the gas clouds absorbing the afterglow light. Indeed, because the GRB afterglow fades rapidly, recombination

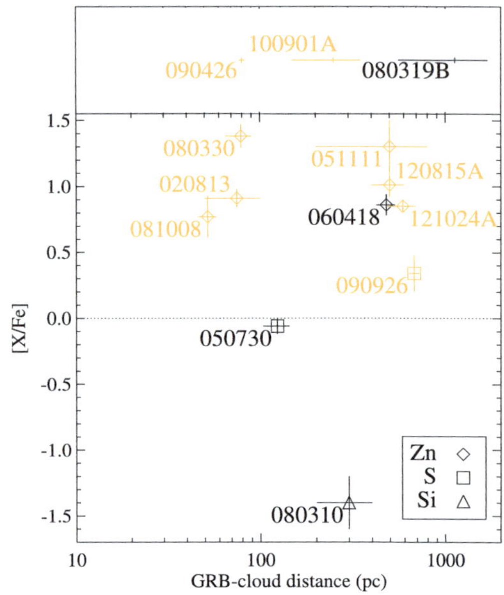

Fig. 8 The abundance ratio of zinc, sulphur or silicon over iron, compared to the GRB-absorber distance inferred from absorption-line modelling (*black points* come from high-resolution spectra and are more reliable). Figure kindly provided by A. De Cia; see also Vreeswijk et al. (2012)

prevails and the populations of the metastable levels change, making these absorptions vary and disappear. The detection of these time-dependent processes, with timescales ranging from seconds to days in the observer frame, can lead to interesting information on the burst itself and on the ISM of the host, in particular on the distance of the absorbing clouds.

To date, this analysis has been possible for $\sim$ 10 GRBs. It demonstrated that the gas is at least at $\sim$ 100 pc from the GRB explosion site (see Fig. 8), and that the GRB ionizes its surrounding gas at least up to 20 pc (Vreeswijk et al. 2008, 2012; Prochaska et al. 2008; D'Elia et al. 2014). Therefore, through optical spectroscopy, we are not probing the close environment of the GRB progenitor, but possibly its star-forming region, the inner region of its host galaxy and its circumgalactic gas.

Many attempts to find signatures of the close (pc-scale) GRB environment have been carried out. Unfortunately, there are no robust identifications to date, and possible evidences have been found only in a couple of cases (Fox et al. 2008; Castro-Tirado et al. 2010, but see Chen et al. 2007). On the other hand, by studying the profile of high ionization lines, some evidence of outflows has been found (Fox et al. 2008).

3.6 Comparison Between ISM and Ionized Gas

Once the afterglow vanishes, it is possible to carry out photometric and spectroscopic campaigns to study the properties of the GRB host galaxy. Most exciting is the combination of the gas properties obtained by the afterglow spectroscopy and those retrieved by the host observations. A systematic study in this sense can provide important information on the physical processes of galaxy evolution and cosmological star-formation. The kinematics and geometry of the gas can be assessed as well, looking for inflow/outflow signatures. The physical state of the gas and its connection with the star-formation activity in the host galaxy can also be modeled. It is possible to obtain unique information to understand the link between the properties of the ISM and star formation activity at any redshift.

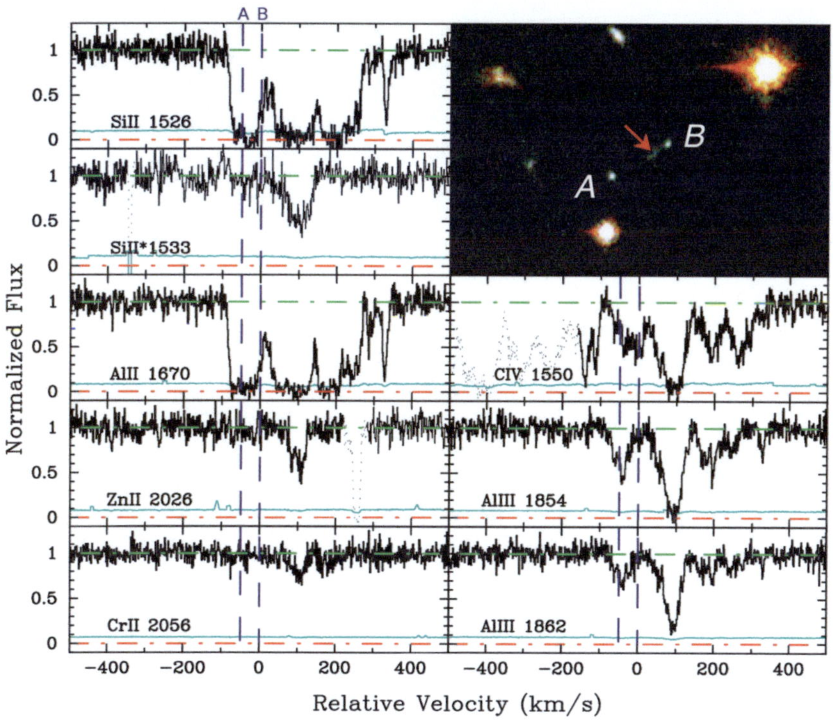

Fig. 9 HST image of the field of GRB050820 at $z = 2.6$, superimposed on some absorption lines from the afterglow spectrum. The *arrow* indicates the afterglow position. The comparison between the velocities of the afterglow absorption lines and the emission lines of interacting galaxies A and B (represented by *blue-dashed lines*) indicates that the GRB occurred in an extension of galaxy B. Image adapted with permission from Figs. 1 and 3 of Chen (2012)

First attempts in this direction have been performed by Vergani et al. (2011) and Chen (2012) (see Fig. 9). With the larger samples of afterglow and host galaxy spectra now available, it will be possible to perform more systematic studies on this topic.

3.7 Extinction

The issue of dust extinction in GRB afterglows has been central in the study of GRBs, their afterglows, host galaxies and birth sites since the first afterglows were detected in 1997 (e.g., Groot et al. 1998; Djorgovski et al. 2001). The advantage of GRB afterglows for the study of extinction is (at least) threefold: (i) afterglows can be extremely bright, allowing for actual detection of the spectra even when affected by substantial dust extinction; (ii) afterglows have intrinsic relatively simple spectra consisting of power-law segments; (iii) afterglows cover a very broad range of frequencies from radio to gamma-rays allowing for a study of the effects of dust over this broad range of frequencies.

A rich literature on the issue of extinction in GRB afterglows has developed over the past 20 years. Early studies focused on the issue of "dark" bursts (Groot et al. 1998; Djorgovski et al. 2001; Fynbo et al. 2001; Lazzati et al. 2002), the name given to bursts that evaded detection in the optical. Evidence then already suggested that these events were fairly common (perhaps as much as 50 % of the sample) and that a majority of these are partially or entirely the result of dust extinction. A study of dust-extinction in pre-Swift afterglows can be found in Kann et al. (2006). However, because follow-up of pre-Swift GRBs was usually delayed until the afterglow faded significantly and because observations at dust-unobscured (e.g. X-ray) wavelengths could not always be obtained, large and systematically-complete analyses of dust extinction were not possible in the early years.

Swift has allowed much larger and more complete studies to be built. First of all the concept of *dark burst* became more precisely defined based on the expected power-law slopes of GRB afterglows between the optical and X-ray bands (Jakobsson et al. 2004; Rol et al. 2005; van der Horst et al. 2009). Examples of single-event studies of well observed and dust extinguished dark bursts are in Watson et al. (2006), Krühler et al. (2008), Elíasdóttir et al. (2009), Prochaska et al. (2009). Efforts to build complete samples in particular allowed for more robust statistics on the amount of extinction towards GRBs: in particular the fraction of dark bursts (as defined, e.g., by Jakobsson et al. 2004) was found to be about ~ 40 %, and about half of these (i.e., ~ 20 % of GRBs) are *very* obscured (e.g., Cenko et al. 2009; Perley et al. 2009; Fynbo et al. 2009; Kann et al. 2010; Greiner et al. 2011; Hjorth et al. 2012; Melandri et al. 2012; Perley et al. 2013; Covino et al. 2013; Littlejohns et al. 2015).

An important, still open issue is that of photoelectric absorption and dust destruction. Galama and Wijers (2001) first noted that there were cases of afterglows with very strong photoelectric absorption in the X-rays and yet apparently rather limited extinction in the optical. This raised the issue of dust destruction (e.g., Fruchter et al. 2001). The issue of photoelectric absorption in the X-rays is further discussed in several dedicated papers (Behar et al. 2011; Watson and Jakobsson 2012; Watson et al. 2013; Starling et al. 2013; Krongold and Prochaska 2013), but without consensus. A possibly related phenomenon is bursts with high dust depletion of refractory elements, but limited extinction in the optical (Savaglio et al. 2003; Savaglio and Fall 2004; Perley et al. 2008; Friis et al. 2015).

More detailed studies of extinction curves derived from afterglow spectra are presented in several sample papers: Starling et al. (2007), Zafar et al. (2011), Schady et al. (2010, 2012), Covino et al. (2013), Japelj et al. (2015). Extinction curves similar to those found towards stars in the Small Magellanic Cloud give the best fit for most GRB sightlines, but the 2175 Å extinction bump is detected towards a handful of bursts.

Some GRB sightlines show evidence for extinction laws with no local analog: for example, the apparently-grey extinction towards GRB 061126 (Perley et al. 2008) or the unusual extinction towards the high-redshift GRB 071025 (Perley et al. 2010). Perhaps the most exotic system yet is that of GRB 140506A, whose extinction curve has been measured using X-shooter spectroscopy (Fynbo et al. 2014). This sightline shows several peculiarities: (i) absorption lines from excited hydrogen and helium, (ii) molecular absorption from CH^+, and most importantly for this discussion, (iii) very strong dust reddening bluewards of about 4000 Å in the rest frame (see Fig. 10). Furthermore, the afterglow was observed on two consecutive nights and significant change in the shape of the reddening was found.

This strong reddening cannot be fitted by extinction curves known from the Local Group and its nature remains somewhat mysterious. It has some similarity to the sightline to SN 2014J in the starburst galaxy M82 (Amanullah et al. 2014). Along this sightline similar absorption line properties are also seen with strong Calcium absorption and absorption from CH^+ (Ritchey et al. 2014). Another place where similar reddening has been observed is towards some QSOs (e.g., Hall et al. 2002; Leighly et al. 2014): the extinction curves for some reddened QSOs appear to be very steep (e.g., Fynbo et al. 2013) and one of the most remarkable QSOs in this regard, CQ0127+0114 (Hall et al. 2002), can be fitted with a similar prescription as the supernova in M82. There is also evidence for different extinction towards the central regions of the Galaxy (e.g., Sumi 2004; Nishiyama et al. 2008, 2009;

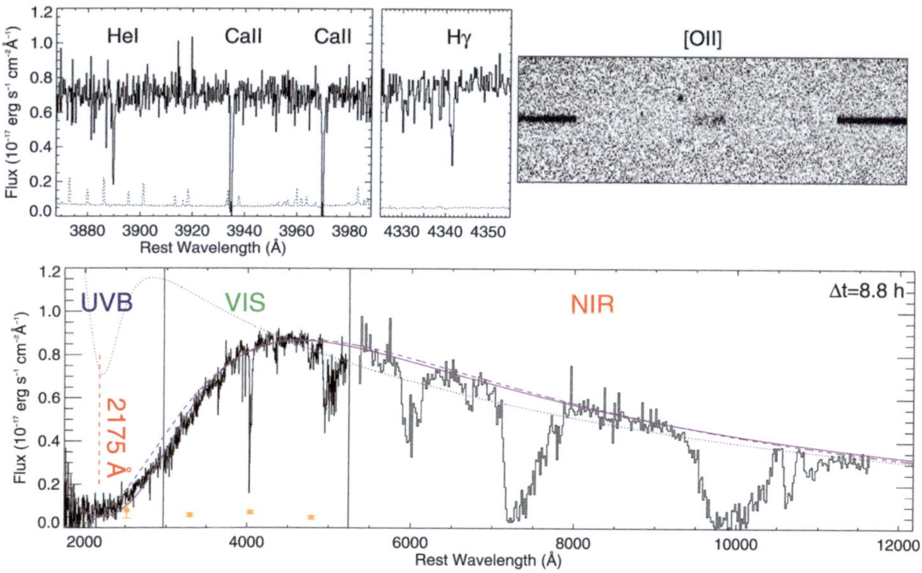

Fig. 10 Spectroscopy of GRB 140506A (Fynbo et al. 2014) and its unusual sightline. *Bottom panel*: An X-shooter spectrum taken 8.8 hr after the burst covering the range from about 2000 to 12000 Å in the rest frame of the $z = 0.889$ GRB is plotted in *black* and late time host photometry is overplotted in orange. The spectrum shows a very steep drop in the blue, which is interpreted as the result of unusual extinction, either an extremely strong 2175 Å extinction bump (the *dotted purple line* shows a fit with the MW extinction curve and the full drawn line an extreme 2175 Å extinction bump) or more likely extinction similar to what is also seen towards some type Ia supernovae and AGN (the *purple dashed line*, see Fynbo et al. 2014 for a full discussion). *Top left:* The spectrum also shows very unusual absorption lines including absorption from excited helium and strong Calcium absorption as well as hydrogen Balmer lines and molecular CH^+ lines (the latter not shown here, but in Fynbo et al. 2014)—never seen before in any afterglow spectra. *Top right:* Emission lines from the underlying host galaxy visible under the light of the afterglow; here [OII]

Gosling et al. 2009; Nataf et al. 2015) and taken all together there seems to be a hint that extinction towards the central regions of galaxies is different and in particular steeper.

GRBs are strongly centered on the light of their host galaxies (Fruchter et al. 2006) and hence often probe the central regions of their hosts. Furthermore, there are still a significant number of bursts every year where we do not detect any afterglow light in the optical or near-IR so it appears quite plausible that this kind of unusual extinction could be more frequent for GRB sightlines than this (so far) only detection suggests. In fact, there is evidence that similar extinction was observed towards GRB 070318 (Watson 2009). Future GRBs may provide additional constraints on unusual extinction of this type and others.

4 Conclusions

GRBs, their afterglows, and their host galaxies provide powerful means of studying both the nearby Universe and the most distant galaxies. Studies of the population in different ways using different techniques (in emission and absorption, photometrically and spectroscopically, and at a wide range of wavelengths) provide unique insight into the factors governing GRB production and the nature and evolution of galaxies in the Universe.

Although the subset of GRBs with bright optical afterglows probes a low-metallicity, low-luminosity, low-mass host population compared to the population of galaxies that provide the bulk of the Universe's star-formation over most of its history, optically-obscured dark GRBs probe a redder and more massive galaxy population and are essential to provide a complete view of the GRB host population. Even with these included, GRBs are not perfect tracers of all of cosmic star-formation: their lower abundance in high-mass and metal-rich galaxies and a tendency towards physically smaller systems point to the existence of physical factors that operate to favor the production of GRBs (relative to other end-states of massive stellar evolution) in some environments over others. Metallicity is a leading candidate, and a model in which the GRB rate relative to star-formation is heavily suppressed above approximately solar metallicity but largely constant at lower metallicity appears to provide good consistency with photometric and spectroscopic observations of the GRB host population. On the other hand, other aspects of the host population are harder to explain by this model alone: the small sizes of GRB hosts, the concentration of GRB sites on the brightest star-forming regions, and the low-mass nature of the most luminous host galaxies may indicate that stellar density may play a supporting role in encouraging GRB production. These possibilities will soon be tested by detailed analysis of the large and uniform samples that have recently been acquired, and by more sophisticated modeling and analysis techniques (including the effects of chemical inhomogeneity within galaxies).

In either case, these environmental influences are expected, and observed, to become less significant at higher redshifts, where galaxies are uniformly small, star-forming, and metal-poor. GRBs are already being used in this way to place independent constraints on the sites of cosmic star-formation and sources of reionization. The nondetection of most $z > 5$ GRB hosts in very deep HST observations is fully consistent with the leading view that reionization was driven by galaxies too faint to be currently observed—as is the recent detection of two $z \sim 6$ GRB hosts.

GRB host studies using afterglow spectroscopy provide information independent of that deriving from the host light. Very detailed information about metal abundances, molecular content, escape of ionizing radiation, dust depletion patterns, and dust extinction curves can be inferred for host galaxies over a wide range of redshifts. Variability of fine structure lines provides unique information about the distance between the burst site and the absorbing material. In addition, information about the ionization state of the IGM can be inferred from the shape of the red damping wing of the Lyman-α line. However, absorption studies still provide only an incomplete view of the overall host population because optical afterglow spectroscopy generally requires a relatively unobscured afterglow. The few exceptions to this limitation (such as GRB 080607) suggest that sight lines towards GRBs in more mature and massive hosts are extremely rich in metals and molecules and there may be much to learn in the future when these sight lines can be studied with the next generation of extremely large telescopes.

For the past decade, Swift (supported by a wide range of ground-based observatories) has been the predominant facility for detecting and locating new GRBs, as well as a source of critical afterglow observations needed to characterize their sightlines and build complete samples. In spite of its achievements, Swift does have some limitations—in particular, difficulty in rapidly confirming the rare but valuable high-redshift ($z > 6$) GRBs. For these purposes, a satellite capable of triggering on a larger number of high-redshift GRBs and earlier identification of the afterglow counterparts of these events through optical and NIR photometry are essential. The SVOM mission[1] will be particularly useful in this sense, thanks to the sensitive Visible Telescope (VT) onboard and to the network of optical-NIR robotic telescopes dedicated to the photometric follow-up of SVOM GRBs. A further step forward

[1] http://smsc.cnes.fr/SVOM/.

towards a large sample of high-redshift GRB DLAs could be made with the proposed space mission THESEUS, specifically developed to detect high-redshift GRBs and observe their afterglows via on-board NIR photometry and spectroscopy.

Acknowledgements The authors would like to thank the organizers of the "Gamma-Ray Bursts—A Tool to Explore the Young Universe" workshop (in particular Diego Götz and Maurizio Falanga) for a stimulating and productive conference and for their excellent hospitality.

Research leading to these results has received funding from the European Research Council under the European Union's Seventh Framework Program (FP7/2007-2013)/ERC Grant agreement no. EGGS-278202.

Susanna D. Vergani would like to thank Patrick Petitjean for accepting the duty of presenting her contribution at the ISSI-BJ meeting.

References

A. Afflerbach, E. Churchwell, M.W. Werner, Galactic abundance gradients from infrared fine-structure lines in compact H II regions. Astrophys. J. **478**, 190 (1997). doi:10.1086/303771

R. Amanullah, A. Goobar, J. Johansson, D.P.K. Banerjee, V. Venkataraman, V. Joshi, N.M. Ashok, Y. Cao, M.M. Kasliwal, S.R. Kulkarni, P.E. Nugent, T. Petrushevska, V. Stanishev, The peculiar extinction law of SN 2014J measured with the Hubble Space Telescope. Astrophys. J. Lett. **788**, 21 (2014). doi:10.1088/2041-8205/788/2/L21

E. Behar, S. Dado, A. Dar, A. Laor, Can the soft X-ray opacity toward high-redshift sources probe the missing baryons? Astrophys. J. **734**, 26 (2011). doi:10.1088/0004-637X/734/1/26

E. Berger, Short-duration gamma-ray bursts. Annu. Rev. Astron. Astrophys. **52**, 43–105 (2014). doi:10.1146/annurev-astro-081913-035926

E. Berger, L.L. Cowie, S.R. Kulkarni, D.A. Frail, H. Aussel, A.J. Barger, A submillimeter and radio survey of gamma-ray burst host galaxies: a glimpse into the future of star formation studies. Astrophys. J. **588**, 99–112 (2003). doi:10.1086/373991

E. Berger, D.B. Fox, S.R. Kulkarni, D.A. Frail, S.G. Djorgovski, The ERO Host Galaxy of GRB 020127: implications for the metallicity of GRB progenitors. Astrophys. J. **660**, 504–508 (2007). doi:10.1086/513007

P.K. Blanchard, E. Berger, W.-f. Fong, The offset and host light distributions of long gamma-ray bursts: a new view from HST observations of swift bursts. ArXiv e-prints (2015)

J.S. Bloom, S.R. Kulkarni, S.G. Djorgovski, The observed offset distribution of gamma-ray bursts from their host galaxies: a robust clue to the nature of the progenitors. Astron. J. **123**, 1111–1148 (2002). doi:10.1086/338893

J.S. Bloom, D.A. Perley, W. Li, N.R. Butler, A.A. Miller, D. Kocevski, D.A. Kann, R.J. Foley, H.-W. Chen, A.V. Filippenko, D.L. Starr, B. Macomber, J.X. Prochaska, R. Chornock, D. Poznanski, S. Klose, M.F. Skrutskie, S. Lopez, P. Hall, K. Glazebrook, C.H. Blake, Observations of the Naked-Eye GRB 080319B: implications of nature's brightest explosion. Astrophys. J. **691**, 723–737 (2009). doi:10.1088/0004-637X/691/1/723

R.C. Bohlin, B.D. Savage, J.F. Drake, A survey of interstellar H I from L-alpha absorption measurements. II. Astrophys. J. **224**, 132–142 (1978). doi:10.1086/156357

R.J. Bouwens, G.D. Illingworth, P.A. Oesch, M. Trenti, I. Labbé, L. Bradley, M. Carollo, P.G. van Dokkum, V. Gonzalez, B. Holwerda, M. Franx, L. Spitler, R. Smit, D. Magee, UV luminosity functions at redshifts $z = 4 - 10 :$ 10,000 galaxies from HST legacy fields. Astrophys. J. **803**, 34 (2015). doi:10.1088/0004-637X/803/1/34

N.R. Butler, Refined astrometry and positions for 179 swift X-ray afterglows. Astron. J. **133**, 1027–1033 (2007). doi:10.1086/510836

M.A. Campisi, G. De Lucia, L. Li, S. Mao, X. Kang, Properties of long gamma-ray burst host galaxies in cosmological simulations. Mon. Not. R. Astron. Soc. **400**, 1613–1624 (2009). doi:10.1111/j.1365-2966.2009.15568.x

C.M. Casey, D. Narayanan, A. Cooray, Dusty star-forming galaxies at high redshift. Phys. Rep. **541**, 45–161 (2014). doi:10.1016/j.physrep.2014.02.009

J.M. Castro Cerón, M.J. Michałowski, J. Hjorth, D. Watson, J.P.U. Fynbo, J. Gorosabel, Star formation rates and stellar masses in $z \sim 1$ gamma-ray burst hosts. Astrophys. J. Lett. **653**, 85–88 (2006). doi:10.1086/510618

J.M. Castro Cerón, M.J. Michałowski, J. Hjorth, D. Malesani, J. Gorosabel, D. Watson, J.P.U. Fynbo, M. Morales Calderón, On the distribution of stellar masses in gamma-ray burst host galaxies. Astrophys. J. **721**, 1919–1927 (2010). doi:10.1088/0004-637X/721/2/1919

A.J. Castro-Tirado, M. Bremer, S. McBreen, J. Gorosabel, S. Guziy, T.A. Fakthullin, V.V. Sokolov, R.M. González Delgado, G. Bihain, S.B. Pandey, M. Jelínek, A. de Ugarte Postigo, K. Misra, R. Sagar, P. Bama, A.P. Kamble, G.C. Anupama, J. Licandro, D. Pérez-Ramírez, D. Bhattacharya, F.J. Aceituno, R. Neri, The dark nature of GRB 051022 and its host galaxy. Astron. Astrophys. **475**, 101–107 (2007). doi:10.1051/0004-6361+20066748

A.J. Castro-Tirado, P. Møller, G. García-Segura, J. Gorosabel, E. Pérez, A. de Ugarte Postigo, E. Solano, D. Barrado, S. Klose, D.A. Kann, J.M. Castro Cerón, C. Kouveliotou, J.P.U. Fynbo, J. Hjorth, H. Pedersen, E. Pian, E. Rol, E. Palazzi, N. Masetti, N.R. Tanvir, P.M. Vreeswijk, M.I. Andersen, A.S. Fruchter, J. Greiner, R.A.M.J. Wijers, E.P.J. van den Heuvel, GRB 021004: tomography of a gamma-ray burst progenitor and its host galaxy. Astron. Astrophys. **517**, 61 (2010). doi:10.1051/0004-6361/200913966

S.B. Cenko, J. Kelemen, F.A. Harrison, D.B. Fox, S.R. Kulkarni, M.M. Kasliwal, E.O. Ofek, A. Rau, A. Gal-Yam, D.A. Frail, D.-S. Moon, Dark bursts in the swift era: the Palomar 60 inch-swift early optical afterglow catalog. Astrophys. J. **693**, 1484–1493 (2009). doi:10.1088/0004-637X/693/2/1484

R. Chary, E.E. Becklin, L. Armus, Are starburst galaxies the hosts of gamma-ray bursts? Astrophys. J. **566**, 229–238 (2002). doi:10.1086/337964

H.-W. Chen, Near-infrared spectroscopy of gamma-ray burst host galaxies at $z > 1.5$: insights into host galaxy dynamics and interpretations of afterglow absorption spectra. Mon. Not. R. Astron. Soc. **419**, 3039–3047 (2012). doi:10.1111/j.1365-2966.2011.19944.x

H.-W. Chen, J.X. Prochaska, E. Ramirez-Ruiz, J.S. Bloom, M. Dessauges-Zavadsky, R.J. Foley, On the absence of wind signatures in GRB afterglow spectra: constraints on the Wolf-Rayet winds of GRB progenitors. Astrophys. J. **663**, 420–436 (2007). doi:10.1086/518111

N.E. Chisari, P.B. Tissera, L.J. Pellizza, Host galaxies of long gamma-ray bursts in the Millennium Simulation. Mon. Not. R. Astron. Soc. **408**, 647–656 (2010). doi:10.1111/j.1365-2966.2010.17169.x

R. Chornock, E. Berger, D.B. Fox, R. Lunnan, M.R. Drout, W.-f. Fong, T. Laskar, K.C. Roth, GRB 130606A as a probe of the intergalactic medium and the interstellar medium in a star-forming galaxy in the first Gyr after the Big Bang. Astrophys. J. **774**, 26 (2013). doi:10.1088/0004-637X/774/1/26

L. Christensen, P.M. Vreeswijk, J. Sollerman, C.C. Thöne, E. Le Floc'h, K. Wiersema, IFU observations of the GRB 980425/SN 1998bw host galaxy: emission line ratios in GRB regions. Astron. Astrophys. **490**, 45–59 (2008). doi:10.1051/0004-6361+200809896

M. Cioni, The metallicity gradient as a tracer of history and structure: the Magellanic Clouds and M33 galaxies. Astron. Astrophys. **506**, 1137–1146 (2009). doi:10.1051/0004-6361/200912138

S. Covino, A. Melandri, R. Salvaterra, S. Campana, S.D. Vergani, M.G. Bernardini, P. D'Avanzo, V. D'Elia, D. Fugazza, G. Ghirlanda, G. Ghisellini, A. Gomboc, Z.P. Jin, T. Krühler, D. Malesani, L. Nava, B. Sbarufatti, G. Tagliaferri, Dust extinctions for an unbiased sample of gamma-ray burst afterglows. Mon. Not. R. Astron. Soc. **432**, 1231–1244 (2013). doi:10.1093/mnras/stt540

P.A. Crowther, L. Dessart, D.J. Hillier, J.B. Abbott, A.W. Fullerton, Stellar and wind properties of LMC WC4 stars. A metallicity dependence for Wolf-Rayet mass-loss rates. Astron. Astrophys. **392**, 653–669 (2002). doi:10.1051/0004-6361:20020941

A. Cucchiara, A.J. Levan, D.B. Fox, N.R. Tanvir, T.N. Ukwatta, E. Berger, T. Krühler, A. Küpcü Yoldaş, X.F. Wu, K. Toma, J. Greiner, F.E. Olivares, A. Rowlinson, L. Amati, T. Sakamoto, K. Roth, A. Stephens, A. Fritz, J.P.U. Fynbo, J. Hjorth, D. Malesani, P. Jakobsson, K. Wiersema, P.T. O'Brien, A.M. Soderberg, R.J. Foley, A.S. Fruchter, J. Rhoads, R.E. Rutledge, B.P. Schmidt, M.A. Dopita, P. Podsiadlowski, R. Willingale, C. Wolf, S.R. Kulkarni, P. D'Avanzo, A photometric redshift of $z \sim 9.4$ for GRB 090429B. Astrophys. J. **736**, 7 (2011). doi:10.1088/0004-637X/736/1/7

R. Davé, The galaxy stellar mass-star formation rate relation: evidence for an evolving stellar initial mass function? Mon. Not. R. Astron. Soc. **385**, 147–160 (2008). doi:10.1111/j.1365-2966.2008.12866.x

A. De Cia, C. Ledoux, S. Savaglio, P. Schady, P.M. Vreeswijk, Dust-to-metal ratios in damped Lyman-α absorbers. Fresh clues to the origins of dust and optical extinction towards γ-ray bursts. Astron. Astrophys. **560**, 88 (2013). doi:10.1051/0004-6361/201321834

V. D'Elia, J.P.U. Fynbo, P. Goldoni, S. Covino, A. de Ugarte Postigo, C. Ledoux, F. Calura, J. Gorosabel, D. Malesani, F. Matteucci, R. Sánchez-Ramírez, S. Savaglio, A.J. Castro-Tirado, O.E. Hartoog, L. Kaper, T. Muñoz-Darias, E. Pian, S. Piranomonte, G. Tagliaferri, N. Tanvir, S.D. Vergani, D.J. Watson, D. Xu, VLT/X-shooter spectroscopy of the GRB 120327A afterglow. Astron. Astrophys. **564**, 38 (2014). doi:10.1051/0004-6361/201323057

S.G. Djorgovski, D.A. Frail, S.R. Kulkarni, J.S. Bloom, S.C. Odewahn, A. Diercks, The afterglow and the host galaxy of the dark burst GRB 970828. Astrophys. J. **562**, 654–663 (2001). doi:10.1086/323845

B.T. Draine, L. Hao, Gamma-ray burst in a molecular cloud: destruction of dust and H_2 and the emergent spectrum. Astrophys. J. **569**, 780–791 (2002). doi:10.1086/339394

Á. Elíasdóttir, J.P.U. Fynbo, J. Hjorth, C. Ledoux, D.J. Watson, A.C. Andersen, D. Malesani, P.M. Vreeswijk, J.X. Prochaska, J. Sollerman, A.O. Jaunsen, Dust extinction in high-z galaxies with gamma-ray burst afterglow spectroscopy: the 2175 Å feature at $z = 2.45$. Astrophys. J. **697**, 1725–1740 (2009). doi:10.1088/0004-637X/697/2/1725

J. Elliott, T. Krühler, J. Greiner, S. Savaglio, F. Olivares, E.A. Rau, A. de Ugarte Postigo, R. Sánchez-Ramírez, K. Wiersema, P. Schady, D.A. Kann, R. Filgas, M. Nardini, E. Berger, D. Fox, J. Gorosabel, S. Klose, A. Levan, A. Nicuesa Guelbenzu, A. Rossi, S. Schmidl, V. Sudilovsky, N.R. Tanvir, C.C. Thöne, The low-extinction afterglow in the solar-metallicity host galaxy of γ-ray burst 110918A. Astron. Astrophys. **556**, 23 (2013). doi:10.1051/0004-6361/201220968

J. Elliott, S. Khochfar, J. Greiner, C. Dalla Vecchia, The first billion years project: gamma-ray bursts at $z > 5$. Mon. Not. R. Astron. Soc. **446**, 4239–4249 (2015). doi:10.1093/mnras/stu2417

D.K. Erb, A.E. Shapley, M. Pettini, C.C. Steidel, N.A. Reddy, K.L. Adelberger, The mass-metallicity relation at $z \gtrsim 2$. Astrophys. J. **644**, 813–828 (2006). doi:10.1086/503623

P.A. Evans, A.P. Beardmore, K.L. Page, J.P. Osborne, P.T. O'Brien, R. Willingale, R.L.C. Starling, D.N. Burrows, O. Godet, L. Vetere, J. Racusin, M.R. Goad, K. Wiersema, L. Angelini, M. Capalbi, G. Chincarini, N. Gehrels, J.A. Kennea, R. Margutti, D.C. Morris, C.J. Mountford, C. Pagani, M. Perri, P. Romano, N. Tanvir, Methods and results of an automatic analysis of a complete sample of Swift-XRT observations of GRBs. Mon. Not. R. Astron. Soc. **397**, 1177–1201 (2009). doi:10.1111/j.1365-2966.2009.14913.x

A.J. Fox, C. Ledoux, P.M. Vreeswijk, A. Smette, A.O. Jaunsen, High-ion absorption in seven GRB host galaxies at $z = 2$–4. Evidence for both circumburst plasma and outflowing interstellar gas. Astron. Astrophys. **491**, 189–207 (2008). doi:10.1051/0004-6361+200810286

M. Friis, A. De Cia, T. Krühler, J.P.U. Fynbo, C. Ledoux, P.M. Vreeswijk, D.J. Watson, D. Malesani, J. Gorosabel, R.L.C. Starling, P. Jakobsson, K. Varela, K. Wiersema, A.P. Drachmann, A. Trotter, C.C. Thöne, A. de Ugarte Postigo, V. D'Elia, J. Elliott, M. Maturi, P. Goldoni, J. Greiner, J. Haislip, L. Kaper, F. Knust, A. LaCluyze, B. Milvang-Jensen, D. Reichart, S. Schulze, V. Sudilovsky, N. Tanvir, S.D. Vergani, The warm, the excited, and the molecular gas: GRB 121024A shining through its star-forming galaxy. Mon. Not. R. Astron. Soc. **451**, 167–183 (2015). doi:10.1093/mnras/stv960

A. Fruchter, J.H. Krolik, J.E. Rhoads, X-Ray destruction of dust along the line of sight to γ-ray bursts. Astrophys. J. **563**, 597–610 (2001). doi:10.1086/323520

A.S. Fruchter, A.J. Levan, L. Strolger, P.M. Vreeswijk, S.E. Thorsett, D. Bersier, I. Burud, J.M. Castro Cerón, A.J. Castro-Tirado, C. Conselice, T. Dahlen, H.C. Ferguson, J.P.U. Fynbo, P.M. Garnavich, R.A. Gibbons, J. Gorosabel, T.R. Gull, J. Hjorth, S.T. Holland, C. Kouveliotou, Z. Levay, M. Livio, M.R. Metzger, P.E. Nugent, L. Petro, E. Pian, J.E. Rhoads, A.G. Riess, K.C. Sahu, A. Smette, N.R. Tanvir, R.A.M.J. Wijers, S.E. Woosley, Long γ-ray bursts and core-collapse supernovae have different environments. Nature **441**, 463–468 (2006). doi:10.1038/nature04787

C.L. Fryer, A. Heger, Binary merger progenitors for gamma-ray bursts and hypernovae. Astrophys. J. **623**, 302–313 (2005). doi:10.1086/428379

J.P.U. Fynbo, P. Jakobsson, P. Möller, J. Hjorth, B. Thomsen, M.I. Andersen, A.S. Fruchter, J. Gorosabel, S.T. Holland, C. Ledoux, H. Pedersen, J. Rhoads, M. Weidinger, R.A.M.J. Wijers, On the Lyalpha emission from gamma-ray burst host galaxies: evidence for low metallicities. Astron. Astrophys. **406**, 63–66 (2003). doi:10.1051/0004-6361+20030931

J.P.U. Fynbo, R.L.C. Starling, C. Ledoux, K. Wiersema, C.C. Thöne, J. Sollerman, P. Jakobsson, J. Hjorth, D. Watson, P.M. Vreeswijk, P. Møller, E. Rol, J. Gorosabel, J. Näränen, R.A.M.J. Wijers, G. Björnsson, J.M. Castro Cerón, P. Curran, D.H. Hartmann, S.T. Holland, B.L. Jensen, A.J. Levan, M. Limousin, C. Kouveliotou, G. Nelemans, K. Pedersen, R.S. Priddey, N.R. Tanvir, Probing cosmic chemical evolution with gamma-ray bursts: GRB 060206 at $z = 4.048$. Astron. Astrophys. **451**, 47–50 (2006). doi:10.1051/0004-6361+20065056

J.P.U. Fynbo, P. Jakobsson, J.X. Prochaska, D. Malesani, C. Ledoux, A. de Ugarte Postigo, M. Nardini, P.M. Vreeswijk, K. Wiersema, J. Hjorth, J. Sollerman, H.-W. Chen, C.C. Thöne, G. Björnsson, J.S. Bloom, A.J. Castro-Tirado, L. Christensen, A. De Cia, A.S. Fruchter, J. Gorosabel, J.F. Graham, A.O. Jaunsen, B.L. Jensen, D.A. Kann, C. Kouveliotou, A.J. Levan, J. Maund, N. Masetti, B. Milvang-Jensen, E. Palazzi, D.A. Perley, E. Pian, E. Rol, P. Schady, R.L.C. Starling, N.R. Tanvir, D.J. Watson, D. Xu, T. Augusteijn, F. Grundahl, J. Telting, P.-O. Quirion, Low-resolution spectroscopy of gamma-ray burst optical afterglows: biases in the swift sample and characterization of the absorbers. Astrophys. J. Suppl. **185**, 526–573 (2009). doi:10.1088/0067-0049/185/2/526

J.P.U. Fynbo, J.-K. Krogager, B. Venemans, P. Noterdaeme, M. Vestergaard, P. Møller, C. Ledoux, S. Geier, Optical/near-infrared selection of red quasi-stellar objects: evidence for steep extinction curves toward galactic centers? Astrophys. J. Suppl. **204**, 6 (2013). doi:10.1088/0067-0049/204/1/6

J.P.U. Fynbo, T. Krühler, K. Leighly, C. Ledoux, P.M. Vreeswijk, S. Schulze, P. Noterdaeme, D. Watson, R.A.M.J. Wijers, J. Bolmer, Z. Cano, L. Christensen, S. Covino, V. D'Elia, H. Flores, M. Friis, P. Goldoni, J. Greiner, F. Hammer, J. Hjorth, P. Jakobsson, J. Japelj, L. Kaper, S. Klose, F. Knust, G. Leloudas, A. Levan, D. Malesani, B. Milvang-Jensen, P. Møller, A. Nicuesa Guelbenzu, S. Oates, E. Pian, P. Schady, M. Sparre, G. Tagliaferri, N. Tanvir, C.C. Thöne, A. de Ugarte Postigo, S. Vergani, K.

Wiersema, D. Xu, T. Zafar, The mysterious optical afterglow spectrum of GRB 140506A at $z = 0.889$. Astron. Astrophys. **572**, 12 (2014). doi:10.1051/0004-6361/201424726

J.U. Fynbo, B.L. Jensen, J. Gorosabel, J. Hjorth, H. Pedersen, P. Møller, T. Abbott, A.J. Castro-Tirado, D. Delgado, J. Greiner, A. Henden, A. Magazzù, N. Masetti, S. Merlino, J. Masegosa, R. Østensen, E. Palazzi, E. Pian, H.E. Schwarz, T. Cline, C. Guidorzi, J. Goldsten, K. Hurley, E. Mazets, T. McClanahan, E. Montanari, R. Starr, J. Trombka, Detection of the optical afterglow of GRB 000630: implications for dark bursts. Astron. Astrophys. **369**, 373–379 (2001). doi:10.1051/0004-6361+20010112

T.J. Galama, R.A.M.J. Wijers, High column densities and low extinctions of gamma-ray bursts: evidence for hypernovae and dust destruction. Astrophys. J. Lett. **549**, 209–213 (2001). doi:10.1086/319162

T.J. Galama, P.M. Vreeswijk, J. van Paradijs, C. Kouveliotou, T. Augusteijn, H. Böhnhardt, J.P. Brewer, V. Doublier, J.-F. Gonzalez, B. Leibundgut, C. Lidman, O.R. Hainaut, F. Patat, J. Heise, J. in't Zand, K. Hurley, P.J. Groot, R.G. Strom, P.A. Mazzali, K. Iwamoto, K. Nomoto, H. Umeda, T. Nakamura, T.R. Young, T. Suzuki, T. Shigeyama, T. Koshut, M. Kippen, C. Robinson, P. de Wildt, R.A.M.J. Wijers, N. Tanvir, J. Greiner, E. Pian, E. Palazzi, F. Frontera, N. Masetti, L. Nicastro, M. Feroci, E. Costa, L. Piro, B.A. Peterson, C. Tinney, B. Boyle, R. Cannon, R. Stathakis, E. Sadler, M.C. Begam, P. Ianna, An unusual supernova in the error box of the γ-ray burst of 25 April 1998. Nature **395**, 670–672 (1998). doi:10.1038/27150

N. Gehrels, G. Chincarini, P. Giommi, K.O. Mason, J.A. Nousek, A.A. Wells, N.E. White, S.D. Barthelmy, D.N. Burrows, L.R. Cominsky, K.C. Hurley, F.E. Marshall, P. Mészáros, P.W.A. Roming, L. Angelini, L.M. Barbier, T. Belloni, S. Campana, P.A. Caraveo, M.M. Chester, O. Citterio, T.L. Cline, M.S. Cropper, J.R. Cummings, A.J. Dean, E.D. Feigelson, E.E. Fenimore, D.A. Frail, A.S. Fruchter, G.P. Garmire, K. Gendreau, G. Ghisellini, J. Greiner, J.E. Hill, S.D. Hunsberger, H.A. Krimm, S.R. Kulkarni, P. Kumar, F. Lebrun, N.M. Lloyd-Ronning, C.B. Markwardt, B.J. Mattson, R.F. Mushotzky, J.P. Norris, J. Osborne, B. Paczynski, D.M. Palmer, H.-S. Park, A.M. Parsons, J. Paul, M.J. Rees, C.S. Reynolds, J.E. Rhoads, T.P. Sasseen, B.E. Schaefer, A.T. Short, A.P. Smale, I.A. Smith, L. Stella, G. Tagliaferri, T. Takahashi, M. Tashiro, L.K. Townsley, J. Tueller, M.J.L. Turner, M. Vietri, W. Voges, M.J. Ward, R. Willingale, F.M. Zerbi, W.W. Zhang, The swift gamma-ray burst mission. Astrophys. J. **611**, 1005–1020 (2004). doi:10.1086/422091

A.J. Gosling, R.M. Bandyopadhyay, K.M. Blundell, The complex, variable near-infrared extinction towards the Nuclear Bulge. Mon. Not. R. Astron. Soc. **394**, 2247–2254 (2009). doi:10.1111/j.1365-2966.2009.14493.x

J.F. Graham, A.S. Fruchter, The metal aversion of long-duration gamma-ray bursts. Astrophys. J. **774**, 119 (2013). doi:10.1088/0004-637X/774/2/119

J. Greiner, T. Krühler, S. Klose, P. Afonso, C. Clemens, R. Filgas, D.H. Hartmann, A. Küpcü Yoldaş, M. Nardini, F.E. Olivares, A. Rau, A. Rossi, P. Schady, A. Updike, The nature of "dark" gamma-ray bursts. Astron. Astrophys. **526**, 30 (2011). doi:10.1051/0004-6361/201015458

J. Greiner, D.B. Fox, P. Schady, T. Krühler, M. Trenti, A. Cikota, J. Bolmer, J. Elliott, C. Delvaux, R. Perna, P. Afonso, D.A. Kann, S. Klose, S. Savaglio, S. Schmidl, T. Schweyer, M. Tanga, K. Varela, Gamma-ray bursts trace UV metrics of star formation over $3 < z < 5$. Astrophys. J. **809**, 76 (2015). doi:10.1088/0004-637X/809/1/76

P.J. Groot, T.J. Galama, J. van Paradijs, C. Kouveliotou, R.A.M.J. Wijers, J. Bloom, N. Tanvir, R. Vanderspek, J. Greiner, A.J. Castro-Tirado, J. Gorosabel, T. von Hippel, M. Lehnert, K. Kuijken, H. Hoekstra, N. Metcalfe, C. Howk, C. Conselice, J. Telting, R.G.M. Rutten, J. Rhoads, A. Cole, D.J. Pisano, R. Naber, R. Schwarz, A search for optical afterglow from GRB 970828. Astrophys. J. Lett. **493**, 27–30 (1998). doi:10.1086/311125

P.B. Hall, S.F. Anderson, M.A. Strauss, D.G. York, G.T. Richards, X. Fan, G.R. Knapp, D.P. Schneider, D.E. Vanden Berk, T.R. Geballe, A.E. Bauer, R.H. Becker, M. Davis, H.-W. Rix, R.C. Nichol, N.A. Bahcall, J. Brinkmann, R. Brunner, A.J. Connolly, I. Csabai, M. Doi, M. Fukugita, J.E. Gunn, Z. Haiman, M. Harvanek, T.M. Heckman, G.S. Hennessy, N. Inada, Ž. Ivezić, D. Johnston, S. Kleinman, J.H. Krolik, J. Krzesinski, P.Z. Kunszt, D.Q. Lamb, D.C. Long, R.H. Lupton, G. Miknaitis, J.A. Munn, V.K. Narayanan, E. Neilsen, P.R. Newman, A. Nitta, S. Okamura, L. Pentericci, J.R. Pier, D.J. Schlegel, S. Snedden, A.S. Szalay, A.R. Thakar, Z. Tsvetanov, R.L. White, W. Zheng, Unusual broad absorption line quasars from the Sloan digital sky survey. Astrophys. J. Suppl. **141**, 267–309 (2002). doi:10.1086/340546

X.H. Han, F. Hammer, Y.C. Liang, H. Flores, M. Rodrigues, J.L. Hou, J.Y. Wei, The Wolf-Rayet features and mass-metallicity relation of long-duration gamma-ray burst host galaxies. Astron. Astrophys. **514**, 24 (2010). doi:10.1051/0004-6361/200912475

O.E. Hartoog, D. Malesani, J.P.U. Fynbo, T. Goto, T. Krühler, P.M. Vreeswijk, A. De Cia, D. Xu, P. Møller, S. Covino, V. D'Elia, H. Flores, P. Goldoni, J. Hjorth, P. Jakobsson, J.-K. Krogager, L. Kaper, C. Ledoux, A.J. Levan, B. Milvang-Jensen, J. Sollerman, M. Sparre, G. Tagliaferri, N.R. Tanvir, A. de Ugarte Postigo, S.D. Vergani, K. Wiersema, J. Datson, R. Salinas, K. Mikkelsen, N. Aghanim, VLT/X-shooter

spectroscopy of the afterglow of the Swift GRB 130606A. Chemical abundances and reionisation at $z \sim 6$. Astron. Astrophys. **580**, 139 (2015). doi:10.1051/0004-6361/201425001
T. Hashimoto, K. Ohta, K. Aoki, I. Tanaka, K. Yabe, N. Kawai, W. Aoki, H. Furusawa, T. Hattori, M. Iye, K.S. Kawabata, N. Kobayashi, Y. Komiyama, G. Kosugi, Y. Minowa, Y. Mizumoto, Y. Niino, K. Nomoto, J. Noumaru, R. Ogasawara, T.-S. Pyo, T. Sakamoto, K. Sekiguchi, Y. Shirasaki, M. Suzuki, A. Tajitsu, T. Takata, T. Tamagawa, H. Terada, T. Totani, J. Watanabe, T. Yamada, A. Yoshida, "Dark" GRB 080325 in a dusty massive galaxy at $z \sim 2$. Astrophys. J. **719**, 378–384 (2010). doi:10.1088/0004-637X/719/1/378
T. Hashimoto, D.A. Perley, K. Ohta, K. Aoki, I. Tanaka, Y. Niino, K. Yabe, N. Kawai, The star formation rate and metallicity of the host galaxy of the Dark GRB 080325 at $z = 1.78$. Astrophys. J. **806**, 250 (2015). doi:10.1088/0004-637X/806/2/250
B. Hatsukade, K. Ohta, A. Endo, K. Nakanishi, Y. Tamura, T. Hashimoto, K. Kohno, Two γ-ray bursts from dusty regions with little molecular gas. Nature **510**, 247–249 (2014). doi:10.1038/nature13325
A. Heger, C.L. Fryer, S.E. Woosley, N. Langer, D.H. Hartmann, How massive single stars end their life. Astrophys. J. **591**, 288–300 (2003). doi:10.1086/375341
R. Hirschi, G. Meynet, A. Maeder, Stellar evolution with rotation. XIII. Predicted GRB rates at various Z. Astron. Astrophys. **443**, 581–591 (2005). doi:10.1051/0004-6361+20053329
R. Hirschi, C. Chiappini, G. Meynet, A. Maeder, S. Ekström, Stellar evolution at low metallicity, in *Proceedings of the International Astronomical Union*, ed. by F. Bresolin, P.A. Crowther, J. Puls IAU Symposium, vol. 250, 2008, pp. 217–230. doi:10.1017/S1743921308020528
J. Hjorth, J.S. Bloom, The gamma-ray burst—supernova connection, in *Gamma-Ray Bursts* (2012), pp. 169–190
J. Hjorth, D. Malesani, P. Jakobsson, A.O. Jaunsen, J.P.U. Fynbo, J. Gorosabel, T. Krühler, A.J. Levan, M.J. Michałowski, B. Milvang-Jensen, P. Møller, S. Schulze, N.R. Tanvir, D. Watson, The Optically Unbiased Gamma-ray burst Host (TOUGH) survey. I. Survey design and catalogs. Astrophys. J. **756**, 187 (2012). doi:10.1088/0004-637X/756/2/187
L.K. Hunt, E. Palazzi, M.J. Michałowski, A. Rossi, S. Savaglio, S. Basa, S. Berta, S. Bianchi, S. Covino, V. D'Elia, P. Ferrero, D. Götz, J. Greiner, S. Klose, D. Le Borgne, E. Le Floc'h, E. Pian, S. Piranomonte, P. Schady, S.D. Vergani, New light on gamma-ray burst host galaxies with Herschel. Astron. Astrophys. **565**, 112 (2014). doi:10.1051/0004-6361/201323340
L. Hunt, E. Palazzi, A. Rossi, S. Savaglio, G. Cresci, S. Klose, M. Michałowski, E. Pian, The extremely red host galaxy of GRB 080207. Astrophys. J. Lett. **736**, 36 (2011). doi:10.1088/2041-8205/736/2/L36
R.G. Izzard, E. Ramirez-Ruiz, C.A. Tout, Formation rates of core-collapse supernovae and gamma-ray bursts. Mon. Not. R. Astron. Soc. **348**, 1215–1228 (2004). doi:10.1111/j.1365-2966.2004.07436.x
P. Jakobsson, J. Hjorth, J.P.U. Fynbo, D. Watson, K. Pedersen, G. Björnsson, J. Gorosabel, Swift identification of dark gamma-ray bursts. Astrophys. J. Lett. **617**, 21–24 (2004). doi:10.1086/427089
P. Jakobsson, G. Björnsson, J.P.U. Fynbo, G. Jóhannesson, J. Hjorth, B. Thomsen, P. Møller, D. Watson, B.L. Jensen, G. Östlin, J. Gorosabel, E.H. Gudmundsson, Ly-α and ultraviolet emission from high-redshift gamma-ray burst hosts: to what extent do gamma-ray bursts trace star formation? Mon. Not. R. Astron. Soc. **362**, 245–251 (2005). doi:10.1111/j.1365-2966.2005.09289.x
J. Japelj, S. Covino, A. Gomboc, S.D. Vergani, P. Goldoni, J. Selsing, Z. Cano, V. D'Elia, H. Flores, J.P.U. Fynbo, F. Hammer, J. Hjorth, P. Jakobsson, L. Kaper, D. Kopač, T. Krühler, A. Melandri, S. Piranomonte, R. Sánchez-Ramírez, G. Tagliaferri, N.R. Tanvir, A. de Ugarte Postigo, D. Watson, R.A.M.J. Wijers, Spectrophotometric analysis of gamma-ray burst afterglow extinction curves with X-Shooter. Astron. Astrophys. **579**, 74 (2015). doi:10.1051/0004-6361/201525665
M. Kajisawa, T. Ichikawa, I. Tanaka, T. Yamada, M. Akiyama, R. Suzuki, C. Tokoku, Y. Katsuno Uchimoto, M. Konishi, T. Yoshikawa, T. Nishimura, K. Omata, M. Ouchi, I. Iwata, T. Hamana, M. Onodera, MOIRCS deep survey. IX. Deep near-infrared imaging data and source catalog. Publ. Astron. Soc. Jpn. **63**, 379 (2011). doi:10.1093/pasj/63.sp2.S379
D.A. Kann, S. Klose, A. Zeh, Signatures of extragalactic dust in pre-swift GRB afterglows. Astrophys. J. **641**, 993–1009 (2006). doi:10.1086/500652
D.A. Kann, S. Klose, B. Zhang, D. Malesani, E. Nakar, A. Pozanenko, A.C. Wilson, N.R. Butler, P. Jakobsson, S. Schulze, M. Andreev, L.A. Antonelli, I.F. Bikmaev, V. Biryukov, M. Böttcher, R.A. Burenin, J.M. Castro Cerón, A.J. Castro-Tirado, G. Chincarini, B.E. Cobb, S. Covino, P. D'Avanzo, V. D'Elia, M. Della Valle, A. de Ugarte Postigo, Y. Efimov, P. Ferrero, D. Fugazza, J.P.U. Fynbo, M. Gålfalk, F. Grundahl, J. Gorosabel, S. Gupta, S. Guziy, B. Hafizov, J. Hjorth, K. Holhjem, M. Ibrahimov, M. Im, G.L. Israel, M. Jelínek, B.L. Jensen, R. Karimov, I.M. Khamitov, Ü. Kiziloğlu, E. Klunko, P. Kubánek, A.S. Kutyrev, P. Laursen, A.J. Levan, F. Mannucci, C.M. Martin, A. Mescheryakov, N. Mirabal, J.P. Norris, J.-E. Ovaldsen, D. Paraficz, E. Pavlenko, S. Piranomonte, A. Rossi, V. Rumyantsev, R. Salinas, A. Sergeev, D. Sharapov, J. Sollerman, B. Stecklum, L. Stella, G. Tagliaferri, N.R. Tanvir, J. Telting, V. Testa, A.C. Updike, A. Volnova, D. Watson, K. Wiersema, D. Xu, The afterglows of swift-era gamma-

ray bursts. I. Comparing pre-swift and swift-era long/soft (type II) GRB optical afterglows. Astrophys. J. **720**, 1513–1558 (2010). doi:10.1088/0004-637X/720/2/1513

P.L. Kelly, R.P. Kirshner, M. Pahre, Long γ-ray bursts and type Ic core-collapse supernovae have similar locations in hosts. Astrophys. J. **687**, 1201–1207 (2008). doi:10.1086/591925

P.L. Kelly, A.V. Filippenko, M. Modjaz, D. Kocevski, The host galaxies of fast-ejecta core-collapse supernovae. Astrophys. J. **789**, 23 (2014). doi:10.1088/0004-637X/789/1/23

L.J. Kewley, M.A. Dopita, Using strong lines to estimate abundances in extragalactic H II regions and starburst galaxies. Astrophys. J. Suppl. **142**, 35–52 (2002). doi:10.1086/341326

H.A. Kobulnicky, L.J. Kewley, Metallicities of $z = 0.3$–1.0 galaxies in the GOODS-north field. Astrophys. J. **617**, 240–261 (2004). doi:10.1086/425299

D. Kocevski, A.A. West, M. Modjaz, Modeling the GRB host galaxy mass distribution: are GRBs unbiased tracers of star formation? Astrophys. J. **702**, 377–385 (2009). doi:10.1088/0004-637X/702/1/377

S.A. Kohn, M.J. Michałowski, N. Bourne, M. Baes, J. Fritz, A. Cooray, I. De Looze, G. De Zotti, H. Dannerbauer, L. Dunne, S. Dye, S. Eales, C. Furlanetto, J. Gonzalez-Nuevo, E. Ibar, R.J. Ivison, S.J. Maddox, D. Scott, D.J.B. Smith, M.W.L. Smith, M. Symeonidis, E. Valiante, Far-infrared observations of an unbiased sample of gamma-ray burst host galaxies. Mon. Not. R. Astron. Soc. **448**, 1494–1503 (2015). doi:10.1093/mnras/stv088

Y. Krongold, J.X. Prochaska, An explanation for the different X-ray to optical column densities in the environments of gamma ray bursts: a progenitor embedded in a dense medium. Astrophys. J. **774**, 115 (2013). doi:10.1088/0004-637X/774/2/115

T. Krühler, A. Küpcü Yoldaş, J. Greiner, C. Clemens, S. McBreen, N. Primak, S. Savaglio, A. Yoldaş, G.P. Szokoly, S. Klose, The 2175 Å dust feature in a gamma-ray burst afterglow at redshift 2.45. Astrophys. J. **685**, 376–383 (2008). doi:10.1086/590240

T. Krühler, J. Greiner, P. Schady, S. Savaglio, P.M.J. Afonso, C. Clemens, J. Elliott, R. Filgas, D. Gruber, D.A. Kann, S. Klose, A. Küpcü-Yoldaş, S. McBreen, F. Olivares, D. Pierini, A. Rau, A. Rossi, M. Nardini, A. Nicuesa Guelbenzu, V. Sudilovsky, A.C. Updike, The SEDs and host galaxies of the dustiest GRB afterglows. Astron. Astrophys. **534**, 108 (2011). doi:10.1051/0004-6361/201117428

T. Krühler, C. Ledoux, J.P.U. Fynbo, P.M. Vreeswijk, S. Schmidl, D. Malesani, L. Christensen, A. De Cia, J. Hjorth, P. Jakobsson, D.A. Kann, L. Kaper, S.D. Vergani, P.M.J. Afonso, S. Covino, A. de Ugarte Postigo, V. D'Elia, R. Filgas, P. Goldoni, J. Greiner, O.E. Hartoog, B. Milvang-Jensen, M. Nardini, S. Piranomonte, A. Rossi, R. Sánchez-Ramírez, P. Schady, S. Schulze, V. Sudilovsky, N.R. Tanvir, G. Tagliaferri, D.J. Watson, K. Wiersema, R.A.M.J. Wijers, D. Xu, Molecular hydrogen in the damped Lyman α system towards GRB 120815A at $z = 2.36$. Astron. Astrophys. **557**, 18 (2013). doi:10.1051/0004-6361/201321772

T. Krühler, D. Malesani, J.P.U. Fynbo, O.E. Hartoog, J. Hjorth, P. Jakobsson, D.A. Perley, A. Rossi, P. Schady, S. Schulze, N.R. Tanvir, S.D. Vergani, K. Wiersema, P.M.J. Afonso, J. Bolmer, Z. Cano, S. Covino, V. D'Elia, A. de Ugarte Postigo, R. Filgas, M. Friis, J.F. Graham, J. Greiner, P. Goldoni, A. Gomboc, F. Hammer, J. Japelj, D.A. Kann, L. Kaper, S. Klose, A.J. Levan, G. Leloudas, B. Milvang-Jensen, A. Nicuesa Guelbenzu, E. Palazzi, E. Pian, S. Piranomonte, R. Sánchez-Ramírez, S. Savaglio, J. Selsing, G. Tagliaferri, P.M. Vreeswijk, D.J. Watson, D. Xu, GRB hosts through cosmic time. VLT/X-shooter emission-line spectroscopy of 96 γ-ray-burst-selected galaxies at $0.1 < z < 3.6$. Astron. Astrophys. **581**, 125 (2015). doi:10.1051/0004-6361/201425561

D.Q. Lamb, D.E. Reichart, Gamma-ray bursts as a probe of the very high redshift universe. Astrophys. J. **536**, 1–18 (2000). doi:10.1086/308918

N. Langer, C.A. Norman, On the collapsar model of long gamma-ray bursts:constraints from cosmic metallicity evolution. Astrophys. J. Lett. **638**, 63–66 (2006). doi:10.1086/500363

A. Lapi, N. Kawakatu, Z. Bosnjak, A. Celotti, A. Bressan, G.L. Granato, L. Danese, Long gamma-ray bursts and their host galaxies at high redshift. Mon. Not. R. Astron. Soc. **386**, 608–618 (2008). doi:10.1111/j.1365-2966.2008.13076.x

J. Larsson, A.J. Levan, M.B. Davies, A.S. Fruchter, A new constraint for gamma-ray burst progenitor mass. Mon. Not. R. Astron. Soc. **376**, 1285–1290 (2007). doi:10.1111/j.1365-2966.2007.11523.x

T. Laskar, E. Berger, R.-R. Chary, Exploring the galaxy mass-metallicity relation at $z \sim 3$–5. Astrophys. J. **739**, 1 (2011). doi:10.1088/0004-637X/739/1/1

D. Lazzati, S. Covino, G. Ghisellini, On the role of extinction in failed gamma-ray burst optical/infrared afterglows. Mon. Not. R. Astron. Soc. **330**, 583–590 (2002). doi:10.1046/j.1365-8711.2002.05076.x

E. Le Floc'h, P.-A. Duc, I.F. Mirabel, D.B. Sanders, G. Bosch, R.J. Diaz, C.J. Donzelli, I. Rodrigues, T.J.-L. Courvoisier, J. Greiner, S. Mereghetti, J. Melnick, J. Maza, D. Minniti, Are the hosts of gamma-ray bursts sub-luminous and blue galaxies? Astron. Astrophys. **400**, 499–510 (2003). doi:10.1051/0004-6361+20030001

E. Le Floc'h, V. Charmandaris, W.J. Forrest, I.F. Mirabel, L. Armus, D. Devost, Probing cosmic star formation using long gamma-ray bursts: new constraints from the Spitzer space telescope. Astrophys. J. **642**, 636–652 (2006). doi:10.1086/501118

K.M. Leighly, D.M. Terndrup, E. Baron, A.B. Lucy, M. Dietrich, S.C. Gallagher, Evidence for active galactic nucleus feedback in the broad absorption lines and reddening of Mrk 231. Astrophys. J. **788**, 123 (2014). doi:10.1088/0004-637X/788/2/123

A. Levan, A. Fruchter, J. Rhoads, B. Mobasher, N. Tanvir, J. Gorosabel, E. Rol, C. Kouveliotou, I. Dell'Antonio, M. Merrill, E. Bergeron, J.M. Castro Cerón, N. Masetti, P. Vreeswijk, A. Antonelli, D. Bersier, A. Castro-Tirado, J. Fynbo, P. Garnavich, S. Holland, J. Hjorth, P. Nugent, E. Pian, A. Smette, B. Thomsen, S.E. Thorsett, R. Wijers, Infrared and optical observations of GRB 030115 and its extremely red host galaxy: implications for dark bursts. Astrophys. J. **647**, 471–482 (2006). doi:10.1086/503595

E.M. Levesque, L.J. Kewley, J.F. Graham, A.S. Fruchter, A High-metallicity Host Environment for the Long-duration GRB 020819. Astrophys. J. Lett. **712**, 26–30 (2010a). doi:10.1088/2041-8205/712/1/L26

E.M. Levesque, A.M. Soderberg, L.J. Kewley, E. Berger, No correlation between host galaxy metallicity and gamma-ray energy release for long-duration gamma-ray bursts. Astrophys. J. **725**, 1337–1341 (2010b). doi:10.1088/0004-637X/725/1/1337

E.M. Levesque, L.J. Kewley, E. Berger, H.J. Zahid, The host galaxies of gamma-ray bursts. II. A mass-metallicity relation for long-duration gamma-ray burst host galaxies. Astron. J. **140**, 1557–1566 (2010c). doi:10.1088/0004-6256/140/5/1557

E.M. Levesque, E. Berger, A.M. Soderberg, R. Chornock, Metallicity in the GRB 100316D/SN 2010bh host complex. Astrophys. J. **739**, 23 (2011). doi:10.1088/0004-637X/739/1/23

O.M. Littlejohns, N.R. Butler, A. Cucchiara, A.M. Watson, O.D. Fox, W.H. Lee, A.S. Kutyrev, M.G. Richer, C.R. Klein, J.X. Prochaska, J.S. Bloom, E. Troja, E. Ramirez-Ruiz, J.A. de Diego, L. Georgiev, J. González, C.G. Román-Zúñiga, N. Gehrels, H. Moseley, A detailed study of the optical attenuation of gamma-ray bursts in the Swift era. Mon. Not. R. Astron. Soc. **449**, 2919–2936 (2015). doi:10.1093/mnras/stv479

A.I. MacFadyen, S.E. Woosley, Collapsars: gamma-ray bursts and explosions in "Failed Supernovae". Astrophys. J. **524**, 262–289 (1999). doi:10.1086/307790

J.T.W. McGuire, N.R. Tanvir, A.J. Levan, M. Trenti, E.R. Stanway, J.M. Shull, K. Wiersema, D.A. Perley, R.L.C. Starling, M. Bremer, J.T. Stocke, J. Hjorth, J.E. Rhoads, E.M. Levesque, B. Robertson, J.P.U. Fynbo, R.S. Ellis, A.S. Fruchter, R. Perna, Detection of three gamma-ray burst host galaxies at $z \sim 6$. ArXiv e-prints (2015)

A. Melandri, B. Sbarufatti, P. D'Avanzo, R. Salvaterra, S. Campana, S. Covino, S.D. Vergani, L. Nava, G. Ghisellini, G. Ghirlanda, D. Fugazza, V. Mangano, M. Capalbi, G. Tagliaferri, The dark bursts population in a complete sample of bright Swift long gamma-ray bursts. Mon. Not. R. Astron. Soc. **421**, 1265–1272 (2012). doi:10.1111/j.1365-2966.2011.20398.x

M.J. Michałowski, J. Hjorth, J.M. Castro Cerón, D. Watson, The nature of GRB-selected submillimeter galaxies: hot and young. Astrophys. J. **672**, 817–824 (2008). doi:10.1086/523891

M.J. Michałowski, A. Kamble, J. Hjorth, D. Malesani, R.F. Reinfrank, L. Bonavera, J.M. Castro Cerón, E. Ibar, J.S. Dunlop, J.P.U. Fynbo, M.A. Garrett, P. Jakobsson, D.L. Kaplan, T. Krühler, A.J. Levan, M. Massardi, S. Pal, J. Sollerman, N.R. Tanvir, A.J. van der Horst, D. Watson, K. Wiersema, The Optically Unbiased GRB Host (TOUGH) survey. VI. Radio observations at $z \lesssim 1$ and consistency with typical star-forming galaxies. Astrophys. J. **755**, 85 (2012). doi:10.1088/0004-637X/755/2/85

M.J. Michałowski, G. Gentile, J. Hjorth, M.R. Krumholz, N.R. Tanvir, P. Kamphuis, D. Burlon, M. Baes, S. Basa, S. Berta, J.M. Castro Ceron, D. Crosby, V. D'Elia, J. Elliott, J. Greiner, L.K. Hunt, S. Klose, M.P. Koprowski, E. Le Floc'h, D. Malesani, T. Murphy, A. Nicuesa Guelbenzu, E. Palazzi, J. Rasmussen, A. Rossi, S. Savaglio, P. Schady, J. Sollerman, A. de Ugarte Postigo, D. Watson, P. van der Werf, S.D. Vergani, D. Xu, Massive stars formed in atomic hydrogen reservoirs: HI observations of gamma-ray burst host galaxies. ArXiv e-prints (2015)

B. Milvang-Jensen, J.P.U. Fynbo, D. Malesani, J. Hjorth, P. Jakobsson, P. Møller, The Optically Unbiased GRB Host (TOUGH) survey. IV. Lyα emitters. Astrophys. J. **756**, 25 (2012). doi:10.1088/0004-637X/756/1/25

M. Modjaz, L. Kewley, R.P. Kirshner, K.Z. Stanek, P. Challis, P.M. Garnavich, J.E. Greene, P.L. Kelly, J.L. Prieto, Measured metallicities at the sites of nearby broad-lined type Ic supernovae and implications for the supernovae gamma-ray burst connection. Astron. J. **135**, 1136–1150 (2008). doi:10.1088/0004-6256/135/4/1136

M. Modjaz, L. Kewley, J.S. Bloom, A.V. Filippenko, D. Perley, J.M. Silverman, Progenitor diagnostics for stripped core-collapse supernovae: measured metallicities at explosion sites. Astrophys. J. Lett. **731**, 4 (2011). doi:10.1088/2041-8205/731/1/L4

D.M. Nataf, O.A. Gonzalez, L. Casagrande, G. Zasowski, C. Wegg, C. Wolf, A. Kunder, J. Alonso-Garcia, D. Minniti, M. Rejkuba, R.K. Saito, E. Valenti, M. Zoccali, R. Poleski, G. Pietrzynski, J. Skowron,

I. Soszynski, M.K. Szymanski, A. Udalski, K. Ulaczyk, L. Wyrzykowski, Interstellar extinction curve variations toward the Inner Milky Way: a challenge to observational cosmology. ArXiv e-prints (2015)

Y. Niino, Revisiting the metallicity of long-duration gamma-ray burst host galaxies: the role of chemical inhomogeneity within galaxies. Mon. Not. R. Astron. Soc. **417**, 567–572 (2011). doi:10.1111/j.1365-2966.2011.19299.x

Y. Niino, K. Nagamine, B. Zhang, Metallicity measurements of gamma-ray burst and supernova explosion sites: lessons from H II regions in M31. Mon. Not. R. Astron. Soc. **449**, 2706–2717 (2015). doi:10.1093/mnras/stv436

Y. Niino, J. Choi, M.A.R. Kobayashi, K. Nagamine, T. Totani, B. Zhang, Luminosity distribution of gamma-ray burst host galaxies at redshift $z = 1$ in cosmological smoothed particle hydrodynamic simulations: implications for the metallicity dependence of GRBs. Astrophys. J. **726**, 88 (2011). doi:10.1088/0004-637X/726/2/88

S. Nishiyama, T. Nagata, M. Tamura, R. Kandori, H. Hatano, S. Sato, K. Sugitani, The interstellar extinction law toward the galactic center. II. V, J, H, and K_S bands. Astrophys. J. **680**, 1174–1179 (2008). doi:10.1086/587791

S. Nishiyama, M. Tamura, H. Hatano, D. Kato, T. Tanabé, K. Sugitani, T. Nagata, Interstellar extinction law toward the galactic center III: J, H, K_S bands in the 2MASS and the MKO systems. Astrophys. J. **696**, 1407–1417 (2009). doi:10.1088/0004-637X/696/2/1407

S.E. Nuza, P.B. Tissera, L.J. Pellizza, D.G. Lambas, C. Scannapieco, M.E. de Rossi, The host galaxies of long-duration gamma-ray bursts in a cosmological hierarchical scenario. Mon. Not. R. Astron. Soc. **375**, 665–672 (2007). doi:10.1111/j.1365-2966.2006.11324.x

D. Perley, Y. Kemper, GRBOX: a new online catalog of GRBs, in *AIP Conf. Proc.*, ed. by M. Galassi, D. Palmer, E. Fenimore American Institute of Physics Conference Series, vol. 1000, 2008, pp. 631–634. doi:10.1063/1.2943549

D.A. Perley, J.S. Bloom, N.R. Butler, L.K. Pollack, J. Holtzman, C.H. Blake, D. Kocevski, W.T. Vestrand, W. Li, R.J. Foley, E. Bellm, H.-W. Chen, J.X. Prochaska, D. Starr, A.V. Filippenko, E.E. Falco, A.H. Szentgyorgyi, J. Wren, P.R. Wozniak, R. White, J. Pergande, The troublesome broadband evolution of GRB 061126: does a gray burst imply gray dust? Astrophys. J. **672**, 449–464 (2008). doi:10.1086/523929

D.A. Perley, S.B. Cenko, J.S. Bloom, H.-W. Chen, N.R. Butler, D. Kocevski, J.X. Prochaska, M. Brodwin, K. Glazebrook, M.M. Kasliwal, S.R. Kulkarni, S. Lopez, E.O. Ofek, M. Pettini, A.M. Soderberg, D. Starr, The host galaxies of swift dark gamma-ray bursts: observational constraints on highly obscured and very high redshift GRBs. Astron. J. **138**, 1690–1708 (2009). doi:10.1088/0004-6256/138/6/1690

D.A. Perley, J.S. Bloom, C.R. Klein, S. Covino, T. Minezaki, P. Woźniak, W.T. Vestrand, G.G. Williams, P. Milne, N.R. Butler, A.C. Updike, T. Krühler, P. Afonso, A. Antonelli, L. Cowie, P. Ferrero, J. Greiner, D.H. Hartmann, Y. Kakazu, A. Küpcü Yoldaş, A.N. Morgan, P.A. Price, J.X. Prochaska, Y. Yoshii, Evidence for supernova-synthesized dust from the rising afterglow of GRB071025 at $z \sim 5$. Mon. Not. R. Astron. Soc. **406**, 2473–2487 (2010). doi:10.1111/j.1365-2966.2010.16772.x

D.A. Perley, A.J. Levan, N.R. Tanvir, S.B. Cenko, J.S. Bloom, J. Hjorth, T. Krühler, A.V. Filippenko, A. Fruchter, J.P.U. Fynbo, P. Jakobsson, J. Kalirai, B. Milvang-Jensen, A.N. Morgan, J.X. Prochaska, J.M. Silverman, A population of massive, luminous galaxies hosting heavily dust-obscured gamma-ray bursts: implications for the use of GRBs as tracers of cosmic star formation. Astrophys. J. **778**, 128 (2013). doi:10.1088/0004-637X/778/2/128

D.A. Perley, R.A. Perley, J. Hjorth, M.J. Michałowski, S.B. Cenko, P. Jakobsson, T. Krühler, A.J. Levan, D. Malesani, N.R. Tanvir, Connecting GRBs and ULIRGs: a sensitive, unbiased survey for radio emission from gamma-ray burst host galaxies at $0 < z < 2.5$. Astrophys. J. **801**, 102 (2015)

D.A. Perley, T. Krühler, S. Schulze, A. de Ugarte Postigo, J. Hjorth, E. Berger, S.B. Cenko, R. Chary, A. Cucchiara, R. Ellis, W. Fong, J.P.U. Fynbo, J. Gorosabel, J. Greiner, P. Jakobsson, T. Laskar, A.J. Levan, M.J. Michałowski, B. Milvang-Jensen, N.R. Tanvir, C.C. Thöne, K. Wiersema, The swift gamma-ray burst host galaxy legacy survey—I. Sample selection and redshift distribution. Astrophys. J. **817**(1), 7 (2016a), 23 pp. doi:10.3847/0004-637X/817/1/7

D.A. Perley, N.R. Tanvir, J. Hjorth, T. Laskar, E. Berger, R. Chary, A. de Ugarte Postigo, J.P.U. Fynbo, T. Krühler, A.J. Levan, M.J. Michałowski, S. Schulze, The swift GRB host galaxy legacy survey—II. Rest-frame NIR luminosity distribution and evidence for a near-solar metallicity threshold. Astrophys. J. **817**(1), 8 (2016b), 17 pp. doi:10.3847/0004-637X/817/1/8

R. Perna, D. Lazzati, Time-dependent photozionization in a dusty medium. I. Code description and general results. Astrophys. J. **570**, 271–277 (2002)

S. Piranomonte, J. Japelj, S.D. Vergani, S. Savaglio, E. Palazzi, S. Covino, H. Flores, P. Goldoni, G. Cupani, T. Krühler, F. Mannucci, F. Onori, A. Rossi, V. D'Elia, E. Pian, P. D'Avanzo, A. Gomboc, F. Hammer, S. Randich, F. Fiore, L. Stella, G. Tagliaferri, GRB host galaxies with VLT/X-shooter: properties at $z = 0.8$–1.3. Mon. Not. R. Astron. Soc. **452**, 3293–3303 (2015). doi:10.1093/mnras/stv1569

P. Podsiadlowski, N. Ivanova, S. Justham, S. Rappaport, Explosive common-envelope ejection: implications for gamma-ray bursts and low-mass black-hole binaries. Mon. Not. R. Astron. Soc. **406**, 840–847 (2010). doi:10.1111/j.1365-2966.2010.16751.x

A. Pontzen, A. Deason, F. Governato, M. Pettini, J. Wadsley, T. Quinn, A. Brooks, J. Bellovary, J.P.U. Fynbo, The nature of HI absorbers in gamma-ray burst afterglows: clues from hydrodynamic simulations. Mon. Not. R. Astron. Soc. **402**, 1523–1535 (2010). doi:10.1111/j.1365-2966.2009.16017.x

J.X. Prochaska, A.M. Wolfe, The UCSD HIRES/Keck I Damped Lyα abundance database. II. The implications. Astrophys. J. **566**, 68–92 (2002). doi:10.1086/338080

J.X. Prochaska, H.-W. Chen, J.S. Bloom, Dissecting the circumstellar environment of γ-ray burst progenitors. Astrophys. J. **648**, 95–110 (2006). doi:10.1086/505737

J.X. Prochaska, M. Dessauges-Zavadsky, E. Ramirez-Ruiz, H.-W. Chen, A survey for N V absorption at $z \sim z_{GRB}$ in GRB afterglow spectra: clues to gas near the progenitor star. Astrophys. J. **685**, 344–353 (2008). doi:10.1086/590529

J.X. Prochaska, Y. Sheffer, D.A. Perley, J.S. Bloom, L.A. Lopez, M. Dessauges-Zavadsky, H.-W. Chen, A.V. Filippenko, M. Ganeshalingam, W. Li, A.A. Miller, D. Starr, The first positive detection of molecular gas in a GRB host galaxy. Astrophys. J. Lett. **691**, 27–32 (2009). doi:10.1088/0004-637X/691/1/L27

C. Raskin, E. Scannapieco, J. Rhoads, M. Della Valle, Using spatial distributions to constrain progenitors of supernovae and gamma-ray bursts. Astrophys. J. **689**, 358–370 (2008). doi:10.1086/592495

A.M. Ritchey, D.E. Welty, J.A. Dahlstrom, D.G. York, Diffuse atomic and molecular gas in the interstellar medium of M82 toward SN 2014J. ArXiv e-prints (2014)

E. Rol, R.A.M.J. Wijers, C. Kouveliotou, L. Kaper, Y. Kaneko, How special are dark gamma-ray bursts: a diagnostic tool. Astrophys. J. **624**, 868–879 (2005). doi:10.1086/429082

E. Rol, A. van der Horst, K. Wiersema, S.K. Patel, A. Levan, M. Nysewander, C. Kouveliotou, R.A.M.J. Wijers, N. Tanvir, D. Reichart, A.S. Fruchter, J. Graham, J.-E. Ovaldsen, A.O. Jaunsen, P. Jonker, W. van Ham, J. Hjorth, R.L.C. Starling, P.T. O'Brien, J. Fynbo, D.N. Burrows, R. Strom, GRB 051022: physical parameters and extinction of a prototype dark burst. Astrophys. J. **669**, 1098–1106 (2007). doi:10.1086/521336

A. Rossi, S. Klose, P. Ferrero, J. Greiner, L.A. Arnold, E. Gonsalves, D.H. Hartmann, A.C. Updike, D.A. Kann, T. Krühler, E. Palazzi, S. Savaglio, S. Schulze, P.M.J. Afonso, L. Amati, A.J. Castro-Tirado, C. Clemens, R. Filgas, J. Gorosabel, L.K. Hunt, A. Küpcü Yoldaş, N. Masetti, M. Nardini, A. Nicuesa Guelbenzu, F.E. Olivares, E. Pian, A. Rau, P. Schady, S. Schmidl, A. Yoldaş, A. de Ugarte Postigo, A deep search for the host galaxies of gamma-ray bursts with no detected optical afterglow. Astron. Astrophys. **545**, 77 (2012). doi:10.1051/0004-6361/201117201

R. Salvaterra, S. Campana, S.D. Vergani, S. Covino, P. D'Avanzo, D. Fugazza, G. Ghirlanda, G. Ghisellini, A. Melandri, L. Nava, B. Sbarufatti, H. Flores, S. Piranomonte, G. Tagliaferri, A complete sample of bright swift long gamma-ray bursts. I. Sample presentation, luminosity function and evolution. Astrophys. J. **749**, 68 (2012). doi:10.1088/0004-637X/749/1/68

R. Salvaterra, U. Maio, B. Ciardi, M.A. Campisi, Simulating high-z gamma-ray burst host galaxies. Mon. Not. R. Astron. Soc. **429**, 2718–2726 (2013). doi:10.1093/mnras/sts541

N.E. Sanders, N. Caldwell, J. McDowell, P. Harding, The metallicity profile of M31 from spectroscopy of hundreds of H II regions and PNe. Astrophys. J. **758**, 133 (2012). doi:10.1088/0004-637X/758/2/133

B.D. Savage, K.R. Sembach, Interstellar gas-phase abundances and physical conditions toward two distant high-latitude Halo stars. Astrophys. J. **470**, 893 (1996). doi:10.1086/177919

S. Savaglio, S.M. Fall, Dust depletion and extinction in a gamma-ray burst afterglow. Astrophys. J. **614**, 293–300 (2004). doi:10.1086/423447

S. Savaglio, S.M. Fall, F. Fiore, Heavy-element abundances and dust depletions in the host galaxies of three gamma-ray bursts. Astrophys. J. **585**, 638–646 (2003). doi:10.1086/346225

S. Savaglio, K. Glazebrook, D. Le Borgne, The galaxy population hosting gamma-ray bursts. Astrophys. J. **691**, 182–211 (2009). doi:10.1088/0004-637X/691/1/182

P. Schady, M.J. Page, S.R. Oates, M. Still, M. de Pasquale, T. Dwelly, N.P.M. Kuin, S.T. Holland, F.E. Marshall, P.W.A. Roming, Dust and metal column densities in gamma-ray burst host galaxies. Mon. Not. R. Astron. Soc. **401**, 2773–2792 (2010). doi:10.1111/j.1365-2966.2009.15861.x

P. Schady, T. Dwelly, M.J. Page, T. Krühler, J. Greiner, S.R. Oates, M. de Pasquale, M. Nardini, P.W.A. Roming, A. Rossi, M. Still, The dust extinction curves of gamma-ray burst host galaxies. Astron. Astrophys. **537**, 15 (2012). doi:10.1051/0004-6361/201117414

P. Schady, S. Savaglio, T. Müller, T. Krühler, T. Dwelly, E. Palazzi, L.K. Hunt, J. Greiner, H. Linz, M.J. Michałowski, D. Pierini, S. Piranomonte, S.D. Vergani, W.K. Gear, Herschel observations of gamma-ray burst host galaxies: implications for the topology of the dusty interstellar medium. Astron. Astrophys. **570**, 52 (2014). doi:10.1051/0004-6361/201424092

S. Schulze, R. Chapman, J. Hjorth, A.J. Levan, P. Jakobsson, G. Björnsson, D.A. Perley, T. Krühler, J. Gorosabel, N.R. Tanvir, A. de Ugarte Postigo, J.P.U. Fynbo, B. Milvang-Jensen, P. Møller, D.J. Watson, The Optically Unbiased GRB Host (TOUGH) survey. VII. The host galaxy luminosity function: probing the relationship between GRBs and star formation to redshift 6. Astrophys. J. **808**, 73 (2015). doi:10.1088/0004-637X/808/1/73

S.J. Smartt, K.A. Venn, P.L. Dufton, D.J. Lennon, W.R.J. Rolleston, F.P. Keenan, Chemical abundances in the inner 5 kpc of the galactic disk. Astron. Astrophys. **367**, 86–105 (2001). doi:10.1051/0004-6361+20000479

A.M. Soderberg, E. Nakar, E. Berger, S.R. Kulkarni, Late-time radio observations of 68 type Ibc supernovae: strong constraints on off-axis gamma-ray bursts. Astrophys. J. **638**, 930–937 (2006). doi:10.1086/499121

M. Sparre, O.E. Hartoog, T. Krühler, J.P.U. Fynbo, D.J. Watson, K. Wiersema, V. D'Elia, T. Zafar, P.M.J. Afonso, S. Covino, A. de Ugarte Postigo, H. Flores, P. Goldoni, J. Greiner, J. Hjorth, P. Jakobsson, L. Kaper, S. Klose, A.J. Levan, D. Malesani, B. Milvang-Jensen, M. Nardini, S. Piranomonte, J. Sollerman, R. Sánchez-Ramírez, S. Schulze, N.R. Tanvir, S.D. Vergani, R.A.M.J. Wijers, The metallicity and dust content of a redshift 5 gamma-ray burst host galaxy. Astrophys. J. **785**, 150 (2014). doi:10.1088/0004-637X/785/2/150

K.Z. Stanek, O.Y. Gnedin, J.F. Beacom, A.P. Gould, J.A. Johnson, J.A. Kollmeier, M. Modjaz, M.H. Pinsonneault, R. Pogge, D.H. Weinberg, Protecting life in the Milky Way: metals keep the GRBs away. Acta Astron. **56**, 333–345 (2006)

E.R. Stanway, M.N. Bremer, N.R. Tanvir, A.J. Levan, L.J.M. Davies, Constraining the molecular gas in the environs of a $z = 8$ gamma-ray burst host galaxy. Mon. Not. R. Astron. Soc. **410**, 1496–1502 (2011). doi:10.1111/j.1365-2966.2010.17534.x

R.L.C. Starling, R.A.M.J. Wijers, K. Wiersema, E. Rol, P.A. Curran, C. Kouveliotou, A.J. van der Horst, M.H.M. Heemskerk, Gamma-ray burst afterglows as probes of environment and blast wave physics. I. Absorption by host-galaxy gas and dust. Astrophys. J. **661**, 787–800 (2007). doi:10.1086/511953

R.L.C. Starling, K. Wiersema, A.J. Levan, T. Sakamoto, D. Bersier, P. Goldoni, S.R. Oates, A. Rowlinson, S. Campana, J. Sollerman, N.R. Tanvir, D. Malesani, J.P.U. Fynbo, S. Covino, P. D'Avanzo, P.T. O'Brien, K.L. Page, J.P. Osborne, S.D. Vergani, S. Barthelmy, D.N. Burrows, Z. Cano, P.A. Curran, M. de Pasquale, V. D'Elia, P.A. Evans, H. Flores, A.S. Fruchter, P. Garnavich, N. Gehrels, J. Gorosabel, J. Hjorth, S.T. Holland, A.J. van der Horst, C.P. Hurkett, P. Jakobsson, A.P. Kamble, C. Kouveliotou, N.P.M. Kuin, L. Kaper, P.A. Mazzali, P.E. Nugent, E. Pian, M. Stamatikos, C.C. Thöne, S.E. Woosley, Discovery of the nearby long, soft GRB 100316D with an associated supernova. Mon. Not. R. Astron. Soc. **411**, 2792–2803 (2011). doi:10.1111/j.1365-2966.2010.17879.x

R.L.C. Starling, R. Willingale, N.R. Tanvir, A.E. Scott, K. Wiersema, P.T. O'Brien, A.J. Levan, G.C. Stewart, X-ray absorption evolution in gamma-ray bursts: intergalactic medium or evolutionary signature of their host galaxies. Mon. Not. R. Astron. Soc. **431**, 3159–3176 (2013). doi:10.1093/mnras/stt400

T. Sumi, Extinction map of the galactic centre: OGLE-II galactic bulge fields. Mon. Not. R. Astron. Soc. **349**, 193–204 (2004). doi:10.1111/j.1365-2966.2004.07482.x

K.M. Svensson, A.J. Levan, N.R. Tanvir, A.S. Fruchter, L.-G. Strolger, The host galaxies of core-collapse supernovae and gamma-ray bursts. Mon. Not. R. Astron. Soc. **405**, 57–76 (2010). doi:10.1111/j.1365-2966.2010.16442.x

K.M. Svensson, A.J. Levan, N.R. Tanvir, D.A. Perley, M.J. Michalowski, K.L. Page, J.S. Bloom, S.B. Cenko, J. Hjorth, P. Jakobsson, D. Watson, P.J. Wheatley, The dark GRB 080207 in an extremely red host and the implications for gamma-ray bursts in highly obscured environments. Mon. Not. R. Astron. Soc. **421**, 25–35 (2012). doi:10.1111/j.1365-2966.2011.19811.x

N.R. Tanvir, V.E. Barnard, A.W. Blain, A.S. Fruchter, C. Kouveliotou, P. Natarajan, E. Ramirez-Ruiz, E. Rol, I.A. Smith, R.P.J. Tilanus, R.A.M.J. Wijers, The submillimetre properties of gamma-ray burst host galaxies. Mon. Not. R. Astron. Soc. **352**, 1073–1080 (2004). doi:10.1111/j.1365-2966.2004.08001.x

N.R. Tanvir, A.J. Levan, A.S. Fruchter, J.P.U. Fynbo, J. Hjorth, K. Wiersema, M.N. Bremer, J. Rhoads, P. Jakobsson, P.T. O'Brien, E.R. Stanway, D. Bersier, P. Natarajan, J. Greiner, D. Watson, A.J. Castro-Tirado, R.A.M.J. Wijers, R.L.C. Starling, K. Misra, J.F. Graham, C. Kouveliotou, Star formation in the early universe: beyond the tip of the iceberg. Astrophys. J. **754**, 46 (2012). doi:10.1088/0004-637X/754/1/46

C.C. Thöne, J.P.U. Fynbo, G. Östlin, B. Milvang-Jensen, K. Wiersema, D. Malesani, D. Della Monica Ferreira, J. Gorosabel, D.A. Kann, D. Watson, M.J. Michałowski, A.S. Fruchter, A.J. Levan, J. Hjorth, J. Sollerman, Spatially resolved properties of the GRB 060505 host: implications for the nature of the progenitor. Astrophys. J. **676**, 1151–1161 (2008). doi:10.1086/528943

C.C. Thöne, J.P.U. Fynbo, P. Goldoni, A.P. de Ugarte, S. Campana, S.D. Vergani, S. Covino, T. Krühler, L. Kaper, N. Tanvir, T. Zafar, V. D'Elia, J. Gorosabel, J. Greiner, P. Groot, F. Hammer, P. Jakobsson, S. Klose, A.J. Levan, B. Milvang-Jensen, A.G. Nicuesa, E. Palazzi, S. Piranomonte, G. Tagliaferri, D. Watson, K. Wiersema, R.A.M.J. Wijers, GRB 100219A with X-shooter—abundances in a galaxy at $z = 4.7$. Mon. Not. R. Astron. Soc. **428**, 3590–3606 (2013). doi:10.1093/mnras/sts303

T. Totani, K. Aoki, T. Hattori, G. Kosugi, Y. Niino, T. Hashimoto, N. Kawai, K. Ohta, T. Sakamoto, T. Yamada, Probing intergalactic neutral hydrogen by the Lyman alpha red damping wing of gamma-ray burst 130606A afterglow spectrum at $z = 5.913$. Publ. Astron. Soc. Jpn. **66**, 63 (2014). doi:10.1093/pasj/psu032

C.A. Tremonti, T.M. Heckman, G. Kauffmann, J. Brinchmann, S. Charlot, S.D.M. White, M. Seibert, E.W. Peng, D.J. Schlegel, A. Uomoto, M. Fukugita, J. Brinkmann, The origin of the mass-metallicity relation: insights from 53,000 star-forming galaxies in the Sloan digital sky survey. Astrophys. J. **613**, 898–913 (2004). doi:10.1086/423264

M. Trenti, R. Perna, R. Jimenez, The luminosity and stellar mass functions of GRB host galaxies: insight into the metallicity bias. Astrophys. J. **802**, 103 (2015). doi:10.1088/0004-637X/802/2/103

M. Trenti, R. Perna, E.M. Levesque, J.M. Shull, J.T. Stocke, Gamma-ray burst host galaxy surveys at redshift $z \gtrsim 4$: probes of star formation rate and cosmic reionization. Astrophys. J. Lett. **749**, 38 (2012). doi:10.1088/2041-8205/749/2/L38

E.P.J. van den Heuvel, S.F. Portegies Zwart, Are superluminous supernovae and long GRBs the products of dynamical processes in young dense star clusters? Astrophys. J. **779**, 114 (2013). doi:10.1088/0004-637X/779/2/114

A.J. van der Horst, C. Kouveliotou, N. Gehrels, E. Rol, R.A.M.J. Wijers, J.K. Cannizzo, J. Racusin, D.N. Burrows, Optical classification of gamma-ray bursts in the swift era. Astrophys. J. **699**, 1087–1091 (2009). doi:10.1088/0004-637X/699/2/1087

S.D. Vergani, S. Piranomonte, P. Petitjean, H. Flores, P. Goldoni, M. Rodrigues, F. Hammer, S. Covino, GRB 021004 host galaxy and environment with X-shooter. Astron. Nachr. **332**, 292 (2011). doi:10.1002/asna.201111538

S.D. Vergani, R. Salvaterra, J. Japelj, E. Le Floc'h, P. D'Avanzo, A. Fernandez-Soto, T. Krühler, A. Melandri, S. Boissier, S. Covino, M. Puech, J. Greiner, L.K. Hunt, D. Perley, P. Petitjean, T. Vinci, F. Hammer, A. Levan, F. Mannucci, S. Campana, H. Flores, A. Gomboc, G. Tagliaferri, Are long gamma-ray bursts biased tracers of star formation? Clues from the host galaxies of the Swift/BAT6 complete sample of LGRBs. I. Stellar mass at $z < 1$. Astron. Astrophys. **581**, 102 (2015). doi:10.1051/0004-6361/201425013

J. Vernet, H. Dekker, S. D'Odorico, L. Kaper, P. Kjaergaard, F. Hammer, S. Randich, F. Zerbi, P.J. Groot, J. Hjorth, I. Guinouard, R. Navarro, T. Adolfse, P.W. Albers, J.-P. Amans, J.J. Andersen, M.I. Andersen, P. Binetruy, P. Bristow, R. Castillo, F. Chemla, L. Christensen, P. Conconi, R. Conzelmann, J. Dam, V. de Caprio, A. de Ugarte Postigo, B. Delabre, P. di Marcantonio, M. Downing, E. Elswijk, G. Finger, G. Fischer, H. Flores, P. François, P. Goldoni, L. Guglielmi, R. Haigron, H. Hanenburg, I. Hendriks, M. Horrobin, D. Horville, N.C. Jessen, F. Kerber, L. Kern, M. Kiekebusch, P. Kleszcz, J. Klougart, J. Kragt, H.H. Larsen, J.-L. Lizon, C. Lucuix, V. Mainieri, R. Manuputy, C. Martayan, E. Mason, R. Mazzoleni, N. Michaelsen, A. Modigliani, S. Moehler, P. Møller, A. Norup Sørensen, P. Nørregaard, C. Péroux, F. Patat, E. Pena, J. Pragt, C. Reinero, F. Rigal, M. Riva, R. Roelfsema, F. Royer, G. Sacco, P. Santin, T. Schoenmaker, P. Spano, E. Sweers, R. Ter Horst, M. Tintori, N. Tromp, P. van Dael, H. van der Vliet, L. Venema, M. Vidali, J. Vinther, P. Vola, R. Winters, D. Wistisen, G. Wulterkens, A. Zacchei, X-shooter, the new wide band intermediate resolution spectrograph at the ESO Very Large Telescope. Astron. Astrophys. **536**, 105 (2011). doi:10.1051/0004-6361/201117752

J.S. Vink, A. de Koter, On the metallicity dependence of Wolf-Rayet winds. Astron. Astrophys. **442**, 587–596 (2005). doi:10.1051/0004-6361+20052862

G. Vladilo, M. Centurión, S.A. Levshakov, C. Péroux, P. Khare, V.P. Kulkarni, D.G. York, Extinction and metal column density of HI regions up to redshift $z \simeq 2$. Astron. Astrophys. **454**, 151–164 (2006). doi:10.1051/0004-6361+20054742

P.M. Vreeswijk, A. Smette, D. Malesani, J.P.U. Fynbo, B. Milvang-Jensen, P. Jakobsson, A.O. Jaunsen, U. Oslo, C. Ledoux, VLT/UVES redshift of GRB 080319B. GRB Coordinates Network **7444** (2008)

P.M. Vreeswijk, C. Ledoux, A. De Cia, A. Smette, Probing the environment of GRBs with absorption-line spectroscopy. Mem. Soc. Astron. Ital. Suppl. **21**, 14 (2012)

C. Wainwright, E. Berger, B.E. Penprase, A morphological study of gamma-ray burst host galaxies. Astrophys. J. **657**, 367–377 (2007). doi:10.1086/510794

F.Y. Wang, Z.G. Dai, An evolving stellar initial mass function and the gamma-ray burst redshift distribution. Astrophys. J. Lett. **727**, 34 (2011). doi:10.1088/2041-8205/727/2/L34

W.-H. Wang, H.-W. Chen, K.-Y. Huang, ALMA submillimeter continuum imaging of the host galaxies of GRB 021004 and GRB 080607. Astrophys. J. Lett. **761**, 32 (2012). doi:10.1088/2041-8205/761/2/L32

D. Watson, Probing dust with gamma-ray bursts, in *Cosmic Dust—Near and Far*, ed. by T. Henning, E. Grün, J. Steinacker Astronomical Society of the Pacific Conference Series, vol. 414, 2009, p. 277

D. Watson, P. Jakobsson, Dust extinction bias in the column density distribution of gamma-ray bursts: high column density, low-redshift GRBs are more heavily obscured. Astrophys. J. **754**, 89 (2012). doi:10.1088/0004-637X/754/2/89

D. Watson, J.P.U. Fynbo, C. Ledoux, P. Vreeswijk, J. Hjorth, A. Smette, A.C. Andersen, K. Aoki, T. Augusteijn, A.P. Beardmore, D. Bersier, J.M. Castro Cerón, P. D'Avanzo, D. Diaz-Fraile, J. Gorosabel, P. Hirst, P. Jakobsson, B.L. Jensen, N. Kawai, G. Kosugi, P. Laursen, A. Levan, J. Masegosa, J. Näränen, K.L. Page, K. Pedersen, A. Pozanenko, J.N. Reeves, V. Rumyantsev, T. Shahbaz, D. Sharapov, J. Sollerman, R.L.C. Starling, N. Tanvir, K. Torstensson, K. Wiersema, A log $N_{HI} = 22.6$ damped Lyα absorber in a dark gamma-ray burst: the environment of GRB 050401. Astrophys. J. **652**, 1011–1019 (2006). doi:10.1086/508049

D. Watson, T. Zafar, A.C. Andersen, J.P.U. Fynbo, J. Gorosabel, J. Hjorth, P. Jakobsson, T. Krühler, P. Laursen, G. Leloudas, D. Malesani, Helium in Natal H II regions: the origin of the x-ray absorption in gamma-ray burst afterglows. Astrophys. J. **768**, 23 (2013). doi:10.1088/0004-637X/768/1/23

E. Waxman, B.T. Draine, Dust sublimation by gamma-ray bursts and its implications. Astrophys. J. **537**, 796–802 (2000). doi:10.1086/309053

C. Wolf, P. Podsiadlowski, The metallicity dependence of the long-duration gamma-ray burst rate from host galaxy luminosities. Mon. Not. R. Astron. Soc. **375**, 1049–1058 (2007). doi:10.1111/j.1365-2966.2006.11373.x

S.E. Woosley, A. Heger, The progenitor stars of gamma-ray bursts. Astrophys. J. **637**, 914–921 (2006). doi:10.1086/498500

S.-C. Yoon, N. Langer, Evolution of rapidly rotating metal-poor massive stars towards gamma-ray bursts. Astron. Astrophys. **443**, 643–648 (2005). doi:10.1051/0004-6361+20054030

S.-C. Yoon, N. Langer, C. Norman, Single star progenitors of long gamma-ray bursts. I. Model grids and redshift dependent GRB rate. Astron. Astrophys. **460**, 199–208 (2006). doi:10.1051/0004-6361+20065912

T. Zafar, D. Watson, The metals-to-dust ratio to very low metallicities using GRB and QSO absorbers; extremely rapid dust formation. Astron. Astrophys. **560**, 26 (2013). doi:10.1051/0004-6361/201321413

T. Zafar, D. Watson, J.P.U. Fynbo, D. Malesani, P. Jakobsson, A. de Ugarte Postigo, The extinction curves of star-forming regions from $z = 0.1$ to 6.7 using GRB afterglow spectroscopy. Astron. Astrophys. **532**, 143 (2011). doi:10.1051/0004-6361/201116663

H.J. Zahid, G.I. Dima, R.-P. Kudritzki, L.J. Kewley, M.J. Geller, H.S. Hwang, J.D. Silverman, D. Kashino, The universal relation of galactic chemical evolution: the origin of the mass-metallicity relation. Astrophys. J. **791**, 130 (2014). doi:10.1088/0004-637X/791/2/130

Space Sci Rev (2016) 202:145–160
DOI 10.1007/s11214-016-0253-4

GRBs as Probes of the IGM

Antonino Cucchiara[1] · Tonomori Totani[2] · Nial Tanvir[3]

Received: 22 October 2015 / Accepted: 9 April 2016 / Published online: 6 May 2016

Abstract Gamma-ray Bursts (GRBs) are the most powerful explosions known, capable of outshining the rest of gamma-ray sky during their short-lived prompt emission. Their cosmological nature makes them the best tool to explore the final stages in the lives of very massive stars up to the highest redshifts. Furthermore, studying the emission from their low-energy counterparts (optical and infrared) via rapid spectroscopy, we have been able to pin down the exact location of the most distant galaxies as well as placing stringent constraints on their host galaxies and intervening systems at low and high-redshift (e.g. metallicity and neutral hydrogen fraction). In fact, each GRB spectrum contains absorption features imprinted by metals in the host interstellar medium (ISM) as well as the intervening intergalactic medium (IGM) along the line of sight. In this chapter we summarize the progress made using a large dataset of GRB spectra in understanding the nature of both these absorbers and how GRBs can be used to study the early Universe, in particular to measure the neutral hydrogen fraction and the escape fraction of UV photons before and during the epoch of re-ionization.

Keywords Gamma-ray bursts · Interstellar medium · Intergalactic medium · Re-ionization · Cosmology

1 Introduction

Gamma-ray Bursts (GRB) have been observed from $z = 0.03$ (Pian et al. 2006) to $z \sim 9$ (Cucchiara et al. 2011; Salvaterra et al. 2009; Tanvir et al. 2009) and, similar to quasars (QSOs), their X-ray/optical/NIR afterglows intersect the interstellar medium (ISM) and intergalactic medium (IGM) along their lines of sight (e.g. Jakobsson et al. 2004, 2012; Behar

✉ A. Cucchiara
acucchiara@stsci.edu

1 Space Telescope Science Institute, 3700 San Martin Drive, Baltimore, United States

2 Department of Astronomy, The University of Tokyo, Hongo, Tokyo 113-0033, Japan

3 Department of Physics and Astronomy, University of Leicester, University Road, Leicester, LE1 7RH, UK

et al. 2011; Starling et al. 2013; Campana et al. 2015): from matter near to the host (local ISM, Zafar et al. 2010) to other galaxies (intervening ISM and/or IGM) at lower redshift (and different impact parameters). The gas located in each of these components leaves an imprint clearly identified in GRB (and QSO) optical and NIR spectra by the presence of absorption lines.

Therefore, absorption spectroscopy allows us to disentangle all these absorption systems allowing us to discern not only their properties (e.g. metal content and gas kinematics of these galaxies), but also the overall characteristics of the cosmic IGM during and after the end of re-ionization in a very similar fashion to QSOs (see Fan et al. 2006, for a review).

In the following sections we summarize the advantage of using GRB afterglows for these studies with respect to QSOs, the current status of studying the chemical evolution of the Universe using GRB intervening systems, and the investigation of the cosmic IGM (e.g. neutral hydrogen fraction) based on recent GRB spectroscopic datasets. We will also emphasize some of the drawback of using GRBs, e.g., their transient nature and the low number of events in comparison with QSOs over most redshifts.

2 Intervening Systems

Quasar lines of sight have been explored in depth in recent years thanks to the availability of large spectroscopic surveys, like the Sloan Digital Sky Surveys (York et al. 2000). Since they are bright background sources, they probe gas and matter located in foreground objects at different impact parameters. These intervening systems can be identified in the QSO spectra based on metal absorption features at redshifts lower than the QSO systemic redshift (usually determined by Ly α, C IV, and Mg II emission lines). However, the complexity of the intrinsic QSO continuum and the decreased numbers of bright QSOs at high-redshift makes the study of the intervening systems challenging. Nevertheless, thanks to several observing campaigns spanning the near-UV to the near-Infrared, a large sample of such objects has been collected: the identification of Ly α lines at the intervening system redshifts has offered a unique opportunity to use them as tracers of the metal enrichment of the Universe from low-redshift (Lehner et al. 2013; Muzahid et al. 2016; Cooper et al. 2015) to $z \sim 5$. Furthermore, Mg II intervening absorbers have been identified in several QSO spectra: in the last decade the presence and metal content of such absorbers have been matter of debate, in particular regarding the role of such absorbers in fuelling star-formation and/or tracing cosmological galaxy build up.

Both types of absorbers, Ly α and Mg II, have been also identified in GRB afterglow spectra, the former indicating the presence of large reservoir of neutral hydrogen in the GRB host galaxy, while the latter are often produced by foreground absorbers along the GRB line of sight. The synchrotron emission responsible for the simple power-law afterglow continuum makes the identification and measurement of absorption lines easier than QSOs, but the rarity and rapidly fading nature of the emission means that much lower numbers of such systems have been identified. Nevertheless, GRBs have been spectroscopically identified up to $z = 8.2$, providing unprecedented tool to study the epoch of re-ionization. Thanks largely to the sample of GRBs discovered by the *Swift* satellite (Gehrels 2004), and the identification of their afterglows in many cases (from X-ray to optical and NIR), we have been able to investigate samples of these absorbers in considerable detail. In the following sub-sections we will briefly discuss these two types of absorbers, though we direct the reader to the *Host galaxy* chapter for a more complete view of the GRB hosts properties.

2.1 Lyman-α Absorbers

Intervening systems for which a Ly α line is identified in absorption are subdivided based on the total hydrogen column density: if $\log N_{\rm HI} \geq 20.3$ cm^{-2} they are called "Damped Lyman-α systems" (DLAs), while if $\log N_{\rm HI} < 20.3$ cm^{-2}, the systems are generically called sub-DLAs (see Wolfe et al. 2005, for a detailed review on such absorbers and further subdivisions). This cold gas usually surrounds galaxies and can be used to characterize the circumgalactic (CGM) medium (Stewart et al. 2011; Stinson et al. 2012; Rudie et al. 2012) testing theoretical predictions (e.g. Nagamine et al. 2004; Berry et al. 2014; Cen 2012; Razoumov et al. 2008; Fumagalli et al. 2011). Because of their connection with star-formation, DLAs represent important laboratories to chart the metal enrichment of the Universe, from the end of re-ionization to the peak of star-formation ($z \sim 2$) and at lower redshifts. Recently, Rafelski et al. (2014) have shown how metallicity decreases with redshift up to $z \sim 4.5$ in a large QSO-DLA sample. Furthermore, these authors suggest that at even higher redshifts there is a steepening in the decline of cosmic metallicity, probably due to an increase of the covering fraction of neutral gas as function of redshift. Ly α absorbers have also been identified in the GRB afterglow spectra and offer a unique possibility to obtain not only higher signal-to-noise ratio spectra (bright GRB afterglows observed early are usually brighter than QSOs at any redshift) but also to trace the stellar content of such DLAs (GRB-DLAs): in fact, GRB-DLAs are associated with the host galaxies enabling, once the afterglow fades away, detailed studies of DLAs itself.

In 2015, Cucchiara et al. (2015) compiled the largest sample of spectroscopic afterglow data available and determined uniformly the metallicities of 54 GRB-DLAs from weak metal absorption lines (like S II Si II). They compared this sample with the QSO-DLA results from Rafelski et al. and, despite having a smaller sample, showed a higher metallicity around GRB-DLAs at $z = 3$–5, roughly 10 % the solar value (a factor of two higher than the one measured for QSO-DLAs). These authors suggest that GRB-DLAs may be the best tracer of the interconnection between DLAs and metal build up, especially at high redshift since usually GRBs occur within their host, in close proximity to star-forming regions. The host galaxy metallicity determination obtained by absorption lines diagnostics is an important aspect of cold gas fuelling star-formation and the advent of NIR spectrographs will enable comparisons with emission line (integrated over the whole galaxy) diagnostics, a critical step forward on our understanding of star-formation in the local Universe and at the highest redshifts. Furthermore, understanding the nature of the galaxies hosting these absorbers will provide insights on galaxy formation and evolution. The data required to perform such analyses (e.g. extensive multiband observations) is demanding because of the faint nature of the sources and the need for space-based observatory time in many cases.

2.2 Mg II Absorbers

One of the most studied and commonly identified feature is the Mg II doublet at $\lambda\lambda 2796, 2803$ Å, thanks to its large rest wavelength (which makes it them detectable at $z = 0.5$–2.2), the relatively high abundance, and its oscillator strength.

The Mg II systems are usually classified in terms of the rest-frame equivalent width, W_r, of the bluer component as "weak" ($W_{2796} < 0.3$ Å), "strong" ($W_{2796} > 0.3$ Å) as in Steidel and Sargent (1992) and Churchill et al. (1999), and "very strong" ($W_{2796} > 1.0$ Å, like in Rodríguez Hidalgo et al. 2012).

In the last few years several surveys have expanded our samples of Mg II intervening systems up to $z = 5.2$, thanks to near-IR spectroscopy (e.g. Steidel and Sargent 1992; Nestor

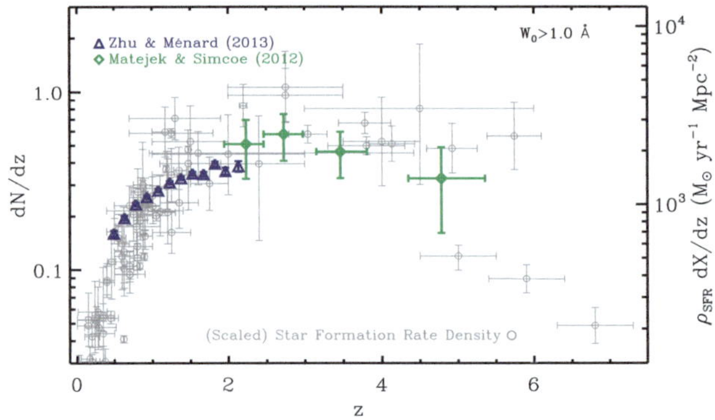

Fig. 1 Incident rate dN/dz of very strong Mg II absorbers ($W_{2796} > 1.0$ Å). *Blue points* are from Zhu and Ménard (2013) based on SDSS data at moderate redshifts, while the *green points* show near-infrared measurements from Matejek and Simcoe (2012). On the *right axes* we indicate the cosmic star formation rate density, $\dot{\rho}_* = dM_*/dtdAdX$, scaled by dX/dz, where A is comoving area and X is comoving distance

et al. 2005; Prochter et al. 2006a; Quider et al. 2011; Simcoe et al. 2011; Zhu and Ménard 2013). These surveys indicate that the very strong Mg II absorbers present an increasing trend up to $z \sim 3$ before declining at higher redshift (see Fig. 1, Prochter et al. 2006a; Matejek and Simcoe 2012). Intriguingly, this behavior closely tracks the cosmic star formation history (Prochter et al. 2006a; Zhu and Ménard 2013), suggesting that some systems may be causally connected to on-going star formation (Ménard et al. 2011; Matejek and Simcoe 2012).

For many years, strong Mg II absorbers have been associated with galaxies ($L \approx L^*$) at modest impact parameter and gas around them (Bergeron 1986; Lanzetta et al. 1987; Steidel 1993), either in the outer disks and/or the circumgalactic medium (CGM).

More recently, QSO lines of sight have been explored in order to probe the baryon content around low-z galaxies (Kacprzak et al. 2012; Chen and Tinker 2008, and reference therein), as well as a diagnostic of the inner part of these galaxies' interstellar medium (Bowen et al. 1995).

In a similar fashion, Gamma-ray Bursts (GRBs) provide both information on their hosts and on the intercepting matter along their lines of sight (Metzger et al. 1997). There are two main advantages of using these powerful sources: first, they can be observed up to very high redshifts (Kawai et al. 2006; Tanvir et al. 2009; Salvaterra et al. 2009; Cucchiara et al. 2011), which allows one to explore a larger redshift path length, and second, their discovery is largely unbiased with respect to intrinsic properties of their hosts (extinction, luminosity or mass). In fact, the detection GRBs, at least in X-ray is independent (or substantially less affected) by the presence of dust extinction along the line of sight (in the host or in the IGM) (Zafar et al. 2010; Watson and Jakobsson 2012; Watson et al. 2013; Schady et al. 2010).

Unfortunately, the number of GRBs discovered and observed (thanks also to the *Swift* satellite, Gehrels et al. 2004, 2009) is several orders of magnitude less than the number of quasars that have been found in large surveys. Nevertheless, in 2006 a survey of Mg II absorbers was performed for an early sample of *Swift* bursts and a heterogeneous sample of pre-existing GRB spectra (Prochter et al. 2006b, P06 hereafter). These authors revealed an extremely puzzling result: the incidence of strong ($W_{2796} \geq 1$ Å) intervening Mg II absorbers was about 4 times higher along GRB sightlines than quasar sightlines. No such excess was found in other common class of absorbers, e.g. C IV features (Tejos et al. 2007; Sudilovsky et al. 2007), or for weak Mg II absorbers (Tejos et al. 2009).

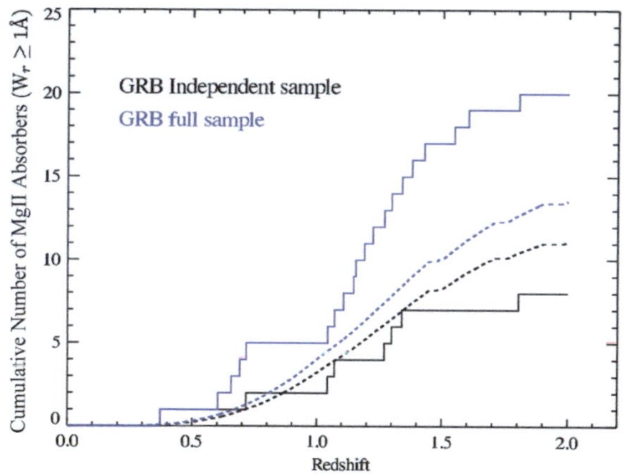

Fig. 2 Cumulative distribution of strong Mg II absorbers along GRB sight lines for an independent sample from Prochter et al. (2006b) (Sample I) and for the full sample (Sample F, *black* and *blue solid curves*, respectively). These are compared to the predicted incidence based on measurement along QSO lines of sight (*dashed curves*). The independent Sample I actually shows fewer absorbers than expected while a modest excess remains in Sample F. Neither result corresponds to a statistically significant difference from the QSO results

Several explanations for this discrepancies were proposed by P06, including: 1) a possible intrinsic origin of these absorbers, which would imply an average escape velocity from the GRB hosts of $\sim$ 10–30 % the speed of light (Cucchiara et al. 2009; Bergeron et al. 2011); 2) a significant dust bias along QSO lines of sight (Ménard et al. 2008; Porciani et al. 2007; Budzynski and Hewett 2011; Sudilovsky et al. 2009); 3) a geometric effect difference due to the sizes of the emitting regions between GRBs and QSOs (Stocke and Rector 1997; Frank et al. 2007; Porciani et al. 2007; Lawther et al. 2012); 4) a gravitational lensing effect (Vergani et al. 2009a; Porciani and Madau 2001; Rapoport et al. 2011, 2013).

Most of these suggestions were later on discarded, but the puzzling result was only solved few years ago: the original P06 work, and even the studies that followed, relied on a small sample of GRB afterglow spectra. Even the largest analysis used only 26 lines of sight (finding 22 absorbers), for a total redshift path of $\Delta z = 31.55$ (Vergani et al. 2009b). Furthermore, no study had analyzed a completely independent set of GRB sightlines from the P06 analysis.

In 2013 Cucchiara et al. (2013) collected the largest compilation of GRB spectroscopic data, exploring a larger redshift path length ($\Delta z = 44.9$, Fig. 2). No excess was identified in an independent sample and further tests, including Monte Carlo simulation and bootstrapping analysis, suggested that the earlier work by P06 was biased by a statistical fluke: in particular, the presence of a small set of lines of sight with multiple absorbers appears to have driven the results (as suggested by Kann et al. 2010). The inclusion of the original P06 dataset does not provide any significant improvement (Fig. 3).

3 Neutral Hydrogen Fraction

The reionization era describes the time when all the intergalactic hydrogen was first ionized by the photons produced by early galaxies and stars. Moreover, the same process heated the IGM gas precluding the formation of sub-haloes, which ultimately would form the smaller

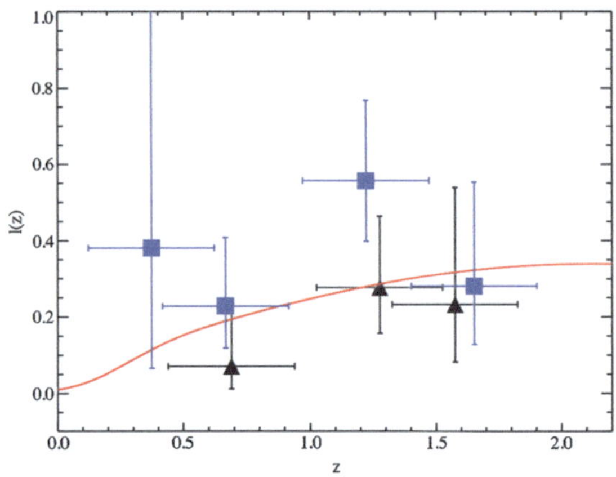

Fig. 3 Incidence of very strong Mg II absorption intervening systems, dN/dz or $\ell(z)$, evolution for the sample of GRB sight lines: *triangles* and *square symbols* refer to the Sample I and Sample F, respectively (see Cucchiara et al. 2013, for more details). The *red curve* shows the evolution of the Mg II incidence along quasar sight lines as recently computed by Zhu and Ménard (2013). We derive an average $\ell(z) = 0.20$ for Sample I, in agreement with the prediction, while $\ell(z) = 0.36$ for Sample F, indicating a slight overabundance of absorbers compared to the QSOs

dwarf galaxies. Therefore, in order to properly describe the evolution of the universe until the present time, it is critical to be able to characterize this important period of its history, for example how long it lasted and how it proceeded (patchy or continuously).

The intergalactic medium (IGM) is known to have been almost completely ionized since a redshift of $z \sim 6$ (Becker et al. 2006, 2015; Fan et al. 2006), while observations of the electron scattering of the cosmic microwave background radiation suggest the universe transitioned from a cold neutral IGM to a hot ionized one at an average redshift of $z \sim 9$ (Ade et al. 2015), but with the possibility that it was a significantly extended and environment dependent process (we will come back to this issue in Sect. 4).

From the detection of the Gunn-Peterson trough (GP, Gunn and Peterson 1965) using high redshift quasars the optical depth of the universe to ionizing photons rapidly decreases from $z \sim 7$ to $z \sim 6$, suggesting that the IGM changed due to re-ionization. Counting the fraction of spectral pixels that are completely absorbed McGreer et al. (2011) were able to place a conservative, almost model independent, lower bound on the filling factor of ionised regions. Specifically, first results showed that the neutral hydrogen fraction should be smaller than $f_{\rm HI} < 0.2$ (1-σ confidence) in the $5 \leq z \leq 5.5$ redshift range. Later, McGreer et al. (2015), refined this result obtaining $f_{\rm HI} < 0.11$ at $z = 5.6$, and $f_{\rm HI} < 0.09$ at $z = 5.9$. Another technique, involves the size distribution of 'dark gaps' in the Ly α forest (see for example, Songaila and Cowie 2010; Gallerani et al. 2008; Mesinger 2010), which shows that reionization is indeed an inhomogeneous process, with $f_{\rm HI} < 0.1$ at $z \sim 5$.

The large number of quasar spectra now available (largely thanks to SDSS, York et al. 2000), offers an opportunity to make such measurements, but unfortunately the signal to noise required to perform accurate measurements is only achieved in a small fraction of cases. Moreover, the GP trough gives only weak lower bounds on the neutral gas fraction and the quasar itself may have altered the ionization of the surrounding IGM (Barkana and Loeb 2004), making obtaining $f_{\rm HI}$ estimates even more complicated.

A new possibility to study the IGM and reionization comes by using high-redshift GRBs: the typical GRB afterglow intrinsic synchrotron emission can be fit by a power-law enabling

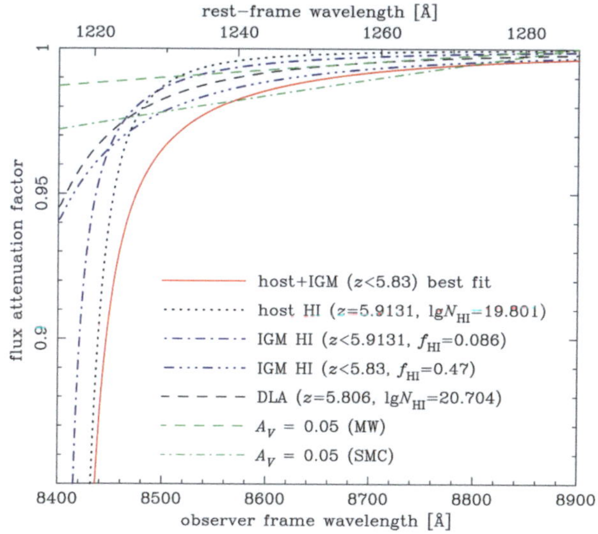

Fig. 4 Flux attenuation factor of the Ly α damping wing by various HI components and extinction at the host. The *red solid curve* is the total attenuation by HI in IGM and the host galaxy. The IGM and host components of this model are also separately shown. The IGM absorption model and the DLA component of the intervening DLA model are also shown for comparison. The *green curves* are the effect of introducing extinction by dust in the host galaxy (normalized to unity at 8900 Å), using the MW or SMC extinction curves (Fitzpatrick 1999; Gordon et al. 2003)

better constraints on the red damping wing of Ly α than is possible for quasars (Miralda-Escudé and Rees 1998; Lamb and Reichart 2000; Ciardi and Loeb 2000; McQuinn et al. 2008). Also GRBs are less biased than QSOs, tracing more typical IGM environments, their radiation has negligible affect on the IGM itself (shorter time-scale radiation than QSOs), and in the first couple of days after their discovery they are much brighter than the average QSO at $z \gtrsim 6$ (e.g. GRB 090423 at $z = 8.2$ was $K = 20.6$ at 4 days post-burst), with the potential for high-resolution, high signal-to-noise ratio spectroscopy.

For several years, only the spectrum of GRB 050904 (at $z = 6.295$, Totani et al. 2006) was able to provide constraints on the neutral hydrogen fraction ($f_{\mathrm{HI}} < 0.6$), based solely on analysis of the red-damping wing. Recently, a newly discovered GRB at $z = 5.912$, GRB 130606A, has provided a great dataset for a similar analysis (Chornock et al. 2013; Totani et al. 2014; Hartoog et al. 2015): the bright optical/NIR afterglow enabled not only a large follow-up effort, but absorption spectroscopy has shown a low HI column density in the host, implying a much better constraint on the actual f_{HI} of the IGM.

GRB 130606A was discovered by the *Swift* and KONUS-*Wind* (Golenetskii et al. 2013) and its redshift was identified based on HI and metal absorption features at the same redshift ($z = 5.912$). The afterglow was observed spectroscopically with several instruments/telescopes: from the low-resolution FOCAS camera on Subaru telescope (resolving power of $R \sim 900$) to the X-Shooter instrument on VLT ($R \sim 10000$). Here we present the results from the former dataset and discuss some discrepancies with other groups' conclusions (Chornock et al. 2013; Totani et al. 2014; Hartoog et al. 2015).

The FOCAS data were obtained during 9.3–16.5 hours after burst and clearly show the presence of a strong Ly α line. Also, flux calibration was performed using the standard star Feige 34, obtained the same night.

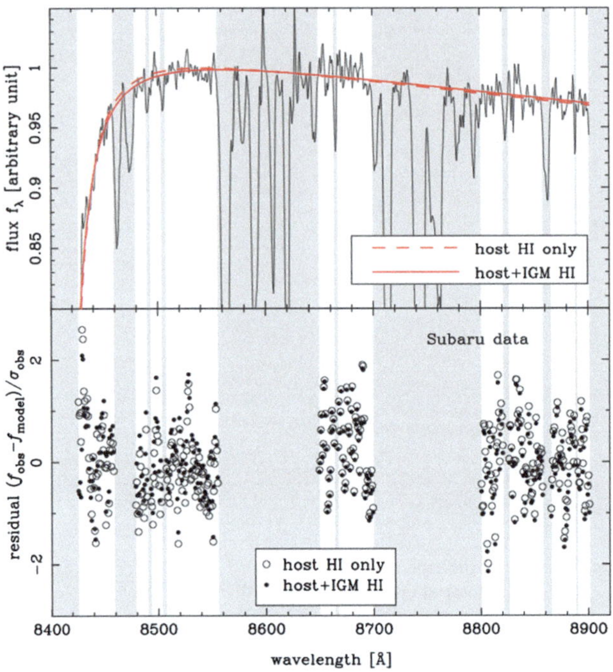

Fig. 5 *Top:* The optical spectrum of GRB 130606A afterglow taken by the SUBARU telescope, around the redshifted Ly α and its red damping wing region. The best-fit curves of the host HI only model and the host + IGM model are shown by *thick dashed* and *solid curves*, respectively. *Bottom:* The fit residuals $(f_{obs} - f_{model})/\sigma_{obs}$ are shown for the two models. The *gray shaded regions* indicate the wavelength ranges removed from the fits because of apparent features of absorption lines or airglow

The observed spectrum was fitted by a model which includes a simple power-law intrinsic continuum ($F_\nu \propto \nu^\beta$) and three components for the HI absorption feature: HI from the host galaxy (host DLA), a diffuse contribution from the IGM, and a possible intervening Damped Ly α system at lower redshift (DLA). For the host component a simple Gaussian profile with radial dispersion σ_v and with a column density of N_{HI}^{host} was used.

For the diffuse IGM we used Lorentzian (or the Voigt profile when convolved with the Gaussian velocity distribution), and the two-level approximation formula by Peebles (1993) and Miralda-Escudé and Rees (1998), with neutral hydrogen fraction f_{HI} in a redshift range from $z_{IGM,l} = 5.67$ and $z_{IGM,u} = [z_{GRB}, free]$. Also, the GP optical depth is $\tau_{GP} = 3.97 \times 10^5 f_{HI}[(1+z)/7]^{3.2}$ calculated from the cosmological parameters of Komatsu et al. (2011) and the primordial helium abundances of Peimbert et al. (2007). These components can be seen in Fig. 4. For this exercise we considered the reionization to be uniform (rather then, e.g., patchy).

Regions of the spectrum cleaned of absorption features like C II or N V, as well as strong skyline emission, were excluded by the fit and simple chi-squared minimization statistics were employed in order to find the best model. The best-fit model parameters are listed in Table 1 of Totani et al. (2014), and provide a neutral hydrogen fraction of $f_{HI} = 0.086^{+0.012}_{-0.011}$. Furthermore, we were able to exclude with high-confidence a model that contains only a host DLA, while one or two components (diffuse IGM only, or IGM + intervening DLA) are equally valuable good fits (Fig. 5). The presence of a second Ly α feature at lower redshift ($z = 5.806$) is supported by several metal lines at such redshift, but further analysis of the possible presence (and strength) of the expected Ly β line allows to exclude that this is another DLA along the line of sight.

We point out that the chances of finding an intervening DLA at lower z is very low: from the $z \sim 5$–6 DLA numbers density the chance of random encountering such system is roughly ~ 3 % (Songaila and Cowie 2010; Castro-Tirado et al. 2013). Finally, the GRB 130606A spectrum shows the power of using high-z GRBs to probe the diffuse IGM down to $f_{\rm HI} \approx 0.1$ for gas close (≤ 5 Mpc in proper distance) to the host galaxy. This value is consistent with the limits obtained by QSO studies (e.g, Mortlock et al. 2011). Such analyses will be even more productive with the upcoming generation of 30 m telescopes, which will be optimized for rapid follow-up of such transients as well as be equipped with powerful NIR spectrographs.

It is interesting that other groups arrived at different results from that of Subaru presented above. Chornock et al. (2013) found no evidence for nonzero IGM HI, though their upper limit is not seriously in contradiction with the best-fit IGM HI of Totani et al. (2014). It should be noted that Chornock et al. (2013) used an intrinsic afterglow spectral index of $\beta = -1.99$, which is not supported by observed afterglow colors or the VLT spectrum that shows $\beta = -1.0$. However, Hartoog et al. (2015) found the best fit $f_{\rm HI}$ to be zero, with a stringent 3σ upper limit of $f_{\rm HI} \leq 0.05$, that is statistically inconsistent with the best-fit $f_{\rm HI}$ of Totani et al. (2014). Totani et al. (2015) examined the origin of this discrepancy by analyzing the VLT data with the code adopted to the Subaru data. It is found that the same result as Totani et al. (2014) is obtained by adopting the same analysis method to the VLT spectrum, and hence the origin of the discrepancy is likely the difference of analysis methods, especially the wavelength ranges used. It is important to carefully optimize an analysis method to minimize systematic uncertainties, when very high precision spectra are obtained for high-z GRBs.

4 Escape Fraction and Re-ionization: First Galaxies

We mentioned in the previous section that a transition from a cold neutral IGM to the hot ionized one must have occurred at some point in the Universe's history. A key question is whether this phase change was predominantly driven by ionizing radiation from early generations of stars (UV emission from star forming galaxies), or whether other mechanisms (e.g. decaying particle ionizing background, contribution from accretion luminosity) must have had a significant effect. Considering the known star formation budget at $z \sim 6$–9, recent calculations confirm that it is plausible that massive stars provided sufficient ionizing photons, providing that a substantial fraction (typical analyses suggest $f_{\rm esc}$ of at least 10–20 %) could escape the host galaxies in which the stars reside (e.g., Bouwens et al. 2015). Unfortunately it is highly challenging observationally to pin down the escape fraction even at lower redshifts, and most recent studies have concluded that it is likely to be rather low, e.g. a few percent or less at $z \sim 3$ (e.g., Grazian et al. 2015).

GRBs provide a novel route to addressing this question, since optical afterglow spectroscopy of $z > 2$ GRBs frequently provides a measurement of the neutral hydrogen column density along the line of sight through the host galaxy. Typically these columns are high, as measured in damped Lyman-α absorbers, and are a lower limit to the column that would have been in front of the GRB progenitor, in as much as hydrogen proximate to the burst may well be ionized by the burst itself. Thus most lines of sight are opaque to ionizing radiation, and in particular, very few GRB afterglows have ever been found with columns low enough to allow even a small fraction of ionizing radiation to escape and therefore to be measured directly (e.g. via far-UV photometry).

There are reasons to think that at higher redshifts, with more star formation occurring in smaller galaxies, the escape fraction may increase, although measurements of $N_{\rm HI}$ from GRBs at $z > 4$ show at best a modest decline (Thöne et al. 2013; Chornock et al. 2014). Nevertheless, recently, thanks to the discovery of many $z > 2$ GRBs with spectroscopic redshift and $N_{\rm HI}$ column measurements, $f_{\rm esc}$ has been measured using an indirect technique, which involves only the determination of $N_{\rm HI}$ distribution and it is not subject to systematics due background subtraction (Chen et al. 2007; Gnedin et al. 2008; Fynbo et al. 2009).

Following the work of Chen et al. (2007), the mean escape fraction of Lyman limit photons averaged over all directions (GRB sightlines) is evaluated according to

$$\langle f_{\rm esc}\rangle = \frac{1}{n}\sum_{i=1}^{i=n} \exp\bigl[-\sigma_{\rm LL}\, N_i({\rm H\,I})\bigr], \tag{1}$$

where the sum extends over the total number of n GRB sightlines in the sample and $\sigma_{\rm LL} = 6.28 \times 10^{-18}$ cm^2 is the photoionization cross section of hydrogen atoms. They find $\langle f_{\rm esc}\rangle = 0.02 \pm 0.02$ (where error is estimated using the bootstrap re-sampling method). They also where able to determine a 95 % c.l. upper limit of $\langle f_{\rm esc}\rangle \le 0.075$.

Fynbo et al. (2009), using a larger number of sight lines selected in a more unbiased way, found very similar results. These results are also consistent with measurements from Lyman Break Galaxies at $z \sim 3$, therefore suggesting that the $f_{\rm esc}$ from QSOs and sub-L^* galaxies, like the ones hosting GRBs, contribute for a comparable amount of ionizing radiation.

The large GRB sample detected by *Swift*, and accurate redshift measurement from ground based follow-up, have sparked deep observations of the highest redshift sample of GRB host galaxies, in particular near the time when re-ionization was complete. Numbers remain small at $z > 6$, so the constraints do not yet provide a critical problem for re-ionization, but samples with good spectroscopy will hopefully improve in the coming years.

References

P.A.R. Ade, N. Aghanim, M. Arnaud, M. Ashdown, J. Aumont, C. Baccigalupi, A.J. Banday, R.B. Barreiro, J.G. Bartlett (Planck Collaboration), et al., Planck 2015 results. XIII. Cosmological parameters. ArXiv e-prints (2015)

R. Barkana, A. Loeb, Gamma-ray bursts versus quasars: Lyα signatures of reionization versus cosmological infall. Astrophys. J. **601**, 64–77 (2004). doi:10.1086/380435

G.D. Becker, W.L.W. Sargent, M. Rauch, R.A. Simcoe, Discovery of excess O I absorption toward the $z = 6.42$ QSO SDSS J1148+5251. Astrophys. J. **640**, 69–80 (2006). doi:10.1086/500079

G.D. Becker, J.S. Bolton, P. Madau, M. Pettini, E.V. Ryan-Weber, B.P. Venemans, Evidence of patchy hydrogen reionization from an extreme Lyα trough below redshift six. Mon. Not. R. Astron. Soc. **447**, 3402–3419 (2015). doi:10.1093/mnras/stu2646

E. Behar, S. Dado, A. Dar, A. Laor, Can the soft X-ray opacity toward high-redshift sources probe the missing baryons? Astrophys. J. **734**, 26 (2011). doi:10.1088/0004-637X/734/1/26

J. Bergeron, The Mg II absorption system in the QSO PKS 2128-12—a galaxy disc/halo with a radius of 65 kpc. Astron. Astrophys. **155**, 8–11 (1986)

J. Bergeron, P. Boissé, B. Ménard, Incidence of Mg II absorbers towards blazars and the GRB/QSO puzzle. Astron. Astrophys. **525**, 51 (2011). doi:10.1051/0004-6361/201015265

M. Berry, R.S. Somerville, M.R. Haas, E. Gawiser, A. Maller, G. Popping, S.C. Trager, Damped Lyα absorption systems in semi-analytic models with multiphase gas. Mon. Not. R. Astron. Soc. **441**, 939–963 (2014). doi:10.1093/mnras/stu613

R.J. Bouwens, G.D. Illingworth, P.A. Oesch, J. Caruana, B. Holwerda, R. Smit, S. Wilkins, Reionization after Planck: The derived growth of the cosmic ionizing emissivity now matches the growth of the galaxy UV luminosity density. Astrophys. J. **811**, 140 (2015). doi:10.1088/0004-637X/811/2/140

D.V. Bowen, J.C. Blades, M. Pettini, Interstellar Mg II absorption lines from low-redshift galaxies. Astrophys. J. **448**, 634 (1995). doi:10.1086/175993

J.M. Budzynski, P.C. Hewett, Dusty Mg II absorbers: Population statistics, extinction curves and gamma-ray burst sightlines. Mon. Not. R. Astron. Soc. **416**, 1871–1889 (2011). doi:10.1111/j.1365-2966.2011.19158.x

S. Campana, R. Salvaterra, A. Ferrara, A. Pallottini, Missing cosmic metals revealed by X-ray absorption towards distant sources. Astron. Astrophys. **575**, 43 (2015). doi:10.1051/0004-6361/201425083

A.J. Castro-Tirado, R. Sánchez-Ramírez, S.L. Ellison, M. Jelínek, A. Martín-Carrillo, V. Bromm, J. Gorosabel, M. Bremer, J.M. Winters, L. Hanlon, S. Meegan, M. Topinka, S.B. Pandey, S. Guziy, S. Jeong, E. Sonbas, A.S. Pozanenko, R. Cunniffe, R. Fernández-Muñoz, P. Ferrero, N. Gehrels, R. Hudec, P. Kubánek, O. Lara-Gil, V.F. Muñoz-Martínez, D. Pérez-Ramírez, J. Štrobl, C. Álvarez-Iglesias, R. Inasaridze, V. Rumyantsev, A. Volnova, S. Hellmich, S. Mottola, J.M. Castro Cerón, J. Cepa, E. Göğüş, T. Güver, Ö. Önal Taş, I.H. Park, L. Sabau-Graziati, A. Tejero, GRB 130606A within a sub-DLA at redshift 5.91. ArXiv e-prints (2013)

R. Cen, The nature of damped Lyα systems and their hosts in the standard cold dark matter universe. Astrophys. J. **748**, 121 (2012). doi:10.1088/0004-637X/748/2/121

H.-W. Chen, J.L. Tinker, The baryon content of dark matter halos: Empirical constraints from Mg II absorbers. Astrophys. J. **687**, 745–756 (2008). doi:10.1086/591927

H.-W. Chen, J.X. Prochaska, N.Y. Gnedin, A new constraint on the escape fraction in distant galaxies using γ-ray burst afterglow spectroscopy. Astrophys. J. Lett. **667**, 125–128 (2007). doi:10.1086/522306

R. Chornock, E. Berger, D.B. Fox, R. Lunnan, M.R. Drout, W.-f. Fong, T. Laskar, K.C. Roth, GRB 130606A as a probe of the intergalactic medium and the interstellar medium in a star-forming galaxy in the first Gyr after the Big Bang. Astrophys. J. **774**, 26 (2013). doi:10.1088/0004-637X/774/1/26

R. Chornock, E. Berger, D.B. Fox, W. Fong, T. Laskar, K.C. Roth, GRB 140515A at $z = 6.33$: Constraints on the end of reionization from a gamma-ray burst in a low hydrogen column density environment. ArXiv e-prints (2014)

C.W. Churchill, J.R. Rigby, J.C. Charlton, S.S. Vogt, The population of weak Mg II absorbers. I. A survey of 26 QSO HIRES/Keck spectra. Astrophys. J. Suppl. Ser. **120**, 51–75 (1999). doi:10.1086/313168

B. Ciardi, A. Loeb, Expected number and flux distribution of gamma-ray burst afterglows with high redshifts. Astrophys. J. **540**, 687–696 (2000). doi:10.1086/309384

T.J. Cooper, R.A. Simcoe, K.L. Cooksey, J.M. O'Meara, P. Torrey, The incidence of low-metallicity Lyman-limit systems at $z \sim 3.5$: Implications for the cold-flow hypothesis of baryonic accretion. Astrophys. J. **812**, 58 (2015). doi:10.1088/0004-637X/812/1/58

A. Cucchiara, T. Jones, J.C. Charlton, D.B. Fox, D. Einsig, A. Narayanan, Testing the possible intrinsic origin of the excess very strong Mg II absorbers along gamma-ray burst lines-of-sight. Astrophys. J. **697**, 345–360 (2009). doi:10.1088/0004-637X/697/1/345

A. Cucchiara, A.J. Levan, D.B. Fox, N.R. Tanvir, T.N. Ukwatta, E. Berger, T. Krühler, A. Küpcü Yoldaş, X.F. Wu, K. Toma, J. Greiner, F.E. Olivares, A. Rowlinson, L. Amati, T. Sakamoto, K. Roth, A. Stephens, A. Fritz, J.P.U. Fynbo, J. Hjorth, D. Malesani, P. Jakobsson, K. Wiersema, P.T. O'Brien, A.M. Soderberg, R.J. Foley, A.S. Fruchter, J. Rhoads, R.E. Rutledge, B.P. Schmidt, M.A. Dopita, P. Podsiadlowski, R. Willingale, C. Wolf, S.R. Kulkarni, P. D'Avanzo, A photometric redshift of $z \sim 9.4$ for GRB 090429B. Astrophys. J. **736**, 7 (2011). doi:10.1088/0004-637X/736/1/7

A. Cucchiara, J.X. Prochaska, G. Zhu, B. Ménard, J.P.U. Fynbo, D.B. Fox, H.-W. Chen, K.L. Cooksey, S.B. Cenko, D. Perley, J.S. Bloom, E. Berger, N.R. Tanvir, V. D'Elia, S. Lopez, R. Chornock, T. de Jaeger, An independent measurement of the incidence of Mg II absorbers along gamma-ray burst sight lines: The end of the mystery? Astrophys. J. **773**, 82 (2013). doi:10.1088/0004-637X/773/2/82

A. Cucchiara, M. Fumagalli, M. Rafelski, D. Kocevski, J.X. Prochaska, R.J. Cooke, G.D. Becker, Unveiling the secrets of metallicity and massive star formation using DLAs along gamma-ray bursts. Astrophys. J. **804**, 51 (2015). doi:10.1088/0004-637X/804/1/51

X. Fan, M.A. Strauss, R.H. Becker, R.L. White, J.E. Gunn, G.R. Knapp, G.T. Richards, D.P. Schneider, J. Brinkmann, M. Fukugita, Constraining the evolution of the ionizing background and the epoch of reionization with $z \sim 6$ quasars. II. A sample of 19 quasars. Astrophys. J. **132**, 117–136 (2006). doi:10.1086/504836

E.L. Fitzpatrick, Correcting for the effects of interstellar extinction. Publ. Astron. Soc. Pac. **111**, 63–75 (1999). doi:10.1086/316293

S. Frank, M.C. Bentz, K.Z. Stanek, S. Mathur, M. Dietrich, B.M. Peterson, D.W. Atlee, Disparate Mg II absorption statistics towards quasars and gamma-ray bursts: A possible explanation. Astrophys. Space Sci. **312**, 325–330 (2007). doi:10.1007/s10509-007-9699-x

M. Fumagalli, J.X. Prochaska, D. Kasen, A. Dekel, D. Ceverino, J.R. Primack, Absorption-line systems in simulated galaxies fed by cold streams. Mon. Not. R. Astron. Soc. **418**, 1796–1821 (2011). doi:10.1111/j.1365-2966.2011.19599.x

J.P.U. Fynbo, P. Jakobsson, J.X. Prochaska, D. Malesani, C. Ledoux, A. de Ugarte Postigo, M. Nardini, P.M. Vreeswijk, K. Wiersema, J. Hjorth, J. Sollerman, H. Chen, C.C. Thöne, G. Björnsson, J.S. Bloom, A.J. Castro-Tirado, L. Christensen, A. De Cia, A.S. Fruchter, J. Gorosabel, J.F. Graham, A.O. Jaunsen, B.L. Jensen, D.A. Kann, C. Kouveliotou, A.J. Levan, J. Maund, N. Masetti, B. Milvang-Jensen, E. Palazzi, D.A. Perley, E. Pian, E. Rol, P. Schady, R.L.C. Starling, N.R. Tanvir, D.J. Watson, D. Xu, T. Augusteijn, F. Grundahl, J. Telting, P. Quirion, Low-resolution spectroscopy of gamma-ray burst optical afterglows: Biases in the Swift sample and characterization of the absorbers. Astrophys. J. Suppl. Ser. **185**, 526–573 (2009). doi:10.1088/0067-0049/185/2/526

S. Gallerani, A. Ferrara, X. Fan, T.R. Choudhury, Glimpsing through the high-redshift neutral hydrogen fog. Mon. Not. R. Astron. Soc. **386**, 359–369 (2008). doi:10.1111/j.1365-2966.2008.13029.x

N. Gehrels, The Swift gamma-ray burst mission. Astrophys. J. **611**, 1005 (2004).

N. Gehrels, G. Chincarini, P. Giommi, K.O. Mason, J.A. Nousek, A.A. Wells, N.E. White, S.D. Barthelmy, D.N. Burrows, L.R. Cominsky, K.C. Hurley, F.E. Marshall, P. Mészáros, P.W.A. Roming, L. Angelini, L.M. Barbier, T. Belloni, S. Campana, P.A. Caraveo, M.M. Chester, O. Citterio, T.L. Cline, M.S. Cropper, J.R. Cummings, A.J. Dean, E.D. Feigelson, E.E. Fenimore, D.A. Frail, A.S. Fruchter, G.P. Garmire, K. Gendreau, G. Ghisellini, J. Greiner, J.E. Hill, S.D. Hunsberger, H.A. Krimm, S.R. Kulkarni, P. Kumar, F. Lebrun, N.M. Lloyd-Ronning, C.B. Markwardt, B.J. Mattson, R.F. Mushotzky, J.P. Norris, J. Osborne, B. Paczynski, D.M. Palmer, H. Park, A.M. Parsons, J. Paul, M.J. Rees, C.S. Reynolds, J.E. Rhoads, T.P. Sasseen, B.E. Schaefer, A.T. Short, A.P. Smale, I.A. Smith, L. Stella, G. Tagliaferri, T. Takahashi, M. Tashiro, L.K. Townsley, J. Tueller, M.J.L. Turner, M. Vietri, W. Voges, M.J. Ward, R. Willingale, F.M. Zerbi, W.W. Zhang, The Swift gamma-ray burst mission. Astrophys. J. **611**, 1005–1020 (2004). doi:10.1086/422091

N. Gehrels, E. Ramirez-Ruiz, D.B. Fox, Gamma-ray bursts in the Swift era. Annu. Rev. Astron. Astrophys. **47**, 567–617 (2009). doi:10.1146/annurev.astro.46.060407.145147

N.Y. Gnedin, A.V. Kravtsov, H.-W. Chen, Escape of ionizing radiation from high-redshift galaxies. Astrophys. J. **672**, 765–775 (2008). doi:10.1086/524007

S. Golenetskii, R. Aptekar, D. Frederiks, V. Pal'Shin, P. Oleynik, M. Ulanov, D. Svinkin, T. Cline, Konus-wind observation of GRB 130606B. GRB Coord. Netw. **14803**, 1 (2013)

K.D. Gordon, G.C. Clayton, K.A. Misselt, A.U. Landolt, M.J. Wolff, A quantitative comparison of the Small Magellanic Cloud, Large Magellanic Cloud, and Milky Way ultraviolet to near-infrared extinction curves. Astrophys. J. **594**, 279–293 (2003). doi:10.1086/376774

A. Grazian, E. Giallongo, R. Gerbasi, F. Fiore, A. Fontana, O. Le Fevre, L. Pentericci, E. Vanzella, G. Zamorani, P. Cassata, B. Garilli, V. Le Brun, D. Maccagni, L.A.M. Tasca, R. Thomas, E. Zucca, R. Amorin, S. Bardelli, L.P. Cassara', M. Castellano, A. Cimatti, O. Cucciati, A. Durkalec, M. Giavalisco, N.P. Hathi, O. Ilbert, B.C. Lemaux, S. Paltani, B. Ribeiro, D. Schaerer, M. Scodeggio, V. Sommariva, M. Talia, L. Tresse, D. Vergani, A. Bonchi, K. Boutsia, P. Capak, S. Charlot, T. Contini, S. de la Torre, J. Dunlop, S. Fotopoulou, L. Guaita, A. Koekemoer, C. Lopez-Sanjuan, Y. Mellier, E. Merlin, D. Paris, J. Pforr, S. Pilo, P. Santini, N. Scoville, Y. Taniguchi, P.W. Wang, The Lyman Continuum escape fraction of galaxies at $z = 3.3$ in the VUDS-LBC/COSMOS field. ArXiv e-prints (2015)

J.E. Gunn, B.A. Peterson Astrophys. J. **142**, 1633 (1965)

O.E. Hartoog, D. Malesani, J.P.U. Fynbo, T. Goto, T. Krühler, P.M. Vreeswijk, A. De Cia, D. Xu, P. Møller, S. Covino, V. D'Elia, H. Flores, P. Goldoni, J. Hjorth, P. Jakobsson, J.-K. Krogager, L. Kaper, C. Ledoux, A.J. Levan, B. Milvang-Jensen, J. Sollerman, M. Sparre, G. Tagliaferri, N.R. Tanvir, A. de Ugarte Postigo, S.D. Vergani, K. Wiersema, J. Datson, R. Salinas, K. Mikkelsen, N. Aghanim, VLT/X-Shooter spectroscopy of the afterglow of the Swift GRB 130606A. Chemical abundances and reionisation at $z \sim 6$. Astron. Astrophys. **580**, 139 (2015). doi:10.1051/0004-6361/201425001

P. Jakobsson, J. Hjorth, J.P.U. Fynbo, M. Weidinger, J. Gorosabel, C. Ledoux, D. Watson, G. Björnsson, E.H. Gudmundsson, R.A.M.J. Wijers, P. Möller, K. Pedersen, J. Sollerman, A.A. Henden, B.L. Jensen, A. Gilmore, P. Kilmartin, A. Levan, J.M. Castro Cerón, A.J. Castro-Tirado, A. Fruchter, C. Kouveliotou, N. Masetti, N. Tanvir, The line-of-sight towards GRB 030429 at $z = 2.66$: Probing the matter at stellar, galactic and intergalactic scales. Astron. Astrophys. **427**, 785–794 (2004). doi:10.1051/0004-6361:20041233

P. Jakobsson, J. Hjorth, D. Malesani, R. Chapman, J.P.U. Fynbo, N.R. Tanvir, B. Milvang-Jensen, P.M. Vreeswijk, G. Letawe, R.L.C. Starling, The optically unbiased GRB host (TOUGH) survey. III. Redshift distribution. Astrophys. J. **752**, 62 (2012). doi:10.1088/0004-637X/752/1/62

G.G. Kacprzak, C.W. Churchill, N.M. Nielsen, Tracing outflows and accretion: A bimodal azimuthal dependence of Mg II absorption. Astrophys. J. Lett. **760**, 7 (2012). doi:10.1088/2041-8205/760/1/L7

D.A. Kann, S. Klose, B. Zhang, D. Malesani, E. Nakar, A. Pozanenko, A.C. Wilson, N.R. Butler, P. Jakobsson, S. Schulze, M. Andreev, L.A. Antonelli, I.F. Bikmaev, V. Biryukov, M. Böttcher, R.A. Burenin, J.M. Castro Cerón, A.J. Castro-Tirado, G. Chincarini, B.E. Cobb, S. Covino, P. D'Avanzo, V. D'Elia, M. Della Valle, A. de Ugarte Postigo, Y. Efimov, P. Ferrero, D. Fugazza, J.P.U. Fynbo, M. Gålfalk, F.

Grundahl, J. Gorosabel, S. Gupta, S. Guziy, B. Hafizov, J. Hjorth, K. Holhjem, M. Ibrahimov, M. Im, G.L. Israel, M. Jelínek, B.L. Jensen, R. Karimov, I.M. Khamitov, Ü. Kiziloğlu, E. Klunko, P. Kubánek, A.S. Kutyrev, P. Laursen, A.J. Levan, F. Mannucci, C.M. Martin, A. Mescheryakov, N. Mirabal, J.P. Norris, J.-E. Ovaldsen, D. Paraficz, E. Pavlenko, S. Piranomonte, A. Rossi, V. Rumyantsev, R. Salinas, A. Sergeev, D. Sharapov, J. Sollerman, B. Stecklum, L. Stella, G. Tagliaferri, N.R. Tanvir, J. Telting, V. Testa, A.C. Updike, A. Volnova, D. Watson, K. Wiersema, D. Xu, The afterglows of Swift-era gamma-ray bursts. I. Comparing pre-Swift and Swift-era long/soft (type II) GRB optical afterglows. Astrophys. J. **720**, 1513–1558 (2010). doi:10.1088/0004-637X/720/2/1513

N. Kawai, G. Kosugi, K. Aoki, T. Yamada, T. Totani, K. Ohta, M. Iye, T. Hattori, W. Aoki, H. Furusawa, K. Hurley, K.S. Kawabata, N. Kobayashi, Y. Komiyama, Y. Mizumoto, K. Nomoto, J. Noumaru, R. Ogasawara, R. Sato, K. Sekiguchi, Y. Shirasaki, M. Suzuki, T. Takata, T. Tamagawa, H. Terada, J. Watanabe, Y. Yatsu, A. Yoshida, An optical spectrum of the afterglow of a γ-ray burst at a redshift of $z = 6.295$. Nature **440**, 184–186 (2006). doi:10.1038/nature04498

E. Komatsu, K.M. Smith, J. Dunkley, C.L. Bennett, B. Gold, G. Hinshaw, N. Jarosik, D. Larson, M.R. Nolta, L. Page, D.N. Spergel, M. Halpern, R.S. Hill, A. Kogut, M. Limon, S.S. Meyer, N. Odegard, G.S. Tucker, J.L. Weiland, E. Wollack, E.L. Wright, Seven-year Wilkinson Microwave Anisotropy Probe (WMAP) observations: Cosmological interpretation. Astrophys. J. Suppl. Ser. **192**, 18 (2011). doi:10.1088/0067-0049/192/2/18

D.Q. Lamb, D.E. Reichart, Gamma-ray bursts as a probe of the very high redshift universe. Astrophys. J. **536**, 1–18 (2000)

K.M. Lanzetta, D.A. Turnshek, A.M. Wolfe, An absorption-line survey of 32 QSOs at red wavelengths—properties of the Mg II absorbers. Astrophys. J. **322**, 739–769 (1987). doi:10.1086/165769

D. Lawther, T. Paarup, M. Schmidt, M. Vestergaard, J. Hjorth, D. Malesani, Constraints on the relative sizes of intervening Mg II-absorbing clouds and quasar emitting regions. Astron. Astrophys. **546**, 67 (2012). doi:10.1051/0004-6361/201219326

N. Lehner, J.C. Howk, T.M. Tripp, J. Tumlinson, J.X. Prochaska, J.M. O'Meara, C. Thom, J.K. Werk, A.J. Fox, J. Ribaudo, The bimodal metallicity distribution of the cool circumgalactic medium at $z \lesssim 1$. Astrophys. J. **770**, 138 (2013). doi:10.1088/0004-637X/770/2/138

M.S. Matejek, R.A. Simcoe, A survey of MgII absorption at $2 \leq z6$ with Magellan/FIRE: I: Sample and evolution of the MgII frequency. ArXiv e-prints (2012)

I.D. McGreer, A. Mesinger, X. Fan, The first (nearly) model-independent constraint on the neutral hydrogen fraction at $z \sim 6$. Mon. Not. R. Astron. Soc. **415**, 3237–3246 (2011). doi:10.1111/j.1365-2966.2011.18935.x

I.D. McGreer, A. Mesinger, V. D'Odorico, Model-independent evidence in favour of an end to reionization by $z \approx 6$. Mon. Not. R. Astron. Soc. **447**, 499–505 (2015). doi:10.1093/mnras/stu2449

M. McQuinn, A. Lidz, M. Zaldarriaga, L. Hernquist, S. Dutta, Probing the neutral fraction of the IGM with GRBs during the epoch of reionization. Mon. Not. R. Astron. Soc. **388**, 1101–1110 (2008). doi:10.1111/j.1365-2966.2008.13271.x

B. Ménard, D. Nestor, D. Turnshek, A. Quider, G. Richards, D. Chelouche, S. Rao, Lensing, reddening and extinction effects of MgII absorbers from $z = 0.4$ to 2. Mon. Not. R. Astron. Soc. **385**, 1053–1066 (2008). doi:10.1111/j.1365-2966.2008.12909.x

B. Ménard, V. Wild, D. Nestor, A. Quider, S. Zibetti, S. Rao, D. Turnshek, Probing star formation across cosmic time with absorption-line systems. Mon. Not. R. Astron. Soc. **417**, 801–811 (2011). doi:10.1111/j.1365-2966.2011.18227.x

A. Mesinger, Was reionization complete by $z \sim 5$–6? Mon. Not. R. Astron. Soc. **407**, 1328–1337 (2010). doi:10.1111/j.1365-2966.2010.16995.x

M.R. Metzger, S.G. Djorgovski, S.R. Kulkarni, C.C. Steidel, K.L. Adelberger, D.A. Frail, E. Costa, F. Frontera, Spectral constraints on the redshift of the optical counterpart to the γ-ray burst of 8 May 1997. Nature **387**, 878–880 (1997). doi:10.1038/43132

J. Miralda-Escudé, M.J. Rees, Searching for the earliest galaxies using the Gunn-Peterson trough and the Ly-α emission line. Astrophys. J. **497**, 21–27 (1998)

D.J. Mortlock, S.J. Warren, B.P. Venemans, M. Patel, P.C. Hewett, R.G. McMahon, C. Simpson, T. Theuns, E.A. Gonzáles-Solares, A. Adamson, S. Dye, N.C. Hambly, P. Hirst, M.J. Irwin, E. Kuiper, A. Lawrence, H.J.A. Röttgering, A luminous quasar at a redshift of $z = 7.085$. Nature **474**, 616–619 (2011). doi:10.1038/nature10159

S. Muzahid, G.G. Kacprzak, J.C. Charlton, C.W. Churchill, Molecular hydrogen absorption from the halo of a $z \sim 0.4$ galaxy. ArXiv e-prints (2016)

K. Nagamine, V. Springel, L. Hernquist, Star formation rate and metallicity of damped Lyman α absorbers in cosmological smoothed particle hydrodynamics simulations. Mon. Not. R. Astron. Soc. **348**, 435–450 (2004). doi:10.1111/j.1365-2966.2004.07180.x

D.B. Nestor, D.A. Turnshek, S.M. Rao, Mg II absorption systems in Sloan Digital Sky Survey QSO spectra. Astrophys. J. **628**, 637–654 (2005). doi:10.1086/427547

P.J.E. Peebles, *Principles of Physical Cosmology* 1993

M. Peimbert, V. Luridiana, A. Peimbert, Revised primordial helium abundance based on new atomic data. Astrophys. J. **666**, 636–646 (2007). doi:10.1086/520571

E. Pian, P.A. Mazzali, N. Masetti, P. Ferrero, S. Klose, E. Palazzi, E. Ramirez-Ruiz, S.E. Woosley, C. Kouveliotou, J. Deng, A.V. Filippenko, R.J. Foley, J.P.U. Fynbo, D.A. Kann, W. Li, J. Hjorth, K. Nomoto, F. Patat, D.N. Sauer, J. Sollerman, P.M. Vreeswijk, E.W. Guenther, A. Levan, P. O'Brien, N.R. Tanvir, R.A.M.J. Wijers, C. Dumas, O. Hainaut, D.S. Wong, D. Baade, L. Wang, L. Amati, E. Cappellaro, A.J. Castro-Tirado, S. Ellison, F. Frontera, A.S. Fruchter, J. Greiner, K. Kawabata, C. Ledoux, K. Maeda, P. Møller, L. Nicastro, E. Rol, R. Starling, An optical supernova associated with the X-ray flash XRF 060218. Nature **442**, 1011–1013 (2006). doi:10.1038/nature05082

C. Porciani, P. Madau, On the association of gamma-ray bursts with massive stars: Implications for number counts and lensing statistics. Astrophys. J. **548**, 522–531 (2001). doi:10.1086/319027

C. Porciani, M. Viel, S.J. Lilly, Strong Mg II systems in quasar and gamma-ray burst spectra. Astrophys. J. **659**, 218–224 (2007). doi:10.1086/512358

G.E. Prochter, J.X. Prochaska, S.M. Burles, On the incidence and kinematics of strong Mg II absorbers. Astrophys. J. **639**, 766–780 (2006a). doi:10.1086/499341

G.E. Prochter, J.X. Prochaska, H. Chen, J.S. Bloom, M. Dessauges-Zavadsky, R.J. Foley, S. Lopez, M. Pettini, A.K. Dupree, P. Guhathakurta, On the incidence of strong Mg II absorbers along gamma-ray burst sight lines. Astrophys. J. Lett. **648**, 93–96 (2006b). doi:10.1086/508061

A.M. Quider, D.B. Nestor, D.A. Turnshek, S.M. Rao, E.M. Monier, A.N. Weyant, J.R. Busche, The Pittsburgh Sloan Digital Sky Survey Mg II quasar absorption-line survey catalog. Astrophys. J. **141**, 137 (2011). doi:10.1088/0004-6256/141/4/137

M. Rafelski, M. Neeleman, M. Fumagalli, A.M. Wolfe, J.X. Prochaska, The rapid decline in metallicity of damped Lyα systems at $z \sim 5$. Astrophys. J. Lett. **782**, 29 (2014). doi:10.1088/2041-8205/782/2/L29

S. Rapoport, C.A. Onken, B.P. Schmidt, J.S.B. Wyithe, B.E. Tucker, A.J. Levan, Testing the gravitational lensing explanation for the MgII problem in GRBs. ArXiv e-prints (2011)

S. Rapoport, C.A. Onken, J.S.B. Wyithe, B.P. Schmidt, A.O. Thygesen, On the significance of the excess number of strong Mg II absorbers observed toward gamma-ray bursts. Astrophys. J. **766**, 23 (2013). doi:10.1088/0004-637X/766/1/23

A.O. Razoumov, M.L. Norman, J.X. Prochaska, J. Sommer-Larsen, A.M. Wolfe, Y.-J. Yang, Can gravitational infall energy lead to the observed velocity dispersion in DLAs? Astrophys. J. **683**, 149–160 (2008). doi:10.1086/589136

P. Rodríguez Hidalgo, K. Wessels, J.C. Charlton, A. Narayanan, A. Mshar, A. Cucchiara, T. Jones, Evolution of the population of very strong Mg II absorbers. Mon. Not. R. Astron. Soc. **427**, 1801–1815 (2012). doi:10.1111/j.1365-2966.2012.21586.x

G.C. Rudie, C.C. Steidel, R.F. Trainor, O. Rakic, M. Bogosavljević, M. Pettini, N. Reddy, A.E. Shapley, D.K. Erb, D.R. Law, The gaseous environment of high-z galaxies: Precision measurements of neutral hydrogen in the circumgalactic medium of $z \sim 2$–3 galaxies in the Keck Baryonic Structure Survey. Astrophys. J. **750**, 67 (2012). doi:10.1088/0004-637X/750/1/67

R. Salvaterra, M. Della Valle, S. Campana, G. Chincarini, S. Covino, P. D'Avanzo, A. Fernández-Soto, C. Guidorzi, F. Mannucci, R. Margutti, C.C. Thöne, L.A. Antonelli, S.D. Barthelmy, M. de Pasquale, V. D'Elia, F. Fiore, D. Fugazza, L.K. Hunt, E. Maiorano, S. Marinoni, F.E. Marshall, E. Molinari, J. Nousek, E. Pian, J.L. Racusin, L. Stella, L. Amati, G. Andreuzzi, G. Cusumano, E.E. Fenimore, P. Ferrero, P. Giommi, D. Guetta, S.T. Holland, K. Hurley, G.L. Israel, J. Mao, C.B. Markwardt, N. Masetti, C. Pagani, E. Palazzi, D.M. Palmer, S. Piranomonte, G. Tagliaferri, V. Testa, GRB090423 at a redshift of $z \sim 8.1$. Nature **461**, 1258–1260 (2009). doi:10.1038/nature08445

P. Schady, M.J. Page, S.R. Oates, M. Still, M. de Pasquale, T. Dwelly, N.P.M. Kuin, S.T. Holland, F.E. Marshall, P.W.A. Roming, Dust and metal column densities in gamma-ray burst host galaxies. Mon. Not. R. Astron. Soc. **401**, 2773–2792 (2010). doi:10.1111/j.1365-2966.2009.15861.x

R.A. Simcoe, K.L. Cooksey, M. Matejek, A.J. Burgasser, J. Bochanski, E. Lovegrove, R.A. Bernstein, J.L. Pipher, W.J. Forrest, C. McMurtry, X. Fan, J. O'Meara, Constraints on the universal C IV mass density at $z \sim 6$ from early infrared spectra obtained with the Magellan FIRE spectrograph. Astrophys. J. **743**, 21 (2011). doi:10.1088/0004-637X/743/1/21

A. Songaila, L.L. Cowie, The evolution of Lyman limit absorption systems to redshift six. Astrophys. J. **721**, 1448–1466 (2010). doi:10.1088/0004-637X/721/2/1448

R.L.C. Starling, R. Willingale, N.R. Tanvir, A.E. Scott, K. Wiersema, P.T. O'Brien, A.J. Levan, G.C. Stewart, X-ray absorption evolution in gamma-ray bursts: Intergalactic medium or evolutionary signature of their host galaxies. Mon. Not. R. Astron. Soc. **431**, 3159–3176 (2013). doi:10.1093/mnras/stt400

C.C. Steidel, The History of Halo Gas in Normal Galaxies, in *Galaxy Evolution. The Milky Way Perspective*, ed. by S.R. Majewski Astronomical Society of the Pacific Conference Series, vol. 49, 1993, p. 227
C.C. Steidel, W.L.W. Sargent, Mg II absorption in the spectra of 103 QSOs—implications for the evolution of gas in high-redshift galaxies. Astrophys. J. Suppl. Ser. **80**, 1–108 (1992). doi:10.1086/191660
K.R. Stewart, T. Kaufmann, J.S. Bullock, E.J. Barton, A.H. Maller, J. Diemand, J. Wadsley, Orbiting circumgalactic gas as a signature of cosmological accretion. Astrophys. J. **738**, 39 (2011). doi:10.1088/0004-637X/738/1/39
G.S. Stinson, C. Brook, J.X. Prochaska, J. Hennawi, S. Shen, J. Wadsley, A. Pontzen, H.M.P. Couchman, T. Quinn, A.V. Macciò, B.K. Gibson, MAGICC haloes: Confronting simulations with observations of the circumgalactic medium at $z=0$. Mon. Not. R. Astron. Soc. **425**, 1270–1277 (2012). doi:10.1111/j.1365-2966.2012.21522.x
J.T. Stocke, T.A. Rector, An excess of Mg II absorbers in BL Lacertae objects. Astrophys. J. Lett. **489**, 17 (1997). doi:10.1086/310962
V. Sudilovsky, S. Savaglio, P. Vreeswijk, C. Ledoux, A. Smette, J. Greiner, Intervening metal systems in GRB and QSO sight lines: The Mg II and C IV question. Astrophys. J. **669**, 741–748 (2007). doi:10.1086/521525
V. Sudilovsky, D. Smith, S. Savaglio, Dusty Mg II absorbers: Implications for the gamma-ray burst/quasar incidence discrepancy. Astrophys. J. **699**, 56–59 (2009). doi:10.1088/0004-637X/699/1/56
N.R. Tanvir, D.B. Fox, A.J. Levan, E. Berger, K. Wiersema, J.P.U. Fynbo, A. Cucchiara, T. Krühler, N. Gehrels, J.S. Bloom, J. Greiner, P.A. Evans, E. Rol, F. Olivares, J. Hjorth, P. Jakobsson, J. Farihi, R. Willingale, R.L.C. Starling, S.B. Cenko, D. Perley, J.R. Maund, J. Duke, R.A.M.J. Wijers, A.J. Adamson, A. Allan, M.N. Bremer, D.N. Burrows, A.J. Castro-Tirado, B. Cavanagh, A. de Ugarte Postigo, M.A. Dopita, T.A. Fatkhullin, A.S. Fruchter, R.J. Foley, J. Gorosabel, J. Kennea, T. Kerr, S. Klose, H.A. Krimm, V.N. Komarova, S.R. Kulkarni, A.S. Moskvitin, C.G. Mundell, T. Naylor, K. Page, B.E. Penprase, M. Perri, P. Podsiadlowski, K. Roth, R.E. Rutledge, T. Sakamoto, P. Schady, B.P. Schmidt, A.M. Soderberg, J. Sollerman, A.W. Stephens, G. Stratta, T.N. Ukwatta, D. Watson, E. Westra, T. Wold, C. Wolf, A γ-ray burst at a redshift of $z\sim 8.2$. Nature **461**, 1254–1257 (2009). doi:10.1038/nature08459
N. Tejos, S. Lopez, J.X. Prochaska, H.-W. Chen, M. Dessauges-Zavadsky, On the incidence of C IV absorbers along the sight lines to gamma-ray bursts. Astrophys. J. **671**, 622–627 (2007). doi:10.1086/523088
N. Tejos, S. Lopez, J.X. Prochaska, J.S. Bloom, H.-W. Chen, M. Dessauges-Zavadsky, M.J. Maureira, Casting light on the "anomalous" statistics of Mg II absorbers toward gamma-ray burst afterglows: The incidence of weak systems. Astrophys. J. **706**, 1309–1315 (2009). doi:10.1088/0004-637X/706/2/1309
C.C. Thöne, J.P.U. Fynbo, P. Goldoni, A.P. de Ugarte, S. Campana, S.D. Vergani, S. Covino, T. Krühler, L. Kaper, N. Tanvir, T. Zafar, V. D'Elia, J. Gorosabel, J. Greiner, P. Groot, F. Hammer, P. Jakobsson, S. Klose, A.J. Levan, B. Milvang-Jensen, A.G. Nicuesa, E. Palazzi, S. Piranomonte, G. Tagliaferri, D. Watson, K. Wiersema, R.A.M.J. Wijers, GRB 100219A with X-shooter—abundances in a galaxy at $z=4.7$. Mon. Not. R. Astron. Soc. **428**, 3590–3606 (2013). doi:10.1093/mnras/sts303
T. Totani, N. Kawai, G. Kosugi, K. Aoki, T. Yamada, M. Iye, K. Ohta, T. Hattori, Implications for cosmic reionization from the optical afterglow spectrum of the gamma-ray burst 050904 at $z=6.3$. Publ. Astron. Soc. Jpn. **58**, 485–498 (2006). doi:10.1093/pasj/58.3.485
T. Totani, K. Aoki, T. Hattori, G. Kosugi, Y. Niino, T. Hashimoto, N. Kawai, K. Ohta, T. Sakamoto, T. Yamada, Probing intergalactic neutral hydrogen by the Lyman alpha red damping wing of gamma-ray burst 130606A afterglow spectrum at $z=5.913$. Publ. Astron. Soc. Jpn. **66**, 63 (2014). doi:10.1093/pasj/psu032
T. Totani, K. Aoki, T. Hattori, N. Kawai, High precision analyses of Lyman alpha damping wing of gamma-ray bursts in the reionization era: On the controversial results from GRB 130606A at $z=5.91$. ArXiv e-prints (2015)
S.D. Vergani, P. Petitjean, C. Ledoux, P. Vreeswijk, A. Smette, E.J.A. Meurs, Statistics and characteristics of MgII absorbers along GRB lines of sight observed with VLT-UVES. Astron. Astrophys. **503**, 771–781 (2009a). doi:10.1051/0004-6361/200911747
S.D. Vergani, P. Petitjean, C. Ledoux, P. Vreeswijk, A. Smette, E.J.A. Meurs, Statistics and characteristics of MgII absorbers along GRB lines of sight observed with VLT-UVES. Astron. Astrophys. **503**, 771–781 (2009b). doi:10.1051/0004-6361/200911747
D. Watson, P. Jakobsson, Dust extinction bias in the column density distribution of gamma-ray bursts: High column density, low-redshift GRBs are more heavily obscured. Astrophys. J. **754**, 89 (2012). doi:10.1088/0004-637X/754/2/89
D. Watson, T. Zafar, A.C. Andersen, J.P.U. Fynbo, J. Gorosabel, J. Hjorth, P. Jakobsson, T. Krühler, P. Laursen, G. Leloudas, D. Malesani, Helium in natal H II regions: The origin of the X-ray absorption in gamma-ray burst afterglows. Astrophys. J. **768**, 23 (2013). doi:10.1088/0004-637X/768/1/23
A.M. Wolfe, E. Gawiser, J.X. Prochaska, Damped Lyα systems. Annu. Rev. Astron. Astrophys. **43**, 861–918 (2005). doi:10.1146/annurev.astro.42.053102.133950

D.G. York, J. Adelman, J.E. Anderson Jr., S.F. Anderson, J. Annis, N.A. Bahcall, J.A. Bakken, R. Barkhouser, S. Bastian, E. Berman, W.N. Boroski, S. Bracker, C. Briegel, J.W. Briggs, J. Brinkmann, R. Brunner, S. Burles, L. Carey, M.A. Carr, F.J. Castander, B. Chen, P.L. Colestock, A.J. Connolly, J.H. Crocker, I. Csabai, P.C. Czarapata, J.E. Davis, M. Doi, T. Dombeck, D. Eisenstein, N. Ellman, B.R. Elms, M.L. Evans, X. Fan, G.R. Federwitz, L. Fiscelli, S. Friedman, J.A. Frieman, M. Fukugita, B. Gillespie, J.E. Gunn, V.K. Gurbani, E. de Haas, M. Haldeman, F.H. Harris, J. Hayes, T.M. Heckman, G.S. Hennessy, R.B. Hindsley, S. Holm, D.J. Holmgren, C.-h. Huang, C. Hull, D. Husby, S.-I. Ichikawa, T. Ichikawa, Ž. Ivezić, S. Kent, R.S.J. Kim, E. Kinney, M. Klaene, A.N. Kleinman, S. Kleinman, G.R. Knapp, J. Korienek, R.G. Kron, P.Z. Kunszt, D.Q. Lamb, B. Lee, R.F. Leger, S. Limmongkol, C. Lindenmeyer, D.C. Long, C. Loomis, J. Loveday, R. Lucinio, R.H. Lupton, B. MacKinnon, E.J. Mannery, P.M. Mantsch, B. Margon, P. McGehee, T.A. McKay, A. Meiksin, A. Merelli, D.G. Monet, J.A. Munn, V.K. Narayanan, T. Nash, E. Neilsen, R. Neswold, H.J. Newberg, R.C. Nichol, T. Nicinski, M. Nonino, N. Okada, S. Okamura, J.P. Ostriker, R. Owen, A.G. Pauls, J. Peoples, R.L. Peterson, D. Petravick, J.R. Pier, A. Pope, R. Pordes, A. Prosapio, R. Rechenmacher, T.R. Quinn, G.T. Richards, M.W. Richmond, C.H. Rivetta, C.M. Rockosi, K. Ruthmansdorfer, D. Sandford, D.J. Schlegel, D.P. Schneider, M. Sekiguchi, G. Sergey, K. Shimasaku, W.A. Siegmund, S. Smee, J.A. Smith, S. Snedden, R. Stone, C. Stoughton, M.A. Strauss, C. Stubbs, M. SubbaRao, A.S. Szalay, I. Szapudi, G.P. Szokoly, A.R. Thakar, C. Tremonti, D.L. Tucker, A. Uomoto, D. Vanden Berk, M.S. Vogeley, P. Waddell, S.-i. Wang, M. Watanabe, D.H. Weinberg, B. Yanny, N. Yasuda (SDSS Collaboration), The Sloan Digital Sky Survey: Technical summary. Astrophys. J. **120**, 1579–1587 (2000). doi:10.1086/301513

T. Zafar, D.J. Watson, D. Malesani, P.M. Vreeswijk, J.P.U. Fynbo, J. Hjorth, A.J. Levan, M.J. Michałowski, No evidence for dust extinction in GRB 050904 at $z \sim 6.3$. Astron. Astrophys. **515**, 94 (2010). doi:10.1051/0004-6361/200913795

G. Zhu, B. Ménard, The JHU-SDSS metal absorption line catalog: Redshift evolution and properties of Mg II absorbers. Astrophys. J. **770**, 130 (2013). doi:10.1088/0004-637X/770/2/130

Space Sci Rev (2016) 202:161–182
DOI 10.1007/s11214-016-0250-7

Gamma-Ray Bursts and Population III Stars

Kenji Toma[1,2] · **Sung-Chul Yoon**[3] · **Volker Bromm**[4]

Received: 18 October 2015 / Accepted: 14 March 2016 / Published online: 13 April 2016

Abstract Gamma-ray bursts (GRBs) are ideal probes of the epoch of the first stars and galaxies. We review the recent theoretical understanding of the formation and evolution of the first (so-called Population III) stars, in light of their viability of providing GRB progenitors. We proceed to discuss possible unique observational signatures of such bursts, based on the current formation scenario of long GRBs. These include signatures related to the prompt emission mechanism, as well as to the afterglow radiation, where the surrounding intergalactic medium might imprint a telltale absorption spectrum. We emphasize important remaining uncertainties in our emerging theoretical framework.

Keywords Gamma-ray bursts · First stars · Cosmology · Dark ages

1 Introduction

With recent progress in observational cosmology, one of the key frontiers is to understand how the first stars and galaxies ended the cosmic dark ages a few hundred million years after the Big Bang (Bromm et al. 2009; Loeb 2010). Prior to their emergence, the universe exhibited a simple state, devoid of complex structure, of any elements heavier than lithium,

✉ K. Toma
toma@astr.tohoku.ac.jp

S.-C. Yoon
yoon@astro.snu.ac.kr

V. Bromm
vbromm@astro.as.utexas.edu

1 Frontier Research Institute for Interdisciplinary Sciences, Tohoku University, Sendai 980-8578, Japan

2 Astronomical Institute, Tohoku University, Sendai 980-8578, Japan

3 Department of Physics and Astronomy, Seoul National University, Gwanak-ro 1, Gwanak-gu, Seoul, 88026, South Korea

4 Department of Astronomy, University of Texas, 2511 Speedway, Austin, TX 78712, USA

and of high-energy radiation fields. Within ΛCDM cosmology, the first stars, the so-called Population III (Pop III), are predicted to form at $z \sim 20$–30 in dark matter minihalos of mass $\sim 10^6$ $M_\odot$. The formation of the first bona-fide galaxies, being able to host long-lived stellar systems, may be delayed until more massive dark matter halos virialize (Bromm and Yoshida 2011). Once the first sources of light have appeared on the cosmic scene, the universe was rapidly transformed through the input of ionizing radiation (Barkana and Loeb 2007) and heavy chemical elements (Karlsson et al. 2013). The character of this fundamental transition, as well as the assembly process of the first galaxies, crucially depended on the feedback exerted by Pop III stars (Ciardi and Ferrara 2005). The feedback in turn is determined by the initial mass function (IMF) of the first stars (Bromm and Larson 2004; Glover 2005). Although important uncertainties remain, the key prediction is that the Pop III IMF is biased towards high mass, implying a top-heavy distribution (Bromm 2013). At least a fraction of the first stars could therefore have collapsed into massive black holes (BHs) at the end of their short lives, and thus provide viable gamma-ray burst (GRB) progenitors.

Upcoming facilities such as the James Webb Space Telescope (JWST), and the next generation of extremely-large telescopes on the ground (GMT, TMT, E-ELT) promise to open up a direct window into the first billion years of cosmic evolution (Gardner et al. 2006). Despite their exquisite sensitivity at near-IR wavelengths, even these observatories may not be able to directly probe the first stars, unless they formed in massive clusters (Pawlik et al. 2011), or were gravitationally lensed (Rydberg et al. 2013). The only opportunity to probe individual Pop III stars may be to catch them at the moment of their explosive death. This could involve extremely energetic supernova (SN) events, such as hypernovae or pair-instability SNe (Hummel et al. 2012; Pan et al. 2012; Whalen et al. 2013), or GRBs. The latter fate depends on whether Pop III stars could give rise to suitable collapsar progenitors, involving rapidly rotating massive stars (MacFadyen and Woosley 1999). Since Pop III stars are predicted to fulfill both requirements (see the discussion below), GRBs are expected to be prevalent at very high redshifts. Indeed, GRBs may play a key role in elucidating primordial star formation, as well as the properties of the early intergalactic medium (IGM), given their extreme intrinsic brightness, both of the prompt γ-ray emission, as well as that of the prolonged afterglow.

A number of features render GRBs ideal probes of the epoch of first light (Loeb 2010): *(i)* Traditional sources to observe the high-z universe, such as quasars and Lyman-α emitting galaxies, severely suffer from the effects of cosmological dimming. It was suggested, on the other hand, that GRB afterglows, if observed at a fixed time after the trigger, may exhibit nearly-flat infrared fluxes out to very high z (Ciardi and Loeb 2000). The argument behind this counter-intuitive effect was as follows: A fixed time interval in the observer frame translates into an increasingly early time in the source frame. Such earlier times in turn would sample the rapidly decaying GRB lightcurve at the moment of maximal brightness, thus compensating for the cosmological dimming (increasing luminosity distance). With the realization that such simple power-law decay may not be established until quite late after the trigger, this distance independence of the high-z burst flux now appears too optimistic. *(ii)* In the hierarchical setting of cosmic structure formation, earlier times are dominated by lower-mass host systems. The massive hosts required for quasars and bright galaxies therefore are "dying out" at the highest redshifts (Mortlock et al. 2011). GRBs, on the other hand, mark the death of individual stars, which can form even in very low-mass systems. *(iii)* Finally, Pop III GRBs would provide very clean background sources to probe the early IGM. Again reflecting the low masses of their hosts, any proximity effect should be much reduced, as ionized bubbles are confined to the immediate vicinity of the Pop III system; the IGM would thus largely remain unperturbed. In addition, since GRB afterglow spectra can be described as featureless, broken power-laws (Vreeswijk et al. 2004), any signature imprinted by absorption and emission events along a given line of sight can be easily discerned.

The outlook for GRB cosmology, therefore, is bright, provided that we can continue to fly wide-field γ-ray trigger instruments in space, beyond the end of the *Swift* satellite. We already have tantalizing examples of high-redshift bursts with the spectroscopically confirmed GRB 090423 at $z \simeq 8.2$ (Salvaterra et al. 2009; Tanvir et al. 2009), and the photometrically detected GRB 090429B at $z \simeq 9.4$ (Cucchiara et al. 2011). Future missions, such as SVOM (Paul et al. 2011), promise to fully unleash the potential of GRBs to probe the early universe. This review is organized as follows. In Sect. 2, we discuss the recent consensus on the formation of the first stars, with a particular focus on assessing their suitability as GRB progenitors. In Sect. 3, we continue by summarizing the key stellar evolution physics of these suggested Pop III progenitors. In Sect. 4, we discuss promising observational avenues to probe the signature of Pop III GRBs, and end with a brief outlook into the future in Sect. 5. In concluding, we would like to refer the reader to two recent reviews, one providing a comprehensive perspective on modern GRB astrophysics (Kumar and Zhang 2015), and the other a concise summary of the lessons from the *Swift* era for GRB cosmology (Salvaterra 2015).

2 Formation of Population III Stars

The longstanding consensus view has been that the conditions in the early universe favored the formation of predominantly massive stars, such that the Pop III IMF was top-heavy (Abel et al. 2002; Bromm et al. 2002; Bromm and Larson 2004). This expectation rests on the much less efficient cooling in pure H/He gas, where the only viable cooling agent is molecular hydrogen. The primordial gas can therefore reach temperatures of only ~ 200 K, compared to the 10 K reached in dust-cooled molecular clouds in the present-day Milky Way (Schneider et al. 2006). The correspondingly enhanced thermal pressure is reflected in a Jeans mass that is larger by one to two orders of magnitude in the Pop III case. Another element of this "standard model" of primordial star formation has been that the first stars formed typically in isolation, one per minihalo (Omukai and Nishi 1999).

In recent years, this traditional paradigm has been refined in important ways (Turk et al. 2009; Stacy et al. 2010; Clark et al. 2011; Greif et al. 2011, 2012). Supercomputing power, as well as algorithmic advances, now enable us to follow the protostellar collapse to densities, $n \sim 10^{22}$ cm^{-3}, where the initial hydrostatic core forms in the center of the cloud (Yoshida et al. 2008). Crucially, the computations can now also be extended into the main accretion phase (Omukai and Palla 2003). An important lesson has been that accretion is mediated through a near-Keplerian disk, similar to present-day star formation. The hot conditions in the surrounding cloud result in extremely large rates of infall onto the disk ($\dot{M} \propto T^{3/2}$); this rapid mass-loading drives the disk inevitably towards gravitational instability, such that a small multiple of Pop III protostars emerges, often dominated by a binary system. It is not yet possible to extend such ab-initio simulations all the way to the completion of the protostellar assembly process; the final mass of Pop III stars and their final IMF are thus still subject to considerable uncertainty. However, first attempts to carry out the radiation-hydrodynamical calculations required to treat the late accretion phase, where protostellar feedback tends to limit further infall, have confirmed the basic prediction: the first stars were typically massive, with masses of a few ~ 10 M$_\odot$, although rarely very massive (> 100 M$_\odot$), as previously thought, forming as a member of small multiple systems (McKee and Tan 2008; Hosokawa et al. 2011; Stacy et al. 2012). There are indications, though, that the Pop III mass could occasionally grow to $> 300\ M_{\odot}$, in cases of unusually weak protostellar feedback (Hirano et al. 2014; Susa et al. 2014).

Are Pop III stars suitable GRB progenitors? To successfully trigger a collapsar event, the leading contender for long-duration GRBs (Woosley 1993; MacFadyen et al. 2001), a number of conditions have to be met: a central BH has to form, a relativistic jet has to escape the stellar envelope before being quenched, and there has to be a sufficient degree of angular momentum close to the center, to delay the accretion of material onto the BH. These are quite stringent, and often difficult to fulfill simultaneously (Zhang and Fryer 2004; Petrovic et al. 2005; Belczynski et al. 2007).

The first requirement for a collapsar central GRB engine, the emergence of BH remnants, is fulfilled because of the top-heavy nature of the primordial IMF. The binary nature of Pop III stars may allow to meet the second requirement of providing an escape channel for the jet, if the binary is sufficiently close to allow for Roche-lobe overflow and a common-envelope phase, to expel the extended hydrogen (and helium?) envelope. This may be crucial to prevent the quenching of the relativistic jet, launched by the central engine (Bromm and Loeb 2006) (but see Suwa and Ioka 2011, and the discussion on this key point below). Simulations have begun to constrain the statistics of Pop III binaries, within a fully cosmological context (Stacy and Bromm 2013). The resulting distribution of semi-major axes is found to be quite broad, with a peak around ~ 300 AU. Those simulations, however, cannot yet resolve the formation and evolution of tight binaries, due to the resolution limit of ~ 20 AU. Such improved simulations would be required to probe the regime of contact binaries, where Roche-lobe overflow or common-envelope evolution could occur during the later red supergiant phase. It appears likely, though, that such tight binaries exist. A fraction of the sink particles that numerically represent Pop III stars in the simulations undergo mergers when they approach to within the resolution limit. With better resolution, some of those sinks/stars may well survive as tight binaries.

What about the additional requirement that the collapsar progenitor retains enough angular momentum? This question ties in with the rate of rotation of Pop III stars, where almost nothing is known yet. A first attempt to address this within a fully cosmological context has recently been carried out (Stacy et al. 2011, 2013), indicating that the first stars may have typically been very fast rotators, with surface rotation speeds of a few 10 percent of the break-up value. Such high rates of rotation would have important consequences for Pop III stellar evolution, possibly enabling strong mixing currents, and for the fate encountered at death (Yoon et al. 2006). Thus, it is plausible that all requirements for a collapsar central engine were in place in the early universe.

An important caveat here is that any effect of possible magnetic braking on the final stellar rotation rate has been neglected so far. Recent studies have argued that turbulent dynamo amplification in the primordial protostellar disks might rapidly build up dynamically significant magnetic fields (e.g., Schober et al. 2012). It is currently not clear whether such small-scale, tangled fields could be organized into larger scale arrangements. If they can, magnetic torques may be responsible for establishing Pop III stellar rotation rates similar to what is known for present-day O stars. Ongoing magneto-hydrodynamical simulations should soon help to clarify this key point.

3 Evolution of Pop III Stars and GRBs

Metallicity is one of the prime factors that determine the evolution of massive stars. Many features of stellar evolution are therefore uniquely found with massive Pop III stars, compared to the case of metal-rich counterparts.

Massive stars on the main sequence in the nearby universe are powered by the CNO cycle. In the early universe, CNO elements are absent and core hydrogen burning starts with the pp chain. Because energy production by the pp-chain is too weak to maintain thermal equilibrium, a massive Pop III star initially undergoes thermal contraction until the central temperature becomes high enough for helium burning to produce carbon. The CNO cycle then becomes active with thus-produced carbon, and the structure of the star is adjusted accordingly until thermal equilibrium is reached. The main sequence evolution begins only thereafter (e.g., Marigo et al. 2001; Ekström et al. 2008; Yoon et al. 2012).

The evolution on the main sequence and in later stages is more critically affected by the lack of heavy elements. In metal-rich environments, massive star evolution is characterized by strong mass loss by radiation-driven winds, for which iron lines play a particularly important role (e.g., Puls et al. 2008). For massive Pop III stars, the radiation force resulting from hydrogen and helium lines is too weak to drive a wind, and the predicted mass loss rate is extremely low (i.e., $\dot{M} \leq 10^{-14}$ $M_\odot$ yr^{-1}; Krtička and Kubát 2006).

This has two important consequences. First, massive Pop III stars would not easily lose their initial angular momentum via mass loss. This implies that the evolution of massive Pop III stars could be dominated by the effects of rotation (Sects. 3.1 and 3.2). Second, even very massive stars ($M \gtrsim 100$ $M_\odot$) that are close to the Eddington limit would retain their hydrogen envelopes until the end of the evolution, This can produce very massive red supergiant stars (see Fig. 1), which are not found in the local universe. Metallicity effects also critically influence the condition for convection inside stars, which plays an important role in the final structure at the pre-collapse stage (Sect. 3.3). All of these effects are closely related to the properties of Pop III GRB progenitors, as discussed below.

3.1 The Role of Rotation

3.1.1 Evolution of Angular Momentum

One of the key necessary conditions for GRB progenitors is rapid rotation. To trigger relativistic jets, a large amount of angular momentum should be retained in the innermost layers of GRB progenitors until the pre-collapse stage. More specifically, the specific angular momentum in the core must be higher than $j > 1.5 \times 10^{16} (\frac{M_{BH}}{3\ M_\odot})$ $cm^2\,s^{-1}$ within the collapsar scenario, where M_{BH} is the black hole mass (Woosley 1993), and $j \simeq 4 \times 10^{15}$ $cm^2\,s^{-1}$ within the magnetar scenario (e.g., Wheeler et al. 2000).

Observations indicate that the amounts of angular momentum are sufficient to meet these criteria for a large fraction of massive main-sequence stars in the nearby universe (e.g., Mokiem et al. 2006; Wolff et al. 2006; Ramírez-Agudelo et al. 2013). Recent numerical studies also indicate that massive Pop III stars would be rapid rotators (Stacy et al. 2011; Rosen et al. 2012). However, numerous observations imply rapid transport of angular momentum inside stars (e.g., Suijs et al. 2008; Charpinet et al. 2009; Eggenberger et al. 2012; Marques et al. 2013), which must have significant impact on the final angular momentum distribution in the stellar core (Hirschi et al. 2004; Heger et al. 2005). While the main mechanism for angular momentum transport is still a matter of debate, the following two mechanisms have been considered most important in stellar evolution models.

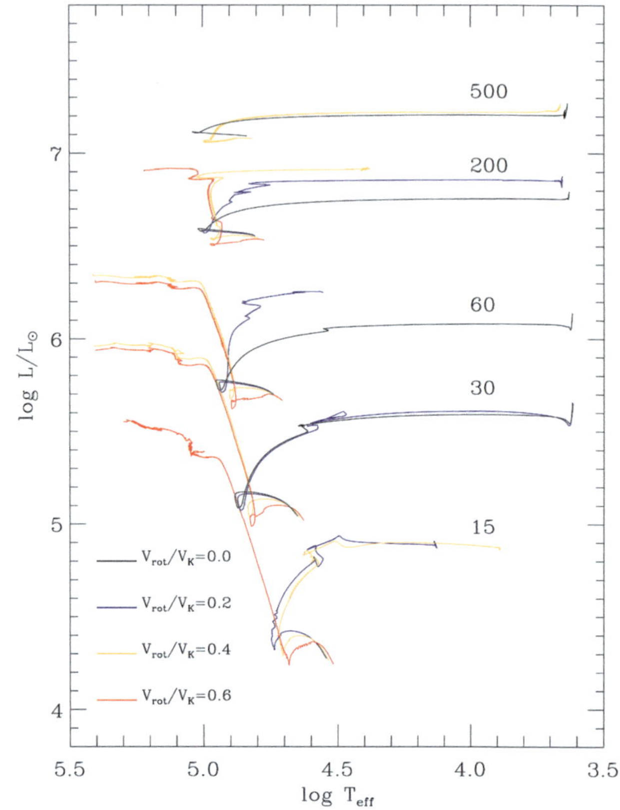

Fig. 1 Evolution of massive Pop III stars (15, 30, 60, 200 and 500 $M_\odot$) with and without rotation. The adopted initial rotational speeds in units of the Keplerian value are marked by *different colors* as indicated by the labels. These evolutionary models are taken from Yoon et al. (2012)

- *Eddington-Sweet circulations:* Thermal equilibrium in rotating stars generally breaks down because the radiative flux along the polar axis becomes excessive compared to that along the equatorial one (the von Zeipel theorem; e.g., Kippenhahn and Weigert 1990). To compensate this thermal imbalance, large-scale meridional circulations are induced in rotating stars, which are nowadays often called 'Eddington-Sweet (ES) circulations' (e.g., Maeder and Meynet 2012). These ES circulations are not only important for the energy flux, but also for the transport of angular momentum. The timescale for ES circulations in a chemically homogeneous star is roughly given by $\tau_{ES} \approx \tau_{KH}(\Omega/\Omega_K)^{-2}$, where τ_{KH} denotes the Kelvin-Helmholtz timescale and Ω/Ω_K the ratio of the angular velocity to the Keplerian value. Although τ_{ES} can be in principle shorter than the nuclear timescale (τ_{nuc}) of a star, ES circulations are severely slowed down once the chemical stratification across the boundary between the core and the hydrogen envelope is built up (Maeder and Meynet 1998). Therefore, it is believed that ES circulations cannot efficiently brake down the stellar core for most cases (Heger et al. 2000; Hirschi et al. 2004).
- *Taylor-Spruit dynamo:* According to Taylor (1973), toroidal fields in radiative layers of stars are always unstable to create poloidal fields (see also Spruit 1999). If differential rotation can wind up thus-created poloidal fields to amplify toroidal fields, the dynamo loop can be closed (Spruit 2002). This so-called Taylor-Spruit (TS) dynamo may lead to magnetic torques that can redistribute angular momentum inside stars much more efficiently than ES circulations. Theoretical studies indeed show that solid-body rotation

can be maintained in main sequence stars with TS dynamo (Heger et al. 2005). This rapid angular momentum transfer by the TS dynamo has been invoked to explain the relatively slow spin rates of young neutron stars and isolated white dwarfs, as well as angular velocity profiles in low-mass stars on various evolutionary stages that are inferred from asteroseismological data (e.g. Eggenberger et al. 2005; Heger et al. 2005; Suijs et al. 2008; Cantiello et al. 2014). However, the validity of the TS dynamo theory is still debated (e.g., Braithwaite 2006; Zahn et al. 2007; Arlt and Rüdiger 2011; Cantiello et al. 2014).

Evidently, the transport mechanisms may not be limited to these ones. Rotationally-induced hydrodynamic instabilities apart from ES circulations include the shear instability and the baroclinic instability. The interplay between ES circulations and the shear instability would be particularly important for the transport process (Maeder and Meynet 2012). Several authors also investigated the role of the magneto-rotational instability (Wheeler et al. 2015) and internal gravity waves in massive stars (Fuller et al. 2014). More progress is certainly needed to have a reliable prescription for angular momentum transport.

3.1.2 *Chemical Mixing*

Rotationally-induced hydrodynamic instabilities can transport not only angular momentum but also chemical species. This may lead to chemical mixing of hydrogen-burning products from the convective core into the radiative envelope in a massive star on the main sequence. Enhanced abundances of nitrogen and helium are indeed found at the surfaces of many massive stars, which provide evidence for such mixing (e.g. Hunter et al. 2008, 2009). The efficiency of chemical mixing due to rotation has been recently calibrated by Brott et al. (2011) using the inferred nitrogen abundances at the surfaces of B-type stars in the Large Magellanic Cloud. But the anomalously high/low nitrogen abundances observed in some slowly/rapidly rotating stars have not been well understood yet (Hunter et al. 2008, 2009; Brott et al. 2011; Aerts et al. 2014).

3.1.3 *Mass Shedding*

Rapidly rotating stars can reach the critical rotation during the course of their evolution if they do not lose a sufficient amount of angular momentum via strong mass loss. Note that the critical rotation speed ($\upsilon_{\rm crit}$) can become lower than the Keplerian limit ($\upsilon_{\rm K} = \sqrt{GM/R}$) if the stellar luminosity approaches the Eddington limit, as the following:

$$\upsilon_{\rm crit} = \upsilon_{\rm K}\sqrt{1 - \varGamma}\ , \tag{1}$$

where $\varGamma$ is the Eddington factor (Heger et al. 2000; see Maeder and Meynet 2000 for an alternative description for the critical rotation speed). Once the star reaches the critical rotation, mechanical mass shedding would occur, even when radiation-driven winds are very weak. This would be the dominant mode of mass-loss from massive Pop III stars (Marigo et al. 2003; Ekström et al. 2008; Yoon et al. 2012). For very massive stars ($M \gtrsim 100\ \mathrm{M}_\odot$), however, pulsationally driven winds during the red supergiant phase might also play an important role (Baraffe et al. 2001; Moriya and Langer 2015).

3.2 Chemically Homogeneous Evolution and GRB Progenitors

Typical timescales about ~ 10 secs of long GRBs imply that their progenitors are generally compact ($R 10\ R_\odot$; Woosley and Bloom 2006). All of the supernovae associated with a long

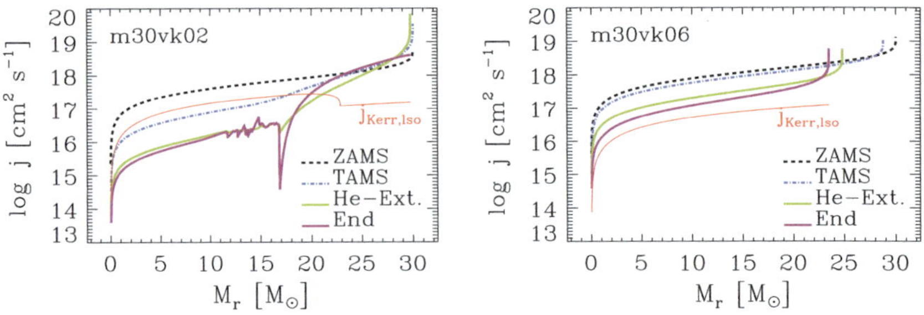

Fig. 2 Mean specific angular momentum profile as a function of the mass coordinate for different evolutionary epochs of 30 $M_\odot$ Pop III star: zero age main sequence (ZAMS), terminal age of the main sequence (TAMS), core helium exhaustion (He-Ext.) and the last calculated model, which corresponds to core neon exhaustion. The TS dynamo was included in the calculations, and the two different initial rotational speeds were adopted: 20 % and 60 % Keplerian rotation speeds for the *left* and the *right panels*, respectively. In the latter case, the star evolves chemically homogeneously. Adapted from Yoon et al. (2012)

GRB turn out to be Type Ic, which provide further evidence for compact progenitors that have lost their hydrogen envelopes (e.g., Woosley and Bloom 2006; D'Elia et al. 2015). It is therefore widely believed that the majority of long GRB progenitors are naked helium stars like Wolf-Rayet (WR) stars. Another necessary condition for long GRB progenitors is rapid rotation as mentioned above.

Mass loss by radiation-driven winds is negligible for Pop III stars, and mass shedding due to rotation is not significant enough to remove the whole hydrogen envelope (Marigo et al. 2003; Ekström et al. 2008; Yoon et al. 2012). This may raise a question of whether massive single Pop III stars can produce an ordinary long GRB.

Probably, the only possible solution for ordinary long GRBs from single Pop III stars would involve the chemically homogeneous evolution (CHE). CHE may occur with a sufficiently high rotation speed if the timescale for chemical mixing due to ES circulations becomes shorter than the nuclear timescale (Maeder 1987). In other words, CHE can be realized if rapid chemical mixing between the hydrogen-burning convective core and the radiative envelope occurs, before nuclear burning builds up a strong degree of chemical stratification across the boundary between them that would dramatically slow down ES circulations. In this case, almost all the hydrogen in the envelope of a star is mixed into the hydrogen-burning core to be fused into helium, which is in turn mixed back into the envelope. The whole star is thus gradually transformed into a helium star by the end of the main sequence. This makes the CHE stars evolve blueward as shown in Fig. 1.

In the case of normal evolution, the helium core during the post-main sequence stage would be significantly braked down by the slowly rotating hydrogen envelope that expands to a red-supergiant phase, which can serve as a large angular momentum reservoir (Heger et al. 2005). By contrast, CHE stars do not undergo the red-supergiant phase and can retain a large fraction of the initial amount of angular momentum until the pre-collapse stage (Fig. 2; Yoon and Langer 2005; Woosley and Heger 2006). CHE can be more easily realized at lower metallicity (Yoon et al. 2006), and therefore CHE is a promising pathway to long GRBs for massive Pop III stars (Yoon et al. 2012).

As summarized in Fig. 3 (see also Yoon et al. 2012), CHE would produce GRB progenitor candidates for the initial mass range of 13–84 $M_\odot$ if the initial rotation speed is higher than about 50 % of the Keplerian value. Another interesting prediction of CHE is that the initial mass range for pair-instability supernovae (PISNe) would significantly decrease to 84–190 $M_\odot$, compared to the non-rotating case (i.e., 140–300 $M_\odot$; see also Chatzopoulos

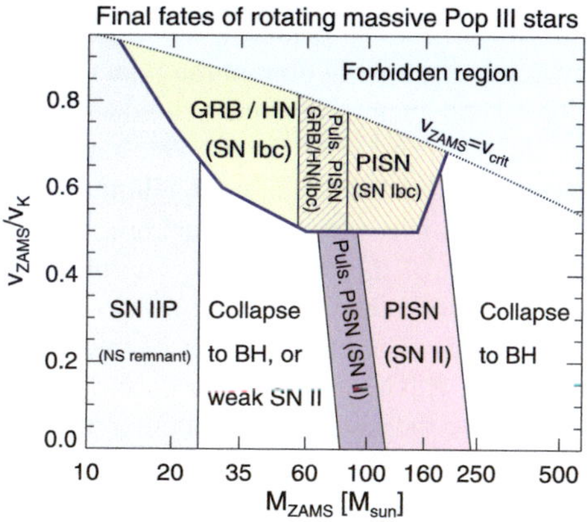

Fig. 3 Final fates of rotating massive Pop III stars, in terms of the initial mass and the initial rotational speed in units of the Keplerian value. The boundary for the CHE regime is marked by the *thick solid line*. The *dotted line* denotes the critical rotational speed for a given ZAMS mass, as defined by Eq. (1). This figure is a reproduction of Fig. 12 in Yoon et al. (2012) with permission from Astronomy & Astrophysics, © ESO

and Wheeler 2012). These PISNe from CHE stars would appear as Type Ic, instead of Type IIP that is expected for non-rotating or slowly rotating PISN progenitors. For the initial mass range of 56–84 $M_\odot$ of GRB progenitors, a pulsational PISN would occur shortly before the core collapses into a black hole, which may produce a very bright optical event via an interaction between the pulsational PISN ejecta and the GRB jet that follows.

3.3 Supergiant Pop III Progenitors for Ultra-Long GRBs

With relatively slow rotation, the typical core-envelope structure is developed in a massive Pop III star after the end of core hydrogen burning. The helium core is expected to be spun down as the core angular momentum is transferred to the expanding hydrogen envelope. At the pre-collapse stage, the inner core would rotate relatively slowly and its specific angular momentum would not exceed the critical limit for producing a GRB.

Interestingly, because Pop III stars do not lose much mass, the total angular momentum can be more or less conserved and the specific angular momentum of the hydrogen envelope increases as a result of the angular momentum transfer from the core (see the left panel of Fig. 2). In this case, the core would directly collapse into a BH, and the envelope material would form a Keplerian disk around it. The consequent accretion rate depends on the free-fall time of the envelope material: it would be $\sim 10^{-4}\ M_\odot\ \mathrm{yr}^{-1}$ for a blue-supergiant (BSG) progenitor, and $\sim 10^{-6}\ M_\odot\ \mathrm{yr}^{-1}$ for a red-supergiant (RSG) progenitor (Woosley and Heger 2012). Therefore, supergiant Pop III stars could produce an ultra-long GRB having a timescale of 10^3–10^7 sec (cf. Gendre et al. 2013; Virgili et al. 2013; Evans et al. 2014; Levan et al. 2014; Piro et al. 2014), depending on the final structure (Suwa and Ioka 2011; Nakauchi et al. 2012; Woosley and Heger 2012; Yoon et al. 2015).

Note, however, that the hydrogen envelope of a RSG is very loosely bound. Recently, Lovegrove and Woosley (2013) showed that most of the RSG envelope may be ejected as a result of the rapid loss of gravitational mass due to neutrino emission from the core while a BH is formed. This makes it difficult for a RSG to produce a long GRB transient even if its envelope is rotating rapidly. By contrast, the binding energy of a BSG envelope is fairly high and it would remain gravitationally bound to the newly formed BH. Therefore, rapidly rotating BSGs should be considered good progenitor candidates for ultra-long GRBs with a timescale of about 10^3–10^4 sec (Woosley and Heger 2012).

The remaining question is whether massive Pop III BH progenitors end their lives as a BSG or a RSG. Although the relatively low-opacity with zero-metallicity helps a massive

Pop III star remain compact, non-linear effects of stellar evolution make it very difficult to make a robust prediction on the final structure. For example, lower opacity in the core of a Pop III star than in a metal-rich counterpart leads to a smaller size of the convective core on the main sequence for a given mass, which in turn results in a more compact carbon-oxygen core at the end of core helium burning. As a consequence, the helium-burning shell source becomes very hot, inducing a more violent convection above it. This often leads to penetration of the convection zone of the helium shell source into the hydrogen-burning shell, boosting the CNO cycle (Heger and Woosley 2010; Limongi and Chieffi 2012; Yoon et al. 2012). The huge amount of nuclear energy produced in this way can make the star become a RSG shortly before core collapse. The onset of this CNO boosting sensitively depends on the adopted overshooting parameter and the initial rotation speed, but it is usually found for Pop III stars in the mass range of about 10–30 $M_\odot$. For more massive Pop III stars, the RSG solution seems to be generally favored (Marigo et al. 2001, 2003; Ekström et al. 2008; Yoon et al. 2012), but for a limited mass range of about 30–60 $M_\odot$, the BSG solution could be obtained with moderate rotation or weak overshooting (Limongi and Chieffi 2012; Yoon et al. 2012; see also Fig. 1).

3.4 Very Massive Pop III Stars and Supercollapsars

For the mass range above the PISN limit and below the general relativistic instability limit (e.g., 260 $M_\odot \lesssim M \lesssim 50000\ M_\odot$ without rotation), Pop III stars would directly collapse to a BH with $M \gtrsim 100\ M_\odot$ (e.g., Fryer et al. 2001a). This would lead to a collapsar that may produce a long GRB having a very high total energy ($\gtrsim 10^{53}$–10^{54} erg) and a very long timescale ($T \sim 10^4$–10^6 sec), depending on the final structure of the progenitor star (the so-called supercollapsar; Komissarov and Barkov 2010; Mészáros and Rees 2010).

In terms of stellar evolution, the following issues should be considered to probe the possibility of Pop III supercollapsars.

- Very massive stars are close to the Eddington limit, and can reach the critical rotation at a relatively low rotation speed. Angular momentum loss through mass shedding becomes significant accordingly, which slows down ES circulations. For this reason, CHE cannot be easily realized for very massive Pop III stars with $M \gtrsim 200\ M_\odot$ (Yoon et al. 2012). This means that they would not be able to avoid developing the typical core-envelope structure during the post-main sequence phase.
- The angular momentum transport from the core to the envelope can easily remove the core angular momentum in this case, if there exists a strong coupling between them via, for example, magnetic torques. The angular momentum condition for a collapsar cannot be fulfilled in this case (Yoon et al. 2012). By contrast, if ES circulations were the dominant mode of angular momentum transfer, the core and/or the envelope would retain an sufficient amount of angular momentum to produce a collapsar for a limited mass range of about 300 $M_\odot \lesssim M \lesssim 700\ M_\odot$ (Yoon et al. 2015).
- The RSG solution is strongly favored for these very massive stars: it is most likely that Pop III stars with $M > 260\ M_\odot$ end their lives as a RSG (Marigo et al. 2003; Yoon et al. 2012, 2015). Therefore, even if the core was rotating rapidly enough to produce a collapsar, the resultant relativistic jet would not be able to penetrate the envelope. The final outcome would be a jet-driven Type IIP supernova, rather than a GRB. The caveat here is that the mass loss rate from such a very massive RSG is not well understood. RSG stars are usually unstable to radial pulsations, and pulsation-driven winds might be significant. Whether or not such winds can completely remove the hydrogen envelope of

a very massive Pop III RSG star is a matter of debate (e.g., Baraffe et al. 2001; Moriya and Langer 2015).

- For a supercollapsar, a neutrino-driven jet cannot be efficient because of the large radius of the last stable orbit around such a massive black hole. Therefore, another important ingredient for a successful supercollapsar is strong large-scale magnetic fields in the progenitor star such that a relativistic jet may be triggered via the Blandford-Znajek (BZ) mechanism (see also Sect. 4 Komissarov and Barkov 2010; Mészáros and Rees 2010). Magnetic field configuration in massive stars at the pre-collapse stage has been hardly investigated so far, and we still do not know if this condition of strong large-scale magnetic fields can be fulfilled in very massive Pop III stars.

In short, a supercollapsar seems to be rather difficult to produce from very massive Pop III single stars, given that they would not follow CHE and that they would become a RSG at the pre-collapse stage, rather than a BSG or a naked He star. However, binary interactions might produce a BSG supercollapsar progenitor more easily, as discussed in Sect. 3.5 below.

3.5 Binary Interactions and GRB Progenitors

Binary interactions would be an alternative way to produce naked helium stars at zero metallicity. However, it is expected that they are slow rotators in general and would not produce a GRB (Yoon et al. 2010): GRB production even from binary stars would require an exotic evolutionary path. Several different scenarios for binary GRB progenitors have been suggested (see Brown et al. 2000; Izzard et al. 2004; Fryer and Heger 2005; van den Heuvel and Yoon 2007; Podsiadlowski et al. 2010), but because of the complexity of the related physical processes like common envelope ejection, the details of these scenarios have not been properly investigated with self-consistent binary evolution models for most cases.

Here we summarize the possibly important aspects of binary stars for Pop III GRBs, which should be carefully studied in the near future.

- Mass exchange in close binary systems can enhance the number fraction of rapidly rotating stars (de Mink et al. 2013). As shown by Cantiello et al. (2007), the Case B mass transfer (i.e., mass transfer during the helium core contraction phase or the early stage of core helium burning) is likely to enhance the possibility for CHE in mass accretors. The parameter space for this solution should be systematically explored for Pop III binaries.
- In the literature, evolutionary channels including the Case C mass transfer (i.e., mass transfer during the late stage of core helium burning or thereafter) has been often invoked to explain GRB progenitors (Brown et al. 2000; van den Heuvel and Yoon 2007; Podsiadlowski et al. 2010). In metal-rich environments, strong mass loss tends to prohibit the Case C mass transfer from massive stars with $M > 30\ M_\odot$ that are potential BH progenitors, but it would occur more frequently for Pop III stars that do not lose much mass. It is therefore possible that the Case C channel for long GRBs would be more important for Pop III stars than in the nearby universe.
- Binary mergers (in particular Case B mergers) may produce a BSG progenitor more easily, because a fairly small mass ratio of the helium core to the hydrogen envelope can be achieved in the merger remnants (e.g., Justham et al. 2014). Although the helium core may not be necessarily rapidly rotating in this case, the hydrogen envelope must have a very high specific angular momentum until the pre-collapse stage because mass loss from Pop III stars is weak. Therefore, Case B mergers at low-metallicity would be a very

promising pathway to ultra-long GRBs. It would be particularly interesting if very massive BSG GRB progenitors ($M \gtrsim 300\ M_{\odot}$) could be produced in this way (see Yoon et al. 2015, for a more detailed discussion).

4 Observational Signatures in Pop III GRBs

4.1 How to Identify Pop III GRBs?

Recent studies on stellar formation in the early universe predict that Pop III stars may be most prominent at $z \sim 20$–30 (e.g. Abel et al. 2002; Yoshida et al. 2008; Bromm and Yoshida 2011, and see Sects. 1 and 2). Observations of GRBs may provide unique probes of the physical conditions of the universe at such redshifts. Due to their high luminosities, GRB prompt emissions and afterglows are expected to be observable at least out to $z \gtrsim 10$ (e.g. Lamb and Reichart 2000; Ciardi and Loeb 2000; Bromm and Loeb 2002; Yonetoku et al. 2004; Gou et al. 2004). This can serve as a tracer of the history of the cosmic star formation rate (Totani 1997; Porciani and Madau 2001; Bromm and Loeb 2006; Kistler et al. 2009; de Souza et al. 2011) and provide invaluable information on the physical conditions in the intergalactic medium (Barkana and Loeb 2004; Ioka and Mészáros 2005; Inoue et al. 2007). Currently, the most distant GRB that has been spectroscopically confirmed is GRB 090423 at $z \simeq 8.2$ (Tanvir et al. 2009; Salvaterra et al. 2009), and GRB 090429B has a photometric redshift $z \simeq 9.4$ (Cucchiara et al. 2011). The detailed spectroscopic observation of GRB 050904 at $z \simeq 6.3$ has put a unique upper bound on the neutral hydrogen fraction in the intergalactic medium at that redshift and estimated the metallicity (Totani et al. 2006; Kawai et al. 2006). See also a recent detailed analysis of GRB 130606A at $z \simeq 5.9$ and related debates (Totani et al. 2016, and references therein). These observations indicate that GRBs are very promising for exploring the high-redshift universe (for a recent review, see Salvaterra 2015).

In the context of the study of Pop III stars, a crucial question is how can we identify GRBs originating from Pop III stars (Pop III GRBs) among bursts whose redshifts are determined as $z \gtrsim 10$ e.g. by observation of the Lyα cutoff at the IR frequencies. An unambiguous way to pinpoint a Pop III progenitor is to examine whether the afterglow spectrum from its surrounding medium is devoid of iron-group elements[1] through high-resolution IR and X-ray spectroscopy by ground-based facilities and/or future space experiments. However, there may also be the case that the surrounding medium is embedded in a region where stellar explosions have already occurred slightly earlier, and the absorption lines of the first heavy elements produced by those explosions are thus imprinted (Wang et al. 2012). A true Pop III origin of the GRB progenitor could then be missed due to this masquerading effect. It is therefore important to explore alternative strategies to identify Pop III GRBs. In this regard, another proposed way is to focus on the total energies and durations of GRBs estimated by the X-ray and gamma-ray observations of prompt emissions and/or afterglows (Komissarov and Barkov 2010; Mészáros and Rees 2010; Toma et al. 2011; Suwa and Ioka 2011). Pop III stars could be very massive stars (VMSs) with $M \gtrsim 300\ M_{\odot}$, and then GRBs originating from Pop III VMSs (Pop III VMS GRBs) could have peculiarly large total energies and long

[1] It may be misleading to state "devoid of metals". Pop III GRB progenitors via CHE, for example, could have ejected CNO elements into the surrounding medium via mass shedding during the late stages of their evolution.

durations.[2] Their values could be orders of magnitude larger than those of ordinary Pop I and Pop II GRBs, providing strong hints for Pop III progenitors.

The recent numerical studies of the evolutions of Pop III stars imply that luminous GRBs are rather difficult to occur from VMSs (Yoon et al. 2012, 2015), as discussed in Sect. 3. However, it may not be interpreted as a robust prediction yet, since theoretical modelings of formation and evolution of Pop III stars are highly non-linear, complex problems, as also noted in Sect. 3. There are several factors to be investigated, such as the multi-dimensional effects related to magnetic field and stellar rotation, the radiative feedback, the mass accretion after the onset of core hydrogen burning, the binary interactions, and so on. If only Pop III stars with $M < 100\ \mathrm{M}_\odot$ could produce GRBs, it would not be simple to identify Pop III GRBs through the estimate of the total energy and duration scales. Such bursts at the high redshifts may not be detectable by the current satellites (Nakauchi et al. 2012), and then persistent efforts on observational and theoretical studies with new high-sensitivity satellites would be needed to find out possible statistical differences between Pop III GRBs and ordinary GRBs. On the other hand, the history of GRB study tells us that some breakthroughs happened by simple observations, such as the serendipitous discovery of GRBs itself, the isotropic angular distribution of GRBs in the sky revealed by BATSE, the discovery of the afterglow by Beppo-SAX, the confirmation of the connection between GRBs and supernovae by HETE-2, and several recent discoveries by Swift and Fermi (for reviews, see e.g. Zhang and Mészáros 2004; Kumar and Zhang 2015). The connection between GRBs and Pop III stars might also be discovered by a serendipitous observation of a source which looks peculiar. Here we consider a case in which Pop III VMSs can produce GRBs as luminous as ordinary GRBs and discuss their unique observational signatures. Searches of such peculiar GRBs will constrain the theoretical models of the formation and evolution of Pop III stars.

4.2 Pop III VMS GRB Model

Let us first make a phenomenological interpretation of the total gamma-ray energies and durations of ordinary long GRBs which are currently observed. Focusing on the most energetic GRBs, which are mainly detected by Fermi satellite, they have $E_{\gamma,iso} \sim 10^{54}$–$10^{55}$ erg. The typical duration is $T \sim 10$–100 s (Ackermann et al. 2013). In the collapsar model, the total gamma-ray energy can be written as

$$E_{\gamma,iso} = \frac{\epsilon_\gamma \eta M_{\mathrm{acc}} c^2}{1 - \cos\theta_j} \sim 6 \times 10^{54} \epsilon_\gamma \left(\frac{\eta}{10^{-2}}\right)\left(\frac{M_{\mathrm{acc}}}{\mathrm{M}_\odot}\right)\left(\frac{\theta_j}{0.1}\right)^{-2} \mathrm{erg}, \tag{2}$$

where ϵ_γ is the gamma-ray radiation efficiency, η the conversion factor from the accretion energy to the jet energy, M_{acc} the total mass accreted by the central black hole (BH), and θ_j the opening half angle of the jet. While $\epsilon_\gamma \gtrsim 0.5$ and $\theta_j \sim 0.1$ are supported by some observational indications (Ioka et al. 2006; Zhang et al. 2007; Frail et al. 2001) (but see Beniamini et al. 2015), the values of η and M_{acc} are highly uncertain. (Furthermore, the conversion factor from the accretion luminosity $\dot{M}_{\mathrm{acc}} c^2$ to the jet luminosity may be time-dependent, so that η in Eq. (2) should be interpreted as a rough temporally-averaged value of the energy conversion factor.) In a scenario in which the progenitor star has an initial mass $\sim 30\ \mathrm{M}_\odot$, ends its life as a Wolf-Rayet star of $\sim 15\ \mathrm{M}_\odot$ after losing a large fraction of

[2] Recently, GRBs from supermassive stars with $M \sim 10^5\ \mathrm{M}_\odot$ which could make seed BHs for supermassive BHs are also discussed (Matsumoto et al. 2015).

mass by the stellar wind, and collapses making a BH of $\sim 3\ \mathrm{M}_\odot$, being accompanied by a supernova explosion with an ejecta mass of $\sim 10\ \mathrm{M}_\odot$ (Iwamoto et al. 2006; Mazzali et al. 2006), the total accreted mass is $M_{\mathrm{acc}} \sim \mathrm{M}_\odot$. If the stellar wind is so weak that the star does not lose much of its mass, we have $M_{\mathrm{acc}} \sim 10\ \mathrm{M}_\odot$. Correspondingly, η can be estimated to be $\sim 10^{-4}$–10^{-2} for $E_{\gamma,iso} \sim 10^{54}$–$10^{55}$ erg.

The conversion factor η should be determined by the jet production mechanism. For driving relativistic jets, the thermal energy injection by $\nu\bar{\nu}$ annihilation (Eichler et al. 1989) and/or the electromagnetic energy injection by magnetic braking of the accretion disk or the BH may be viable. In the former thermal model, η is estimated to be $\sim 10^{-4}$–10^{-2} for $\dot{M}_{\mathrm{acc}} \sim (0.1\text{–}1) \times \mathrm{M}_\odot\,\mathrm{s}^{-1}$ and $M_{\mathrm{BH}} = 3\ \mathrm{M}_\odot$ (Zalamea and Beloborodov 2011). However, it depends on the condition of the BH accretion such as the strength of the disk wind (see also Suwa 2013). In the latter magnetic model, BZ process, i.e. the magnetic braking of the BH or the energy extraction of the BH rotational energy (Blandford and Znajek 1977), can give rise to even $\eta > 1$ as demonstrated by MHD numerical simulations (Tchekhovskoy et al. 2011). There have been some observational indications supporting such a high value of η in relativistic jets of active galactic nuclei (e.g. Fernandes et al. 2011; Ghisellini et al. 2014). In this model also, η depends on the condition of the BH accretion as well as on the behavior of the magnetic field (e.g. Tchekhovskoy and Giannios 2015). There are also some issues relating to the physics of the BZ process itself (Komissarov 2009; Toma and Takahara 2014).

The total duration scale may be estimated as the disk lifetime. If the progenitor star is rotating at half of the break-up speed, then the initial outer edge of the disk is at $R_{\mathrm{d}} \simeq R/4$, where R is the stellar radius. The disk lifetime is given by its viscous accretion timescale $\sim R_{\mathrm{d}}^2/(\alpha c_s H)$, where α is the effective viscous stress parameter of the α-disk model (Shakura and Sunyaev 1973), c_s the sound speed, and H the scale height of the disk. The balance equation between the thermal pressure gradient and the gravitational force gives $c_s/H \sim \Omega_K$, where Ω_K is the Keplerian angular velocity, so that we have in the thick disk case ($H \sim R_{\mathrm{d}}$)

$$T \sim \frac{1}{\alpha \Omega_K}(1+z) \simeq 200 \left(\frac{\alpha}{0.1}\right)^{-1} \left(\frac{R}{10^{10}\ \mathrm{cm}}\right)^{\frac{3}{2}} \left(\frac{M_{\mathrm{BH}}}{3\ \mathrm{M}_\odot}\right)^{-\frac{1}{2}} \left(\frac{1+z}{3}\right)\ \mathrm{s}. \tag{3}$$

We should note that the duration of the bright part of the prompt emission may be shorter than this estimate since the conversion factor from the accretion luminosity to the jet luminosity and the accretion luminosity itself are time-dependent (Komissarov and Barkov 2010; Kumar et al. 2008; Tchekhovskoy and Giannios 2015).[3]

We apply the above scalings of the total gamma-ray energy and duration to a VMS progenitor with $M_* \sim 10^3\ \mathrm{M}_\odot$, although those scalings include many uncertainties. If we substitute $M_{\mathrm{acc}} \sim 300\ \mathrm{M}_\odot$, $M_{\mathrm{BH}} \sim 300\ \mathrm{M}_\odot$, $R \sim 10^{12}$ cm, and $z \sim 20$ for Eqs. (2) and (3), we obtain

$$E_{\gamma,iso} \sim 10^{57} \left(\frac{\eta}{10^{-2}}\right)\ \mathrm{erg}, \qquad T \sim 1\ \mathrm{day}. \tag{4}$$

Therefore, if we were to observe a burst at redshift $z \gtrsim 10$ with such large $E_{\gamma,iso}$ and T, this would very likely be a Pop III VMS GRB (Mészáros and Rees 2010; Toma et al. 2011).

[3] Mizuta and Ioka (2013) propose an alternative scenario that the duration of the bright part of the prompt emission is determined by an increase of θ_j on a timescale of the pressure decay of the cocoon emerging from the progenitor star.

For such large BH masses, the density and temperature of the accretion disk are too low for neutrino cooling to be important, and then the thermal energy from the $\nu\bar{\nu}$ annihilation is insufficient to power strong jets (Fryer et al. 2001b; Komissarov and Barkov 2010; Zalamea and Beloborodov 2011). The electromagnetic effects such as BZ process may instead power the jets, which should be dominated by Poynting flux (see also Suwa et al. 2007).

Pop III VMS GRBs share the property of the extremely long durations with Ultra-long GRBs, which were recently detected by Swift with low redshifts (e.g. Gendre et al. 2013; Levan et al. 2014). Ultra-long GRBs have durations as long as $T \sim 10^4$ s, although their total gamma-ray energies are $E_{\gamma,iso} \sim 10^{53}$–$10^{54}$ erg, which are comparable to those of ordinary GRBs. Thus the property of the extremely high $E_{\gamma,iso}$ would still be a unique property of Pop III VMS GRBs.

Given the propagation speed of the Poynting flux dominated jet inside the progenitor star, $\sim 0.2c$, deduced from MHD simulations (Barkov and Komissarov 2008; Bromberg and Tchekhovskoy 2016), the intrinsic duration of Pop III VMS GRBs $T/(1+z) \sim 10^4$ s is sufficient to break through the star (see also Mészáros and Rees 2001). Suwa and Ioka (2011) performed a semi-analytic calculation of the jet propagation inside the star with $R \sim 10^{13}$ cm, which is an order of magnitude larger than our assumption, and showed that the jets can successfully break through the star (see also Nagakura et al. 2012). While the jet is propagating inside the star, the energy of the jet is dissipated at the jet head and makes a thermally dominated cocoon between the jet and the stellar envelope (Begelman and Cioffi 1989; Mészáros and Rees 2001; Matzner 2003; Bromberg et al. 2011; Bromberg and Tchekhovskoy 2016). The cocoon can also be ejected after the jet breakout, and it can release thermal emission, which provide information about the progenitor star (Kashiyama et al. 2013). Indeed, the luminous supernova-like emission associated with Ultra-long GRB 111209A can be explained by the cocoon thermal emission (Nakauchi et al. 2013) (but see Greiner et al. 2015).

4.3 Prompt Emission

We may roughly estimate the isotropic gamma-ray luminosity from Eqs. (2) and (3) by $\sim E_{\gamma,iso}/[T/(1+z)]$ as

$$L_{\gamma,iso} \sim 3 \times 10^{53} \epsilon_\gamma \eta_{-2} \alpha_{-1} \theta_{j,-1} R_{12}^{-\frac{3}{2}} \left(\frac{M_{\rm acc}}{300\,{\rm M_\odot}}\right) \left(\frac{M_{\rm BH}}{300\,{\rm M_\odot}}\right)^{\frac{1}{2}} {\rm erg\,s^{-1}}, \tag{5}$$

where the notation $Q_x = Q/10^x$ in cgs units is adopted. This value is only one order of magnitude larger than that of ordinary GRBs.

We require the spectral model to examine the detectability of the prompt emission of Pop III VMS GRBs, although the prompt emission mechanism of GRBs is still under debate. Particularly for the magnetically dominated jets, it is very uncertain.

Even the magnetically dominated jets may have a subdominant thermal component, which may be released at the photosphere of the jets. The spectrum of this thermal emission can be estimated in a standard fireball model (Toma et al. 2011). For the photospheric thermal luminosity as 3×10^{52} erg s^{-1} and $1 + z = 20$, the spectral peak energy and the flux are estimated as $E_{p,obs} \simeq 70$ keV and $F \simeq 7 \times 10^{-9}$ erg cm^{-2} s^{-1}. This emission is marginally detectable by the BAT detector on Swift.

An alternative way to estimate the spectral peak energy of the prompt emission is based on empirical laws claimed for observed ordinary GRBs (Nakauchi et al. 2012). If the Poynting flux of the jets are efficiently converted to the radiation energy flux, the observed flux is $F \sim 7 \times 10^{-8}\,\mathrm{erg\,cm^{-2}\,s^{-1}}$ for $L_{\gamma,iso} \sim 3 \times 10^{53}\,\mathrm{erg\,s^{-1}}$ and $1 + z = 20$. Then if the spectral property obeys the empirical $E_p - L_{\gamma,iso}$ relation (Yonetoku et al. 2010), we have $E_{p,obs} \sim 80$ keV, or if the spectral property obeys the empirical $E_p - E_{\gamma,iso}$ relation (Amati 2006), we have $E_{p,obs} \sim 4$ MeV. In both cases the prompt emission is detectable by currently operating satellites.

However, we have adopted $\eta \sim 10^{-2}$ for the above estimates, which correspond to the high end of the estimate of η (see Sect. 4.2). For conservative values $\eta \sim 10^{-4}$–10^{-3}, the radiation luminosity is orders of magnitude smaller, so that their detections by the current satellites are difficult. Next-generation satellites will be needed in such cases (Suwa and Ioka 2011; Nakauchi et al. 2012).

4.4 Afterglow

The external shock driven by the jet in the circumburst medium powers the afterglow, which can be studied independently of the prompt emission (Mészáros and Rees 1997; Sari et al. 1998). The external shock amplifies the magnetic field in the shocked region via plasma and/or magnetohydrodynamic instabilities and accelerates the electrons in the shocked region to a power-law energy distribution. The accelerated electrons produce synchrotron and synchrotron-self-Compton (SSC) radiation as an afterglow. This afterglow model seems to be robust, since it can explain many of the late-time (i.e. since several hours after the burst trigger) multi-band afterglows so far, and simple extension of this model (e.g. continuous energy injection into the external shock) may explain many of the early-time afterglows (Liang et al. 2007).

Calculations of the external shock emission spectrum involve the parameters E_{iso} (i.e. the total jet energy minus $E_{\gamma,iso}$), the number density n of the circumburst medium, the fractions ϵ_B and ϵ_e of the thermal energy in the shocked region that are carried by the magnetic field and the accelerated electrons, respectively, and the index p of the energy spectrum of the accelerated electrons. The latter three microphysical parameters have been constrained by model-fitting the late-time afterglows as $10^{-5} \lesssim \epsilon_B \lesssim 10^{-1}$, $\epsilon_e \sim 10^{-1}$, and $p \sim 2.3$ (e.g. Panaitescu and Kumar 2002; Wijers and Galama 1999).

Figure 4 is an afterglow spectrum at the time $t \sim T$ from the burst trigger (i.e. the early brightest phase of the afterglow) calculated for Pop III VMS GRBs by Toma et al. (2011). The parameters are $E_{iso} = 4 \times 10^{57}$ erg, $T = 2.3$ day, $1 + z = 20$, $n = 1$ cm^{-3}, $\epsilon_B = 10^{-2}$, $\epsilon_e = 10^{-1}$, and $p = 2.3$. We found that the flux at the IR frequency is sufficient to estimate the redshift. (Macpherson et al. 2013 discuss the afterglow detectability with JWST and SPICA more systematically for broad range of parameters.) The Swift XRT and Fermi LAT can detect the X-ray and gamma-ray radiation of this afterglow, which may constrain the total energy scale E_{iso} and the number density n of the circumburst medium. The constraint on n would provide invaluable information about the environment and the radiation feedback of Pop III VMSs at the phase prior to their explosions (Kitayama et al. 2004; Whalen et al. 2004; Wang et al. 2012). A caveat is that in order to perform X-ray observation, one of the wide-field detectors, e.g. Swift BAT and Fermi LAT, has to be triggered by the prompt emission or the afterglow itself.

Toma et al. (2011) also found that the afterglows in the radio band are so bright that they might be detected by survey observations with current radio telescopes. Figure 5 is the light curves at various radio frequencies of the afterglow with the same physical parameters as Fig. 4. This shows that the radio afterglows of Pop III VMS GRBs can be very bright with very long durations. These may be point sources with very long variability times in the radio sky. Deep radio surveys might detect them or can constrain the rate of the Pop III

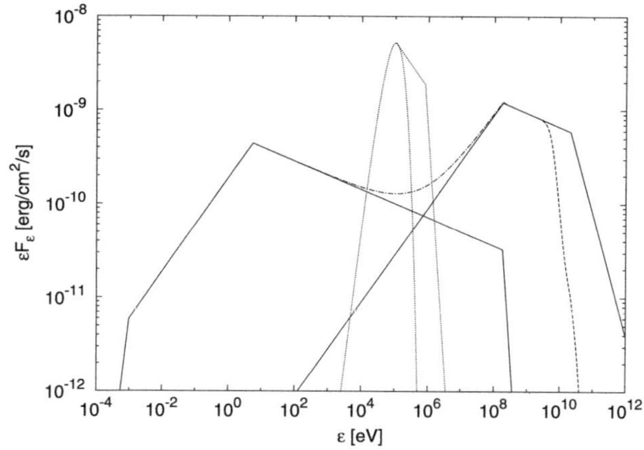

Fig. 4 Example of the spectrum of a Pop III VMS GRB at the time when the jet activity ends. The physical parameters are $E_{iso} = 4 \times 10^{57}$ erg, $T = 2.3$ day, $1 + z = 20$, $n = 1$ cm^{-3}, $\epsilon_B = 10^{-2}$, $\epsilon_e = 10^{-1}$, and $p = 2.3$. The *dot-dashed line* is the external shock spectrum, which consists of the synchrotron and SSC components (*solid lines*). The absorption due to extragalactic background light is expected to become significant above 7 GeV, as shown by the *dashed line*. The *dotted line* represents the prompt emission as a subdominant photospheric thermal component. See Toma et al. (2011) for more details

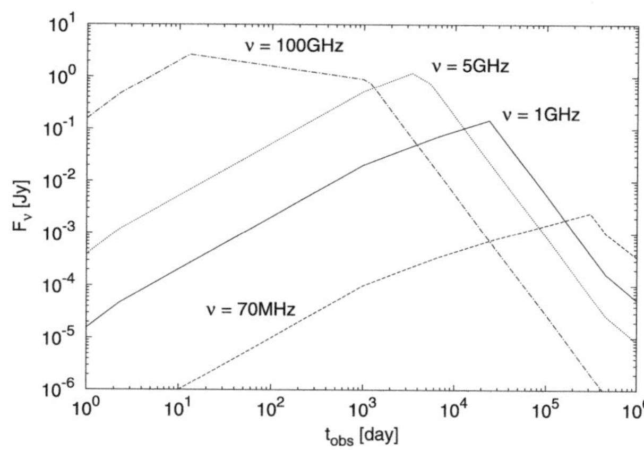

Fig. 5 Radio light curves at frequencies, 100 GHz, 5 GHz, 1 GHz, 70 MHz, of a Pop III VMS GRB at $1 + z = 20$ with the same physical parameters as Figure 4. See Toma et al. (2011) for more details

VMS GRBs (Toma et al. 2011; de Souza et al. 2011; Ghirlanda et al. 2013; Macpherson and Coward 2015). Furthermore, such bright radio sources could be useful for 21 cm absorption line searches (Furlanetto and Loeb 2002; Ioka and Mészáros 2005; Toma et al. 2011; Ciardi et al. 2015), although we should note that the brightness of the afterglow highly depends on the uncertain parameters discussed in Sect. 4.2, similarly to the prompt emission.

The external shock can also accelerate protons to the non-thermal energy distribution, which can produce high-energy neutrinos via $p\gamma$ interaction. The flux of these neutrinos can be high in the 10–100 PeV energy range (Gao et al. 2011) (see also Berezinsky and Blasi 2012).

5 Outlook

What then are the prospects for probing the end of the cosmic dark ages with GRBs, and what are key challenges ahead? For the time being, the *Swift* satellite, with its on-board X-ray and optical telescopes for the rapid localization of gamma-ray transients in the sky, will remain a good facility for catching high-redshift GRBs with $z \gtrsim 6$–9. The next mission, SVOM, with similar capabilities as *Swift*, is now being prepared to follow it. Several other satellite missions targeting higher-redshift GRBs, such as JANUS and Lobster, have also been proposed. Increasing the statistics of high-redshift GRBs, and detecting peculiar sources suggestive of VMS progenitors, will constrain theories of the formation and evolution of Pop III stars, and their GRB production mechanism, as discussed in this article. Wide-field, space-based gamma-ray/X-ray detectors are required to keep operating not only for high-redshift observations but also for upcoming gravitational wave astronomy, which requires electromagnetic counterparts for robust confirmation. The deep surveys in the radio wavelengths may also be helpful for Pop III GRB searches.

From the theory side, we cannot yet make any robust predictions for the final masses (i.e., the IMF), radii, rotation speeds, magnetic field strengths and its configurations, and the binarity of Pop III stars either at the time of formation or at the final evolution stage before the explosion. Similarly, during the GRB explosion phase, the mechanisms of driving relativistic jets and releasing high-energy emissions are still subject to considerable uncertainty. However, numerical simulations have been steadily extending our understanding of some of those ingredients, and they will keep playing a crucial role in theoretical study.

It is clearly important to improve our understanding of the Pop III GRB formation channel, in particular regarding the rotation state of the progenitor stars, and the characteristics of tight binaries. The challenge here is two-fold. We need simulations with even higher resolution, to push into the regime of possible binary overflow phenomena (Roche-lobe and common envelope), while maintaining the large-scale cosmological boundary conditions. The second challenge is improved physics, in particular inclusion of magnetohydrodynamic effects. The latter may crucially impact both the fragmentation properties of the primordial gas, and the rotation rates of the resulting Pop III stars.

The effects of rotation, magnetic fields and binary interactions are also critical in the evolution of Pop III GRB progenitor stars. Furthermore, the rapid rotation and strong magnetic field are key ingredients for producing relativistic jets via the BZ process. Our understanding of these physical processes in massive star evolutions has been greatly improved during the last decade both observationally and theoretically, but uncertainties regarding the transport processes of angular momentum and chemical species inside stars still remain large. In particular, we still lack direct observations of very metal poor massive stars, which makes it difficult to test stellar evolution theory of massive Pop III stars. We hope to overcome this difficulty with systematic studies on massive star populations in very metal poor environments like in I Zwicky 18 (e.g. Szécsi et al. 2015).

Another key uncertainty relates to theoretical predictions for the rate of Pop III bursts. Even if we can narrow down the uncertainty in the Pop III GRB formation efficiency, as outlined in the previous paragraph, we still have to contend with the poorly constrained Pop III star formation rate density. This rate is only known within 2 to 3 orders of magnitude, mostly because of our incomplete understanding of the (radiative and SN) feedback effects that regulate the primordial star formation process. Again, improved cosmological simulations with ever more realistic feedback implementations promise to provide more robust determinations in the next few years. To give a rough ballpark impression of current, state-of-the-art predictions for the Pop III GRB rate at $z \gtrsim 6$: a robust upper limit should be $\sim 10\ \mathrm{yr}^{-1}\ \mathrm{sr}^{-1}$

(Salvaterra et al. 2011), and a more realistic limit may be 0.1 $yr^{-1}sr^{-1}$ (Campisi et al. 2011). Again, the hope is that progress in our understanding of both the Pop III GRB progenitor physics and the global star formation rate density is achievable in the near future.

Acknowledgements K.T. thank P. Mészáros, K. Ioka, Y. Suwa, and D. Nakauchi for useful comments. This work is partly supported by JSPS Grants-in-Aid for Scientific Research 15H05437 (K.T.), JST grant "Building of Consortia for the Development of Human Resources in Science and Technology" (K.T.) and NSF grant AST-1211729 (V.B.). SCY was supported by Basic Science Research (2013R1A1A2061842) program through the National Research Foundation of Korea (NRF).

References

T. Abel, G.L. Bryan, M.L. Norman, Science **295**, 93 (2002)
M. Ackermann et al., Astrophys. J. Suppl. Ser. **209**, 11 (2013)
C. Aerts, G. Molenberghs, M.G. Kenward, C. Neiner, Astrophys. J. **781**, 88 (2014)
L. Amati, Mon. Not. R. Astron. Soc. **372**, 233 (2006)
R. Arlt, G. Rüdiger, Mon. Not. R. Astron. Soc. **412**, 107 (2011)
I. Baraffe, A. Heger, S.E. Woosley, Astrophys. J. **550**, 890 (2001)
R. Barkana, A. Loeb, Astrophys. J. **601**, 64 (2004)
R. Barkana, A. Loeb, Rep. Prog. Phys. **70**, 627 (2007)
M.V. Barkov, S.S. Komissarov, Mon. Not. R. Astron. Soc. **385**, L28 (2008)
M.C. Begelman, D.F. Cioffi, Astrophys. J. **345**, L21 (1989)
K. Belczynski, T. Bulik, A. Heger, C.L. Fryer, Astrophys. J. **664**, 986 (2007)
P. Beniamini, L. Nava, R. Barniol Duran, T. Piran Mon. Not. R. Astron. Soc. **454**, 1073 (2015)
V. Berezinsky, P. Blasi, Phys. Rev. D, Part. Fields **85**, 3003 (2012)
R.D. Blandford, R.L. Znajek, Mon. Not. R. Astron. Soc. **179**, 433 (1977)
J. Braithwaite, Astron. Astrophys. **449**, 451 (2006)
O. Bromberg, E. Nakar, T. Piran, R. Sari, Astrophys. J. **740**, 100 (2011)
O. Bromberg, A. Tchekhovskoy Mon. Not. R. Astron. Soc. **456**, 1739 (2016)
V. Bromm, R.B. Larson, Annu. Rev. Astron. Astrophys. **42**, 79 (2004)
V. Bromm, A. Loeb, Astrophys. J. **575**, 111 (2002)
V. Bromm, A. Loeb, Astrophys. J. **642**, 382 (2006)
V. Bromm, P.S. Coppi, R.B. Larson, Astrophys. J. **564**, 23 (2002)
V. Bromm, N. Yoshida, L. Hernquist, C.F. McKee, Nature **459**, 49 (2009)
V. Bromm, N. Yoshida, Annu. Rev. Astron. Astrophys. **49**, 373 (2011)
V. Bromm, Rep. Prog. Phys. **76**, 112901 (2013)
I. Brott, C.J. Evans, I. Hunter et al., Astron. Astrophys. **530**, 116 (2011)
G.E. Brown, C-H. Lee, R.A.M.J. Wijers, H.K. Lee, G. Istraelian, H.A. Bethe, New Astron. **5**, 191 (2000)
M.A. Campisi, U. Maio, R. Salvaterra, B. Ciardi, Mon. Not. R. Astron. Soc. **416**, 2760 (2011)
M. Cantiello, S-C. Yoon, N. Langer, M. Livio, Astron. Astrophys. **465**, L29 (2007)
M. Cantiello, C. Mankovich, L. Bildsten, J. Christensen-Dalsgaard, B. Paxton, Astrophys. J. **788**, 93 (2014)
S. Charpinet, G. Fontaine, P. Brassard, Nature **461**, 501 (2009)
E. Chatzopoulos, J.C. Wheeler, Astrophys. J. **748**, 42 (2012)
B. Ciardi, S. Inoue, F.B. Abdalla et al., Mon. Not. R. Astron. Soc. **453**, 101 (2015)
B. Ciardi, A. Ferrara, Space Sci. Rev. **116**, 625 (2005)
B. Ciardi, A. Loeb, Astrophys. J. **540**, 687 (2000)
P.C. Clark, S.C.O. Glover, R.J. Smith et al., Science **331**, 1040 (2011)
A. Cucchiara, A.J. Levan, D.B. Fox et al., Astrophys. J. **736**, 7 (2011)
V. D'Elia, E. Pian, A. Melandri et al., Astron. Astrophys. **577**, A116 (2015)
R.S. de Souza, N. Yoshida, K. Ioka, Astron. Astrophys. **533**, A32 (2011)
S.E. de Mink, N. Langer, R.G. Izzard, H. Sana, A. de Koter, Astrophys. J. **764**, 166 (2013)
P. Eggenberger, A. Maeder, G. Meynet, Astron. Astrophys. **440**, L9 (2005)
P. Eggenberger, J. Montalbán, A. Miglio, Astron. Astrophys. **544**, 4 (2012)
D. Eichler, M. Livio, T. Piran, D. Schramm, Nature **340**, 126 (1989)
S. Ekström, G. Meynet, C. Chiappini, R. Hirschi, A. Maeder, Astron. Astrophys. **489**, 685 (2008)
P.A. Evans, R. Willingale, J.P. Osborne et al., Mon. Not. R. Astron. Soc. **444**, 250 (2014)
C.A.C. Fernandes, M.J. Jarvis, S. Rawlings et al., Mon. Not. R. Astron. Soc. **411**, 1909 (2011)
D.A. Frail, S.R. Kulkarni, R. Sari et al., Astrophys. J. **562**, L55 (2001)

C.L. Fryer, A. Heger, Astrophys. J. **623**, 302 (2005)
C.L. Fryer, S.E. Woosley, A. Heger, Astrophys. J. **550**, 372 (2001a)
C.L. Fryer, S.E. Woosley, A. Heger, Astrophys. J. **550**, 372 (2001b)
S.R. Furlanetto, A. Loeb, Astrophys. J. **579**, 1 (2002)
J. Fuller, D. Lecoanet, M. Cantiello, B. Brown, Astrophys. J. **769**, 17 (2014)
S. Gao, K. Toma, P. Mészáros, Phys. Rev. D, Part. Fields **83**, 3004 (2011)
J.P. Gardner, J.C. Mather, M. Clampin et al., Space Sci. Rev. **123**, 485 (2006)
B. Gendre, G. Stratta, L.L. Atteia et al., Astrophys. J. **766**, 30 (2013)
G. Ghirlanda et al., Mon. Not. R. Astron. Soc. **435**, 2543 (2013)
G. Ghisellini, F. Tavecchio, L. Maraschi et al., Nature **515**, 376 (2014)
S.C.O. Glover, Space Sci. Rev. **117**, 445 (2005)
L.J. Gou, P. Mészáros, T. Abel, B. Zhang, Astrophys. J. **604**, 508 (2004)
T.H. Greif, V. Springel, S.D.M. White et al., Astrophys. J. **737**, 75 (2011)
T.H. Greif, V. Bromm, P.C. Clark et al., Mon. Not. R. Astron. Soc. **424**, 399 (2012)
J. Greiner et al., Nature **523**, 189 (2015)
A. Heger, N. Langer, S.E. Woosley, Astrophys. J. **528**, 368 (2000)
A. Heger, S.E. Woosley, Astrophys. J. **724**, 341 (2010)
A. Heger, S.E. Woosley, H.C. Spruit, Astrophys. J. **626**, 350 (2005)
R. Hirschi, G. Meynet, A. Maeder, Astron. Astrophys. **425**, 649 (2004)
S. Hirano, T. Hosokawa, N. Yoshida et al., Astrophys. J. **781**, 60 (2014)
T. Hosokawa, K. Omukai, N. Yoshida, H.W. Yorke, Science **334**, 1250 (2011)
J.A. Hummel, A.H. Pawlik, M. Milosavljevic, V. Bromm, Astrophys. J. **755**, 72 (2012)
I. Hunter, I. Brott, D.J. Lennon et al., Astrophys. J. **676**, 29 (2008)
I. Hunter, I. Brott, N. Langer et al., Astron. Astrophys. **496**, 841 (2009)
S. Inoue, K. Omukai, B. Ciardi, Mon. Not. R. Astron. Soc. **380**, 1715 (2007)
K. Ioka, K. Toma, R. Yamazaki, T. Nakamura, Astron. Astrophys. **458**, 7 (2006)
K. Ioka, P. Mészáros, Astrophys. J. **619**, 684 (2005)
K. Iwamoto, P.A. Mazzali, K. Nomoto et al., Nature **395**, 672 (2006)
R.G. Izzard, E. Ramirez-Ruiz, C.A. Tout, Mon. Not. R. Astron. Soc. **348**, 1215 (2004)
S. Justham, Ph. Podsiadlowski, J.S. Vink, Astrophys. J. **796**, 121 (2014)
T. Karlsson, V. Bromm, J. Bland-Hawthorn, Rev. Mod. Phys. **85**, 809 (2013)
K. Kashiyama, D. Nakauchi, Y. Suwa et al., Astrophys. J. **770**, 8 (2013)
N. Kawai et al., Nature **440**, 184 (2006)
R. Kippenhahn, R. Weigert, *Stellar Structure and Evolution* (Springer, Berlin, 1990)
M.D. Kistler et al., Astrophys. J. **705**, L104 (2009)
T. Kitayama, N. Yoshida, H. Susa, M. Umemura, Astrophys. J. **613**, 631 (2004)
S.S. Komissarov, J. Korean Phys. Soc. **54**, 2503 (2009)
S.S. Komissarov, M.V. Barkov, Mon. Not. R. Astron. Soc. **402**, L25 (2010)
J. Krtička, J. Kubát, Astron. Astrophys. **446**, 1039 (2006)
P. Kumar, R. Narayan, J.L. Johnson, Mon. Not. R. Astron. Soc. **388**, 1729 (2008)
P. Kumar, B. Zhang, Phys. Rep. **561**, 1 (2015)
D.Q. Lamb, D.E. Reichart, Astrophys. J. **536**, 1 (2000)
A.J. Levan, N.R. Tanvir, R.L.C. Starling et al., Astrophys. J. **781**, 13 (2014)
E.W. Liang, B.B. Zhang, B. Zhang, Astrophys. J. **670**, 565 (2007)
M. Limongi, A. Chieffi, Astrophys. J. Suppl. Ser. **199**, 38 (2012)
A. Loeb, *How Did the First Stars and Galaxies Form?* (Princeton University Press, Princeton, 2010)
E. Lovegrove, S.E. Woosley, Astrophys. J. **769**, 109 (2013)
A.I. MacFadyen, S.E. Woosley, Astrophys. J. **524**, 262 (1999)
A.I. MacFadyen, S.E. Woosley, A. Heger, Astrophys. J. **550**, 410 (2001)
D. Macpherson, D.M. Coward, M.G. Zadnik, Astrophys. J. **779**, 73 (2013)
D. Macpherson, D.M. Coward, Mon. Not. R. Astron. Soc. **453**, 2144 (2015)
A. Maeder, Astron. Astrophys. **178**, 159 (1987)
A. Maeder, G. Meynet, Astron. Astrophys. **334**, 1000 (1998)
A. Maeder, G. Meynet, Astron. Astrophys. **361**, 159 (2000)
A. Maeder, G. Meynet, Rev. Mod. Phys. **84**, 25 (2012)
J.P. Marques, M.J. Goupil, Y. Lebreton et al., Astron. Astrophys. **549**, A74 (2013)
P. Marigo, L. Girardi, C. Chiosi, P.R. Wood, Astron. Astrophys. **371**, 152 (2001)
P. Marigo, C. Chiosi, R-P. Kudritzki, Astron. Astrophys. **399**, 617 (2003)
T. Matsumoto, N. Nakauchi, K. Ioka, A. Heger, T. Nakamura, Astrophys. J. **810**, 64 (2015)
C.D. Matzner, Mon. Not. R. Astron. Soc. **345**, 575 (2003)
P.A. Mazzali, J. Deng, E. Pian et al., Astrophys. J. **645**, 1323 (2006)

C.F. McKee, J.C. Tan, Astrophys. J. **681**, 771 (2008)
M.R. Mokiem, A. de Koter, C.J. Evans et al., Astron. Astrophys. **456**, 1131 (2006)
P. Mészáros, M.J. Rees, Astrophys. J. **476**, 232 (1997)
P. Mészáros, M.J. Rees, Astrophys. J. **556**, L37 (2001)
P. Mészáros, M.J. Rees, Astrophys. J. **715**, 967 (2010)
A. Mizuta, K. Ioka, Astrophys. J. **777**, 162 (2013)
D.J. Mortlock, S.J. Warren, B. Venemans et al., Nature **474**, 616 (2011)
T.J. Moriya, N. Langer, Astron. Astrophys. **573**, 18 (2015)
H. Nagakura, Y. Suwa, K. Ioka, Astrophys. J. **754**, 85 (2012)
D. Nakauchi, K. Kashiyama, Y. Suwa, T. Nakamura, Astrophys. J. **778**, 67 (2013)
D. Nakauchi, Y. Suwa, T. Sakamoto, K. Kashiyama, T. Nakamura, Astrophys. J. **759**, 128 (2012)
K. Omukai, F. Palla, Astrophys. J. **589**, 677 (2003)
K. Omukai, R. Nishi, Astrophys. J. **518**, 64 (1999)
T. Pan, D. Kasen, A. Loeb, Mon. Not. R. Astron. Soc. **422**, 2701 (2012)
A. Panaitescu, P. Kumar, Astrophys. J. **571**, 779 (2002)
J. Paul, J. Wei, S. Basa, S-N. Zhang, C. R. Phys. **12**, 298 (2011)
A.H. Pawlik, M. Milosavljevic, V. Bromm, Astrophys. J. **731**, 54 (2011)
J. Petrovic, N. Langer, S-C. Yoon, A. Heger, Astron. Astrophys. **435**, 247 (2005)
L. Piro, E. Troja, B. Gendre, Astron. Astrophys. **790**, 15 (2014)
Ph. Podsiadlowski, N. Ivanova, S. Justham, S. Rappaport, Mon. Not. R. Astron. Soc. **406**, 840 (2010)
C. Porciani, P. Madau, Astrophys. J. **548**, 522 (2001)
J. Puls, J.S. Vink, F. Najarro, Astron. Astrophys. Rev. **16**, 209 (2008)
O.H. Ramírez-Agudelo, S. Simón-Díaz, H. Sana et al., Astron. Astrophys. **560**, A29 (2013)
A.L. Rosen, M.R. Krumholz, E. Ramirez-Ruiz, Astrophys. J. **748**, 97 (2012)
C-E. Rydberg, E. Zackrisson, P. Lundqvist, P. Scott, Mon. Not. R. Astron. Soc. **429**, 3658 (2013)
R. Salvaterra, J. High Energy Astrophys. **7**, 35 (2015)
R. Salvaterra, M. Della Valle, S. Campana et al., Nature **461**, 1258 (2009)
R. Salvaterra, A. Ferrara, P. Dayal, Mon. Not. R. Astron. Soc. **414**, 847 (2011)
R. Sari, T. Piran, R. Narayan, Astrophys. J. **497**, L17 (1998)
R. Schneider, K. Omukai, A.K. Inoue, A. Ferrara, Mon. Not. R. Astron. Soc. **369**, 1437 (2006)
J. Schober, D. Schleicher, C. Federrath et al., Astrophys. J. **754**, 99 (2012)
N.I. Shakura, R.A. Sunyaev, Astron. Astrophys. **24**, 337 (1973)
H.C. Spruit, Astron. Astrophys. **349**, 189 (1999)
H.C. Spruit, Astron. Astrophys. **381**, 923 (2002)
A. Stacy, V. Bromm, Mon. Not. R. Astron. Soc. **433**, 1094 (2013)
A. Stacy, T.H. Greif, V. Bromm, Mon. Not. R. Astron. Soc. **403**, 45 (2010)
A. Stacy, V. Bromm, A. Loeb, Mon. Not. R. Astron. Soc. **413**, 543 (2011)
A. Stacy, T.H. Greif, V. Bromm, Mon. Not. R. Astron. Soc. **422**, 290 (2012)
A. Stacy, T.H. Greif, R.S. Klessen, V. Bromm, A. Loeb, Mon. Not. R. Astron. Soc. **431**, 1470 (2013)
M.P.L. Suijs, N. Langer, A-J. Poelarends et al., Astron. Astrophys. **481**, 87 (2008)
H. Susa, K. Hasegawa, N. Tominaga, Astrophys. J. **792**, 32 (2014)
Y. Suwa, K. Ioka, Astrophys. J. **726**, 107 (2011)
Y. Suwa, Mon. Not. R. Astron. Soc. **428**, 2443 (2013)
Y. Suwa, T. Takiwaki, K. Kotake, K. Sato, Publ. Astron. Soc. Jpn. **59**, 771 (2007)
D. Szécsi, N. Langer, S-C. Yoon, D. Sanyal, S. de Mink, C.J. Evans, T. Dermine, Astron. Astrophys. **581**, 15 (2015)
N.R. Tanvir, D.B. Fox, A.J. Levan et al., Nature **461**, 1254 (2009)
R.J. Taylor, Mon. Not. R. Astron. Soc. **161**, 365 (1973)
A. Tchekhovskoy, D. Giannios, Mon. Not. R. Astron. Soc. **447**, 327 (2015)
A. Tchekhovskoy, R. Narayan, J.C. McKinney, Mon. Not. R. Astron. Soc. **418**, L79 (2011)
K. Toma, T. Sakamoto, P. Mészáros, Astrophys. J. **731**, 127 (2011)
K. Toma, F. Takahara, Mon. Not. R. Astron. Soc. **442**, 2855 (2014)
T. Totani, N. Kawai, G. Kosugi et al., Publ. Astron. Soc. Jpn. **58**, 485 (2006)
T. Totani, Astrophys. J. **486**, L71 (1997)
T. Totani, K. Aoki, T. Hattori, N. Kawai Publ. Astron. Soc. Jpn. **68**, 15 (2016)
M.J. Turk, T. Abel, B.W. O'Shea, Science **325**, 601 (2009)
E.P.J. van den Heuvel, S-C. Yoon, Astrophys. Space Sci. **311**, 177 (2007)
F.J. Virgili, C.G. Mundell, V. Pal'shin et al., Astrophys. J. **778**, 54 (2013)
P.M. Vreeswijk, S.L. Ellison, C. Ledoux et al., Astron. Astrophys. **419**, 927 (2004)
F.Y. Wang, V. Bromm, T.H. Greif et al., Astrophys. J. **760**, 27 (2012)
D. Whalen, T. Abel, M.L. Norman, Astrophys. J. **610**, 14 (2004)

D. Whalen, W. Even, L.H. Frey et al., Astrophys. J. **777**, 110 (2013)
J.C. Wheeler, D. Kagan, E. Chatzopoulos, Astrophys. J. **799**, 85 (2015)
J.C. Wheeler, I. Yi, P. Höflich, L. Wang, Astrophys. J. **537**, 810 (2000)
R.A.M. Wijers, T.J. Galama, Astrophys. J. **523**, 177 (1999)
S.C. Wolff, S.E. Strom, D. Dror, L. Lanz, K. Venn, Astrophys. J. **132**, 749 (2006)
S.E. Woosley, Astrophys. J. **405**, 273 (1993)
S.E. Woosley, J.S. Bloom, Annu. Rev. Astron. Astrophys. **44**, 507 (2006)
S.E. Woosley, A. Heger, Astrophys. J. **637**, 914 (2006)
S.E. Woosley, A. Heger, Astrophys. J. **752**, 32 (2012)
D. Yonetoku, T. Murakami, T. Nakamura et al., Astrophys. J. **609**, 935 (2004)
D. Yonetoku, T. Murakami, R. Tsutsui et al., Astrophys. J. **62**, 1495 (2010)
S-C. Yoon, A. Dierks, N. Langer, Astron. Astrophys. **542**, A113 (2012)
S-C. Yoon, N. Langer, Astron. Astrophys. **443**, 643 (2005)
S-C. Yoon, N. Langer, C. Norman, Astron. Astrophys. **460**, 199 (2006)
S-C. Yoon, S.E. Woosley, N. Langer, Astrophys. J. **725**, 940 (2010)
S-C. Yoon, J. Kang, A. Kozyreva, Astrophys. J. **802**, 16 (2015)
N. Yoshida, K. Omukai, L. Hernquist, Science **321**, 669 (2008)
J-P. Zahn, A.S. Brun, S. Mathis, Astron. Astrophys. **474**, 145 (2007)
I. Zalamea, A.M. Beloborodov, Mon. Not. R. Astron. Soc. **410**, 2302 (2011)
B. Zhang, P. Mészáros, Int. J. Mod. Phys. A **15**, 2385 (2004)
B. Zhang, E. Liang, K.L. Page et al., Astrophys. J. **655**, 989 (2007)
W. Zhang, C.L. Fryer, in *Stellar Collapse*, ed. by C.L. Fryer (Kluwer Academic, Dordrecht, 2004), p. 327

Space Sci Rev (2016) 202:183–196
DOI 10.1007/s11214-016-0288-6

Gamma-Ray Bursts and the Early Star-Formation History

GRBs and $z > 4$ Star-Formation

R. Chary[1] · P. Petitjean[2] · B. Robertson[3] · M. Trenti[4] · E. Vangioni[2]

Received: 16 January 2016 / Accepted: 2 September 2016 / Published online: 19 October 2016

Abstract We review the uncertainties in high-z star-formation rate (SFR) measures and the constraints that one obtains from high-z gamma-ray burst (GRB) rates on them. We show that at the present time, the GRB rates per unit star-formation at $z > 3$ are higher than at lower redshift. There could be a multitude of reasons for this: a stellar metallicity bias for GRB production, a top-heavy initial mass function (IMF) and/or missing a significant fraction of star-formation in field galaxy surveys due to incompleteness, surface brightness limitations and cosmic variance. We also compare metallicity predictions made using a hierarchical model of cosmic chemical evolution based on two recently proposed SFRs, one based on the observed galaxy luminosity function at high redshift and one based on the GRB rate and find that within the considerable scatter in metal abundance measures, they both are consistent with the data. Analyzing the ensemble of different measurements together, we conclude that despite metallicity biases, GRBs may be a less biased probe of star-formation at $z > 3$ than at $z < 2$. There is likely to be a common origin to the high GRB rate per unit star-formation and the high observed Lyman-continuum production rate in high redshift galaxies and that this may be due to a relative overabundance of stars with mass $>25\ M_\odot$ which are likely GRB progenitors. We also find that to reconcile these measure-

✉ R. Chary
rchary@caltech.edu

P. Petitjean
petitjean@iap.fr

B. Robertson
brant@ucsc.edu

M. Trenti
mtrenti@unimelb.edu.au

E. Vangioni
vangioni@iap.fr

1 U.S. Planck Data Center, Caltech, MS314-6, Pasadena, CA 91125, USA

2 Institut d'Astrophysique de Paris, Paris, France

3 University of California, Santa Cruz, CA, USA

4 University of Melbourne, Victoria, Melbourne, Australia

ments with the Thomson scattering cross section of cosmic microwave background (CMB) photons measured by Planck, the escape fraction of Lyman-continuum photons from galaxies must be low, about ∼15 % or less and that the clumping factor of the IGM is likely to be small, ∼3. Finally, we demonstrate that GRBs are unique probes of metallicity evolution in low-mass galaxy samples and that GRB hosts likely lost a significant fraction of metals to the intergalactic medium (IGM) due to feedback processes such as stellar winds and supernovae.

Keywords Gamma ray bursts · Star formation history · Reionization · Chemical abundances

1 Introduction

Measuring the evolution of the co-moving star-formation rate density (SFRD) with redshift has been one of the major goals of galaxy evolution studies. Primary tracers of star-formation include the rest-frame ultraviolet continuum, the mid-infrared through sub-millimeter continuum from thermal dust emission, 1.4 GHz and 30 GHz radio emission which traces the synchrotron and free-free components respectively and nebular lines (e.g. Hα and Hβ). Each of these tracers have different systematics associated with them. Ultraviolet wavelengths are affected by dust attenuation and also require uncertain extrapolations for the faint end of the galaxy luminosity function. Mid-infrared and sub-millimeter tracers are biased towards the bright end of the galaxy luminosity function and suffer from AGN contamination. Current radio surveys are insensitive to all but extreme, ultra-luminous galaxies. Nebular lines are powerful but require extensive spectroscopic follow-up which is limited to the bright end of the LF. Furthermore, they too suffer from attenuation and stellar absorption which needs to be corrected for.

The goal has motivated a multitude of deep surveys; they have been undertaken at visible/ultraviolet (UV) wavelengths with *Hubble*, in the mid- and far-infrared with *ISO*, *Spitzer*, *Herschel* and more recently *ALMA* and also with narrow-band filters and grisms (e.g. *Hubble*/WFC3 and NICMOS) to detect strong nebular emission lines. These surveys have resulted in a generally broad agreement whereby the SFR rises from redshift 0, peaks at $z \sim 1$–2 and declines with increasing redshift out to $z \sim 10$. Our understanding of this has been based on observations of the bright end of the luminosity function with significant assumptions needed to extrapolate to the faint end and to derive a co-moving star-formation rate density. For example, the UV SFR measurements are sensitive down to few times 10^8 $L_\odot$ while the mid- and far-infrared surveys only detect galaxies down to 10^{11} $L_\odot$. As a result, it is unclear how much attenuation correction, if any, is required at the faint end of the galaxy LF and whether even the slope of the faint-end of the galaxy LF is robustly constrained.

This is particularly true at the epoch of reionization at $z > 6$, where considerable uncertainties exist at both the bright end of the ultraviolet luminosity function and the faint end. Although 1000s of candidate galaxies have been detected out to $z \sim 10$, these have been through deep, pencil-beam surveys or through observations of galaxies lensed by foreground clusters, both of which trace relatively small comoving volumes at the redshifts of interest. Since these surveys are done with 4 arcmin^2 instrumental fields of view, there is the issue of cosmic variance. A comparison with semi-analytical models indicate that similar fields of view could show as much as a factor of 3 scatter in luminosity density at $z > 6$, depending on the large scale structure.

Long-duration GRBs,[1] since they are thought to be the evolutionary end-states of massive stars, are an alternate technique for tracing the comoving SFRD (Wijers et al. 1998; Blain and Natarajan 2000; Bromm and Loeb 2002; Daigne et al. 2006; Ishida et al. 2011). Provided the selection effects associated with GRBs can be understood, measuring the co-moving rate density of GRBs and/or comparing the properties of their host galaxies with field galaxies populations can allow a calibration of GRBs as a star-formation rate metric. With rates of ~1/day, they are quite abundant with the biggest limitation arising in identification and spectroscopic follow-up of the afterglow (e.g. Krühler et al. 2015). In addition, since GRB afterglows are bright, the spectroscopic follow-up detects absorption by metal lines that arise both in the host galaxy and in the intervening IGM. This would hopefully allow the characterization of star-forming environments in distant galaxies, and their evolution with redshift, which is otherwise a challenge.

However, the question remains whether GRBs are unbiased tracers of star-formation. Are there metallicity/luminosity biases in the kinds of galaxies that GRBs occur in? Are there environment biases in the sense that GRBs only occur in stars with large angular momentum (Woosley and Heger 2006)? Although accurate answers to these questions are unclear at the present time, in this article, we review the current state of using GRB rates as a star-formation rate metric and outline possible ways forward.

2 GRB Rates as a Tracer of Star Formation

Long-duration gamma-ray bursts are thought to arise from the collapse of the core of a massive star through a black hole-accretion disk system. This knowledge has come about due to the association between GRBs and star-forming galaxies (e.g. Bloom et al. 2002; Le Floc'h et al. 2003; Chary et al. 2002) and the detection of a late-time core-collapse supernova light curve underlying many GRB afterglows (Woosley and Bloom 2006). Since the lifetime of a massive star is relatively short $\lesssim$10 Myr relative to typical timescales for star-formation in a galaxy, measuring the rates of GRBs can in principle be a tracer of high-mass star-formation. If the stellar initial mass function is known, a significant source of uncertainty, they could even provide a global tracer of the star-formation rate density.

The primary disadvantage of using GRBs is there is no mechanism to calibrate the mass range of the progenitor star and its physical properties such as metallicity, angular momentum and binarity to the GRB. For example, do high metallicity stars have large enough opacity to radiation pressure and blow off most of their envelope through stellar winds inhibiting core-collapse? Do stars with large angular momentum lose too much of their mass through magnetic torques? Models of stellar evolution have provided the first crucial insights into this process (Heger et al. 2003), yet observational data is limited. Some theoretical attempts have been made to quantify the efficiency of GRB production depending on metallicity and rotation of progenitors (see e.g. Yoon et al. 2006), but a systematic and comprehensive study is still missing.

In the absence of such a physically motivated relationship between the formation of a massive star and the occurrence of a gamma-ray burst, one has to rely on empirical relationships between the GRB-rate and star-formation rate while factoring in observational selection effects such as the fraction of GRBs that are detected above a particular gamma-ray

[1] Throughout this paper, we refer to long-duration GRBs whose gamma-ray emission typically lasts >2 s and has a soft spectrum with hardness ratios of ~0.5 compared to the short GRBs which last <2 s, have hardness ratios of ~1.0 and that are thought to arise from merging double-degenerate systems.

fluence limit, the efficiency of spectroscopic follow-up of the optical/near-infrared afterglow and the nature of galaxies in which the bulk of star-formation is taking place (Vergani et al. 2015).

2.1 Uncertainties in High Redshift Star Formation

The high redshift ($z > 3$) star-formation rate has, almost exclusively, been measured by using the rest-frame ultraviolet light from Lyman-break galaxies detected in deep optical surveys. This has been combined with corrections for dust attenuation derived from the slope of the ultraviolet continuum to yield comoving star-formation rate densities. These measurements, using ground-based (e.g. *Keck*/LRIS and *Subaru*/SuprimeCam) and *Hubble* deep surveys have revealed that the extinction corrected star-formation rate density appears to monotonically decline as $\sim (1+z)^{-3.75}$ with increasing redshift between $3 < z < 8$ (Oesch et al. 2014b). There are even hints of a steeper decline, $\sim (1+z)^{-10.5}$ at $8 < z < 10$ down to the observed luminosity limit (Bouwens et al. 2015; Oesch et al. 2014b).[2]

There are however significant uncertainties associated with this measurement. Primary among these are:

- the contribution from galaxies at the faint end of the UV-luminosity function since the surveys typically reach depths of $0.04\mathrm{L}_{*,\mathrm{UV}}$ (Trenti et al. 2013; Robertson et al. 2013);
- dust obscuration which although constrained by the UV slope, are uncertain, due to the significant scatter in the relationship between UV slope and attenuation correction (Battisti et al. 2016);
- surface brightness limitations, with individual high-z galaxies appearing to be small, typically $\lesssim$1–2 kpc (Oesch et al. 2010) which suggests that current surveys may be picking up the brightest star-forming knots and missing extended, low surface-brightness star-formation;
- cosmic variance since the amount of area covered is much smaller than the correlation length of large scale structure (Trenti and Stiavelli 2008; Robertson 2010);
- stellar initial mass function which affects the translation from ultraviolet luminosity density to a star-formation rate density (Chary 2008).

To illustrate, Kistler et al. (2013) showed that integrating the ultraviolet luminosity function down to an absolute magnitude $M_{\mathrm{UV}} = -10$ (significantly below the lowest measured absolute magnitude value of $M_{\mathrm{UV}} \simeq -18$) would lead to an order of magnitude increase in the estimated SFR at redshifts $z \sim 4$. However, it is unclear what the lower luminosity limit is for the integral over the luminosity function. This is because the luminosity of the lowest mass dark matter halo which can sustain star-formation is uncertain due to the inefficiency of gas cooling in halos lower than mass $\sim 10^7\ \mathrm{M}_\odot$. Since sub-$\mathrm{L}_{*,\mathrm{UV}}$ galaxies may contribute the bulk of the SFRD, measurement of the faint-end slope of the UV luminosity function, where completeness corrections and surface brightness dimming issues are significant, needs to be undertaken carefully (Steidel et al. 1999; Bouwens et al. 2006). This uncertainty is further manifested in the change in derived luminosity function parameters obtained from data presented in Bouwens et al. (2006, 2015) where deeper data seem to show significant scatter in luminosity function parameters, even at the bright end.

Similarly, if dust extinction were a significant issue, the galaxies that dominate the SFRD would be UV-faint or undetected in magnitude-limited rest-frame UV surveys. Since dust can be produced efficiently, within a few years, in the shells of supernovae (e.g. Matsuura

[2] These exponents are just parametric representations and are currently, not physically motivated.

et al. 2011; Marassi et al. 2015), the fraction of dust obscured star-formation could be quite significant even at high redshift. Far-infrared/submillimeter observations would be required to identify these galaxies and measure their contribution to the bolometric luminosity density. Millimeter observations with ALMA, ACT and SPT are indeed finding significant populations of lensed, dust obscured galaxies at $z \sim 5$–6, yet these are rare enough that their contribution to the bolometric luminosity density appears to be small, at the present time (Weiß et al. 2013; Su et al. 2015). Deep, forthcoming ALMA observations will help clarify the uncertainties due to dust obscuration at least in the limited areas of sky that can be observed with the small ALMA primary beam.

At $z > 6$, these uncertainties only get enlarged further. Although the UV measured SFR from Lyman-break galaxies, the optical depth to electron scattering (τ) measured for the CMB by *WMAP* and *Planck*, and other ancillary probes of the IGM neutrality at $z > 6$ are arguably consistent with each other, each of these has very significant uncertainties (e.g. Robertson et al. 2015). For example, the measure of τ from *Planck* is 0.058 ± 0.012. Since almost 80 % of the optical depth (~0.04) arises from the ionized IGM at $z < 6$, it is merely a $\sim 1\sigma$ constraint on the ionizing photon production rate at $z > 6$. Furthermore, translating the star-formation rate to a measure of the τ requires knowing the highly uncertain escape fraction of ionizing photons. While Robertson et al. (2015) assumed a value of 0.2 which is high, recent constraints suggest it may be lower than 0.02 (Grazian et al. 2016) resulting in a lower τ for a particular star-formation history.

GRBs are relatively insensitive to these limitations. If the GRB rate density were correlated with the comoving SFRD at lower redshifts, where the UV and IR have been brought into agreement (e.g. Reddy et al. 2015), measurement of the GRB rate density at $z > 3$ could provide an independent pathway to quantifying the SFRD (see also, e.g., Price et al. 2006). Complementary measures of star-formation using probes such as long duration GRBs could thereby help address these uncertainties that persist.

2.2 Implications from Gamma-Ray Burst Rates

Using a small sample of ~50 spectroscopically-confirmed GRBs over the first two years of the *Swift* mission, Chary et al. (2007) found that the long duration GRB rate was constant over $0.5 < z < 3$ at a rate of $3.7 \pm 1.1 \times 10^{-11}\ \mathrm{Mpc}^{-3}\,\mathrm{yr}^{-1}$. This number is undoubtedly an underestimate of the true rate obtained by integrating over a GRB luminosity function, since it requires factoring in the efficiency of optical afterglow detection and spectroscopic confirmation (see e.g. Sun et al. 2015). However, as long as the follow-up efficiency is constant over this redshift range, it can be considered a reasonable lower limit. After cross calibrating against the SFR at $z < 3$, the measured GRB rate can yield a measure of the SFR at $z > 3$, assuming a constant ratio between the GRB rate and star-formation rate. Although the uncertainties are large, due to the small sample of high-z GRBs, the amount of SFR was a factor of ~2–3 higher than what had been measured through field galaxies at $z \sim 6$.

This calculation however did not take into account evolution of the GRB luminosity function which more recent work has investigated (Lien et al. 2015; Salvaterra et al. 2012; Qin et al. 2010). This is equivalent to factoring in some of the GRB production efficiencies discussed in Sect. 2. For example, by taking a sample of spectroscopically confirmed *Swift* GRBs, Salvaterra et al. (2012) showed that there is evidence for a $\sim (1+z)^2$ luminosity or density evolution with redshift in the GRB rates. Similarly, Qin et al. (2010) argued that the measured GRB rates require possibly both a metallicity threshold for GRB production and a redshift evolution with respect to the star-formation rate density. Regardless, the conclusion that the GRB rate per unit star-formation increases with increasing redshift appears to be robust although the magnitude of enhancement suffers from uncertainties in both the GRB rates and the SFRD.

More recently, Lien et al. (2015) have simulated the trigger algorithm adopted by the *SWIFT*/BAT instrument. They compared the simulated peak flux and redshift distribution of GRBs with those observed. Their sample of 66 bursts only included bursts whose redshifts arise from afterglow measurements rather than host galaxy measurements and thereby rejected a number of low-z GRBs. They however obtained an order of magnitude higher GRB rate (0.4 $\mathrm{Gpc}^{-3}\ \mathrm{yr}^{-1}$) even at $z \sim 0$. The origin of this difference is far from clear and is most likely folded into the detection efficiency as a function of peak energy of *Swift*. Regardless of the absolute comoving rate of GRBs, they too arrive at the same general conclusion that the GRB rate per unit star-formation increases with increasing redshift (see also Kistler et al. 2008; Virgili et al. 2011).

The presence of a significant amount of star formation below typical Lyman break galaxy survey limits at high z can be probed independently through observations of GRB locations at rest-frame UV wavelengths (~0.15 μm) once the afterglow has faded. In fact, the fraction of undetected hosts provides a direct estimate of the amount of missed star formation to a given luminosity limit (Trenti et al. 2012). Based on ultradeep searches for a small sample of GRB hosts at $z > 5$, which only yield upper limits, it is possible to infer that the galaxy luminosity function must extend with a steep faint-end slope ($\alpha \sim -1.8$) to at least $M_{AB} = -15$, about two mag fainter than the Hubble UDF observational limit for direct detection of galaxies (Trenti et al. 2012; Tanvir et al. 2012). One caveat is the small sample size (six targets) of the study. Indeed, more recent observations yielded the first rest-frame UV detections of GRB hosts at $z > 5$ (McGuire et al. 2016), so future data on GRB hosts might lead to a revision of the fraction of missing star formation from Lyman break galaxy surveys.

Robertson and Ellis (2012) used a larger sample of 112 from a *Swift* luminosity-limited catalog to argue that relative to the star-formation rate, there seemed to be evidence for a slight evolution for the GRB rate of $(1 + z)^{0.5}$ although the exponent strongly depends on the rate of optically-faint or "dark" GRBs as a function of redshift (Fig. 1). Since GRB hosts appear to be sub-luminous, low-metallicity galaxies (e.g Modjaz et al. 2008), that would imply a large population of faint (sub-L_*) Lyman-break galaxies below the detection limit of field galaxy surveys. Chary et al. (2007) estimated that this would imply a slope of the UV luminosity function of galaxies to be $\alpha = -1.9$ at $z \sim 6$, compared to the measured value of -1.73 ± 0.21 derived by Bouwens et al. (2006). Although the field galaxy slope was revised to -1.87 ± 0.1 in Bouwens et al. (2015), corresponding changes to the Schechter luminosity function fit parameters (ϕ_* and M_*, the number density and characteristic luminosity respectively), have not significantly modified the total UV luminosity density and the discrepancy between the scaled GRB rate and the Lyman-break galaxy luminosity density remains a factor of ~2. Unlike the redshift evolution for the SFRD quoted earlier, the SFR inferred from GRBs would imply a much slower fall-off as a function of redshift, $\propto (1 + z)^{-3}$ for $z > 4$ (Kistler et al. 2013; Wang 2013) or $\propto (1 + z)^{-2}$ as shown in Fig. 1.

Robertson and Ellis (2012) also argued that the high SFRD inferred from GRBs would violate the stellar mass density at $z \sim 4 - 6$ for canonical shapes of the IMF and produce enough ionizing photons to result in a τ of 0.09 compared to the *Planck* derived value of 0.058 ± 0.012. The former can be easily reconciled if the stellar IMF is more top-heavy ($dn/dM \propto M^{-1.7\pm0.1}$) at high redshift, as seems to be indicated by nebular line observation of some $z > 4$ galaxies (Shim et al. 2011) and has been previously argued at $z \sim 6$ by Chary (2008). Such an IMF has a factor of 5 more massive stars above 25 $M_\odot$ and would produce a factor of 5.4 more Lyman-continuum ionizing photons per second but only a factor of 3.4

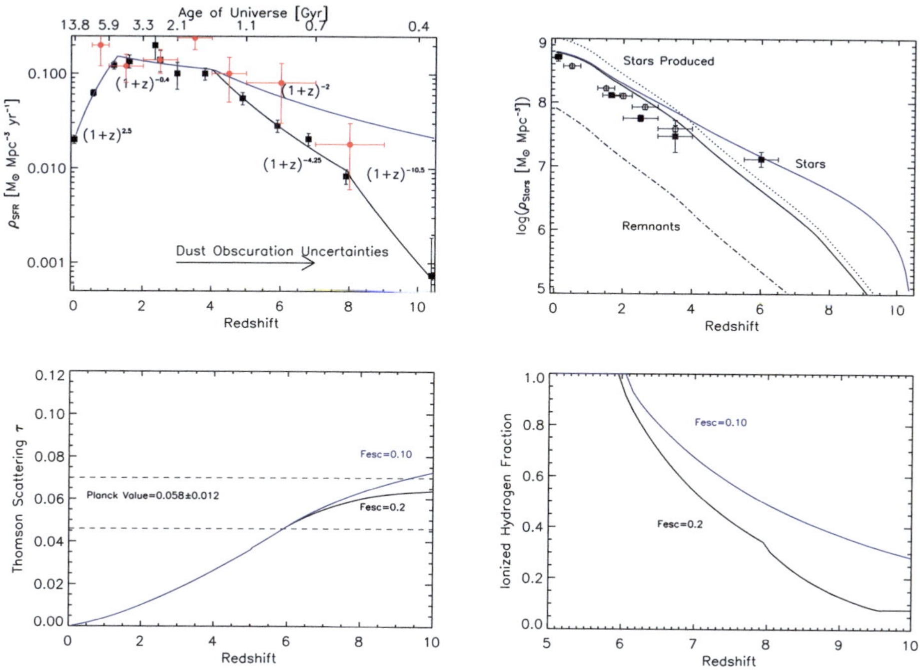

Fig. 1 The *top left panel* shows the comparison between the scaled GRB rate (*red colored points*) and the attenuation-corrected star-formation rate density from deep fields (*black points*) illustrating that the GRB rate per unit star-formation increases with increasing redshift and at $z > 4$ tends to lie above that inferred from the rest-frame ultraviolet observations. The redshift parameterizations of the black and blue lines are shown with the blue line an elevated star-formation history that matches the GRB rates better. Black points are from (Murphy et al. 2011; Reddy and Steidel 2009; Bouwens et al. 2015) while *red points* are from Chary et al. (2007). The *top right panel* shows a comparison between the integral of the star-formation rate density from field galaxies (*black line*) and that from GRBs (*blue line*) with the stellar mass density inferred from rest-frame visible light observations, assuming a standard Salpeter IMF. $z < 4$ stellar mass densities are estimates from Marchesini et al. (2011) and Elsner et al. (2008) and at $z \sim 6$ from Chary (2008). The *bottom left panel* illustrates the Thomson scattering cross section derived for CMB photons from *Planck* and that inferred from the SFRs displayed in *panel 1*, assuming an escape fraction of ionizing photons of 10–20 % and assuming the higher Lyman-continuum production rate inferred in Shim et al. (2011). The *bottom right panel* compares the ionized fraction of the IGM inferred from star-forming galaxies (*black line*) with that inferred from the GRB rate (*blue line*) with the biggest uncertainty in this translation arising from the unknown clumping factor of the IGM; here we adopt a value of 3 for the clumping factor. In this multi-panel figure, we highlight the implications of GRB-derived measures of high-z star-formation and other measures of star-formation. The current data favor a higher far-ultraviolet luminosity density at $z > 4$ than inferred from field galaxy surveys, a high Lyman continuum production rate probably due to a top-heavy IMF, an escape fraction of ~10–15 % and a lower clumping factor for the IGM than current simulations indicate (Trac and Cen 2007)

more far-ultraviolet photons at 1500 Å. The star-formation rate in $M_\odot\ yr^{-1}$ corresponding to a particular measured far-ultraviolet luminosity density is reduced by the latter factor. Such an IMF would imply that a large fraction of the sources of ionizing photons quickly evolve into stellar remnants like neutron stars and black holes, which would not be contributing to the rest-frame visible light luminosity density of galaxies at $z \sim 6$ from which stellar mass densities are derived (Fig. 1). Since GRBs are thought to arise from $>25\ M_\odot$ stars, such an IMF would imply a 1.6 times higher GRB rate per unit star-formation and thus be a convenient solution to the observed discrepancy between the SFRD observed from the UV light of galaxies and the GRB rate at high-z. Intriguingly it also reconciles the discrepancy with the $z \sim 6$ stellar mass density as described in (Chary 2008) and matches the slope of the extragalactic transient luminosity function at high-z seen by Sun et al. (2015).

The discrepancy with the Thomson scattering optical depth can also be addressed. First, if the Lyman continuum production rate is as high as inferred in Shim et al. (2011, assuming the Hα to FUV continuum ratio is 0.08 rather than the 0.008 seen in the local Universe) from observations of Hα in field galaxies, an escape fraction of 10–15 % would reduce the τ corresponding to a particular star-formation history to be within *Planck* derived values (Fig. 1). Second, as explained earlier, τ is only a weak constraint on star-formation at $z > 6$. With such a weak $\sim 1\sigma$ constraint, the current discrepancy with Planck Collaboration (2015) is hardly significant. Reducing the uncertainty in the Thomson scattering optical depth by a factor of two with improved calibration of the *Planck* polarization data on large angular scales will help assess if there is a tension with CMB polarization measurements.

Two other possibilities exist to explain the high GRB rate at $z > 6$. One scenario would argue for a population of dust-obscured galaxies that have thus far not been detected. This argument is disfavored since dusty infrared luminous galaxies are high mass, high metallicity systems at least at $z < 3$ where they have been extensively studied and dominate the star-formation rate density. At these low redshifts, GRB hosts do not sample the infrared luminous galaxy population, even including the significant contribution of dark bursts. Thus it is unlikely that at high-z, where metallicity and dust content is lower, the GRBs would preferentially occur in such systems. Imminent deep *ALMA* surveys which will be sensitive to detecting L_* galaxies at $z \sim 6$ will help constrain the dust obscured star-formation at these early cosmic times better (e.g. Dunlop et al. 2016).

A second possibility would imply that there may be some evolution in the efficiency of GRB production in early stellar populations (e.g., Yüksel et al. 2008; Kistler et al. 2008; Campisi et al. 2010). Work is ongoing to determine possible biases in GRB production observationally (e.g., Schulze et al. 2015; Vergani et al. 2015, Perley et al. 2016) with one such analysis favoring a metallicity ceiling for GRB production of solar, similar to the original work at low-z by Modjaz et al. (2008). Various papers (Levesque et al. 2010; Krühler et al. 2015; Graham and Fruchter 2013), have shown that GRB hosts at $z < 1$ preferentially seem to be low-mass, low-metallicity systems. The presence of hosts with super-solar metallicity in this redshift range argues against a metallicity ceiling, instead arguing in favor of a decreased efficiency for GRB production at high metallicities. The absence of high metallicity galaxies in the distant Universe would therefore increase the fraction of SFR in low metallicity galaxies and thus the increase in GRB rates relative to the SFRD could be explained by such a bias as argued by Perley et al. (2016), Trenti et al. (2015). In other words, GRBs are a *less* biased tracer of SFR at high-z than they are at low-z!

Galaxy-integrated metallicity estimates however need to be interpreted with some caution. In the Milky Way itself, there is abundant evidence for low metallicity stars in the halo. While they are predominantly low-mass stars which do not dominate the galaxy light, a range of metallicities may be present in a galaxy even with sub-solar gas phase metallicity (or vice-versa) making it challenging to associate a metallicity threshold/bias for GRB production. Having a larger sample of gas-phase and absorption-line metallicity estimates in high-z GRB hosts compared with the distribution of metallicities in field galaxies at the same mass, both of which will only become possible with *JWST* spectroscopy, will help to highlight correlations between galaxy-integrated and local-to-the-GRB metallicities and disentangle the role of metallicity as a biasing factor for GRB production (Li 2008).

Clearly, the uncertainties in the star-formation rate density at $z > 3$ are significant. Larger samples of GRBs and a better understanding of the dust obscuration properties of the field galaxy sample are crucial for assessing whether the GRB rates are consistent with the star-formation rate measured from the field galaxy sample or whether the higher GRB rate at $z > 3$ is providing a window into metallicity/IMF trends that evolve with redshift.

Looking into the future, understanding the progenitors of the high-redshift GRB population is critical for characterizing them as a robust star-formation tracer. It has recently become clear that the ionization parameter (i.e. ionizing photon rate per baryon) in high-z star-forming galaxies is much higher than that seen in galaxies in the local Universe as inferred from the strength of nebular lines (Shim et al. 2011). It is however unclear what the implications of the high ionization parameter for the underlying stellar population in those galaxies are. Is this a result of a top-heavy IMF? Is this due to a higher binary fraction of stars in compact star-forming galaxies? Or is the interstellar medium in the distant galaxies so turbulent that the stars have higher angular momentum resulting in a higher production rate of ionizing photons? While the top-heavy IMF argument can be a panacea for both the high GRB rate and the high ionization parameter, tieing the production efficiency of GRBs with the angular momentum distribution of stars is crucial for understanding whether stellar rotation may be playing a role in understanding the observational results.

A further important consideration is the potential identification of GRBs without a high-energy trigger. Now that this feat has been accomplished by the Intermediate Palomar Transient Factory (iPTF; Cenko et al. 2015), we know that with a suitable cadence such a discovery is possible from the GRB afterglow alone. The WFIRST space telescope, with a 0.3 deg^2 FOV, may have the ability to both discover and follow-up high-redshift GRBs from their afterglows and continue the study of new GRBs into the late 2020's beyond the time when current high-energy GRB detection experiments are guaranteed to be operating. This in particular, motivates the development of next-generation GRB missions such as SVOM so that constraints on the X-ray and γ-ray properties of transients seen at optical/near-infrared wavelengths can be obtained. Such a multi-wavelength approach to a large sample of optical transients would help constrain the mass range of the progenitor stars and help improve the mapping from GRB rate to the star-formation rate.

3 Metallicity Constraints on GRB-Derived SFR

In the previous section, we have outlined the results and uncertainties on star-formation rate densities from field galaxy surveys and from GRBs. Here we discuss our current understanding of how observations of metallicities in damped Ly-α systems could constrain the star-formation rate inferred from GRBs.

We consider two star-formation rate histories that span the range that is consistent with the observational constraints presented earlier. In one, we consider the Behroozi et al. (2013) SFR which ties the stellar mass of galaxies with the corresponding dark matter halo mass and uses merger rates and mass accretion histories to derive average star-formation histories (see also Behroozi and Silk 2014). In another, we take the GRB rate derived SFR based on the analysis of Kistler et al. (2013) at $z > 4$ folded in with the Behroozi et al. (2013) SFR at $z < 4$. The difference is the second model is 0.3 deg higher than the first model at high redshifts which would also be consistent with the analysis of Trenti et al. (2013). This latter study also favors a metallicity origin for the increase in the GRB-to-SFR ratio with redshift out to $z \lesssim 4$, while at higher redshift their model predicts the ratio evolves primarily because the SFR from Lyman-break galaxy surveys is missing faint galaxies.

We have used a cosmic evolution model in the context of hierarchical structure formation that follows the interplay between the ISM and IGM due to star formation. This is coupled

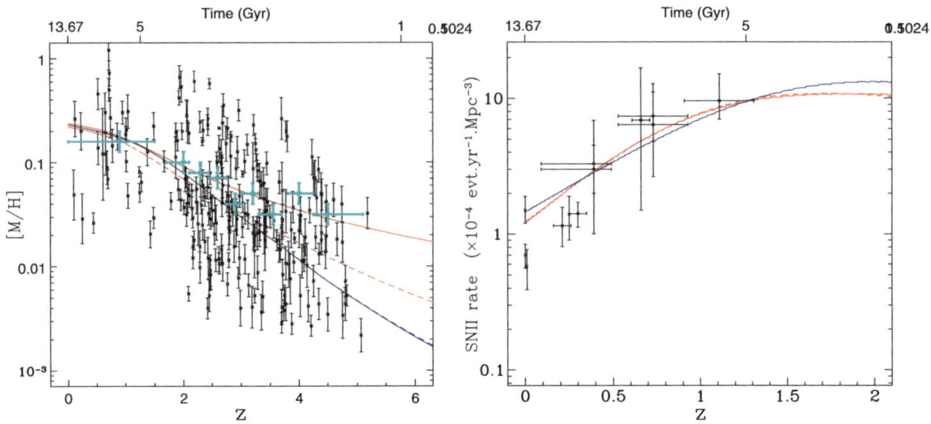

Fig. 2 *Left*: The derived (log of the) metallicity abundance relative to the solar metallicity for different star-formation histories. The *red curve* corresponds to the Kistler et al. (2013) SFR derived from GRBs, the *blue curve* corresponds to the SFR of Behroozi and Silk (2014) while the *dashed red curve* corresponds to the Trenti et al. (2013) rate. Data points are from damped Lyman-α systems (Rafelski et al. 2012) with the *cyan points* showing the average in redshift bins. These metallicities are lower than those plotted in Fig. 3 since quasar DLAs tend to probe lower metallicity environments than the nuclei of GRB hosts which are star-forming galaxies. *Right*: the derived core-collapse supernova rate (SNII) using the same color coding as in the *left panel*. All quantities are shown as functions of redshift. See Vangioni et al. (2015) for more details. Clearly, the observed scatter in both metallicities and Type II SNe rates is large enough that different star-formation histories cannot be distinguished; the preference from the metallicity measurements is however for a high SFR similar to that derived from GRBs

to a detailed model for cosmic chemical evolution (Daigne et al. 2006; Rollinde et al. 2009), and allows us to keep track of the ionization history of the Universe, the abundances of the important elements produced in massive and low/intermediate mass stars as a function of redshift, as well as the rate of core-collapse supernovae (SNII), the specific SFR (sSFR), and the stellar mass density (Vangioni et al. 2015).

We show in Vangioni et al. (2015) that using the SFR from Kistler et al. (2013), the derived Thomson scattering optical depth is too high compared to the WMAP optical depth by a factor of two if we use the usual escape fraction value $f_{esc} = 0.2$. This would imply that compared to the recent *Planck*-derived value of τ, the tension would be even greater. Lowering the escape fraction by a factor of two or more, addresses this issue since it reduces the number of ionizing photons escaping the star-forming galaxy and thereby reduces the number density of electrons that scatter the CMB photons. This result is similar to that seen by Robertson and Ellis (2012) and discussed earlier.

When instead we compare the metal yields with the metallicity of damped Ly-α systems reported in Rafelski et al. (2012), we find that the two models cannot be discriminated using current observational data. Although the average metallicities are consistent with the higher SFR of Kistler et al. (2013) and Trenti et al. (2013), the SNII rate which is primarily measured at $z < 1.5$, is consistent with the full range of models. This can be seen by comparing the red solid and dashed curves in Fig. 2 with the solid blue curve.

It can therefore be concluded that the different data sets considered as a whole favor a higher SFR at $z > 3$ similar to that derived from GRBs and a low escape fraction ($f_{esc} < 0.2$) of ionizing photons. However, the scatter in both the derived metallicities and measured τ is too large for this to be a robust conclusion. Since the current uncertainty on τ provides only a weak constraint on star-formation at $z > 6$, we think that using the optical depth as a constraint on star-formation is only useful once these uncertainties are improved in future analysis of *Planck* polarization data.

4 Metallicity Evolution of the ISM in GRB Hosts: Probing Feedback

Spectroscopy of GRB afterglows can provide a unique window into the gas column density and metallicity of the ISM of their hosts, the intensity of the radiation field therein and the distribution of metals along the line of sight, in some ways the after-effects of star-formation in the galaxies. Although these are measurements along a single line of sight through the host, by obtaining measurements of several GRBs at similar redshifts, we can obtain an average measure of the metal abundances in sub-L_* galaxy populations at high-z, a population which is challenging to characterize in any detail due to their intrinsic faintness (e.g Laskar et al. 2011). Even with the advent of multiplexed infrared spectrographs on 10 m class telescopes, such studies on field galaxies have been limited to the bright end of the galaxy luminosity function and typically at $z < 3$ (Shapley et al. 2015). However, upcoming work by Faisst et al. (private communication) which leverages the entrance of nebular lines into the *Spitzer*/IRAC bandpasses calibrated against rest-frame UV spectroscopy will help push the envelope on mass-metallicity studies out to $z \sim 6$, prior to *JWST*.

By comparing the redshift evolution of metallicity in GRB hosts with the corresponding evolution in stellar mass, one can thereby characterize the nature of stellar winds, supernovae and feedback processes which may drive metals out of the host galaxy. For example, if galaxies were a closed-box, then as long as they were star-forming, the metallicity should increase in direct proportion to their stellar mass density. Since galaxies are not closed-box systems, the infall of pristine gas at high redshift can dilute metallicity, while the corresponding star-formation fueled by that gas supply may be able to compensate for that dilution, depending on the star-formation efficiency. Furthermore, if feedback processes are very efficient, then the gas-phase metallicity of the ISM will not increase at the same rate as the stellar mass density since metal enrichment through star-formation is lost through galactic winds. This effect is best seen at $z \sim 0$ in the mass-metallicity relationship of galaxies seen in the Sloan Digital Sky Survey (Tremonti et al. 2004). In that relationship, dwarf galaxies are five times more metal depleted that L_* galaxies and faint galaxies show significantly larger scatter than bright ones. This suggests that stochastic star-formation in faint galaxies combined with galactic winds efficiently remove metals from galaxy potential wells.

Figure 3 shows the comparison between the average metallicity of GRB hosts derived from line of sight spectroscopy of the afterglow in 3 different redshift bins compared to the stellar mass density in those bins obtained from field galaxy surveys. We find that the best-fit relation has an exponent of 0.7 (i.e. $Z(z) \propto \rho_*(z)^{0.7\pm0.2}$), in the sense that metal abundances in faint galaxies increase at a slower rate than the stellar mass density (Chary et al. 2007). If this is due to the effect of feedback processes, a similar plot generated for bright galaxies would fall closer to a line which has an exponent of unity. Thus, through spectroscopy of the bright afterglows of GRBs, we can obtain insights into feedback processes from star-formation in faint galaxies which would otherwise be challenging to obtain.

5 Conclusions

In this summary, we have highlighted the complementary role that GRBs can play to field galaxy surveys in constraining the star-formation rate density at high-z. The GRB rate per

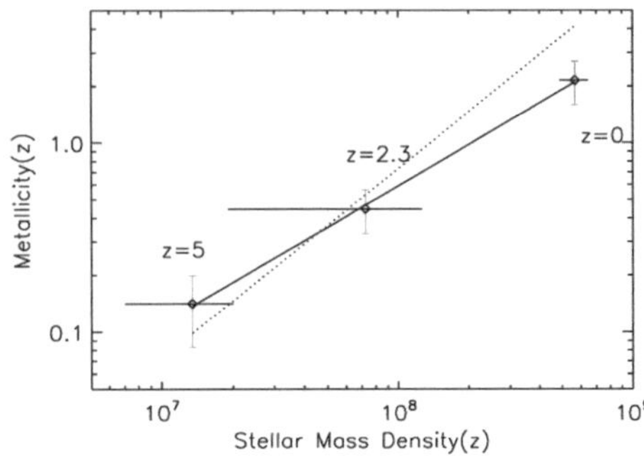

Fig. 3 Evolution of metallicity in GRB hosts at three different redshift bins plotted against the comoving stellar mass density ($M_\odot$ Mpc^{-3}) at those redshifts. The *solid line* is the best fit of the functional form $Z(z) \propto \rho_*(z)^{0.7}$ where $Z(z)$ is the metallicity as a function of redshift averaged over a multitude of GRB hosts and $\rho_*(z)$ is the stellar mass density dominated by faint, field galaxies. The *dotted line* has an exponent of unity. The fact that the best-fit exponent is less than one suggests that metals are blown out of low-mass galaxies very efficiently by stellar winds, an effect which is crucial for the difference in the faint end slope of the UV luminosity function of galaxies and the visible light luminosity function. GRBs are therefore a unique probe of feedback in low luminosity galaxies. However, better metallicity measurements of a larger sample of high-z GRBs are required to constrain this relationship better and to derive it as a function of halo mass

unit star-formation derived from field galaxy surveys appears to be increasing with increasing redshift. This may be due to a top-heavy IMF which increases the fraction of massive stars for a particular UV luminosity density. Alternately, a metallicity bias for GRBs whereby galaxies above roughly solar metallicity are inefficient GRB producers could explain this observational result. Since a larger fraction of star-formation at low-z takes place in such galaxies, than at high-z, if the GRB rate per unit low-metallicity-star-formation is constant, a normalization between the GRB and total SFR at low redshift would result in the apparent GRB rate per unit total star-formation rate increasing with increasing redshift. A large contribution from faint/dusty galaxies below the detection limit of field galaxy surveys could also explain the observed trend but is disfavored based on the sparsity of dust obscured high-z galaxies in current surveys. While the Thomson scattering optical depth of CMB photons (τ) could help constrain the high-z SFR, it too is plagued by both measurement uncertainty and the uncertainty in the escape fraction of ionizing photons from galaxies. However, the ensemble of current data favor the high SFR (assuming a conversion from far-ultraviolet luminosity density to SFR using a Salpeter IMF) inferred from GRB measurements, a high ionizing photon production rate probably due to a top-heavy IMF, a low escape fraction of $\sim$10–15 % and a low clumping factor for the IGM. Future breakthroughs in high-z star-formation will come from detecting significant populations of GRBs at $z > 4$–10 with existing missions such as *Swift* and future missions such as SVOM and targeting them through multi-wavelength spectroscopy. This would yield a better measurement of the luminosity and metallicity distribution of high-z GRB hosts which can robustly discriminate between these scenarios and provide insights into galactic feedback processes at the low-mass end. Better measurements of high-z GRB rates and τ could also provide a unique constraint into the escape fraction of ionizing photons at $z \sim 6$. Finally, improved GRB progenitor models which provide greater insights on the role of magnetic torques, angular momentum and metallicity of massive stars in the formation of GRBs will also help reduce the scatter in the mapping of GRB rates to star-formation rates.

Acknowledgements The work of EV and PPJ has been carried out at the ILP LABEX (under reference ANR-10-LABX-63) supported by French state funds managed by the ANR within the Investissements d'Avenir programme under reference ANR-11-IDEX-0004-02. It was also sponsored by the French Agence Nationale pour la Recherche (A.N.R.) via the grant VACOUL (ANR-2010-Blan-0510-01). We thank the referees for their feedback and clarifying remarks.

References

A.J. Battisti, D. Calzetti, R. Chary, Astrophys. J. **818**, 13 (2016). 10.3847/0004-637X/818/1/13
P.S. Behroozi, J. Silk, Preprint arXiv:1404.5299 (2014)
P.S. Behroozi, R.H. Wechsler, C. Conroy, Astrophys. J. **770**, 57 (2013)
A.W. Blain, P. Natarajan, Mon. Not. R. Astron. Soc. **312**, L35 (2000)
J.S. Bloom, S.R. Kulkarni, S.G. Djorgovski, Astron. J. **123**, 1111 (2002)
R.J. Bouwens, G.D. Illingworth, J.P. Blakeslee, M. Franx, Astrophys. J. **653**, 53 (2006)
R.J. Bouwens, G.D. Illingworth, P.A. Oesch et al., Astrophys. J. **803**, 34 (2015)
V. Bromm, A. Loeb, Astrophys. J. **575**, 111 (2002)
M.A. Campisi, L.-X. Li, P. Jakobsson, Mon. Not. R. Astron. Soc. **407**, 1972 (2010)
R.-R. Chary, Astrophys. J. **680**, 32 (2008)
R. Chary, E.E. Becklin, L. Armus, Astrophys. J. **566**, 229 (2002)
R. Chary, E. Berger, L. Cowie, Astrophys. J. **671**, 272 (2007)
F. Daigne, E.M. Rossi, R. Mochkovitch, Mon. Not. R. Astron. Soc. **372**, 1024 (2006)
J.S. Dunlop, R.J. McLure, A.D. Biggs et al., arXiv:1606.00227 (2016)
F. Elsner, G. Feulner, U. Hopp, Astron. Astrophys. **477**, 503 (2008)
J.F. Graham, A.S. Fruchter, Astrophys. J. **774**, 119 (2013)
A. Grazian, E. Giallongo, R. Gerbasi et al., Astron. Astrophys. **585**, A48 (2016)
A. Heger, C.L. Fryer, S.E. Woosley, N. Langer, D.H. Hartmann, Astrophys. J. **591**, 288 (2003)
E.E.O. Ishida, R.S. de Souza, A. Ferrara, Mon. Not. R. Astron. Soc. **418**, 500 (2011)
M.D. Kistler, H. Yüksel, J.F. Beacom, K.Z. Stanek, Astrophys. J. **673**, L119 (2008)
M.D. Kistler, H. Yuksel, A.M. Hopkins, arXiv:1305.1630 (2013)
T. Krühler, D. Malesani, J.P.U. Fynbo et al., Astron. Astrophys. **581**, A125 (2015)
T. Laskar, E. Berger, R.-R. Chary, Astrophys. J. **739**, 1 (2011)
E. Le Floc'h, P.-A. Duc, I.F. Mirabel et al., Astron. Astrophys. **400**, 499 (2003)
E.M. Levesque, L.J. Kewley, E. Berger, H.J. Zahid, Astron. J. **140**, 1557 (2010)
L.-X. Li, Mon. Not. R. Astron. Soc. **388**, 1487 (2008)
A. Lien, T. Sakamoto, N. Gehrels et al., Astrophys. J. **806**, 276 (2015)
S. Marassi, R. Schneider, M. Limongi et al., Mon. Not. R. Astron. Soc. **454**, 4250 (2015)
D. Marchesini et al., Astrophys. J. **2009**, 1765 (2011)
M. Matsuura, E. Dwek, M. Meixner et al., Science **333**, 1258 (2011)
J.T.W. McGuire et al., arXiv:1512.07808 (2016)
M. Modjaz, L. Kewley, R.P. Kirshner et al., Astron. J. **135**, 1136 (2008)
E.J. Murphy, R.-R. Chary, M. Dickinson et al., Astrophys. J. **732**, 126 (2011)
P.A. Oesch, R.J. Bouwens, C.M. Carollo et al., Astrophys. J. **709**, L21 (2010)
P.A. Oesch, R.J. Bouwens, G.D. Illingworth et al., arXiv:1409.1228 (2014b)
Planck Collaboration, P.A.R. Ade, N. Aghanim et al., arXiv:1502.01589 (2015)
D.A. Perley et al., Astrophys. J. **817**, 8 (2016)
P.A. Price et al., Astrophys. J. **645**, 851 (2006)
S.-F. Qin, E.-W. Liang, R.-J. Lu, J.-Y. Wei, S.-N. Zhang, Mon. Not. R. Astron. Soc. **406**, 558 (2010)
M. Rafelski, A.M. Wolfe, J.X. Prochaska, M. Neeleman, A.J. Mendez, Astrophys. J. **755**, 89 (2012)
N.A. Reddy, C.C. Steidel, Astrophys. J. **692**, 778 (2009)
N.A. Reddy, M. Kriek, A.E. Shapley et al., Astrophys. J. **806**, 259 (2015)
B.E. Robertson, Astrophys. J. **716**, L229 (2010)
B.E. Robertson, R.S. Ellis, Astrophys. J. **744**, 95 (2012)
B.E. Robertson, S.R. Furlanetto, E. Schneider et al., Astrophys. J. **768**, 71 (2013)
B.E. Robertson, R.S. Ellis, S.R. Furlanetto, J.S. Dunlop, Astrophys. J. **802**, L19 (2015)
E. Rollinde, E. Vangioni, D. Maurin, K.A. Olive, F. Daigne, J. Silk, F.H. Vincent, Mon. Not. R. Astron. Soc. **398**, 1782 (2009)
R. Salvaterra, S. Campana, S.D. Vergani et al., Astrophys. J. **749**, 68 (2012)
C.C. Steidel et al., Astrophys. J. **519**, 1 (1999)
A.E. Shapley, N.A. Reddy, M. Kriek et al., Astrophys. J. **801**, 88 (2015)
H. Shim, R.-R. Chary, M. Dickinson et al., Astrophys. J. **738**, 69 (2011)
S. Schulze et al., Astrophys. J. **808**, 73 (2015)

T. Su, T.A. Marriage, V. Asboth et al., arXiv:1511.06770 (2015)
H. Sun, B. Zhang, Z. Li, Astrophys. J. **812**, 33 (2015)
N.R. Tanvir, A.J. Levan, A.S. Fruchter et al., Astrophys. J. **754**, 46 (2012)
H. Trac, R. Cen, Astrophys. J. **671**, 1 (2007)
C.A. Tremonti, T.M. Heckman, G. Kauffmann et al., Astrophys. J. **613**, 898 (2004)
M. Trenti, M. Stiavelli, Astrophys. J. **676**, 767 (2008)
M. Trenti, R. Perna, E.M. Levesque, J.M. Shull, J.T. Stocke, Astrophys. J. **749**, L38 (2012)
M. Trenti, R. Perna, S. Tacchella, Astrophys. J. **773**, L22 (2013)
M. Trenti, R. Perna, R. Jimenez, Astrophys. J. **802**, 103 (2015)
E. Vangioni, K.A. Olive, T. Prestegard, J. Silk, P. Petitjean, V. Mandic, Mon. Not. R. Astron. Soc. **447**, 2575 (2015)
S.D. Vergani, R. Salvaterra, J. Japelj et al., Astron. Astrophys. **581**, A102 (2015)
F.J. Virgili, B. Zhang, K. Nagamine, J.-H. Choi, Mon. Not. R. Astron. Soc. **417**, 3025 (2011)
F.Y. Wang, Astron. Astrophys. **556**, 90 (2013)
A. Weiß, C. De Breuck, D.P. Marrone et al., Astrophys. J. **767**, 88 (2013)
R.A.M.J. Wijers, J.S. Bloom, J.S. Bagla, P. Natarajan, Mon. Not. R. Astron. Soc. **294**, L13 (1998)
S.E. Woosley, J.S. Bloom, Annu. Rev. Astron. Astrophys. **44**, 507 (2006)
S.E. Woosley, A. Heger, Astrophys. J. **637**, 914 (2006)
S.C. Yoon, N. Langer, C. Norman, Astron. Astrophys. **460**, 199 (2006)
H. Yüksel, M.D. Kistler, J.F. Beacom, A.M. Hopkins, Astrophys. J. **683**, L5 (2008)

Space Sci Rev (2016) 202:197–236
DOI 10.1007/s11214-016-0235-6

GRBs and Fundamental Physics

Patrick Petitjean[1] · F.Y. Wang[2,3] · X.F. Wu[4] · J.J. Wei[4]

Received: 20 October 2015 / Accepted: 5 January 2016 / Published online: 8 February 2016

Abstract Gamma-ray bursts (GRBs) are short and intense flashes at the cosmological distances, which are the most luminous explosions in the Universe. The high luminosities of GRBs make them detectable out to the edge of the visible universe. So, they are unique tools to probe the properties of high-redshift universe: including the cosmic expansion and dark energy, star formation rate, the reionization epoch and the metal evolution of the Universe. First, they can be used to constrain the history of cosmic acceleration and the evolution of dark energy in a redshift range hardly achievable by other cosmological probes. Second, long GRBs are believed to be formed by collapse of massive stars. So they can be used to derive the high-redshift star formation rate, which can not be probed by current observations. Moreover, the use of GRBs as cosmological tools could unveil the reionization history and metal evolution of the Universe, the intergalactic medium (IGM) properties and the nature of first stars in the early universe. But beyond that, the GRB high-energy photons can be applied to constrain Lorentz invariance violation (LIV) and to test Einstein's Equivalence Principle (EEP). In this paper, we review the progress on the GRB cosmology and fundamental physics probed by GRBs.

✉ F.Y. Wang
fayinwang@nju.edu.cn

P. Petitjean
petitjean@iap.fr

X.F. Wu
xfwu@pmo.ac.cn

J.J. Wei
jjwei@pmo.ac.cn

1 Institut d'Astrophysique de Paris, 98bis Boulevard Arago, 75014 Paris, France

2 School of Astronomy and Space Science, Nanjing University, Nanjing 210093, China

3 Key Laboratory of Modern Astronomy and Astrophysics, Nanjing University, Ministry of Education, Nanjing 210093, China

4 Purple Mountain Observatory, Chinese Academy of Sciences, Nanjing 210008, China

1 Introduction

Gamma-ray bursts (GRBs) are among the most powerful explosions in the Universe (Mészáros 2006; Zhang 2007; Gehrels et al. 2009; Kumar and Zhang 2015; Wang et al. 2015), which can be classified into two classes: long GRBs and short GRBs (Kouveliotou et al. 1993). Long GRBs are formed by collapses of massive stars, meanwhile the progenitors of short GRBs are mergers of compact stars (Woosley and Bloom 2006). The high luminosities of GRBs make them detectable out to high redshifts. Similar as type Ia supernova (SNe Ia), long GRBs can be treated as "relative standard candles" (i.e., Amati et al. 2002; Ghirlanda et al. 2004a; Liang and Zhang 2005). Due to the collapsar model, long GRBs can probe the high-redshift star formation rate. They also offer the exciting opportunity to detect the Population III (Pop III) stars, because massive Pop III stars can die as GRBs. The smooth spectra of GRB afterglow allow that properties of the intergalactic (IGM) absorption features could be extracted. The metal absorption lines make GRBs powerful sources to study the metal enrichment history. The clean damping wings of GRBs make them ideal tools to study the reionization of IGM and interstellar medium (ISM) properties of their hosts.

First, GRBs can act as the complementary tools to study dark energy and cosmic accelerating expansion. The SNe Ia can be observed when accreting white dwarf stars exceed the mass of the Chandrasekhar limit and explode. Thus they can be treated as ideal standard candles. However, the double white dwarfs merger model of SNe Ia challenges the precision of this standard candle. The accelerating expansion of Universe is discovered by two supernova groups (Riess et al. 1998; Perlmutter et al. 1999), which is attributed to the mysterious component—dark energy. Other observations, such as the cosmic microwave background (CMB) (Spergel et al. 2003), baryonic acoustic oscillations (BAO) (Eisenstein et al. 2005), X-ray gas mass fraction in galaxy clusters (Allen et al. 2004), and Hubble parameters (Jimenez et al. 2003; Wang and Wang 2014), also indicate that the expansion of universe is speeding up. But hitherto the highest redshift of SNe Ia is 1.914 (Jones et al. 2013). GRBs are promising tools to fill the gap between SNe Ia and CMB (Dai et al. 2004; Ghirlanda et al. 2004b, 2006; Liang and Zhang 2005; Wang and Dai 2006b; Wang et al. 2007; Schaefer 2007; Wang 2012). Fortunately, some luminosity correlations of GRBs have been used to standardize GRB energetics (i,e., Ghirlanda et al. 2004a; Xu et al. 2005; Liang and Zhang 2006; Firmani et al. 2005; Amati 2006; Amati et al. 2008; Liang et al. 2008).

The high-redshift ($z > 6$) star formation rate (SFR) measurement beyond the reach of present instruments, particularly at the faint end of the galaxy luminosity function. Long GRBs formed by the collapse of massive stars, provide a complementary tools for measuring the SFR (Totani 1997; Wijers et al. 1998; Bromm et al. 2002). Because the lifetimes of massive stars are short, the formation rate can be treated as their death rate. Surprisingly, the Swift data reveals that long GRBs are not tracing the SFR directly, instead implying some kind of additional evolution (Daigne et al. 2006; Le and Dermer 2007; Yüksel and Kistler 2007; Salvaterra and Chincarini 2007; Guetta and Piran 2007; Kistler et al. 2008; Campisi et al. 2010; Salvaterra et al. 2012). But the conversion factor between GRB rate and SFR is hard to determine. So the common method is relating the GRBs observed at low redshift to the SFR measurements and also considering the additional evolution of the GRB rate relative to the SFR. The SFR derived from GRBs seems to be much higher than that obtained from high-redshift galaxy surveys (Kistler et al. 2008, 2009; Wang and Dai 2009; Wang 2013). The additional evolution term can be expressed as $(1+z)^{\delta}$ ($\delta \sim 0.5$–1.5) (Kistler et al. 2008; Robertson and Ellis 2012; Wang 2013). Many models have been proposed to explain this additional evolution, including metallicity evolution (Li 2008; Qin et al. 2010), GRBs from cosmic strings (Cheng et al. 2010), evolution of initial mass function (Wang and Dai 2011a; Xu and Wei 2008), evolution of luminosity function (Virgili et al. 2011; Yu et al. 2012). However, some studies claimed that there is no discrepancy between GRB rate and SFR (Elliott et al. 2012; Hao and Yuan 2013). The sample selection and adopted SFR are important. Theoretical model and observations of host galaxies both support that long GRBs could occur in low-metallicity environment (Woosley and Heger 2006; Mészáros 2006; Langer and Norman 2006; Stanek et al. 2006; Levesque 2014; Wang and Dai 2014b).

What are the dominant sources of metal enrichment? What is the metal enrichment history? The above two questions are the key questions of metal enrichment. The metal enrichment can change the star formation mode, from a high-mass (Pop III) mode to a low-mass dominated (Pop I/II) one, if the metal exceeds a 'critical metallicity' of $Z_{\rm crit} \sim 10^{-4} Z_\odot$ (Bromm et al. 2001; Schneider et al. 2002, 2006). The metal enrichment history is also connected with the reionization history. Absorption lines of distant bright sources, such as GRBs or quasars, are main sources of information about the chemical properties of high-redshift Universe (Oh 2002; Furlanetto and Loeb 2003; Oppenheimer et al. 2009). GRBs have several advantages compared to traditional quasars (Bromm and Loeb 2007). First, because their progenitors are stellar mass black holes, so the number density drops much less precipitously than quasars at $z > 6$ (Fan et al. 2006). Second, the spectra of GRBs are smooth power laws, which allows to extract absorption lines. Several absorption lines have been extracted from the GRB spectra, such as two absorption lines (Si IV and Fe II) in the spectrum of GRB 090423 (Salvaterra et al. 2009).

In this paper, we review the cosmological implications of GRBs. Then in the following sections the gamma-ray bursts cosmology is reviewed. The second section is dedicated to luminosity correlations and cosmological constraints from GRBs. Section 3 discusses the capability of GRBs to reveal the high-redshift SFR. In Sect. 4, the capability of GRBs to probe the metal enrichment history is discussed. The tests on fundamental physics with GRBs are reviewed in Sects. 5 and 6. The last section provides a summary and future prospect.

2 The Luminosity Correlations of GRBs and Dark Energy

In order to investigate the properties of dark energy, the relation between distance and redshift is needed, i.e., the expansion history of our universe. The best way to measure the redshift-distance relation is using standard candles, such as SNe Ia. To this end, many projects have been proposed to determine the distances of SNe Ia with exquisite accuracy. However, due to the limited luminosity of SN Ia, they can only be detected at low redshifts, i.e., $z < 2.0$. If we study the evolution of dark energy at large redshift range, high-redshift standard candle is needed. GRBs could be complementary probes of dark energy at high redshifts. Many attempts have been performed to standardize GRBs. But the observed energies of GRBs span a wide range. After considering the collimation effect, Frail et al. (2001) found that the collimated energies of GRBs clustered around 5×10^{50} erg, which was confirmed by Bloom et al. (2003). Meanwhile, the collimated jet predicts that the appearance of an achromatic break in the afterglow light curve of GRBs (Rhoads 1997; Sari 1999), which is important to standardize the energetic of GRBs.

In this section, we first review luminosity correlations of GRBs. Then the progress on dark energy revealed by GRBs is discussed. Some reviews have discussed this topic (i.e., Ghirlanda et al. 2006; Dai and Wang 2007; Amati and Della Valle 2013).

2.1 The Luminosity Correlations of GRBs

The isotropic luminosity L and the energy E of GRBs with redshifts can be calculated by the peak flux and the fluence. For example, the isotropic luminosity is

$$L = 4\pi d_L^2 P_{\rm bolo} \tag{1}$$

and the total isotropic energy is

$$E_{\rm iso} = 4\pi d_L^2 S_{\rm bolo}(1+z)^{-1}. \tag{2}$$

Here, $P_{\rm bolo}$ and $S_{\rm bolo}$ are the bolometric peak flux and fluence, respectively. In the classical scenario, the presence of a jet affects the afterglow light curve which presents an achromatic break. So the observation of the afterglow allows to estimate the jet opening angle $\theta_{\rm jet}$. So after the geometric correction, the collimated energy $E_\gamma = E_{\rm iso} F_{\rm beam}$, where $F_{\rm beam} = 1 - \cos\theta_{\rm jet}$ is the beaming factor. The peak flux and fluence are given over a wide variety of observed bandpasses, and with the wide range of redshifts which correspond to different range of energy bands in the rest frame of GRBs. So the K-correction is important (Bloom et al. 2001). $P_{\rm bolo}$ and $S_{\rm bolo}$ are derived from the differential energy spectrum $\Phi(E)$ as follows:

$$P_{\rm bolo} = P \times \frac{\int_{1/(1+z)}^{10^4/(1+z)} E\Phi(E)dE}{\int_{E_{\rm min}}^{E_{\rm max}} \Phi(E)dE}, \tag{3}$$

$$S_{\rm bolo} = S \times \frac{\int_{1/(1+z)}^{10^4/(1+z)} E\Phi(E)dE}{\int_{E_{\rm min}}^{E_{\rm max}} E\Phi(E)dE}, \tag{4}$$

with P and S being the observed peak energy and fluence in units of photons/cm^2/s and erg/cm^2, respectively, and $(E_{\rm min}, E_{\rm max})$ the detection thresholds of the observing instrument. In general, the differential energy spectrum is described by band function (Band et al. 1993),

$$\Phi(E) = \begin{cases} AE^\alpha {\rm e}^{-(2+\alpha)E/E_{\rm peak}} & E \le \frac{\alpha-\beta}{2+\alpha} E_{\rm peak} \\ BE^\beta & \text{otherwise} \end{cases} \tag{5}$$

where α is the power-law index for photon energies below the break and β is the power-law index for photon energies above the break. Some differential energy spectra of GRBs also show power-law spectra plus an exponential cutoff. In the standard flat ΛCDM model, the luminosity distance d_L can be expressed as

$$d_L(z) = (1+z)\frac{c}{H_0}\int_0^z \frac{dz'}{\sqrt{\Omega_M(1+z')^3 + \Omega_\Lambda}}. \tag{6}$$

There are several intrinsic luminosity correlations of GRBs are found, such as $L_{\rm iso} - \tau_{\rm lag}$ correlation (Norris et al. 2000), $L_{\rm iso} - V$ correlation (Fenimore and Ramirez-Ruiz 2000), Amati correlation (Amati et al. 2002), Yonetoku correlation (Yonetoku et al. 2004), Ghirlanda correlation (Ghirlanda et al. 2004a), Liang-Zhang correlation (Liang and Zhang 2005) and Combo-correlation (Izzo et al. 2015).

Because GRBs cover large redshift range, the redshift-evolution effect on correlations should be discussed. It's found that the slope of Amati correlation may vary with redshift significantly using a small sample of GRBs (Li 2007). Basilakos and Perivolaropoulos (2008) found no statistically significant evidence for redshift dependence of slopes in five correlations using 69 GRBs compiled by Schaefer (2007). Wang et al. (2011b) enlarge the GRB sample and test six GRB correlations. There is no statistically significant evidence for the evolution of the luminosity correlations with redshift is found. The slopes of correlations versus redshift are all consistent with zero at the 2σ confidence level. But Lin et al. (2015) found that the Amati correlation of low-redshift GRBs differs from that of high-redshift GRBs at more than 3σ confidence level, which is insensitive to cosmological models.

2.2 Constraints on Dark Energy and Cosmological Parameters

The classical method to constrain dark energy is through its influence on the expansion history of the universe, which can be extracted from the luminosity distance $d_L(z)$ and the angular diameter distance $d_A(z)$. In addition, the weak gravitational lensing, growth of large-scale structure, and redshift space distortion can also provide useful constraints on dark energy. Theoretical models can be tested using the classical χ^2 statistic. The typical way to probe dark energy from standard candles is as follows. With luminosity distance d_L in units of megaparsecs, the theoretically predicted distance modulus is

$$\mu = 5\log(d_L) + 25. \tag{7}$$

The likelihood functions for the cosmological parameters can be determined from χ^2 statistics,

$$\chi^2(\Omega_M, \Omega_{DE}) = \sum_{i=1}^{N} \frac{[\mu_i(z_i, H_0, \Omega_M, \Omega_{DE}) - \mu_{0,i}]^2}{\sigma^2_{\mu_{0,i}}}, \tag{8}$$

where $\mu_{0,i}$ is the observed distance modulus, and $\sigma_{\mu_{0,i}}$ is the standard deviation. The confidence regions in the $\Omega_M - \Omega_{DE}$ plane can be derived through marginalizing the likelihood functions over H_0 (i.e., integrating the probability density $p \propto \exp(-\chi^2/2)$ for all values of H_0).

A lot of effort had been made to constrain cosmological parameters using GRBs since their cosmological origin was confirmed. Schaefer et al. (2003) obtained the first GRB Hubble diagram based on $L_{\rm iso} - V$ correlation, and found the mass density $\Omega_M < 0.35$ at the 1σ confidence level. After Ghirlanda et al. (2004a) found the Ghirlanda correlation, Dai et al. (2004) first used this correlation with 12 bursts and found the mass density $\Omega_M = 0.35 \pm 0.15$ at the 1σ confident level for a flat universe by assuming that some physical explanation comes into existence. Ghirlanda et al. (2004b) using 14 GRBs and SNe Ia obtained $\Omega_M = 0.37 \pm 0.10$ and $\Omega_\Lambda = 0.87 \pm 0.23$. Assuming a flat universe, the cosmological parameters were constrained to be $\Omega_M = 0.29 \pm 0.04$ and $\Omega_\Lambda = 0.71 \pm 0.05$ (Ghirlanda et al. 2004b). Wang and Dai (2006a) using the Liang-Zhang correlation to investigate the transition redshifts in different dark energy models via GRBs and SNe Ia, see also Wang and Dai (2006a). Di Girolamo et al. (2005) simulated different samples of gamma-ray bursts and found that Ω_M could be determined with accuracy $\sim 7\ \%$ with data from 300 GRBs. Meanwhile, many works have been done in this field, such as Mörtsell and Sollerman (2005), Bertolami and Silva (2006), Firmani et al. (2006), Li et al. (2008), Basilakos and Perivolaropoulos (2008), Wang (2008), Yu et al. (2009), Cardone et al. (2009), Qi and Lu (2010), Demianski and Piedipalumbo (2011), Demianski et al. (2011), and Wei et al. (2013).

Unfortunately, because of lack of low-z GRBs, the luminosity correlations has been obtained only from moderate-z GRBs. So the correlations are cosmology-dependent, i.e., the isotropic energy $E_{\rm iso}$, collimation-corrected energy E_γ, and luminosity $L_{\rm iso}$ are as functions of cosmological parameters. This is the so-called "circularity problem" of GRBs. In the following, we discuss different methods to overcome this problem.

The first method is fitting the cosmological parameters and luminosity correlation simultaneously. Schaefer (2007) used 69 GRBs and five correlations to constrain cosmological parameters. Wang et al. (2007) also used 69 GRBs and other cosmological probes to constrain cosmological parameters. They make simultaneous use of five luminosity indicators, which are correlations of $L_{\rm iso} - \tau_{\rm lag}$, $L_{\rm iso} - V$, $E_{\rm peak} - L_{\rm iso}$, $E_{\rm peak} - E_\gamma$, and $\tau_{\rm RT} - L_{\rm iso}$. After obtaining the distance modulus of each burst using one of these correlations, the real distance modulus can be calculated,

$$\mu_{\rm fit} = \Big(\sum_i \mu_i/\sigma_{\mu_i}^2\Big)\Big/\Big(\sum_i \sigma_{\mu_i}^{-2}\Big), \tag{9}$$

where the summation runs from 1–5 over the correlations with available data, μ_i is the best estimated distance modulus from the i-th relation, and σ_{μ_i} is the corresponding uncertainty. The uncertainty of the distance modulus is

$$\sigma_{\mu_{\rm fit}} = \Big(\sum_i \sigma_{\mu_i}^{-2}\Big)^{-1/2}. \tag{10}$$

When calculating constraints on cosmological parameters and dark energy, the normalizations and slopes of the five correlations are marginalized. The marginalization method is to integrate over some parameters for all of its possible values. The χ^2 value is

$$\chi^2_{\rm GRB} = \sum_{i=1}^{N} \frac{[\mu_i(z_i, H_0, \Omega_M, \Omega_{DE}) - \mu_{{\rm fit},i}]^2}{\sigma^2_{\mu_{{\rm fit},i}}}, \tag{11}$$

where $\mu_{{\rm fit},i}$ and $\sigma_{\mu_{{\rm fit},i}}$ are the fitted distance modulus and its error. In addition to GRBs, SNe Ia, CMB, BAO, X-ray gas mass fraction in galaxy clusters and growth rate data are also ideal cosmological probes. In Fig. 1, the constraint on the ΛCDM model is shown. Different color contours represent constraints from different data. The best fitted parameters are consistent with a flat geometry. Li et al. (2008) also performed a Markov Chain Monte Carlo (MCMC) global fitting analysis to overcome the circularity problem. The Ghirlanda correlation and 27 GRBs are used. They treated the slopes of Ghirlanda correlation and cosmological parameters as free parameters and determine them simultaneously through MCMC analysis on GRB data together with other observational data, such as SNe Ia, CMB and large-scale structure (LSS). Amati et al. (2008) measured the cosmological parameters using Amati correlation using global fitting method. The extrinsic scatter was assumed on the parameter of $E_{\rm peak}$, but the cosmological-dependent value is $E_{\rm iso}$ (Ghirlanda 2009).

The second method is to calibrate the correlations of GRBs using SNe Ia data at low redshifts. The principle of this method is that objects at the same redshift should have the same luminosity distance in any cosmology model. Therefore, the luminosity distance at any redshift in the redshift range of GRBs can be obtained by interpolating (or by other approaches) directly from the SNe Ia Hubble diagram. Then if further assuming these calibrated GRB correlations to be valid for all long GRB data, the standard Hubble diagram method to constrain the cosmological parameters from the GRB data at high redshifts obtained by utilizing the correlations. Liang et al. (2008) first calibrated the GRB correlations

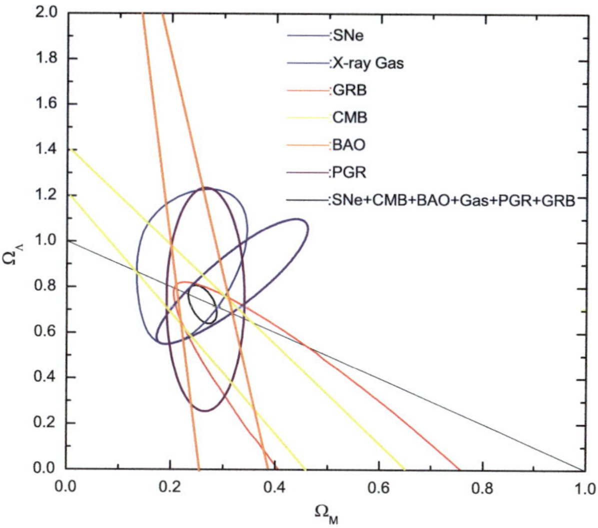

Fig. 1 Joint confidence intervals of 1σ for $(\Omega_M, \Omega_\Lambda)$ from the observational datasets. The *thick black line contour*, the *blue contour*, the *red contour*, the *yellow contour*, the *violet contour*, the *orange contour*, and the *purple contour* show constraints from all the datasets, 26 galaxy clusters, 69 GRBs, CMB shift parameter, 182 SNe Ia, BAO, and 2dF Galaxy Redshift Survey, respectively. The *thin solid line* represents a flat universe (adapted from Fig. 2 in Wang et al. 2007)

using an interpolation method. The error of distance modulus of linear interpolation can be calculated as (Liang et al. 2008; Wei 2010; Wei et al. 2015)

$$\sigma_\mu = \left(\left[(z_{i+1} - z)/(z_{i+1} - z_i)\right]^2 \epsilon_{\mu,i}^2 + \left[(z - z_i)/(z_{i+1} - z_i)\right]^2 \epsilon_{\mu,i+1}^2\right)^{1/2}, \tag{12}$$

where $\epsilon_{\mu,i}$ and $\epsilon_{\mu,i+1}$ are errors of the SNe Ia, μ_i and μ_{i+1} are the distance moduli of the SNe Ia at z_i and z_{i+1} respectively. Similar to the interpolation method, Cardone et al. (2009) constructed an updated GRBs Hubble diagram on six correlations calibrated by local regression from SNe Ia. Kodama et al. (2008) presented that the $L_{\rm iso}$–$E_{\rm peak}$ correlation can be calibrated with the empirical formula fitted from the luminosity distance of SNe Ia.

However, it must be noted that this calibration procedure depends seriously on the choice of the formula and various possible formulae can be fitted from the SNe Ia data that could give different calibration results of GRBs. As the cosmological constraints from GRBs are sensitive to GRBs calibration results (Wang 2008), the reliability of this method should be tested carefully. Moreover, as pointed out by Wang (2008), the GRB luminosity correlations which are calibrated by this way are no longer completely independent of all the SNe Ia data points. Therefore these GRB data can not be used to directly combine with the whole SNe Ia dataset to constrain cosmological parameters and dark energy. In order to search a unique expression of the fitting formula, Wang et al. (2009) used the cosmographic parameters (Capozziello and Izzo 2008; Vitagliano et al. 2010; Xia et al. 2012; Gao et al. 2012). The luminosity distance can be expanded as (Visser 2004)

$$d_L = \frac{c}{H_0}\left\{z + \frac{1}{2}(1 - q_0)z^2 - \frac{1}{6}\left(1 - q_0 - 3q_0^2 + j_0\right)z^3 \right.$$
$$\left. + \frac{1}{24}\left[2 - 2q_0 - 15q_0^2 - 15q_0^3 + 5j_0 + 10q_0 j_0 + s_0\right]z^4 + O\left(z^5\right)\right\}, \tag{13}$$

where q is the deceleration parameter, j is the so-called "jerk", and s is the so-called "snap" parameter. These quantities are defined as

$$q = -\frac{1}{H^2}\frac{\ddot{a}}{a}; \tag{14}$$

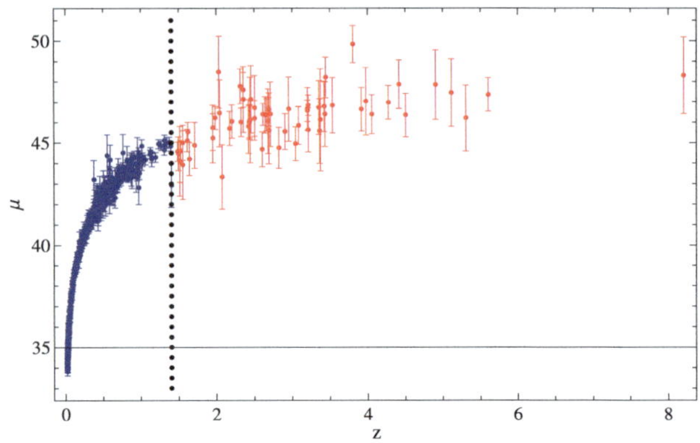

Fig. 2 The Hubble diagram of 557 SNe Ia (*blue*) and 66 high-redshift GRBs (*red*) (adapted from Fig. 2 in Wang and Dai 2011b)

$$j = \frac{1}{H^3}\frac{\dddot{a}}{a}; \tag{15}$$

$$s = \frac{1}{H^4}\frac{\ddddot{a}}{a}. \tag{16}$$

Equation (13) is only dependent on the cosmological principles and FRW metric, so the expansion is model-independent. But the Taylor-expansion of d_L is not valid at $z > 1$. So expansion d_L as a function of $y = z/1 + z$ is much useful (Wang and Dai 2011b). The calibrated Hubble diagram of GRBs is shown in Fig. 2 (Wang and Dai 2011b).

The third method is to calibrate the standard candles using GRBs in a narrow redshift range (δz) near a fiducial redshift (Liang and Zhang 2006; Ghirlanda et al. 2006). Liang and Zhang (2006) proposed a procedure to calibrate the Liang-Zhang correlation with a sample of GRBs in a narrow redshift range. No low-redshift GRB sample is needed in this method. The calibration procedure can be described as follows. First, calibrate the power-law index of Liang-Zhang correlation using a sample of GRBs that satisfy this correlation and are distributed in a narrow redshift range. The power-law index can be derived using a multiple regression method. Second, marginalize the coefficient value over a reasonable range.

However, the gravitational lensing by random fluctuations in the intervening matter distribution induces a dispersion in GRB brightness (Oguri and Takahashi 2006; Schaefer 2007), degrading their value as standard candles as well as SNe Ia (Holz 1998). GRBs can be magnified (or reduced) by the gravitational lensing produced by the structure of the Universe. The gravitational lensing has sometimes a great impact on high-redshift GRBs. First, the probability distribution functions (PDFs) of gravitational lensing magnification have much higher dispersions and are markedly different from the Gaussian distribution (Valageas 2000; Oguri and Takahashi 2006; Wang and Dai 2011b). Figure 3 shows the magnification probability distribution functions of gravitational lensing at different redshifts (Wang and Dai 2011b). Second, there is effectively a threshold for the detection in the burst apparent brightness. With gravitational lensing, bursts just below this threshold might be magnified in brightness and detected, whereas bursts just beyond this threshold might be reduced in brightness and excluded. Wang and Dai (2011b) considered the weak lensing effect on cosmological parameters derived from GRBs, and found that the most probable value of the observed matter density Ω_M is slightly lower than its actual value, see Fig. 4. The weak gravitational lensing also affects the dark energy equation of state by shifting it to a more negative value.

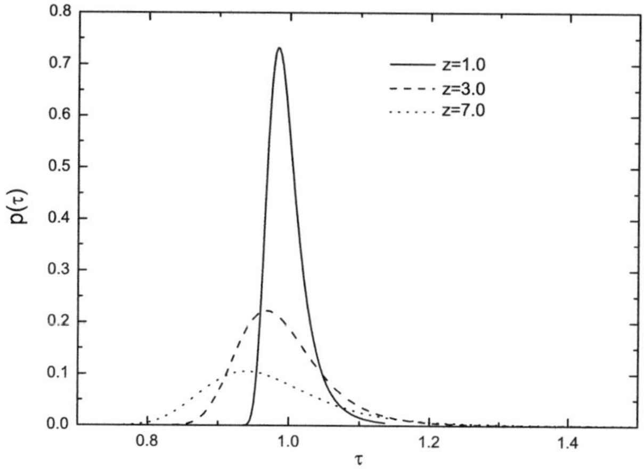

Fig. 3 Magnification probability distribution functions of gravitational lensing at redshifts $z = 1$, $z = 3$ and $z = 7$ (adapted from Fig. 5 in Wang and Dai 2011b)

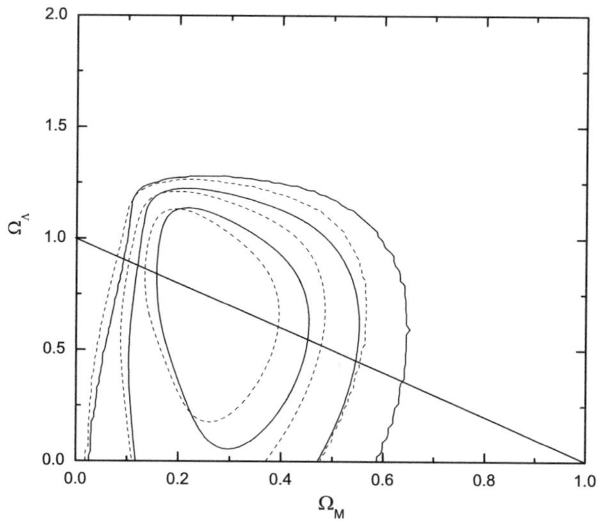

Fig. 4 Confidence contours of likelihood from 1σ to 3σ in the ΛCDM model. The *black line contours* from 116 GRBs and the *dotted contours* from 116 GRBs including magnification bias (adapted from Fig. 7 in Wang and Dai 2011b)

2.3 The Equation of State of Dark Energy

The dark energy equation of state w is the most important parameter that describes the properties of dark energy. Whether and how it evolves with time is crucial for revealing the physics of dark energy. GRBs can provide the high-redshift evolution property of dark energy. The procedure is to bin w in z, and fit the w in each bin to observational data by assuming that w is constant in each bin. w_i is the EOS parameter in the ith redshift bin defined by an upper boundary at z_i, and the zeroth bin is defined as $z_0 = 0$. Qi et al. (2008a) used GRBs and other cosmological observations to construct evolution of the equation of state, and found that the equation of state w is consistent with the cosmological constant (also see Qi et al. 2008b). The confidence interval of the uncorrelated equation of state parameter can be significantly reduced by adding GRBs. After calibrating the GRB correlations using cosmographic parameters, Wang and Dai (2011b) found that the high-redshift ($1.4 < z < 8.2$) equation of state is consistent with the cosmological constant. But some studies found that the equation of state w may deviate from -1 (i.e., Qi et al. 2009; Zhao et al. 2012). In light

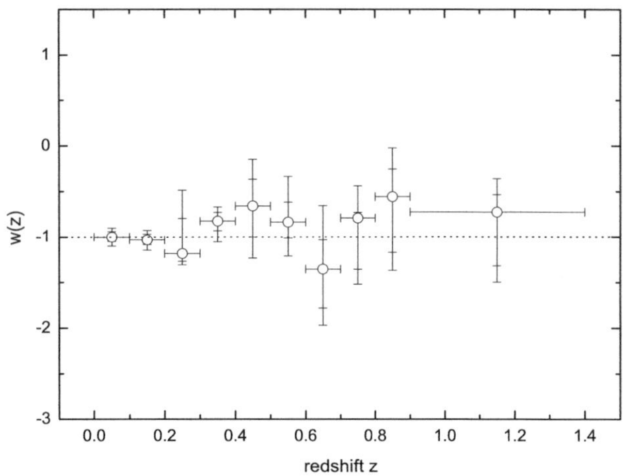

Fig. 5 Estimation of the uncorrelated dark energy EOS parameters at different redshift bins ($w_1, w_2, \ldots, w_{10}$) from SNe Ia+BAO+WMAP9+H(z)+GRB data. The *open points* show the best fit value. The *error bars* are 1σ and 2σ confidence levels. The *dotted line* shows the cosmological constant (adapted from Fig. 3 in Wang and Dai 2014a)

of the Planck CMB data, Wang and Dai (2014a) found that the EOS is consistent with the cosmological constant at the 2σ confidence level, not preferring to a dynamical dark energy, which is shown in Fig. 5.

3 The Star Formation Rate and GRBs

3.1 Star Formation Rate Derived from GRBs

The association of long GRBs with core-collapse supernovae has been confirmed from observations in recent years (Stanek et al. 2003; Hjorth et al. 2003), which provides a complementary technique for measuring the high-redshift SFR (Totani 1997; Wijers et al. 1998; Lamb and Reichart 2000; Porciani and Madau 2001; Bromm et al. 2002). The selection effects should be considered (for a review, see Coward 2007). But one crucial problem appears, i.e., how to calibrate the GRB event rate to the SFR. The luminosity function may play an important role (Natarajan et al. 2005; Daigne et al. 2006; Salvaterra and Chincarini 2007; Salvaterra et al. 2009; Campisi et al. 2010; Wanderman and Piran 2010; Cao et al. 2011). Before the launch of *Swift* (Gehrels et al. 2004), the luminosity function is determined by fitting the observed $\log N - \log P$ distribution (Schmidt 1999; Porciani and Madau 2001; Guetta et al. 2005; Natarajan et al. 2005). Thanks to the *Swift*, more redshifts of GRBs are measured. This makes it possible to give more information on the luminosity function (Wanderman and Piran 2010; Cao et al. 2011; Tan et al. 2013). Because the form of luminosity function should be assumed and the model parameters of luminosity function is degenerate, it is not easy to determine the luminosity function. A straightforward way to estimate the luminosity function is proposed by Lynden-Bell (1971) and then further developed by Efron and Petrosian (1992). This method has been used for GRBs (Lloyd-Ronning et al. 2002; Yonetoku et al. 2004; Wu et al. 2012). They found that the GRB rate is tracing SFR in a wide redshift range. But Yu et al. (2015) for the first time found that the GRB rate shows an unexpectedly low-redshift excess comparing to the observed SFR. In Fig. 6, the blue stepwise line represents the comoving cosmic formation of GRBs as a function of redshift, and the error bar gives the 1σ confidence level. The best-fitting power laws for different

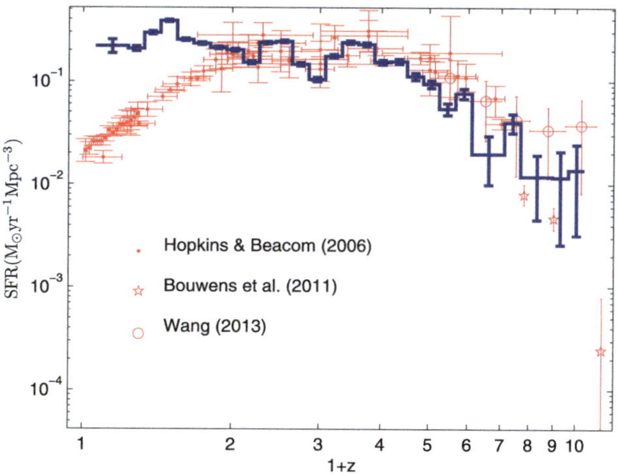

Fig. 6 The comparison between GRB formation rate $\rho(z)$ (*blue*) and the observed SFR. The SFR data are taken from Hopkins and Beacom (2006), which are shown as *red dots*. The SFR data from Bouwens et al. (2011) (*stars*) and Wang (2013) (*open circles*) are also used

segments are

$$\rho(z) \propto \begin{cases} (1+z)^{0.02\pm1.47} & z < 1.0, \\ (1+z)^{-0.35\pm0.70} & 1.0 < z < 4.0, \\ (1+z)^{-3.04\pm1.53} & z > 4.0. \end{cases} \tag{17}$$

The error bars are in the 95 % confidence level. This result has been confirmed by later study by Petrosian et al. (2015). There are some possible reasons for this low-redshift excess.

The first one is that the definition of long GRBs is not clear. In classical method, the long GRBs are defined by $T_{90} > 2$ s (Kouveliotou et al. 1993). There is no clear boundary line in this diagram to separate the long and short GRBs. Moreover, T_{90} is an observed time scale, which represents different time for GRBs at different redshifts. Meanwhile, the observations of low-redshift long GRBs, such as GRB 060614 at $z = 0.125$ and GRB 060505 at $z = 0.089$, show no association of supernovae (Gal-Yam et al. 2006; Gehrels et al. 2006). So more physical criterions are required to classify GRBs. Because only a subclass of GRBs can trace the SFR. Some attempts have been performed (Zhang 2006; Zhang et al. 2009; Bloom et al. 2008). It has been suggested that GRBs can be classified physically into Type I (compact star origin) and Type II (massive star origin) (Zhang 2006, 2007).

The second one is that some selection effects have not been included in analysis. For example, it is easier to measure the redshift of those GRBs which are in lower redshift and therefore create a bias toward low redshift GRBs. It means that we lose some high redshift GRBs, so the formation rate of GRBs at low redshift we calculated will larger than the SFR.

The third one is that there may exist a subclass GRBs, i.e., low-luminosity GRBs (Cobb et al. 2006; Pian et al. 2006; Soderberg et al. 2006; Liang et al. 2007). The local rate of low-luminosity GRBs may be high, i.e., $\rho(0) = 100$–$1000\ \mathrm{yr}^{-1}\,\mathrm{Gpc}^{-3}$ (Soderberg et al. 2006; Liang et al. 2007), much higher than high-luminosity GRBs. The progenitors of low-luminosity GRBs may be different with those of high-luminosity GRBs (Mazzali et al. 2006; Soderberg et al. 2006). The contamination from low-luminosity GRBs could lead to the low-redshift excess.

The expected redshift distribution of GRBs is

$$\frac{dN}{dz} = F(z)\frac{\varepsilon(z)\dot{\rho}_*(z)}{\langle f_{\mathrm{beam}}\rangle}\frac{dV_{\mathrm{com}}/dz}{1+z}, \tag{18}$$

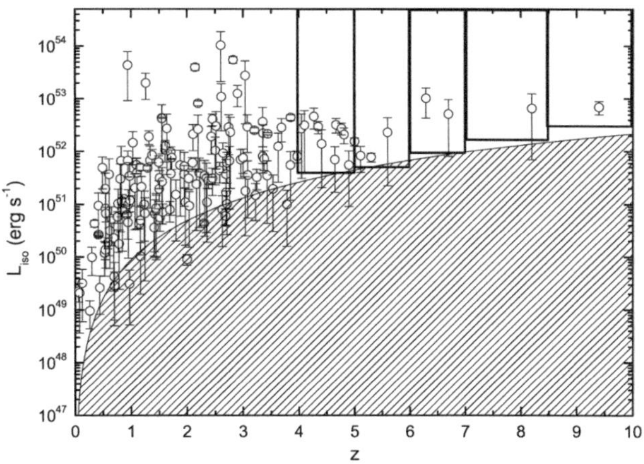

Fig. 7 Distribution of the isotropic-equivalent luminosity for 157 long-duration *Swift* GRBs. The *shaded area* approximates the detection threshold of *Swift* BAT (adapted from Fig. 1 in Wang 2013)

where $F(z)$ represents the ability to obtain the redshift, $\varepsilon(z)$ accounts for the fraction of stars producing GRBs, and $\dot{\rho}_*(z)$ is the SFR density. The $F(z)$ can be treated as constant when we consider the bright bursts with luminosities sufficient to be detected within an entire redshift range. GRBs that are unobservable due to beaming are accounted for through $\langle f_{\rm beam}\rangle$. The $\varepsilon(z)$ can be parameterized as $\varepsilon(z) = \varepsilon_0(1+z)^\delta$, where ε_0 is an unknown constant that includes the absolute conversion from the SFR to the GRB rate in a given GRB luminosity range. Kistler et al. (2008) found the index $\delta = 1.5$ from 63 *Swift* GRBs. A little smaller value $\delta \sim 0.5$–1.2 has been inferred from update *Swift* GRBs (Kistler et al. 2009; Wang 2013). In a flat universe, the comoving volume is calculated by

$$\frac{dV_{\rm com}}{dz} = 4\pi D_{\rm com}^2 \frac{dD_{\rm com}}{dz}, \tag{19}$$

where the comoving distance is

$$D_{\rm com}(z) \equiv \frac{c}{H_0}\int_0^z \frac{dz'}{\sqrt{\Omega_m(1+z')^3 + \Omega_\Lambda}}. \tag{20}$$

In the calculations, the ΛCDM model with $\Omega_m = 0.27$, $\Omega_\Lambda = 0.73$ and $H_0 = 71\ {\rm km\,s^{-1}\,Mpc^{-1}}$ from the WMAP seven-year data is used (Komatsu et al. 2011).

Figure 7 shows the isotropic luminosity distribution of 157 *Swift* GRBs. The isotropic luminosity can be obtained by

$$L_{\rm iso} = E_{\rm iso}(1+z)/T_{90}, \tag{21}$$

where T_{90} is the duration time. The shaded area approximates the detection threshold of *Swift* BAT, which has a flux limit $\sim F_{\rm lim} = 1.2 \times 10^{-8} {\rm erg\,cm^{-2}\,s^{-1}}$. So the selection effect is important. In order to exclude faint low -redshift GRBs that could not be visible at high redshifts, we only select luminous bursts. The luminosity cut $L_{\rm iso} > 10^{51}\ {\rm erg\,s^{-1}}$ is chosen in the redshift bin 0–4 (Yüksel et al. 2008), which removes many low-redshift, low-luminosity bursts that could not be detected at higher redshift. The cumulative distribution of GRB redshift can be expressed as

$$\frac{N(<z)}{N(<z_{\max})} = \frac{N(0,z)}{N(0,z_{\max})}. \tag{22}$$

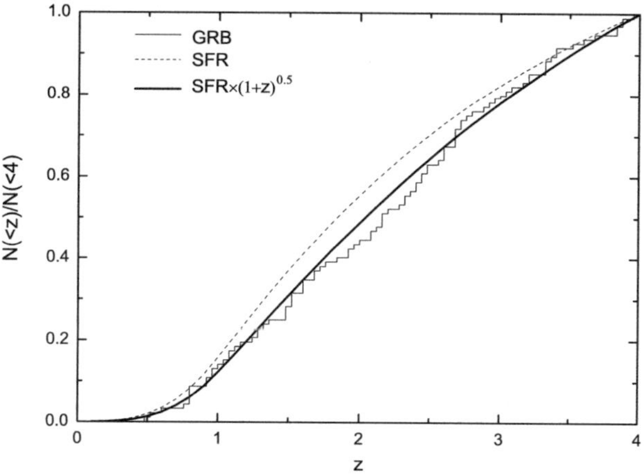

Fig. 8 Cumulative distribution of 92 *Swift* GRBs with $L_{\rm iso} > 10^{51}$ erg s^{-1} in $z = 0$–4 (*stepwise solid line*). The *dashed line* shows the GRB rate inferred from the star formation history of Hopkins and Beacom (2006). The *solid line* shows the GRB rate inferred from the star formation history including $(1+z)^{0.5}$ evolution (adapted from Fig. 2 in Wang 2013)

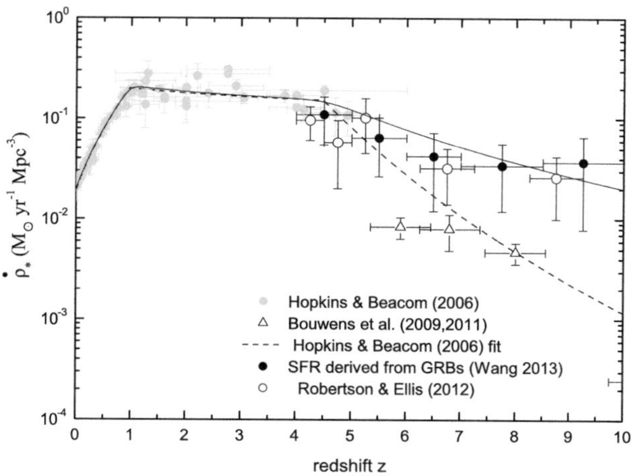

Fig. 9 The cosmic star formation history. The *gray points* are taken from Hopkins and Beacom (2006), the *dashed line* shows their fitting result. The *triangular points* are from Bouwens et al. (2009, 2011). The *open circles* are taken from Robertson and Ellis (2012). The *filled circles* are the SFR derived from GRBs in Wang (2013) (adapted from Fig. 3 in Wang 2013)

The value of $z_{\max}$ is taken as 4.0. Because the SFR has been well measured at $z < 4.0$ (Hopkins and Beacom 2006). The theory predicted and observed cumulative GRB distributions is shown in Fig. 8. The Kolmogorov-Smirnov statistic gives the minimization for $\delta = 0.5$ (Wang 2013).

The derived SFR from GRBs are shown as filled circles in Fig. 9. Error bars correspond to 68 % Poisson confidence intervals for the binned events (Gehrels 1986).

The formation rate of short GRBs is also extensively investigated. From the host galaxy and afterglow observations, it is believed that short GRBs originate from neutron star-neutron star or neutron star-black hole mergers (Berger 2014). So there is a delay time between the SFR and the merger rate, because the stellar evolution of binary stars and the binary orbits to spiral in. The delay time is about 3.0 Gyr (Wanderman and Piran 2015). The coalescence of a neutron star-neutron star is the mainly source of gravitational waves (GW). The simultaneous detection of a short GRB with a GW signal can constrain the nature of short GRBs. The approximate detection horizon of neutron star-neutron star merger by Advanced LIGO/Virgo and other planned advanced GW detectors is about 300 Mpc.

3.2 Possible Origins of High-Redshift GRB Rate Excess

Recent studies show that the rate of GRBs does not strictly follow the SFH but is actually enhanced by some mechanism at high redshift (Le and Dermer 2007; Salvaterra and Chincarini 2007; Kistler et al. 2008; Yüksel et al. 2008; Wang and Dai 2009; Robertson and Ellis 2012; Wang 2013). The SFR inferred from the high-redshift ($z > 6$) GRBs seems to be too high in comparison with the SFR obtained from some high-redshift galaxy surveys (Bouwens et al. 2009, 2011).

3.2.1 Metallicity Evolution

A natural origin of the high-redshift GRB rate excess is the metallicity evolution. Theory and observation both support that long GRBs prefer to occurring in low-metallicity environment. Some theoretical studies of long GRBs progenitors using stellar evolution models suggest that low metallicity may be a necessary condition for a long GRB to occur. For popular collapse models of long GRBs, stars with masses $> 30M_\odot$ can be able to create a black hole (BH) remnant (Woosley 1993; Hirschi et al. 2005). The preservation of high angular momentum and high-stellar mass at the time of collapse (Woosley 1993; MacFadyen and Woosley 1999) is crucial for producing a relativistic jet and high luminosity. Low-metallicity (0.1–$0.3Z_\odot$) progenitors can theoretically retain more of their mass due to smaller line-driven stellar winds (Kudritzki and Puls 2000; Vink and de Koter 2005), and hence preserve their angular momentum (Yoon and Langer 2005; Yoon et al. 2006), because the wind-driven mass loss of massive stars is proportional to the metallicity. Observations of long GRB host galaxies also show that they are typically in low metallicity environment, for several local long GRB host galaxies (Sollerman et al. 2005; Stanek et al. 2006), as well as in distant long GRB hosts (i.e., Fruchter et al. 2006; Prochaska et al. 2007).

Li (2008) studied the possibility of interpreting the observed discrepancy between the GRB rate history and the star formation rate history using cosmic metallicity evolution (Kistler et al. 2008). Under the assumption that the formation of long GRBs follows the cosmic star formation history and form preferentially in low-metallicity galaxies, the rate of GRB is given by

$$R_{\mathrm{GRB}}(z) = k_{\mathrm{GRB}} \Sigma(Z_{\mathrm{th}}, z) \rho_*(z), \tag{23}$$

where k_{GRB} is the GRB formation efficiency, $\Sigma(Z_{\mathrm{th}}, z)$ is the fraction of galaxies at redshift z with metallicity below Z_{th} (Langer and Norman 2006) and $\rho_*(z)$ is the observed SFR. The function $\Sigma(Z_{\mathrm{th}}, z)$ is (Langer and Norman 2006)

$$\Sigma(Z_{\mathrm{th}}, z) = \frac{\hat{\Gamma}[\alpha_1 + 2, (Z_{\mathrm{th}}/Z_\odot)^2 10^{0.15\beta z}]}{\Gamma(\alpha_1 + 2)}, \tag{24}$$

where $\hat{\Gamma}$ and Γ are the incomplete and complete gamma functions, $\alpha_1 = -1.16$ and $\beta = 2$ (Savaglio et al. 2005). Li (2008) found that the distribution of luminosity and cumulative distribution of redshift could be well fitted if $Z_{\mathrm{th}} = 0.3Z_\odot$ is adopted. Wang and Dai (2009) studied the high-redshift SFR by considering the GRBs tracing the star formation history and the cosmic metallicity evolution. They found the SFR derived from GRBs is marginal consistent with that from traditional way (i.e., Hopkins and Beacom 2006). Wei et al. (2014) examined the influence on the GRB distribution due to the background cosmology, i.e., $R_h = ct$ Universe. However, a few GRB hosts with high metallicity are observed

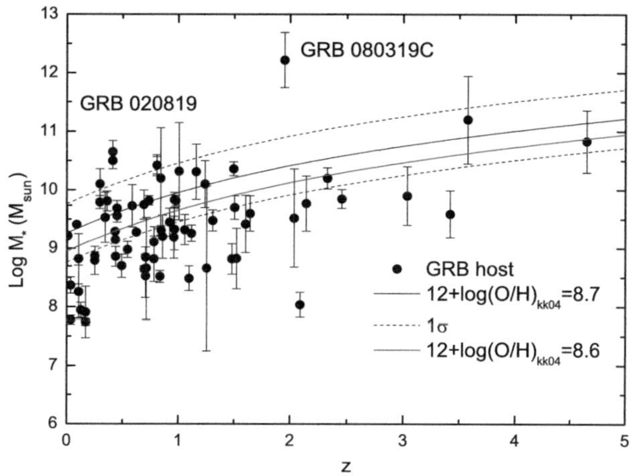

Fig. 10 GRB host galaxy mass distribution. The *solid lines* represent the *upper limits* of the stellar mass of a GRB host galaxy given a metallicity cutoff of $12 + \log(\mathrm{O/H})_{\mathrm{KK04}} = 8.7$ (*black*), and $12 + \log(\mathrm{O/H})_{\mathrm{KK04}} = 8.6$ (*red*). The *dashed lines* represent the 1σ scatter (adapted from Fig. 4 in Wang and Dai 2014b)

(i.e. GRB 020819), so that the role of metallicity in driving the GRB phenomena remains unclear and it is still debated (Price et al. 2007; Wolf and Podsiadlowski 2007; Kocevski et al. 2009; Graham et al. 2009; Svensson et al. 2010). For excellent reviews, see Fynbo et al. (2012) and Levesque (2014). But there are some uncertainties when measure the metallicities of GRBs' explosion region at high-redshifts, such as chemical inhomogeneity (Levesque et al. 2010; Niino 2011). Wang and Dai (2014b) studied the metallicity role from two aspects, the GRB host galaxies and redshift distribution. They found that the observed GRB host galaxy masses and the cumulative redshift distribution can fit the predicted distributions well if GRBs occur in low-metallicity $12 + \log(\mathrm{O/H})_{\mathrm{KK04}} < 8.7$, which is shown in Fig. 10. Trenti et al. (2015) found that there is clear evidence for a relation between SFR and GRB (Jimenez and Piran 2013). But a sharp cut-off of metallicity is ruled out.

3.2.2 Evolving Star Initial Mass Function

Wang and Dai (2011b) proposed that the GRB rate excess may be due to the evolution of star initial mass function (IMF), also see Xu and Wei (2008). Because an "top-heavy" IMF will lead to more massive stars at high-redshift which can result in much more GRBs. Considering long GRBs trace SFR, the rate of GRBs in an evolving IMF is

$$R_{\mathrm{GRB}} \propto \frac{N_{m>30M_\odot}}{V} = K\left(\frac{c}{H_0}\right)^{-3} \frac{\int_{30M_\odot}^{m_l} \xi(m)d\log m}{\int_{m_s}^{m_l} m\xi(m)d\log m}\rho_*(z), \tag{25}$$

where K is a constant to be constrained and R_{GRB} is the rate of GRBs, representing the number of GRBs per unit time per unit volume at redshift z. The evolving IMF proposed by Davé (2008) is

$$\frac{dN}{d\log m} = \xi(m) \propto \begin{cases} m^{-0.3} & \text{for } m < \hat{m}_{\mathrm{IMF}} \\ m^{-1.3} & \text{for } m > \hat{m}_{\mathrm{IMF}}, \end{cases} \tag{26}$$

where $\hat{m}_{\mathrm{IMF}} = 0.5(1+z)^2 M_\odot$, which has been constrained by requiring non-evolving star formation activity parameter. Figure 11 shows that the observed cumulative distribution of GRBs can be well produced by this model.

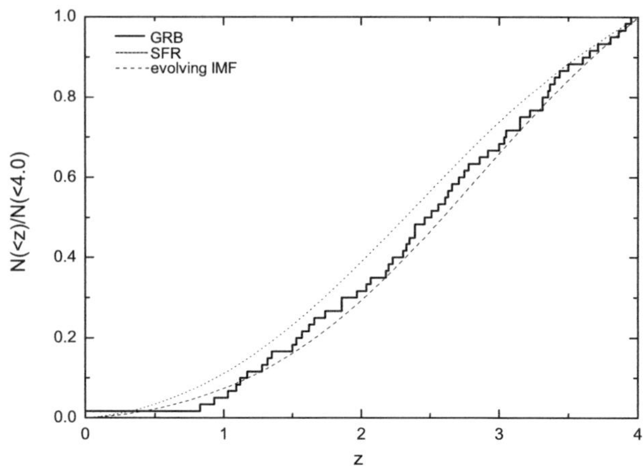

Fig. 11 The cumulative distribution of 72 *Swift* long GRBs with $L_{iso} > 0.8 \times 10^{51}$ erg s^{-1} (*stepwise solid line*). The *dotted line* shows the GRB rate inferred from the star formation history of Hopkins and Beacom (2006). The *dashed line* shows the GRB rate inferred from star formation history including an evolving IMF (adapted from Fig. 2 in Wang and Dai 2011b)

3.2.3 Evolving Luminosity Function Break

Virgili et al. (2011) found that if the break of luminosity function evolves with redshift, the distributions of luminosity, redshift and peak photon flux from the BATSE and Swift data can be reproduced from simulations. The break luminosity function evolution can be in a moderate way $\propto L_b \times (1+z)^{\sim 0.8-1.2}$. Campisi et al. (2010) studied the luminosity function, the rate of long GRBs at high redshift, using high-resolution N-body simulations. A strongly evolving luminosity function with no metallicity cut may well explain the $\log N - \log P$ distribution of BATSE and Swift data.

3.2.4 Superconducting Cosmic String

Cosmic strings are thought to be linear topological defects that could be formed at a phase transition in very early Universe. By considering that high-redshift GRBs 080913 and 090423 are electromagnetic bursts of superconducting cosmic strings, Cheng et al. (2010) showed the high-redshift GRB excess can be reconciled. But Wang et al. (2011a) claimed that GRBs from cosmic string have a very small angle, about 10^{-3}, which could be in contradiction with the opening angle of the GRB outflow. Cheng et al. (2011) pointed out that the angle is not the opening angle of the GRB outflow, but is just the collimation angle of the radiation of the corresponding string segment. We must caution that the existence of cosmic string is only speculative.

4 Probing the Pop III Stars and High-Redshift IGM

4.1 Observational Signature of Pop III GRBs

The first stars, also called Population III (Pop III) stars, are predicted to have formed in minihaloes with virial temperatures $T_{vir} \leq 10^4$ K at $z \geq 15$ (Tegmark et al. 1997; Yoshida et al. 2003; Bromm and Larson 2004). Numerical simulations show that Pop III stars forming in primordial minihaloes, were predominantly very massive stars with typical masses $M_* \geq 100 M_\odot$ (Bromm et al. 1999, 2002), for recent reviews, see Bromm et al. (2009). They had likely played a crucial role in early universe evolution, including reionization, metal enrichment history. Some studies shows that some Pop III stars will end as GRBs, called Pop III GRBs (Heger et al. 2003; Bromm and Loeb 2006; Komissarov and Barkov 2010; Stacy et al. 2011), which will be brighter and more energetic than any GRB yet detected (Toma et al. 2011; Nagakura et al. 2011; Campisi et al. 2011; Mészáros and Rees 2010; Nakauchi et al. 2012). Direct observations of the Pop III stars have so far been out of reach. The properties of Pop III stars may be revealed by their remanents, Pop III GRBs.

In particular, the afterglow emission of Pop III GRBs depend on the circumburst density (Ciardi and Loeb 2000; Gou et al. 2004; Wang et al. 2012). In particular, the central minihalo environments just before a massive star die and a GRB bursts out can be understood as follows. The number of ionizing photons depends strongly on the central stellar mass, which is determined by a accretion flow onto the growing protostar (e.g., McKee and Tan 2008; Hosokawa et al. 2011; Stacy et al. 2012). Meanwhile the accretion is also affected by this radiation field. So the assembly of the Pop III stars and the development of an H II region around them proceed simultaneously, and affect each other. The shallow potential wells of minihalos are unable to maintain photo-ionized gas, so that the gas is effectively blown out of the minihalo. The resulting photo-evaporation has been studied (Alvarez et al. 2006; Abel et al. 2002; Greif et al. 2009).

The photoevaporation from minihalos can be described as the self-similar solution for a champagne flow (Shu et al. 2002). Assuming a $\rho \propto r^{-2}$ density profile, the spherically symmetric continuity and Euler equations for isothermal gas can be described as follows:

$$\left[(v-x)^2-1\right]\frac{1}{\alpha}\frac{d\alpha}{dx}=\left[\alpha-\frac{2}{x}(x-v)\right](x-v), \tag{27}$$

$$\left[(v-x)^2-1\right]\frac{dv}{dx}=\left[(x-v)\alpha-\frac{2}{x}\right](x-v), \tag{28}$$

where $x = r/c_s t$, and $\rho(r,t) = \alpha(x)/4\pi G t^2 = m_{\rm H} n(r)/X$ and $u(r,t) = c_s v(x)$ are the reduced density and velocity, respectively. c_s is the sound speed and $X = 0.75$ the hydrogen mass fraction. We set the typical lifetime of a massive Pop III star as $t = t_* \simeq 3 \times 10^6$ yr.

In the left panel of Fig. 12, we show the density profiles at the end of the Pop III progenitor's life in the minihalo case. The circumstellar densities are nearly uniform at small radii. Such a flat density profile is markedly different from that created by stellar winds. But in the atomic cooling halo case, star formation and radiative feedback is not well understood (Johnson et al. 2009), such as the masses of stars, and stellar multiplicity (Clark et al. 2011). So we also use the formalism of the Shu solution as above. We assume that either one Pop III star or a small stellar cluster forms. The densities are shown in the right panel of Fig. 12. Similar to the minihalo case, densities are nearly constant at small radii, but overall values are much higher, which is due to the deep potential wells, so that photoheated gas can easily be retained. Typical circumburst densities are $n \sim 100$ cm^{-3}. Pop III GRBs originating in atomic cooling halos may be extremely bright.

The typical parameters of the afterglow emission are adopted, $\Gamma_0 = 300$, $E_{\rm iso} = 10^{53}$ erg, $\Delta_0 = 10^{12}$ cm, $\epsilon_e = 0.3$, $\epsilon_B = 0.1$, and $p = 2.5$. As an example, in Fig. 13, the M-band ($\nu = 6.3 \times 10^{13}$ Hz) light curve is shown. Figure 14 gives the observed flux at $\nu = 1.36 \times 10^{14}$ Hz as a function of redshift in the minihalo case. The lines with filled dots, black triangles and open dots correspond to an observed time of 6 minutes, 1 hour, and 1 day respectively. The straight line marks the K-band sensitivity for the near-infrared spectrograph

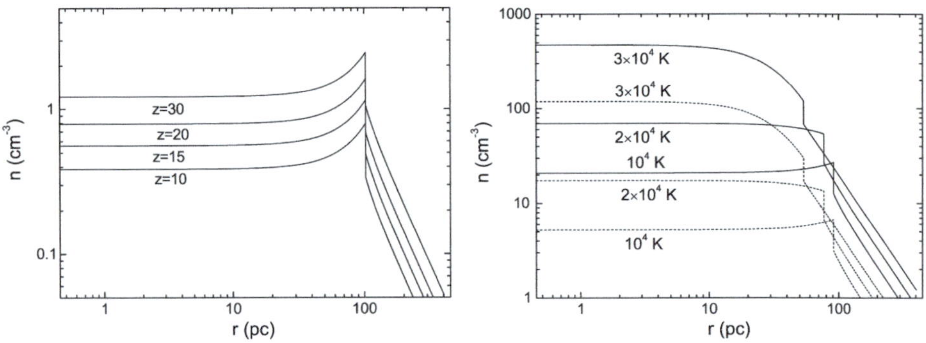

Fig. 12 *Left panel*: minihalo circumburst density. Shown is the hydrogen number density as a function of distance from the central Pop III star at the moment of its death. Typical circumburst densities are ~ 1 cm^{-3}. *Right panel*: atomic-cooling-halo circumburst density. The density profiles are calculated from the Shu solution. The case of photoheating from only a single Pop III star (*solid lines*), and that from a stellar cluster (*dotted lines*) (adapted from Figs. 1 and 2 in Wang et al. 2012)

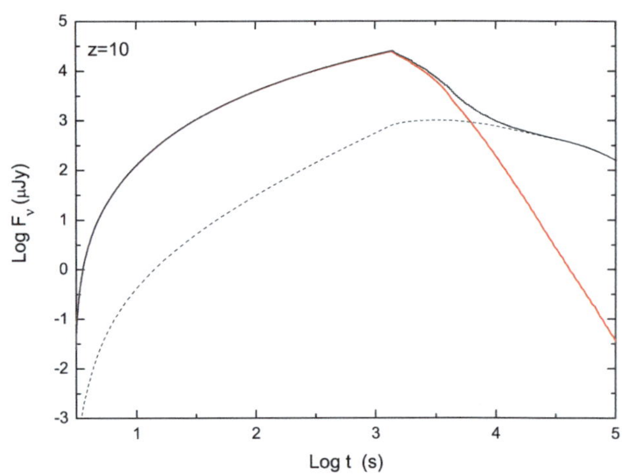

Fig. 13 *Light curve* at $\nu = 6.3 \times 10^{13}$ Hz (*M* band) of Pop III GRBs. The emission from the forward shock (*dashed line*), the reverse shock (*solid red line*), and their combination (*solid black line*) are shown (adapted from Fig. 5 in Wang et al. 2012)

(NIRSpec) on James Webb Space Telescope (Gardner et al. 2006). The high-redshift cut-off is due to the Lyα absorption. The flux will be completely absorbed by the intervening neutral IGM. At these frequencies, the flux of afterglow is weakly dependent on redshift of GRB. There are two reasons. First, the time dilation effect implies that the high redshift means the earlier emission times, where the afterglow are much brighter (Ciardi and Loeb 2000; Bromm and Loeb 2007). Second, circumburst densities of GRBs modestly increase with redshift.

4.2 Metal Enrichment History

The metal enrichment history has several important influences for cosmic structure formation. For example, the metal injection change the mode of star formation (Bromm et al. 2001; Schneider et al. 2002). The transition between Pop III star formation and "normal" (Pop I/II) star formation has important implications, e.g., the expected GRB redshift distribution (Bromm et al. 2002), reionization (Wyithe and Loeb 2003), and the chemical abundance

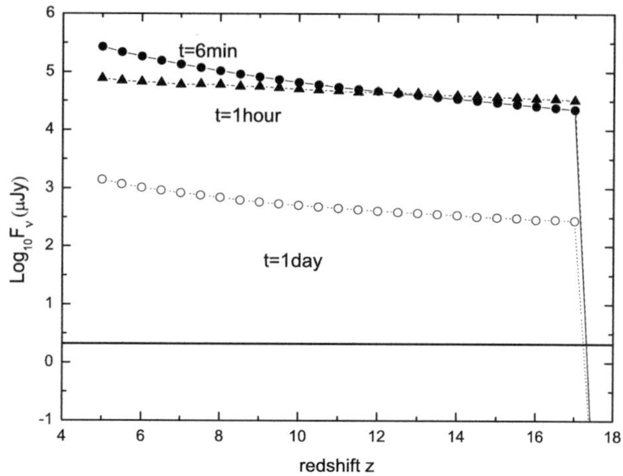

Fig. 14 Observed flux at $\nu = 1.36 \times 10^{14}$ Hz (K band) as a function of redshift at different observed times, as labelled. The K-band sensitivity of the NIRSpec instrument on board the *JWST* is shown as a *horizontal line*. The sharp cut-off at $z \simeq 17$ is due to Lyα absorption in the IGM (adapted from Fig. 6 in Wang et al. 2012)

patterns of stars. So it is important to map the topology of pre-galactic metal enrichment. Ten or thirty meter-class telescopes have been proposed to measure the $z > 6$ IGM metallicity with the GRB afterglow (Oh 2002). Meanwhile, the relative gas column density from metal absorption lines can reflect the enrichment history (Hartmann 2008; Wang et al. 2012).

Absorption processes and absorption lines imprinted on the spectra of GRBs or quasars are the main sources of information about the chemical and physical properties of high-redshift universe. But the bright QSO number is very low at $z > 6$ (Fan et al. 2006). Meanwhile, there are several high-redshift GRBs: GRB 050904 at $z = 6.29$, GRB 080913 at $z = 6.7$, GRB 090423 at $z = 8.3$ and GRB 090429B at $z = 9.4$. The progenitors of long GRBs are thought to be massive stars, so the number of high-redshift GRBs does not decrease significantly. The density, temperature, kinematics and chemical abundances can be extracted from absorption lines (Oh 2002; Furlanetto and Loeb 2003). For instance, Kawai et al. (2006) have identified several metal absorption lines in the afterglow spectrum of GRB 050904 and found that this GRB occur in metal-enriched regions. Two absorption lines have been observed in the spectrum of GRB 090423 at $z = 8.2$ (Salvaterra et al. 2009). These lines are due mainly to absorption metal elements in low ionization stages.

Wang et al. (2012) studied the ability of metal absorption lines in the spectra of Pop III GRBs to probe the pre-galactic metal enrichment. The first galaxy simulation carried out by Greif et al. (2010) was used. The simulation allowed one Pop III progenitor star to explode as an energetic supernova, then the IGM was polluted by the ejected metals. The simulation box size is 1 Mpc (comoving), and is initialized at $z = 99$ according to the ΛCDM model with parameters: $\Omega_\mathrm{m} = 1 - \Omega_\Lambda = 0.3$, $\Omega_\mathrm{b} = 0.04$, $h = H_0/(100\ \mathrm{km\,s^{-1}\,Mpc^{-1}}) = 0.7$, spectral index $n_\mathrm{s} = 1.0$, and normalization $\sigma_8 = 0.9$ (Spergel et al. 2003).

In Fig. 15, the hydrogen number density and metallicity averaged along the line of sight are shown within the central $\simeq 100$ kpc closer to the virialization of the first galaxy at $z = 16.4$. The distribution of metals produced by the first SN explosion is highly inhomogeneous, and the metallicity can reach up to $Z \sim 10^{-2.5} Z_\odot$, which is already larger than the critical metallicity, $Z_\mathrm{crit} \le 10^{-4} Z_\odot$. Therefore, both Pop III and Pop I/II stars will form during the assembly of the first galaxies (Johnson et al. 2008; Maio et al. 2010), so simultaneous occurrence of Pop III and normal GRBs at a given redshift (Bromm and Loeb 2006; de Souza et al. 2011). We consider a Pop III burst exploding in one of the (still metal-free) first galaxy progenitor minihalos at $z \simeq 16.4$.

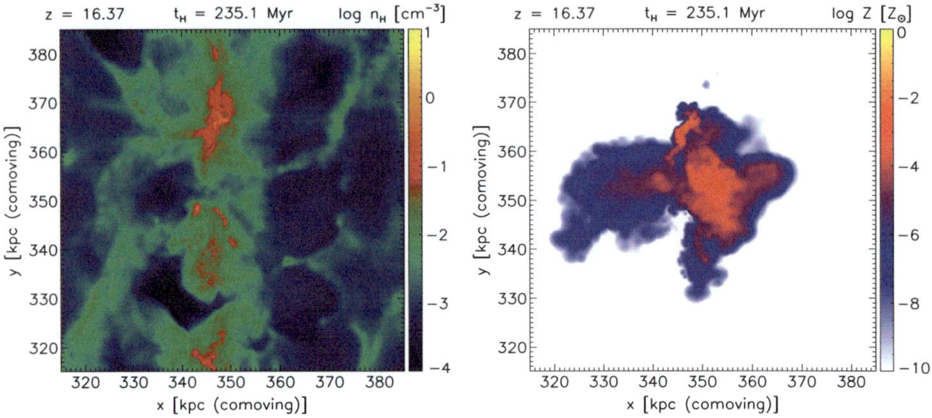

Fig. 15 Possible explosion sites for high-redshift GRBs. Shown are the hydrogen number density (*left panel*) and metallicity contours (*right panel*) averaged along the line of sight at $z \sim 16.37$, when the first galaxy forms. The topology of metal enrichment is highly inhomogeneous (adapted from Fig. 3 in Wang et al. 2012)

Fig. 16 Pop III GRB spectrum observed at the reverse shock crossing time $t_\oplus = 16.7 \times (1 + 16.4)$ s. Metal absorption lines are imprinted according to the Pop III SN event, PISN vs. core-collapse. The former originates from a very massive star (VMS) progenitor, whereas the latter from a less massive one. In each case, the cutoff at short wavelengths is due to Lyman-α scattering in the neutral IGM (adapted from Fig. 10 in Wang et al. 2012)

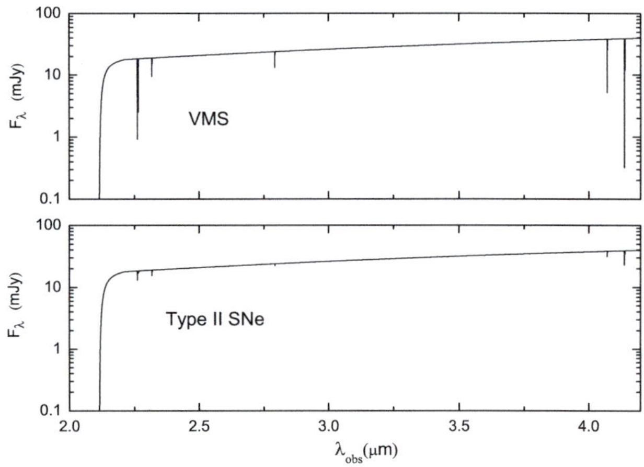

For simplicity, we consider that prior to the GRB only one nearby SN exploded beforehand, dispersing its heavy elements into the pristine IGM. Two nucleosynthetic metal yields for Type II core-collapse SNe (Woosley and Weaver 1995), and for pair-instability supernovae (PISNe; Heger and Woosley 2002, 2010) are considered. Because the hydrogen is substantially neutral, metals will reside in states typical of C II, O I, Si II, and Fe II, because high-energy photons able to further ionize these elements will be absorbed by H I (Furlanetto and Loeb 2003).

Figure 16 shows two spectra of afterglow at the reverse shock crossing time. Top panel is for the top-heavy (Very Massive Star) initial mass function (PISN case) and bottom for normal initial mass function (Type II SNe case). The cutoff is due to Lyman-α absorption in the IGM which is expected to be still completely neutral at $z > 10$. In the two cases, the metal lines are markedly different. The metal yields could be obtained from metal lines. So the initial mass function of Pop III stars can be derived from the metal absorption lines.

5 Using High Resolution Spectroscopy of GRBs to Probe Fundamental Physics

5.1 Measuring the Temperature of the Cosmic Microwave Background Radiation

The existence of the cosmic microwave background (CMB) radiation is a fundamental prediction of the hot Big-Bang theory. If gravitation is described by general relativity and electromagnetism by the Maxwell theory, then photons propagate along null geodesics and the CMB black-body temperature must follow the relation $T_{\mathrm{CMB}}(z) = T_{\mathrm{CMB}}(0) \times (1+z)^{1-\beta}$, with $\beta = 0$ and $T_{\mathrm{CMB}}(0) = 2.725 \pm 0.002$ K for the temperature measured locally. This relation, which is a theoretical consequence of the adiabatic expansion of the Universe, needs to be verified by direct measurements. This has deeper theoretical implications as well (Uzan et al. 2004). A non-zero β would indicate either a violation of the hypothesis of local position invariance (and thus of the equivalence principle) or that the number of photons is not conserved with the constraint that the energy injection does not induce spectral distortion of the CMB. In the first case, this should be associated with a variation of the fundamental constants (see next section). There are currently two methods to measure T_{CMB} at redshifts $z > 0$. The first one relies on the measurement of a small change in the spectral intensity of the CMB towards clusters of galaxies owing to inverse Compton scattering of photons by the hot intra-cluster gas: the so-called Sunyaev-Zeldovich (S-Z) effect. Although this technique permits precise measurements $\Delta T \sim 0.3$ K (Luzzi et al. 2009; Hurier et al. 2014), the method is essentially limited to $z < 1$ because of the scarcity of known clusters at higher redshifts. The other technique uses the excitation of interstellar atomic or molecular species that have transition energies in the sub-millimeter range and can be excited by CMB photons. When the relative population of the different energy levels are in radiative equilibrium with the CMB radiation, the excitation temperature of the species equals that of the black-body radiation at that redshift. Therefore, the detection of these species in diffuse gas, where collisional excitation is negligible, provides one of the best thermometers for determining the black-body temperature of the CMB in the distant Universe (Bahcall and Wolf 1968; Molaro et al. 2002; Srianand et al. 2008).

One of the best thermometers is the molecule CO because it has a non-zéro electromagnetic dipole which implies that in usual conditions of the *diffuse* interstellar medium (ISM), the excitation of the molecule is low (~5 K). Therefore at high redshift, the excitation is easily dominated by the CMB radiation. However, up to very recently, detection of the diffuse CO medium remained elusive and this is only very recently, thanks to the advent of large spectroscopic surveys of quasars that a dozen of such systems have been detected (Noterdaeme et al. 2010) resulting in the best constrain on β shown in Fig. 17.

Up to now only one CO detection has been reported in the spectrum of a GRB afterglow (Prochaska et al. 2009). This is unfortunately a case where the afterglow is highly obscured. However, due to the location of the GRB close to ISM of the host galaxy, the probability to detect diffuse molecular gas (and therefore to avoid very high extinction) is much higher in GRBs than in quasars. With the increased number of GRB detections at high redshift ($z > 3$) there is no doubt that such observations will be performed in the future. It is adament however that to be successful, a high resolution spectrograph must be available at the extremely large telescopes to be built in the near future.

5.2 Testing Variations of Fundamental Constants of Physics

Metal lines of absorption systems due to intervening galaxies along the line of sight towards distant sources provide insight into the atomic structure at the cosmic time and location of

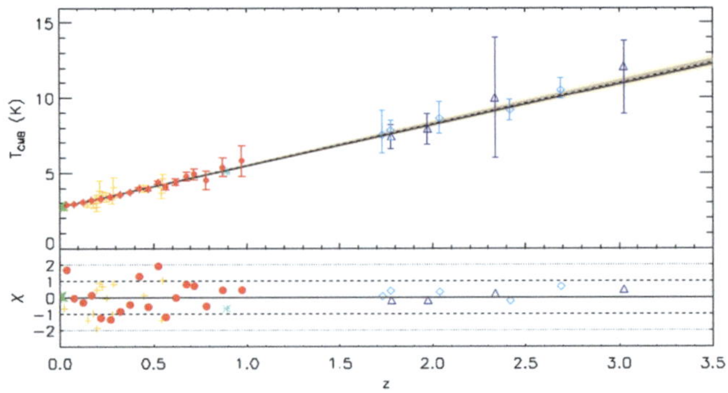

Fig. 17 *Top panel*: $T_{\rm CMB}$ as a function of redshift. The *red filled circles* represent SZ measurements in clusters by the Planck mission (Hurier et al. 2014). *Blue diamonds* show the measurements from CO absorption lines. The *solid black line* presents the standard evolution for $T_{\rm CMB}$ and the *dashed black line* represents the best-fitting model combining all the measurements. The 1 and 2σ envelopes are displayed as *shaded dark* and *light-gray regions*. *Bottom panel*: deviation from the standard evolution in units of standard deviation. The *dashed* and *dotted black lines* correspond to the 1 and 2σ levels. Figure is from Hurier et al. (2014) Note that only one high precision measurement using CO at high redshift could yield much stronger constraint on β

the intervening object. All atomic transitions depend on the fine-structure constant, offering a way to probe possible variations in its value in space and time. These absorption lines are readily observed in the spectra of GRB afterglows and could be used to probe these possible variations.

A first analysis using the multi multiplet method on quasar absorption spectra obtained at the Keck telescope revealed hints that in the early universe the fine structure constant was smaller than today by ~6 parts per million (ppm; Murphy et al. 2003). However, the analysis of smaller data sets obtained from the Very Large Telescope (VLT) in Chile by other groups did not confirm the variation (Srianand et al. 2004, 2007). A recent analysis has been performed on a new sample built up by merging 141 measurements from Keck with 154 measurements from VLT (King et al. 2012). The merged sample confirms the results by Srianand et al. (2007) but indicates a spatial variation across the sky at the 4.1σ level with amplitude ~10 ppm. Since the Keck telescope in Hawaii is at a latitude of 20 degree North, and the VLT in Chile is at latitude of 25 degrees South, the two subsamples survey relatively different hemispheres, and the study of the merged sample provides a much more complete sky coverage. Remarkably, the two sets of measurements are consistent along the region of the sky covered by the two telescopes. So far, this spatial variation has not been confirmed by other groups (e.g. Agafonova et al. 2011) but the lines-of-sight used by these groups fall into a region with minimal reported variation (see however Srianand et al. 2012). One important weakness in the evidence for variations in α is that it derives mostly from archival Keck and VLT spectra.

A VLT/UVES large program has been dedicated to reach the ultimate precision and therefore track systematics. For this, twelves quasar lines-of-sight have been observed at the highest spectral resolution ($R \sim 60\,000$) and highest SNR (> 100) possible. Along each line-of-sight there is at least one absorber in which we expect a precision approaching 2 ppm on the variation of α based on the variety of metal-line transitions available and the shape/structure of their absorption profiles (see Molaro et al. 2013). Some of the lines-of-sight are also dedicated to constraining the variations of μ, the electron-to-proton mass ratio (Rahmani et al. 2013).

GRB afterglows, if well selected, can be unique targets in this field after the advent of extremely large telescopes (TMT and ELT) since GRB afterglows can be very bright even at high redshift and yield spectra of the highest SNR; and the probability that the line-of-sight intersects a diffuse molecular cloud is much higher in the case of GRBs compared to QSOs. However, a fast response mode on a high resolution spectrograph has to be available (e.g. Vreeswijk et al. 2007) in order to collect a high SNR spectrum. It will also be important to follow a well defined procedure for wavelength calibration.

6 Using High Energetic Photons of GRBs to Test Fundamental Physics

6.1 Constraining Lorentz Invariance Violation

Lorentz invariance is a famous postulate of Einstein's special relativity, which indicates that the relevant physics laws of a non-accelerated physical system are not affected when this system undergoes Lorentz transformation. However, many Quantum Gravity (QG) theories have suggested that quantum fluctuations in the space-time would make it appears discontinuous and chaotic, namely the foamy structure (Amelino-Camelia et al. 1997). And the propagation of light through the space-time with the foamy structure would show a non-trivial dispersion relation in vacuum (Amelino-Camelia et al. 1998), which could leads to the violation of Lorentz invariance. Constraining the Lorentz invariance violation (LIV) effect has been thought as an effective method to test the accuracy of QG theories, such as loop quantum gravity (Gambini and Pullin 1999; Alfaro et al. 2002), string theory (Kostelecký and Samuel 1989), double special relativity (Amelino-Camelia and Ahluwalia 2002), and so on. Since it is generally expected for QG to manifest itself fully at the Planck scale, the Planck energy scale ($E_{\rm QG} \approx E_{\rm Pl} = \sqrt{\hbar c^5/G} \simeq 1.22 \times 10^{19}$) being a natural one at which Lorentz invariance is predicted to be broken (see, e.g., Amelino-Camelia 2013, and references therein).

Due to the LIV effect, the speed for the propagation of photons could become energy-dependent, instead of a constant speed of light in vacuum (see Amelino-Camelia et al. 1998; Ellis and Mavromatos 2013). Normally, the modified dispersion relation of photons can be approximatively described as the leading term of the Taylor expansion (see Ellis et al. 2003; Jacob and Piran 2008)

$$E^2 \simeq p^2c^2\left[1 - s_n\left(\frac{pc}{E_{{\rm QG},n}}\right)^n\right], \tag{29}$$

where the n-th order expansion of leading term corresponds to linear ($n = 1$) or quadratic ($n = 2$), $E_{\rm QG}$ denotes the QG energy scale, and $s_n = \pm 1$ represents the sign of the LIV correction. If $s_n = -1$ ($s_n = +1$), the low energy photons travel slower (faster) than the high energy photons. The speed of light derived from Eq. (29) is

$$v = \frac{\partial E}{\partial p} \approx c\left[1 - s_n\frac{n+1}{2}\left(\frac{E}{E_{{\rm QG},n}}\right)^n\right]. \tag{30}$$

Because of the energy dependence of the light speed, two photons with different energies emitted simultaneously from the source will reach us with a time delay Δt. For a cosmic source, the cosmological expansion should be taken into account when calculating the LIV

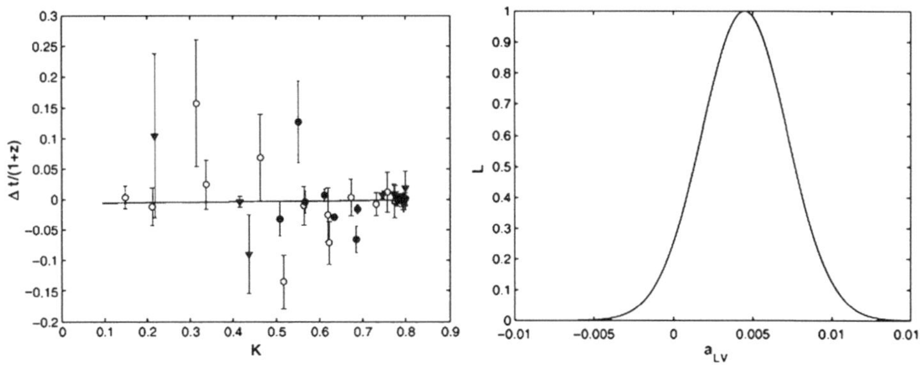

Fig. 18 *Left panel*: the observed spectral time-delays between the arrival times of pairs of genuine high-intensity sharp features detected in the *light curves* of the full set of 35 GRBs with measured redshifts observed by BATSE (*closed circles*), HETE (*open circles*) and SWIFT (*triangles*). *Right panel*: the likelihood function for the slope parameter (adapted from Figs. 1 and 2 in Ellis et al. 2008)

induced time delay (Ellis et al. 2008; Jacob and Piran 2008; Zhang and Ma 2015)

$$\Delta t_{\rm LV} = t_{\rm h} - t_{\rm l} = s_n \frac{1+n}{2H_0} \frac{E_{\rm h}^n - E_{\rm l}^n}{E_{{\rm QG},n}^n} \int_0^z \frac{(1+z')^n dz'}{\sqrt{\Omega_m (1+z')^3 + \Omega_\Lambda}}, \tag{31}$$

where $t_{\rm h}$ and $t_{\rm l}$ correspond to the arrival times of the high energy photon and the low energy one, with $E_{\rm h}$ and $E_{\rm l}$ ($E_{\rm h} > E_{\rm l}$) being the photon energies.

Since photons with different energies will arrive on earth at different time, GRBs with detected high energetic photons and measured redshifts have been proposed to test the LIV effect (see, e.g., Amelino-Camelia et al. 1998; Ellis et al. 2003, 2006, 2008; Jacob and Piran 2008; Abdo et al. 2009a,b; Vasileiou et al. 2013; Zhang and Ma 2015). The first attempt that take the advantage of cosmological distances compared the arrival time of different energetic photons from GRB to constrain the LIV effect was presented in Amelino-Camelia et al. (1998). Ellis et al. (2003, 2006, 2008) developed a method to measure LIV by analyzing a large sample of GRBs with known redshifts. This method has the advantage that it can precisely extract the observed time lag from the GRB light curves in different energy bands. Taking into account the unknown intrinsic time lags $b_{\rm sf}$, the observed time delays should have two contributions $\Delta t_{\rm obs} = \Delta t_{\rm LV} + b_{\rm sf}(1+z)$ (Ellis et al. 2006). A simple linear fitting function can be written as

$$\frac{\Delta t_{\rm obs}}{1+z} = a_{\rm LV} K + b_{\rm sf}, \tag{32}$$

where $K = (1+z)^{-1} \int_0^z dz'(1+z')/\sqrt{\Omega_m(1+z')^3 + \Omega_\Lambda}$ is a function of redshift and the slope $a_{\rm LV} = \Delta E/(H_0 E_{\rm QG})$ is related to the scale of Lorentz violation. Note that the linear term ($n = 1$) is considered here. The result of a linear fit to the observed time delays extracted from 35 GRB light curves is presented in the left panel of Fig. 18. The best fit corresponds to $\frac{\Delta t_{\rm obs}}{1+z} = (0.0068 \pm 0.0067)K - (0.0065 \pm 0.0046)$ and the likelihood function for the slope $a_{\rm LV}$ is shown in the right panel of Fig. 18. The 95 % confidence-level lower limit on the QG energy scale derived from $a_{\rm LV}$ is $E_{\rm QG} \geq 1.4 \times 10^{16}$ GeV (Ellis et al. 2008).

However, there is a limitation of Ellis et al. treatment. They extracted spectral time-lags in the light curves recorded in the MeV energy band relative to those in the keV energy band. The value of the energy band plays an important role in constraining LIV, a higher energy band leading to better constraints on LIV. Therefore, it is necessary to consider using the GeV lags to constrain LIV. Abdo et al. (2009b) used the highest energy (13.2 GeV) photon of GRB 080916C to study the LIV effect, which was detected 16.5 s after the trigger time.[1] Compared with previous estimates, their limit on the linear term ($E_{\rm QG,1}$) of 1.3×10^{18} GeV represented an improvement of one order of magnitude. But, it is still one order of magnitude below the Planck energy scale $E_{\rm Pl}$. With an observed time delay ~ 0.86 s of GRB 090510 (i.e., the time lag between the trigger time and the arrival time of 31 GeV photon), Abdo et al. (2009a) set the most stringent limits on both the linear and quadratic term up to now. The limits set are $E_{\rm QG,1} > 9.1 \times 10^{19}$ GeV $> 1.2E_{\rm Pl}$ and $E_{\rm QG,2} > 1.3 \times 10^{11}$ GeV. We can see that the lincar ($n = 1$) LIV can be excluded by GRB 090510.

6.2 Testing Einstein's Equivalence Principle

The Einstein Equivalence Principle (EEP) is the heart and soul of general relativity and other gravitational theories, which states that the trajectory of any uncharged "test" body is independent of its internal structure and composition. An alternative statement of the EEP is that spacetime is given a symmetric metric and uncharged test bodies follow geodesics of that metric. A lot of methods have been performed to quantify the possible violations of the EEP. Among the most famous are the Eötvös-type experiments, which compare the accelerations of two laboratory-sized bodies made of different composition in a gravitational field (see, e.g., Will 2014, and references therein). If the EEP is violated, the accelerations of these two bodies will be different. The validity of the EEP can be well tested by the comparison of laboratory-sized bodies in a Newtonian context. Nevertheless, Newtonian dynamics is inappropriate for describing particles' (like photons or neutrinos) motion in a gravitational field. In order to measure the accuracy of the EEP with particles, the parameterized post-Newtonian (PPN) formalism has been developed for precisely describing their motion. Most metric theories of gravity satisfying the EEP are contained by the PPN formalism, and each theory is specified by the numerical values of some coefficients (PPN parameters), which have been well reviewed by Will (2006, 2014). The accuracy of the EEP can therefore be tested through the numerical values of PPN parameters.

For instance, Shapiro (1964) has suggested that the time interval for photons or neutrinos to pass through a given distance is longer within an external gravitational potential $U(r)$, the so-called Shapiro time delay, which is given by

$$\delta t = -\frac{1+\gamma}{c^3}\int_{r_e}^{r_o} U(r)dr, \tag{33}$$

where r_e and r_o are locations of source and observation, the PPN parameter γ accounts for how much space-curvature is produced by unit rest mass. General relativity predict that the value of γ should be unity (i.e., $\gamma = 1$), but not all gravity theories predict the same values for γ. There are many ways in which one can test general relativity by measuring the value of γ. Measurements of the parameter γ have reached high precision with the results of solar-system experiments, including the deflection of light and the time delay of light. The most stringent limits on the γ value from the light deflection near the sun is based on the very long baseline radio interferometry measurement, with a result $\gamma - 1 = (-0.8 \pm 1.2) \times 10^{-4}$ (Lambert and Le Poncin-Lafitte 2009, 2011). With the travel time delay of a radar signal

[1] Since the energies of photons arrived at the trigger time (~ 100 keV) are several orders of magnitude lower than those of high energy photons ($\sim$ GeV), $E_{\rm h}^n - E_{\rm l}^n$ in Eq. (31) can be approximated as $E_{\rm h}^n$.

of the Cassini spacecraft, Bertotti et al. (2003) set the most precise limit on the value of γ, i.e., $\gamma - 1 = (2.1 \pm 2.3) \times 10^{-5}$. These results showed that γ is very close to 1, which is consistent with the expectation of general relativity.

However, it is important to note that all gravity theories satisfying the EEP also predict $\gamma_1 = \gamma_2 \equiv \gamma$, where the subscripts represent two different test particles. To test the EEP in general relativity, therefore, the issue is not only whether the value of γ is very nearly 1, but also whether it is the same for different types of particles, or for the same type of particle with different energies. Krauss and Tremaine (1988) and Longo (1988) suggested that the small arrival time delay of the photon and neutrino emitted from supernova 1987A in the Large Magellanic Cloud provides a new precision test of the EEP. By analyzing the arrival time delay between the photons (eV) and neutrinos (MeV) from supernova 1987A, Longo (1988) obtained a strong limit on the differences of γ to the level of 0.34 %. In addition, using the time delay for two neutrinos with different energies, Longo (1988) set a more precise limit on γ differences, yielding $[\gamma(40\ \mathrm{MeV}) - \gamma(7.5\ \mathrm{MeV})] \leq 1.6 \times 10^{-6}$. Besides the extragalactic source (supernova 1987A), the nearly simultaneous arrival of photons with different energies from cosmic transients, such as GRBs, have also been applied to test the EEP (Sivaram 1999; Gao et al. 2015).

For a cosmic source, the observed time delays between correlated photons should have four contributions (Gao et al. 2015):

$$\Delta t_{\mathrm{obs}} = \Delta t_{\mathrm{int}} + \Delta t_{\mathrm{LIV}} + \Delta t_{\mathrm{spe}} + \Delta t_{\mathrm{gra}}, \tag{34}$$

where Δt_{int} represents the unknown intrinsic time lag, Δt_{LIV} is the LIV induced time delay, and Δt_{spe} denotes the potential time delay caused by the non-zero mass of photons in special relativity. With Eq. (33), the time delay (Δt_{gra}) of two photons with energy E_1 and E_2, due to the external gravitational potential $U(r)$, which can be written down as

$$\Delta t_{\mathrm{gra}} = \frac{\gamma_1 - \gamma_2}{c^3} \int_{r_o}^{r_e} U(r) dr. \tag{35}$$

Since both Δt_{LIV} and Δt_{spe} are negligible for the purposes of testing the EEP, they can be ignored in the analysis (see Gao et al. 2015, for more explanations). With the assumption that $\Delta t_{\mathrm{int}} > 0$, one can derive

$$\Delta t_{\mathrm{obs}} > \frac{\gamma_1 - \gamma_2}{c^3} \int_{r_o}^{r_e} U(r) dr. \tag{36}$$

Generally speaking, $U(r)$ should has three contributions: the gravitational potential of our galaxy $U_{\mathrm{MW}}(r)$, the intergalactic potential $U_{\mathrm{IG}}(r)$ between our galaxy and the transient host galaxy, and the gravitational potential of the transient host galaxy $U_{\mathrm{host}}(r)$. Unfortunately, we know nothing about the potential functions for $U_{\mathrm{IG}}(r)$ and $U_{\mathrm{host}}(r)$, but we stand a good chance of considering the combination of these two terms is much larger than the potential of $U_{\mathrm{MW}}(r)$. Hence, it would be reasonable to derive

$$\Delta t_{\mathrm{obs}} > \frac{\gamma_1 - \gamma_2}{c^3} \int_{r_o}^{r_e} U_{\mathrm{MW}}(r) dr. \tag{37}$$

If we adopt the Keplerian potential $U_{\mathrm{MW}}(r) = -GM/r$ for our galaxy, then we have

$$\gamma_1 - \gamma_2 < \Delta t_{\mathrm{obs}} \left(\frac{GM_{\mathrm{MW}}}{c^3} \right)^{-1} \ln^{-1}\left(\frac{d}{b} \right), \tag{38}$$

where $M_{\rm MW} \simeq 6 \times 10^{11} M_{\odot}$ represents the mass of our galaxy (McMillan 2011; Kafle et al. 2012), d is the distance from the transient source to the observer, and b denotes the impact parameters of the light rays relative to our galaxy center. Assuming that the cosmic transient source located in the direction (R.A. $= \beta_S$, Dec. $= \delta_S$), the impact parameter b can be calculated by the expression

$$b = r_G \sqrt{1 - \left(\sin\delta_S \sin\delta_G + \cos\delta_S \cos\delta_G \cos(\beta_S - \beta_G)\right)^2}, \tag{39}$$

where $r_G = 8.3$ kpc is the distance from the Sun to our galaxy center, and the coordinates (J2000) of our galaxy center correspond to $\beta_G = 17^{\rm h}45^{\rm m}40.04^{\rm s}$ and $\delta_G = -29^{\circ}00'28.1''$ (Gillessen et al. 2009).

With the assumption that the observed time delays for photons with different energies are caused dominantly by the gravitational potential of our galaxy, for the GeV and MeV photons of GRB 090510 (Gao et al. 2015) set a severe limit of $\gamma_{\rm GeV} - \gamma_{\rm MeV} \le 2 \times 10^{-8}$; and for the optical and MeV photons from GRB 080319B they obtained $\gamma_{\rm eV} - \gamma_{\rm MeV} \le 1.2 \times 10^{-7}$. Comparing these results with the limits of supernovae 1987 shows that GRBs can increase the accuracy level of the EEP constraints by at least one order of magnitude, to $\sim 10^{-7}$, while expanding the tested EEP energy range to the MeV-eV and GeV-MeV range. Sivaram (1999) also derived the same upper limit on the EEP of 4×10^{-7} for gamma-ray and optical photons from GRB 990123, but his result was restricted to the MeV and eV energy bands.

7 Summary and Future Prospect

GRBs are observed throughout the whole electromagnetic spectrum, from radio waves to γ-rays, which have been observed in distant universe. Recently, GRBs have attracted a lot of attention as promising standardizable candles to construct the Hubble diagram to high redshift, as complementarity to other cosmological probes, such as SNe Ia, CMB and BAO. However, a lot of work is needed to be sure that GRBs can hold this promise in future. The most important thing is to search for a correlation similar to that used to standardize SNe Ia. In order to obtain the correlation, the classification of GRBs may be crucial. We must remind that only SNe Ia are standard candles among all SNe. The classical classification method is based on the prompt emission properties (duration, hardness, and spectral lag). The physics of prompt emission are not fully understood (Zhang 2014), and some new clues from other objects are found (Wang and Dai 2013; Wang et al. 2014). But observations of some GRBs are challenging the standard classification (Zhang 2006; Zhang et al. 2009; Lü et al. 2010). So more physical nature of GRBs is needed (Zhang et al. 2009). The circularity problem could be partially solved by analyzing a sample of GRBs within a small redshift bin (Ghirlanda et al. 2006; Liang and Zhang 2006).

In order to measure high-redshift SFR from GRBs, the relation between long GRB rate and SFR must be known. Besides, theoretical models of the SFR have several free parameters, such as the efficiency of star formation and the chemical feedback strength. From the theoretical SFR, the predicted GRB redshift distribution can be derived. So one can use the GRB redshift distribution observed by Swift (or future missions such as SVOM and EXIST), to calibrate the free parameters. More GRB red damping wing with low HI column density are required to study properties of IGM.

Metal absorption lines in the GRB afterglow spectrum, giving rise to EWs of a few tens of Å, which may allow us to distinguish whether the first heavy elements were produced in a Pop III star died as a PISN or a core-collapse SN. To this extent, the spectrum needs to be obtained sufficiently early, within the first few hours after the trigger. Upcoming JWST would detect much more high-redshift GRBs (properly Pop III GRBs) with high resolution NIR spectra including metal absorption lines, which allow one to measure the cosmic metallicity evolution.

In the future, the French-Chinese satellite Space-based multi-band astronomical Variable Objects Monitor (SVOM) and JWST, have been optimized to increase the number of GRB and the synergy with the ground-based facilities. There are a combination of multi-wavelength detectors on board of SVOM (Paul et al. 2011). ECLAIRs wide-field camera will detect GRBs in the energy range of 4–150 keV. The spectral information of prompt emission will be measured by Gamma-Ray Monitor (GRM). The afterglow can be obtained by the Micro channel X-ray Telescope (MXT; 0.3–10 keV) and the Visible Telescope (VT; 400–900 nm). SVOM can detect about 80 GRBs per year, and more than 50 % of GRBs have redshift measurement (Petitjean and Vergani 2011). JWST is a large, infrared-optimized space telescope with 6.6 m diameter aperture. It has four scientific instruments: a Near-IR Camera (NIRCam), a Near-IR Spectrograph (NIRSpec), a near-IR Tunable Filter Imager (TFI), and a Mid-IR Instrument (MIRI) (Gardner et al. 2006). But the direct detection of a single Pop III star is not feasible even for JWST, i.e., the AB magnitude of a $M = 1000 M_\odot$ star is only 36 at $z \sim 30$. Meanwhile, the Pop III GRBs can be detectable by JWST (Wang et al. 2012; Mesler et al. 2014; Macpherson et al. 2013). This will boost the amount of information available to tackle the important issues revealed by this exciting field of research.

Because of the much larger detector area and higher sensitivity, the Water Cerenkov Detector Array (WCDA) of LHAASO will has the ability to detect much more high-energy photons from GRBs and high quality high-energy light curves will be possible, making it easy to extract the observed time delays from the light curves with different energy bands. We can expect that the high energy GRB observations provided by LHAASO/WCDA will set much more competitive limits on fundamental physics.

Acknowledgements This work is supported by the National Basic Research Program ("973" Program) of China (grants 2014CB845800 and 2013CB83490), the National Natural Science Foundation of China (grants 11422325, 11373022, 11033002, 11322328, and 11433009), the Excellent Youth Foundation of Jiangsu Province (BK20140016), and the Program for New Century Excellent Talents in University (grant No. NCET-13-0279). X.F. Wu was also partially supported by the One-Hundred-Talent Program, the Youth Innovation Promotion Association (2011231), and the Strategic Priority Research Program "The Emergence of Cosmological Structure" (grant No. XDB09000000) of the Chinese Academy of Sciences.

References

A.A. Abdo, M. Ackermann, M. Ajello, K. Asano, W.B. Atwood, M. Axelsson, L. Baldini, J. Ballet, G. Barbiellini, M.G. Baring et al., A limit on the variation of the speed of light arising from quantum gravity effects. Nature **462**, 331–334 (2009a). doi:10.1038/nature08574

A.A. Abdo, M. Ackermann, M. Arimoto, K. Asano, W.B. Atwood, M. Axelsson, L. Baldini, J. Ballet, D.L. Band, G. Barbiellini et al., Fermi observations of high-energy gamma-ray emission from GRB 080916C. Science **323**, 1688 (2009b). doi:10.1126/science.1169101

T. Abel, G.L. Bryan, M.L. Norman, The formation of the first star in the Universe. Science **295**, 93–98 (2002). doi:10.1126/science.295.5552.93

I.I. Agafonova, P. Molaro, S.A. Levshakov, J.L. Hou, First measurement of Mg isotope abundances at high redshifts and accurate estimate of $\Delta\alpha/\alpha$. Astron. Astrophys. **529**, 28 (2011). doi:10.1051/0004-6361/201016194

J. Alfaro, H.A. Morales-Técotl, L.F. Urrutia, Loop quantum gravity and light propagation. Phys. Rev. D **65**(10), 103509 (2002). doi:10.1103/PhysRevD.65.103509

S.W. Allen, R.W. Schmidt, H. Ebeling, A.C. Fabian, L. van Speybroeck, Constraints on dark energy from Chandra observations of the largest relaxed galaxy clusters. Mon. Not. R. Astron. Soc. **353**, 457–467 (2004). doi:10.1111/j.1365-2966.2004.08080.x
M.A. Alvarez, V. Bromm, P.R. Shapiro, The H II region of the first star. Astrophys. J. **639**, 621–632 (2006). doi:10.1086/499578
L. Amati, The $E_{\mathrm{p,i}}$–E_{iso} correlation in gamma-ray bursts: updated observational status, re-analysis and main implications. Mon. Not. R. Astron. Soc. **372**, 233–245 (2006). doi:10.1111/j.1365-2966.2006.10840.x
L. Amati, M. Della Valle, Measuring cosmological parameters with gamma ray bursts. Int. J. Mod. Phys. D **22**, 30028 (2013). doi:10.1142/S0218271813300280
L. Amati, F. Frontera, M. Tavani, J.J.M. in't Zand, A. Antonelli, E. Costa, M. Feroci, C. Guidorzi, J. Heise, N. Masetti, E. Montanari, L. Nicastro, E. Palazzi, E. Pian, L. Piro, P. Soffitta, Intrinsic spectra and energetics of BeppoSAX gamma-ray bursts with known redshifts. Astron. Astrophys. **390**, 81–89 (2002). doi:10.1051/0004-6361:20020722
L. Amati, C. Guidorzi, F. Frontera, M. Della Valle, F. Finelli, R. Landi, E. Montanari, Measuring the cosmological parameters with the $E_{\mathrm{p,i}}$–E_{iso} correlation of gamma-ray bursts. Mon. Not. R. Astron. Soc. **391**, 577–584 (2008). doi:10.1111/j.1365-2966.2008.13943.x
G. Amelino-Camelia, Quantum-spacetime phenomenology. Living Rev. Relativ. **16**, 5 (2013). doi:10.12942/lrr-2013-5
G. Amelino-Camelia, D.V. Ahluwalia, Relativity in spacetimes with short-distance structure governed by an observer-independent (Planckian) length scale. Int. J. Mod. Phys. D **11**, 35–59 (2002). doi:10.1142/S0218271802001330
G. Amelino-Camelia, J. Ellis, N.E. Mavromatos, D.V. Nanopoulos, Distance measurement and wave dispersion in a Liouville-string approach to quantum gravity. Int. J. Mod. Phys. A **12**, 607–623 (1997). doi:10.1142/S0217751X97000566
G. Amelino-Camelia, J. Ellis, N.E. Mavromatos, D.V. Nanopoulos, S. Sarkar, Tests of quantum gravity from observations of γ-ray bursts. Nature **393**, 763–765 (1998). doi:10.1038/31647
J.N. Bahcall, R.A. Wolf, Fine-structure transitions. Astrophys. J. **152**, 701 (1968). doi:10.1086/149589
D. Band, J. Matteson, L. Ford, B. Schaefer, D. Palmer, B. Teegarden, T. Cline, M. Briggs, W. Paciesas, G. Pendleton, G. Fishman, C. Kouveliotou, C. Meegan, R. Wilson, P. Lestrade, BATSE observations of gamma-ray burst spectra. I—Spectral diversity. Astrophys. J. **413**, 281–292 (1993). doi:10.1086/172995
S. Basilakos, L. Perivolaropoulos, Testing gamma-ray bursts as standard candles. Mon. Not. R. Astron. Soc. **391**, 411–419 (2008). doi:10.1111/j.1365-2966.2008.13894.x
E. Berger, Short-duration gamma-ray bursts. Annu. Rev. Astron. Astrophys. **52**, 43–105 (2014). doi:10.1146/annurev-astro-081913-035926
O. Bertolami, P.T. Silva, Gamma-ray bursts as dark energy-matter probes in the context of the generalized Chaplygin gas model. Mon. Not. R. Astron. Soc. **365**, 1149–1159 (2006). doi:10.1111/j.1365-2966.2005.09765.x
B. Bertotti, L. Iess, P. Tortora, A test of general relativity using radio links with the Cassini spacecraft. Nature **425**, 374–376 (2003). doi:10.1038/nature01997
J.S. Bloom, D.A. Frail, R. Sari, The prompt energy release of gamma-ray bursts using a cosmological k-correction. Astron. J. **121**, 2879–2888 (2001). doi:10.1086/321093
J.S. Bloom, D.A. Frail, S.R. Kulkarni, Gamma-ray burst energetics and the gamma-ray burst Hubble diagram: promises and limitations. Astrophys. J. **594**, 674–683 (2003). doi:10.1086/377125
J.S. Bloom, N.R. Butler, D.A. Perley, Gamma-ray bursts, classified physically, in *AIP Conference Proceedings*, ed. by M. Galassi, D. Palmer, E. Fenimore American Institute of Physics Conference Series, vol. 1000, 2008, pp. 11–15. doi:10.1063/1.2943423
R.J. Bouwens, G.D. Illingworth, M. Franx, R.-R. Chary, G.R. Meurer, C.J. Conselice, H. Ford, M. Giavalisco, P. van Dokkum, UV continuum slope and dust obscuration from $z \sim 6$ to $z \sim 2$: the star formation rate density at high redshift. Astrophys. J. **705**, 936–961 (2009). doi:10.1088/0004-637X/705/1/936
R.J. Bouwens, G.D. Illingworth, I. Labbe, P.A. Oesch, M. Trenti, C.M. Carollo, P.G. van Dokkum, M. Franx, M. Stiavelli, V. González, D. Magee, L. Bradley, A candidate redshift $z \sim 10$ galaxy and rapid changes in that population at an age of 500 Myr. Nature **469**, 504–507 (2011). doi:10.1038/nature09717
V. Bromm, R.B. Larson, The first stars. Annu. Rev. Astron. Astrophys. **42**, 79–118 (2004). doi:10.1146/annurev.astro.42.053102.134034
V. Bromm, A. Loeb, High-redshift gamma-ray bursts from population III progenitors. Astrophys. J. **642**, 382–388 (2006). doi:10.1086/500799
V. Bromm, A. Loeb, GRB cosmology. ArXiv e-prints (2007)
V. Bromm, P.S. Coppi, R.B. Larson, Forming the first stars in the Universe: the fragmentation of primordial gas. Astrophys. J. Lett. **527**, 5–8 (1999). doi:10.1086/312385
V. Bromm, A. Ferrara, P.S. Coppi, R.B. Larson, The fragmentation of pre-enriched primordial objects. Mon. Not. R. Astron. Soc. **328**, 969–976 (2001). doi:10.1046/j.1365-8711.2001.04915.x

V. Bromm, P.S. Coppi, R.B. Larson, The formation of the first stars. I. The primordial star-forming cloud. Astrophys. J. **564**, 23–51 (2002). doi:10.1086/323947

V. Bromm, N. Yoshida, L. Hernquist, C.F. McKee, The formation of the first stars and galaxies. Nature **459**, 49–54 (2009). doi:10.1038/nature07990

M.A. Campisi, L.-X. Li, P. Jakobsson, Redshift distribution and luminosity function of long gamma-ray bursts from cosmological simulations. Mon. Not. R. Astron. Soc. **407**, 1972–1980 (2010). doi:10.1111/j.1365-2966.2010.17044.x

M.A. Campisi, U. Maio, R. Salvaterra, B. Ciardi, Population III stars and the long gamma-ray burst rate. Mon. Not. R. Astron. Soc. **416**, 2760–2767 (2011). doi:10.1111/j.1365-2966.2011.19238.x

X.-F. Cao, Y.-W. Yu, K.S. Cheng, X.-P. Zheng, The luminosity function of Swift long gamma-ray bursts. Mon. Not. R. Astron. Soc. **416**, 2174–2181 (2011). doi:10.1111/j.1365-2966.2011.19194.x

S. Capozziello, L. Izzo, Cosmography by gamma ray bursts. Astron. Astrophys. **490**, 31–36 (2008). doi:10.1051/0004-6361:200810337

V.F. Cardone, S. Capozziello, M.G. Dainotti, An updated gamma-ray bursts Hubble diagram. Mon. Not. R. Astron. Soc. **400**, 775–790 (2009). doi:10.1111/j.1365-2966.2009.15456.x

K.S. Cheng, Y.-W. Yu, T. Harko, High-redshift gamma-ray bursts: observational signatures of superconducting cosmic strings? Phys. Rev. Lett. **104**(24), 241102 (2010). doi:10.1103/PhysRevLett.104.241102

K.S. Cheng, Y.W. Yu, T. Harko, Cheng, Yu, and Harko reply. Phys. Rev. Lett. **106**(25), 259002 (2011). doi:10.1103/PhysRevLett.106.259002

B. Ciardi, A. Loeb, Expected number and flux distribution of gamma-ray burst afterglows with high redshifts. Astrophys. J. **540**, 687–696 (2000). doi:10.1086/309384

P.C. Clark, S.C.O. Glover, R.S. Klessen, V. Bromm, Gravitational fragmentation in turbulent primordial gas and the initial mass function of population III stars. Astrophys. J. **727**, 110 (2011). doi:10.1088/0004-637X/727/2/110

B.E. Cobb, C.D. Bailyn, P.G. van Dokkum, P. Natarajan, SN 2006aj and the nature of low-luminosity gamma-ray bursts. Astrophys. J. Lett. **645**, 113–116 (2006). doi:10.1086/506271

D. Coward, Open issues with the gamma-ray burst redshift distribution. New Astron. Rev. **51**, 539–546 (2007). doi:10.1016/j.newar.2007.03.003

Z.G. Dai, F.Y. Wang, Recent progress on GRB cosmology. Adv. Space Res. **40**, 1244–1249 (2007). doi:10.1016/j.asr.2007.01.009

Z.G. Dai, E.W. Liang, D. Xu, Constraining Ω_M and dark energy with gamma-ray bursts. Astrophys. J. Lett. **612**, 101–104 (2004). doi:10.1086/424694

F. Daigne, E.M. Rossi, R. Mochkovitch, The redshift distribution of Swift gamma-ray bursts: evidence for evolution. Mon. Not. R. Astron. Soc. **372**, 1034–1042 (2006). doi:10.1111/j.1365-2966.2006.10837.x

R. Davé, The galaxy stellar mass-star formation rate relation: evidence for an evolving stellar initial mass function? Mon. Not. R. Astron. Soc. **385**, 147–160 (2008). doi:10.1111/j.1365-2966.2008.12866.x

R.S. de Souza, N. Yoshida, K. Ioka, Populations III.1 and III.2 gamma-ray bursts: constraints on the event rate for future radio and X-ray surveys. Astron. Astrophys. **533**, 32 (2011). doi:10.1051/0004-6361/201117242

M. Demianski, E. Piedipalumbo, Standardizing the gamma-ray bursts with the Amati $E_{\rm p,i}$–$E_{\rm iso}$ relation: the updated Hubble diagram and implications for cosmography. Mon. Not. R. Astron. Soc. **415**, 3580–3590 (2011). doi:10.1111/j.1365-2966.2011.18975.x

M. Demianski, E. Piedipalumbo, C. Rubano, The gamma-ray bursts Hubble diagram in quintessential cosmological models. Mon. Not. R. Astron. Soc. **411**, 1213–1222 (2011). doi:10.1111/j.1365-2966.2010.17751.x

T. Di Girolamo, R. Catena, M. Vietri, G. Di Sciascio, Can gamma-ray bursts constrain quintessence? J. Cosmol. Astropart. Phys. **4**, 8 (2005). doi:10.1088/1475-7516/2005/04/008

B. Efron, V. Petrosian, A simple test of independence for truncated data with applications to redshift surveys. Astrophys. J. **399**, 345–352 (1992). doi:10.1086/171931

D.J. Eisenstein, I. Zehavi, D.W. Hogg, R. Scoccimarro, M.R. Blanton, R.C. Nichol, R. Scranton, H.-J. Seo, M. Tegmark, Z. Zheng, S.F. Anderson, J. Annis, N. Bahcall, J. Brinkmann, S. Burles, F.J. Castander, A. Connolly, I. Csabai, M. Doi, M. Fukugita, J.A. Frieman, K. Glazebrook, J.E. Gunn, J.S. Hendry, G. Hennessy, Z. Ivezić, S. Kent, G.R. Knapp, H. Lin, Y.-S. Loh, R.H. Lupton, B. Margon, T.A. McKay, A. Meiksin, J.A. Munn, A. Pope, M.W. Richmond, D. Schlegel, D.P. Schneider, K. Shimasaku, C. Stoughton, M.A. Strauss, M. SubbaRao, A.S. Szalay, I. Szapudi, D.L. Tucker, B. Yanny, D.G. York, Detection of the baryon acoustic peak in the large-scale correlation function of SDSS luminous red galaxies. Astrophys. J. **633**, 560–574 (2005). doi:10.1086/466512

J. Elliott, J. Greiner, S. Khochfar, P. Schady, J.L. Johnson, A. Rau, The long γ-ray burst rate and the correlation with host galaxy properties. Astron. Astrophys. **539**, 113 (2012). doi:10.1051/0004-6361/201118561

J. Ellis, N.E. Mavromatos, Probes of Lorentz violation. Astropart. Phys. **43**, 50–55 (2013). doi:10.1016/j.astropartphys.2012.05.004

J. Ellis, N.E. Mavromatos, D.V. Nanopoulos, A.S. Sakharov, Quantum-gravity analysis of gamma-ray bursts using wavelets. Astron. Astrophys. **402**, 409–424 (2003). doi:10.1051/0004-6361:20030263

J. Ellis, N.E. Mavromatos, D.V. Nanopoulos, A.S. Sakharov, E.K.G. Sarkisyan, Robust limits on Lorentz violation from gamma-ray bursts. Astropart. Phys. **25**, 402–411 (2006). doi:10.1016/j.astropartphys.2006.04.001

J. Ellis, N.E. Mavromatos, D.V. Nanopoulos, A.S. Sakharov, E.K.G. Sarkisyan, Corrigendum to "Robust limits on Lorentz violation from gamma-ray bursts" [Astropart. Phys. **25**, 402 (2006)]. Astropart. Phys. **29**, 158–159 (2008). doi:10.1016/j.astropartphys.2007.12.003

X. Fan, C.L. Carilli, B. Keating, Observational constraints on cosmic reionization. Annu. Rev. Astron. Astrophys. **44**, 415–462 (2006). doi:10.1146/annurev.astro.44.051905.092514

E.E. Fenimore, E. Ramirez-Ruiz, Redshifts for 220 BATSE gamma-ray bursts determined by variability and the cosmological consequences. ArXiv Astrophysics e-prints (2000)

C. Firmani, G. Ghisellini, G. Ghirlanda, V. Avila-Reese, A new method optimized to use gamma-ray bursts as cosmic rulers. Mon. Not. R. Astron. Soc. **360**, 1–5 (2005). doi:10.1111/j.1745-3933.2005.00023.x

C. Firmani, G. Ghisellini, V. Avila-Reese, G. Ghirlanda, Discovery of a tight correlation among the prompt emission properties of long gamma-ray bursts. Mon. Not. R. Astron. Soc. **370**, 185–197 (2006). doi:10.1111/j.1365-2966.2006.10445.x

D.A. Frail, S.R. Kulkarni, R. Sari, S.G. Djorgovski, J.S. Bloom, T.J. Galama, D.E. Reichart, E. Berger, F.A. Harrison, P.A. Price, S.A. Yost, A. Diercks, R.W. Goodrich, F. Chaffee, Beaming in gamma-ray bursts: evidence for a standard energy reservoir. Astrophys. J. Lett. **562**, 55–58 (2001). doi:10.1086/338119

A.S. Fruchter, A.J. Levan, L. Strolger, P.M. Vreeswijk, S.E. Thorsett, D. Bersier, I. Burud, J.M. Castro Cerón, A.J. Castro-Tirado, C. Conselice, T. Dahlen, H.C. Ferguson, J.P.U. Fynbo, P.M. Garnavich, R.A. Gibbons, J. Gorosabel, T.R. Gull, J. Hjorth, S.T. Holland, C. Kouveliotou, Z. Levay, M. Livio, M.R. Metzger, P.E. Nugent, L. Petro, E. Pian, J.E. Rhoads, A.G. Riess, K.C. Sahu, A. Smette, N.R. Tanvir, R.A.M.J. Wijers, S.E. Woosley, Long γ-ray bursts and core-collapse supernovae have different environments. Nature **441**, 463–468 (2006). doi:10.1038/nature04787

S.R. Furlanetto, A. Loeb, Metal absorption lines as probes of the intergalactic medium prior to the reionization epoch. Astrophys. J. **588**, 18–34 (2003). doi:10.1086/374045

J.P.U. Fynbo, D. Malesani, P. Jakobsson, Long gamma-ray burst host galaxies and their environments, in *Gamma-Ray Bursts*. Cambridge Astrophysics Series, vol. 51 (2012), pp. 269–301. doi:10.1017/CBO9780511980336.014

A. Gal-Yam, D.B. Fox, P.A. Price, E.O. Ofek, M.R. Davis, D.C. Leonard, A.M. Soderberg, B.P. Schmidt, K.M. Lewis, B.A. Peterson, S.R. Kulkarni, E. Berger, S.B. Cenko, R. Sari, K. Sharon, D. Frail, D.-S. Moon, P.J. Brown, A. Cucchiara, F. Harrison, T. Piran, S.E. Persson, P.J. McCarthy, B.E. Penprase, R.A. Chevalier, A.I. MacFadyen, A novel explosive process is required for the γ-ray burst GRB 060614. Nature **444**, 1053–1055 (2006). doi:10.1038/nature05373

R. Gambini, J. Pullin, Nonstandard optics from quantum space-time. Phys. Rev. D **59**(12), 124021 (1999). doi:10.1103/PhysRevD.59.124021

H. Gao, N. Liang, Z.-H. Zhu, Calibration of GRB luminosity relations with cosmography. Int. J. Mod. Phys. B **21**, 50016 (2012). doi:10.1142/S0218271812500162

H. Gao, X.-F. Wu, P. Mészáros, Cosmic transients test Einstein's equivalence principle out to GeV energies. Astrophys. J. **810**, 121 (2015). doi:10.1088/0004-637X/810/2/121

J.P. Gardner, J.C. Mather, M. Clampin, R. Doyon, M.A. Greenhouse, H.B. Hammel, J.B. Hutchings, P. Jakobsen, S.J. Lilly, K.S. Long, J.I. Lunine, M.J. McCaughrean, M. Mountain, J. Nella, G.H. Rieke, M.J. Rieke, H.-W. Rix, E.P. Smith, G. Sonneborn, M. Stiavelli, H.S. Stockman, R.A. Windhorst, G.S. Wright, The James Webb space telescope. Space Sci. Rev. **123**, 485–606 (2006). doi:10.1007/s11214-006-8315-7

N. Gehrels, Confidence limits for small numbers of events in astrophysical data. Astrophys. J. **303**, 336–346 (1986). doi:10.1086/164079

N. Gehrels, G. Chincarini, P. Giommi, K.O. Mason, J.A. Nousek, A.A. Wells, N.E. White, S.D. Barthelmy, D.N. Burrows, L.R. Cominsky, K.C. Hurley, F.E. Marshall, P. Mészáros, P.W.A. Roming, L. Angelini, L.M. Barbier, T. Belloni, S. Campana, P.A. Caraveo, M.M. Chester, O. Citterio, T.L. Cline, M.S. Cropper, J.R. Cummings, A.J. Dean, E.D. Feigelson, E.E. Fenimore, D.A. Frail, A.S. Fruchter, G.P. Garmire, K. Gendreau, G. Ghisellini, J. Greiner, J.E. Hill, S.D. Hunsberger, H.A. Krimm, S.R. Kulkarni, P. Kumar, F. Lebrun, N.M. Lloyd-Ronning, C.B. Markwardt, B.J. Mattson, R.F. Mushotzky, J.P. Norris, J. Osborne, B. Paczynski, D.M. Palmer, H.-S. Park, A.M. Parsons, J. Paul, M.J. Rees, C.S. Reynolds, J.E. Rhoads, T.P. Sasseen, B.E. Schaefer, A.T. Short, A.P. Smale, I.A. Smith, L. Stella, G. Tagliaferri, T. Takahashi, M. Tashiro, L.K. Townsley, J. Tueller, M.J.L. Turner, M. Vietri, W. Voges, M.J. Ward, R.

Willingale, F.M. Zerbi, W.W. Zhang, The Swift gamma-ray burst mission. Astrophys. J. **611**, 1005–1020 (2004). doi:10.1086/422091

N. Gehrels, J.P. Norris, S.D. Barthelmy, J. Granot, Y. Kaneko, C. Kouveliotou, C.B. Markwardt, P. Mészáros, E. Nakar, J.A. Nousek, P.T. O'Brien, M. Page, D.M. Palmer, A.M. Parsons, P.W.A. Roming, T. Sakamoto, C.L. Sarazin, P. Schady, M. Stamatikos, S.E. Woosley, A new γ-ray burst classification scheme from GRB060614. Nature **444**, 1044–1046 (2006). doi:10.1038/nature05376

N. Gehrels, E. Ramirez-Ruiz, D.B. Fox, Gamma-ray bursts in the Swift era. Annu. Rev. Astron. Astrophys. **47**, 567–617 (2009). doi:10.1146/annurev.astro.46.060407.145147

G. Ghirlanda, Advances on GRB as cosmological tools, in *AIP Conference Proceedings*, ed. by G. Giobbi, A. Tornambe, G. Raimondo, M. Limongi, L.A. Antonelli, N. Menci, E. Brocato American Institute of Physics Conference Series, vol. 1111, 2009, pp. 579–586. doi:10.1063/1.3141613

G. Ghirlanda, G. Ghisellini, D. Lazzati, The collimation-corrected gamma-ray burst energies correlate with the peak energy of their νF_ν spectrum. Astrophys. J. **616**, 331–338 (2004a). doi:10.1086/424913

G. Ghirlanda, G. Ghisellini, D. Lazzati, C. Firmani, Gamma-ray bursts: new rulers to measure the Universe. Astrophys. J. Lett. **613**, 13–16 (2004b). doi:10.1086/424915

G. Ghirlanda, G. Ghisellini, C. Firmani, Gamma-ray bursts as standard candles to constrain the cosmological parameters. New J. Phys. **8**, 123 (2006). doi:10.1088/1367-2630/8/7/123

S. Gillessen, F. Eisenhauer, S. Trippe, T. Alexander, R. Genzel, F. Martins, T. Ott, Monitoring stellar orbits around the massive black hole in the galactic center. Astrophys. J. **692**, 1075–1109 (2009). doi:10.1088/0004-637X/692/2/1075

L.J. Gou, P. Mészáros, T. Abel, B. Zhang, Detectability of long gamma-ray burst afterglows from very high redshifts. Astrophys. J. **604**, 508–520 (2004). doi:10.1086/382061

J.F. Graham, A.S. Fruchter, L.J. Kewley, E.M. Levesque, A.J. Levan, N.R. Tanvir, D.E. Reichart, M. Nysewander, Unusually high metallicity host of the dark LGRB 051022, in *AIP Conference Proceedings*, ed. by C. Meegan, C. Kouveliotou, N. Gehrels American Institute of Physics Conference Series, vol. 1133, 2009, pp. 269–272. doi:10.1063/1.3155900

T.H. Greif, J.L. Johnson, R.S. Klessen, V. Bromm, The observational signature of the first HII regions. Mon. Not. R. Astron. Soc. **399**, 639–649 (2009). doi:10.1111/j.1365-2966.2009.15336.x

T.H. Greif, S.C.O. Glover, V. Bromm, R.S. Klessen, The first galaxies: chemical enrichment, mixing, and star formation. Astrophys. J. **716**, 510–520 (2010). doi:10.1088/0004-637X/716/1/510

D. Guetta, T. Piran, Do long duration gamma ray bursts follow star formation? J. Cosmol. Astropart. Phys. **7**, 3 (2007). doi:10.1088/1475-7516/2007/07/003

D. Guetta, T. Piran, E. Waxman, The luminosity and angular distributions of long-duration gamma-ray bursts. Astrophys. J. **619**, 412–419 (2005). doi:10.1086/423125

J.-M. Hao, Y.-F. Yuan, Is the metallicity of the progenitor of long gamma-ray bursts really low? Astrophys. J. **772**, 42 (2013). doi:10.1088/0004-637X/772/1/42

D.H. Hartmann, Probing cosmic chemical evolution with GRBs. New Astron. Rev. **52**, 450–453 (2008). doi:10.1016/j.newar.2008.06.028

A. Heger, S.E. Woosley, The nucleosynthetic signature of population III. Astrophys. J. **567**, 532–543 (2002). doi:10.1086/338487

A. Heger, S.E. Woosley, Nucleosynthesis and evolution of massive metal-free stars. Astrophys. J. **724**, 341–373 (2010). doi:10.1088/0004-637X/724/1/341

A. Heger, C.L. Fryer, S.E. Woosley, N. Langer, D.H. Hartmann, How massive single stars end their life. Astrophys. J. **591**, 288–300 (2003). doi:10.1086/375341

R. Hirschi, G. Meynet, A. Maeder, Stellar evolution with rotation. XIII. Predicted GRB rates at various Z. Astron. Astrophys. **443**, 581–591 (2005). doi:10.1051/0004-6361:20053329

J. Hjorth, J. Sollerman, P. Møller, J.P.U. Fynbo, S.E. Woosley, C. Kouveliotou, N.R. Tanvir, J. Greiner, M.I. Andersen, A.J. Castro-Tirado, J.M. Castro Cerón, A.S. Fruchter, J. Gorosabel, P. Jakobsson, L. Kaper, S. Klose, N. Masetti, H. Pedersen, K. Pedersen, E. Pian, E. Palazzi, J.E. Rhoads, E. Rol, E.P.J. van den Heuvel, P.M. Vreeswijk, D. Watson, R.A.M.J. Wijers, A very energetic supernova associated with the γ-ray burst of 29 March 2003. Nature **423**, 847–850 (2003). doi:10.1038/nature01750

D.E. Holz, Lensing and high-z supernova surveys. Astrophys. J. Lett. **506**, 1–5 (1998). doi:10.1086/311631

A.M. Hopkins, J.F. Beacom, On the normalization of the cosmic star formation history. Astrophys. J. **651**, 142–154 (2006). doi:10.1086/506610

T. Hosokawa, K. Omukai, N. Yoshida, H.W. Yorke, Protostellar feedback halts the growth of the first stars in the Universe. Science **334**, 1250 (2011). doi:10.1126/science.1207433

G. Hurier, N. Aghanim, M. Douspis, E. Pointecouteau, Measurement of the $T_{\rm CMB}$ evolution from the Sunyaev-Zel'dovich effect. Astron. Astrophys. **561**, 143 (2014). doi:10.1051/0004-6361/201322632

L. Izzo, M. Muccino, E. Zaninoni, L. Amati, M. Della Valle, New measurements of Ω_m from gamma-ray bursts. ArXiv e-prints (2015)

U. Jacob, T. Piran, Lorentz-violation-induced arrival delays of cosmological particles. J. Cosmol. Astropart. Phys. **1**, 31 (2008). doi:10.1088/1475-7516/2008/01/031

R. Jimenez, T. Piran, Reconciling the gamma-ray burst rate and star formation histories. Astrophys. J. **773**, 126 (2013). doi:10.1088/0004-637X/773/2/126

R. Jimenez, L. Verde, T. Treu, D. Stern, Constraints on the equation of state of dark energy and the Hubble constant from stellar ages and the cosmic microwave background. Astrophys. J. **593**, 622–629 (2003). doi:10.1086/376595

J.L. Johnson, T.H. Greif, V. Bromm, Occurrence of metal-free galaxies in the early Universe. Mon. Not. R. Astron. Soc. **388**, 26–38 (2008). doi:10.1111/j.1365-2966.2008.13381.x

J.L. Johnson, T.H. Greif, V. Bromm, R.S. Klessen, J. Ippolito, The first galaxies: signatures of the initial starburst. Mon. Not. R. Astron. Soc. **399**, 37–47 (2009). doi:10.1111/j.1365-2966.2009.15158.x

D.O. Jones, S.A. Rodney, A.G. Riess, B. Mobasher, T. Dahlen, C. McCully, T.F. Frederiksen, S. Casertano, J. Hjorth, C.R. Keeton, A. Koekemoer, L.-G. Strolger, T.G. Wiklind, P. Challis, O. Graur, B. Hayden, B. Patel, B.J. Weiner, A.V. Filippenko, P. Garnavich, S.W. Jha, R.P. Kirshner, H.C. Ferguson, N.A. Grogin, D. Kocevski, The discovery of the most distant known type Ia supernova at redshift 1.914. Astrophys. J. **768**, 166 (2013). doi:10.1088/0004-637X/768/2/166

P.R. Kafle, S. Sharma, G.F. Lewis, J. Bland-Hawthorn, Kinematics of the stellar halo and the mass distribution of the Milky Way using blue horizontal branch stars. Astrophys. J. **761**, 98 (2012). doi:10.1088/0004-637X/761/2/98

N. Kawai, G. Kosugi, K. Aoki, T. Yamada, T. Totani, K. Ohta, M. Iye, T. Hattori, W. Aoki, H. Furusawa, K. Hurley, K.S. Kawabata, N. Kobayashi, Y. Komiyama, Y. Mizumoto, K. Nomoto, J. Noumaru, R. Ogasawara, R. Sato, K. Sekiguchi, Y. Shirasaki, M. Suzuki, T. Takata, T. Tamagawa, H. Terada, J. Watanabe, Y. Yatsu, A. Yoshida, An optical spectrum of the afterglow of a γ-ray burst at a redshift of $z = 6.295$. Nature **440**, 184–186 (2006). doi:10.1038/nature04498

J.A. King, J.K. Webb, M.T. Murphy, V.V. Flambaum, R.F. Carswell, M.B. Bainbridge, M.R. Wilczynska, F.E. Koch, Spatial variation in the fine-structure constant—new results from VLT/UVES. Mon. Not. R. Astron. Soc. **422**, 3370–3414 (2012). doi:10.1111/j.1365-2966.2012.20852.x

M.D. Kistler, H. Yüksel, J.F. Beacom, K.Z. Stanek, An unexpectedly Swift rise in the gamma-ray burst rate. Astrophys. J. Lett. **673**, 119–122 (2008). doi:10.1086/527671

M.D. Kistler, H. Yüksel, J.F. Beacom, A.M. Hopkins, J.S.B. Wyithe, The star formation rate in the reionization era as indicated by gamma-ray bursts. Astrophys. J. Lett. **705**, 104–108 (2009). doi:10.1088/0004-637X/705/2/L104

D. Kocevski, A.A. West, M. Modjaz, Modeling the GRB host galaxy mass distribution: are GRBs unbiased tracers of star formation? Astrophys. J. **702**, 377–385 (2009). doi:10.1088/0004-637X/702/1/377

Y. Kodama, D. Yonetoku, T. Murakami, S. Tanabe, R. Tsutsui, T. Nakamura, Gamma-ray bursts in suggest that the time variation of the dark energy is small. Mon. Not. R. Astron. Soc. **391**, 1–4 (2008). doi:10.1111/j.1745-3933.2008.00508.x

E. Komatsu, K.M. Smith, J. Dunkley, C.L. Bennett, B. Gold, G. Hinshaw, N. Jarosik, D. Larson, M.R. Nolta, L. Page, D.N. Spergel, M. Halpern, R.S. Hill, A. Kogut, M. Limon, S.S. Meyer, N. Odegard, G.S. Tucker, J.L. Weiland, E. Wollack, E.L. Wright, Seven-year Wilkinson microwave anisotropy probe (WMAP) observations: cosmological interpretation. Astrophys. J. Suppl. Ser. **192**, 18 (2011). doi:10.1088/0067-0049/192/2/18

S.S. Komissarov, M.V. Barkov, Supercollapsars and their X-ray bursts. Mon. Not. R. Astron. Soc. **402**, 25–29 (2010). doi:10.1111/j.1745-3933.2009.00792.x

V.A. Kostelecký, S. Samuel, Spontaneous breaking of Lorentz symmetry in string theory. Phys. Rev. D **39**, 683–685 (1989). doi:10.1103/PhysRevD.39.683

C. Kouveliotou, C.A. Meegan, G.J. Fishman, N.P. Bhat, M.S. Briggs, T.M. Koshut, W.S. Paciesas, G.N. Pendleton, Identification of two classes of gamma-ray bursts. Astrophys. J. Lett. **413**, 101–104 (1993). doi:10.1086/186969

L.M. Krauss, S. Tremaine, Test of the weak equivalence principle for neutrinos and photons. Phys. Rev. Lett. **60**, 176 (1988). doi:10.1103/PhysRevLett.60.176

R.-P. Kudritzki, J. Puls, Winds from hot stars. Annu. Rev. Astron. Astrophys. **38**, 613–666 (2000). doi:10.1146/annurev.astro.38.1.613

P. Kumar, B. Zhang, The physics of gamma-ray bursts and relativistic jets. Phys. Rep. **561**, 1–109 (2015). doi:10.1016/j.physrep.2014.09.008

D.Q. Lamb, D.E. Reichart, Gamma-ray bursts as a probe of the very high redshift universe. Astrophys. J. **536**, 1–18 (2000). doi:10.1086/308918

S.B. Lambert, C. Le Poncin-Lafitte, Determining the relativistic parameter γ using very long baseline interferometry. Astron. Astrophys. **499**, 331–335 (2009). doi:10.1051/0004-6361/200911714

S.B. Lambert, C. Le Poncin-Lafitte, Improved determination of γ by VLBI. Astron. Astrophys. **529**, 70 (2011). doi:10.1051/0004-6361/201016370

N. Langer, C.A. Norman, On the collapsar model of long gamma-ray bursts: constraints from cosmic metallicity evolution. Astrophys. J. Lett. **638**, 63–66 (2006). doi:10.1086/500363

T. Le, C.D. Dermer, On the redshift distribution of gamma-ray bursts in the Swift era. Astrophys. J. **661**, 394–415 (2007). doi:10.1086/513460

E.M. Levesque, The host galaxies of long-duration gamma-ray bursts. Publ. Astron. Soc. Pac. **126**, 1–14 (2014). doi:10.1086/674531

E.M. Levesque, L.J. Kewley, E. Berger, H.J. Zahid, The host galaxies of gamma-ray bursts. II. A mass-metallicity relation for long-duration gamma-ray burst host galaxies. Astron. J. **140**, 1557–1566 (2010). doi:10.1088/0004-6256/140/5/1557

L.-X. Li, Variation of the Amati relation with cosmological redshift: a selection effect or an evolution effect? Mon. Not. R. Astron. Soc. **379**, 55–59 (2007). doi:10.1111/j.1745-3933.2007.00333.x

L.-X. Li, Star formation history up to $z = 7.4$: implications for gamma-ray bursts and cosmic metallicity evolution. Mon. Not. R. Astron. Soc. **388**, 1487–1500 (2008). doi:10.1111/j.1365-2966.2008.13488.x

H. Li, J.-Q. Xia, J. Liu, G.-B. Zhao, Z.-H. Fan, X. Zhang, Overcoming the circular problem for gamma-ray bursts in cosmological global-fitting analysis. Astrophys. J. **680**, 92–99 (2008). doi:10.1086/529582

E. Liang, B. Zhang, Model-independent multivariable gamma-ray burst luminosity indicator and its possible cosmological implications. Astrophys. J. **633**, 611–623 (2005). doi:10.1086/491594

E. Liang, B. Zhang, Calibration of gamma-ray burst luminosity indicators. Mon. Not. R. Astron. Soc. **369**, 37–41 (2006). doi:10.1111/j.1745-3933.2006.00169.x

E. Liang, B. Zhang, F. Virgili, Z.G. Dai, Low-luminosity gamma-ray bursts as a unique population: luminosity function, local rate, and beaming factor. Astrophys. J. **662**, 1111–1118 (2007). doi:10.1086/517959

N. Liang, W.K. Xiao, Y. Liu, S.N. Zhang, A cosmology-independent calibration of gamma-ray burst luminosity relations and the Hubble diagram. Astrophys. J. **685**, 354–360 (2008). doi:10.1086/590903

H.-N. Lin, X. Li, S. Wang, Z. Chang, Are long gamma-ray bursts standard candles? Mon. Not. R. Astron. Soc. **453**, 128–132 (2015). doi:10.1093/mnras/stv1624

N.M. Lloyd-Ronning, C.L. Fryer, E. Ramirez-Ruiz, Cosmological aspects of gamma-ray bursts: luminosity evolution and an estimate of the star formation rate at high redshifts. Astrophys. J. **574**, 554–565 (2002). doi:10.1086/341059

M.J. Longo, New precision tests of the Einstein equivalence principle from SN1987A. Phys. Rev. Lett. **60**, 173–175 (1988). doi:10.1103/PhysRevLett.60.173

H.-J. Lü, E.-W. Liang, B.-B. Zhang, B. Zhang, A new classification method for gamma-ray bursts. Astrophys. J. **725**, 1965–1970 (2010). doi:10.1088/0004-637X/725/2/1965

G. Luzzi, M. Shimon, L. Lamagna, Y. Rephaeli, M. De Petris, A. Conte, S. De Gregori, E.S. Battistelli, Redshift dependence of the cosmic microwave background temperature from Sunyaev-Zeldovich measurements. Astrophys. J. **705**, 1122–1128 (2009). doi:10.1088/0004-637X/705/2/1122

D. Lynden-Bell, A method of allowing for known observational selection in small samples applied to 3CR quasars. Mon. Not. R. Astron. Soc. **155**, 95 (1971)

A.I. MacFadyen, S.E. Woosley, Collapsars: gamma-ray bursts and explosions in "Failed Supernovae". Astrophys. J. **524**, 262–289 (1999). doi:10.1086/307790

D. Macpherson, D.M. Coward, M.G. Zadnik, The potential for detecting gamma-ray burst afterglows from population III stars with the next generation of infrared telescopes. Astrophys. J. **779**, 73 (2013). doi:10.1088/0004-637X/779/1/73

U. Maio, B. Ciardi, K. Dolag, L. Tornatore, S. Khochfar, The transition from population III to population II-I star formation. Mon. Not. R. Astron. Soc. **407**, 1003–1015 (2010). doi:10.1111/j.1365-2966.2010.17003.x

P.A. Mazzali, J. Deng, K. Nomoto, D.N. Sauer, E. Pian, N. Tominaga, M. Tanaka, K. Maeda, A.V. Filippenko, A neutron-star-driven X-ray flash associated with supernova SN 2006aj. Nature **442**, 1018–1020 (2006). doi:10.1038/nature05081

C.F. McKee, J.C. Tan, The formation of the first stars. II. Radiative feedback processes and implications for the initial mass function. Astrophys. J. **681**, 771–797 (2008). doi:10.1086/587434

P.J. McMillan, Mass models of the Milky Way. Mon. Not. R. Astron. Soc. **414**, 2446–2457 (2011). doi:10.1111/j.1365-2966.2011.18564.x

R.A. Mesler, D.J. Whalen, J. Smidt, C.L. Fryer, N.M. Lloyd-Ronning, Y.M. Pihlström, The first gamma-ray bursts in the Universe. Astrophys. J. **787**, 91 (2014). doi:10.1088/0004-637X/787/1/91

P. Mészáros, Gamma-ray bursts. Rep. Prog. Phys. **69**, 2259–2321 (2006). doi:10.1088/0034-4885/69/8/R01

P. Mészáros, M.J. Rees, Population III gamma-ray bursts. Astrophys. J. **715**, 967–971 (2010). doi:10.1088/0004-637X/715/2/967

P. Molaro, S.A. Levshakov, M. Dessauges-Zavadsky, S. D'Odorico, The cosmic microwave background radiation temperature at $z_{\rm abs} = 3.025$ toward QSO 0347-3819. Astron. Astrophys. **381**, 64–67 (2002). doi:10.1051/0004-6361:20011698

P. Molaro, M. Centurión, J.B. Whitmore, T.M. Evans, M.T. Murphy, I.I. Agafonova, P. Bonifacio, S. D'Odorico, S.A. Levshakov, S. Lopez, C.J.A.P. Martins, P. Petitjean, H. Rahmani, D. Reimers, R. Srianand, G. Vladilo, M. Wendt, The UVES large program for testing fundamental physics I. Bounds on a change in α towards quasar HE 2217-2818. Astron. Astrophys. **555**, 68 (2013). doi:10.1051/0004-6361/201321351

E. Mörtsell, J. Sollerman, On the future of gamma-ray burst cosmology. J. Cosmol. Astropart. Phys. **6**, 9 (2005). doi:10.1088/1475-7516/2005/06/009

M.T. Murphy, J.K. Webb, V.V. Flambaum, Further evidence for a variable fine-structure constant from Keck/HIRES QSO absorption spectra. Mon. Not. R. Astron. Soc. **345**, 609–638 (2003). doi:10.1046/j.1365-8711.2003.06970.x

H. Nagakura, H. Ito, K. Kiuchi, S. Yamada, Jet propagations, breakouts, and photospheric emissions in collapsing massive progenitors of long-duration gamma-ray bursts. Astrophys. J. **731**, 80 (2011). doi:10.1088/0004-637X/731/2/80

D. Nakauchi, Y. Suwa, T. Sakamoto, K. Kashiyama, T. Nakamura, Long-duration X-ray flash and X-ray-rich gamma-ray bursts from low-mass population III stars. Astrophys. J. **759**, 128 (2012). doi:10.1088/0004-637X/759/2/128

P. Natarajan, B. Albanna, J. Hjorth, E. Ramirez-Ruiz, N. Tanvir, R. Wijers, The redshift distribution of gamma-ray bursts revisited. Mon. Not. R. Astron. Soc. **364**, 8–12 (2005). doi:10.1111/j.1745-3933.2005.00094.x

Y. Niino, Revisiting the metallicity of long-duration gamma-ray burst host galaxies: the role of chemical inhomogeneity within galaxies. Mon. Not. R. Astron. Soc. **417**, 567–572 (2011). doi:10.1111/j.1365-2966.2011.19299.x

J.P. Norris, G.F. Marani, J.T. Bonnell, Connection between energy-dependent lags and peak luminosity in gamma-ray bursts. Astrophys. J. **534**, 248–257 (2000). doi:10.1086/308725

P. Noterdaeme, P. Petitjean, C. Ledoux, S. López, R. Srianand, S.D. Vergani, A translucent interstellar cloud at $z = 2.69$ CO, H_2, and HD in the line-of-sight to SDSS J123714.60 + 064759.5. Astron. Astrophys. **523**, 80 (2010). doi:10.1051/0004-6361/201015147

M. Oguri, K. Takahashi, Gravitational lensing effects on the gamma-ray burst Hubble diagram. Phys. Rev. D **73**(12), 123002 (2006). doi:10.1103/PhysRevD.73.123002

S.P. Oh, Probing the dark ages with metal absorption lines. Mon. Not. R. Astron. Soc. **336**, 1021–1029 (2002). doi:10.1046/j.1365-8711.2002.05859.x

B.D. Oppenheimer, R. Davé, K. Finlator, Tracing the re-ionization-epoch intergalactic medium with metal absorption lines. Mon. Not. R. Astron. Soc. **396**, 729–758 (2009). doi:10.1111/j.1365-2966.2009.14771.x

J. Paul, J. Wei, S. Basa, S.-N. Zhang, The Chinese-French SVOM mission for gamma-ray burst studies. C. R. Phys. **12**, 298–308 (2011). doi:10.1016/j.crhy.2011.01.009

S. Perlmutter, G. Aldering, G. Goldhaber, R.A. Knop, P. Nugent, P.G. Castro, S. Deustua, S. Fabbro, A. Goobar, D.E. Groom, I.M. Hook, A.G. Kim, M.Y. Kim, J.C. Lee, N.J. Nunes, R. Pain, C.R. Pennypacker, R. Quimby, C. Lidman, R.S. Ellis, M. Irwin, R.G. McMahon, P. Ruiz-Lapuente, N. Walton, B. Schaefer, B.J. Boyle, A.V. Filippenko, T. Matheson, A.S. Fruchter, N. Panagia, H.J.M. Newberg, W.J. Couch, T.S.C. Project, Measurements of Ω and Λ from 42 high-redshift supernovae. Astrophys. J. **517**, 565–586 (1999). doi:10.1086/307221

P. Petitjean, S.D. Vergani, Gamma-ray bursts as probes of the distant Universe. C. R. Phys. **12**, 288–297 (2011). doi:10.1016/j.crhy.2011.01.007

V. Petrosian, E. Kitanidis, D. Kocevski, Cosmological evolution of long gamma-ray bursts and star formation rate. Astrophys. J. **806**, 44 (2015). doi:10.1088/0004-637X/806/1/44

E. Pian, P.A. Mazzali, N. Masetti, P. Ferrero, S. Klose, E. Palazzi, E. Ramirez-Ruiz, S.E. Woosley, C. Kouveliotou, J. Deng, A.V. Filippenko, R.J. Foley, J.P.U. Fynbo, D.A. Kann, W. Li, J. Hjorth, K. Nomoto, F. Patat, D.N. Sauer, J. Sollerman, P.M. Vreeswijk, E.W. Guenther, A. Levan, P. O'Brien, N.R. Tanvir, R.A.M.J. Wijers, C. Dumas, O. Hainaut, D.S. Wong, D. Baade, L. Wang, L. Amati, E. Cappellaro, A.J. Castro-Tirado, S. Ellison, F. Frontera, A.S. Fruchter, J. Greiner, K. Kawabata, C. Ledoux, K. Maeda, P. Møller, L. Nicastro, E. Rol, R. Starling, An optical supernova associated with the X-ray flash XRF 060218. Nature **442**, 1011–1013 (2006). doi:10.1038/nature05082

C. Porciani, P. Madau, On the association of gamma-ray bursts with massive stars: implications for number counts and lensing statistics. Astrophys. J. **548**, 522–531 (2001). doi:10.1086/319027

P.A. Price, A. Songaila, L.L. Cowie, J. Bell Burnell, E. Berger, A. Cucchiara, D.B. Fox, I. Hook, S.R. Kulkarni, B. Penprase, K.C. Roth, B. Schmidt, Properties of a gamma-ray burst host galaxy at $z \sim 5$. Astrophys. J. Lett. **663**, 57–60 (2007). doi:10.1086/520047

J.X. Prochaska, H.-W. Chen, M. Dessauges-Zavadsky, J.S. Bloom, Probing the interstellar medium near star-forming regions with gamma-ray burst afterglow spectroscopy: gas, metals, and dust. Astrophys. J. **666**, 267–280 (2007). doi:10.1086/520042

J.X. Prochaska, Y. Sheffer, D.A. Perley, J.S. Bloom, L.A. Lopez, M. Dessauges-Zavadsky, H.-W. Chen, A.V. Filippenko, M. Ganeshalingam, W. Li, A.A. Miller, D. Starr, The first positive detection of molecular gas in a GRB host galaxy. Astrophys. J. Lett. **691**, 27–32 (2009). doi:10.1088/0004-637X/691/1/L27

S. Qi, T. Lu, A new luminosity relation for gamma-ray bursts and its implication. Astrophys. J. **717**, 1274–1278 (2010). doi:10.1088/0004-637X/717/2/1274

S. Qi, F.-Y. Wang, T. Lu, Constraining the evolution of dark energy with type Ia supernovae and gamma-ray bursts. Astron. Astrophys. **483**, 49–55 (2008a). doi:10.1051/0004-6361:20079329

S. Qi, F.-Y. Wang, T. Lu, Gamma-ray burst contributions to constraining the evolution of dark energy. Astron. Astrophys. **487**, 853–855 (2008b). doi:10.1051/0004-6361:200809954

S. Qi, T. Lu, F.-Y. Wang, Deviation from the cosmological constant or systematic errors? Mon. Not. R. Astron. Soc. **398**, 78–82 (2009). doi:10.1111/j.1745-3933.2009.00713.x

S.-F. Qin, E.-W. Liang, R.-J. Lu, J.-Y. Wei, S.-N. Zhang, Simulations on high-z long gamma-ray burst rate. Mon. Not. R. Astron. Soc. **406**, 558–565 (2010). doi:10.1111/j.1365-2966.2010.16691.x

H. Rahmani, M. Wendt, R. Srianand, P. Noterdaeme, P. Petitjean, P. Molaro, J.B. Whitmore, M.T. Murphy, M. Centurion, H. Fathivavsari, S. D'Odorico, T.M. Evans, S.A. Levshakov, S. Lopez, C.J.A.P. Martins, D. Reimers, G. Vladilo, The UVES large program for testing fundamental physics—II. Constraints on a change in μ towards quasar HE 0027-1836. Mon. Not. R. Astron. Soc. **435**, 861–878 (2013). doi:10.1093/mnras/stt1356

J.E. Rhoads, How to tell a jet from a balloon: a proposed test for beaming in gamma-ray bursts. Astrophys. J. Lett. **487**, 1–4 (1997). doi:10.1086/310876

A.G. Riess, A.V. Filippenko, P. Challis, A. Clocchiatti, A. Diercks, P.M. Garnavich, R.L. Gilliland, C.J. Hogan, S. Jha, R.P. Kirshner, B. Leibundgut, M.M. Phillips, D. Reiss, B.P. Schmidt, R.A. Schommer, R.C. Smith, J. Spyromilio, C. Stubbs, N.B. Suntzeff, J. Tonry, Observational evidence from supernovae for an accelerating universe and a cosmological constant. Astron. J. **116**, 1009–1038 (1998). doi:10.1086/300499

B.E. Robertson, R.S. Ellis, Connecting the gamma ray burst rate and the cosmic star formation history: implications for reionization and galaxy evolution. Astrophys. J. **744**, 95 (2012). doi:10.1088/0004-637X/744/2/95

R. Salvaterra, G. Chincarini, The gamma-ray burst luminosity function in the light of the Swift 2 year data. Astrophys. J. Lett. **656**, 49–52 (2007). doi:10.1086/512606

R. Salvaterra, M. Della Valle, S. Campana, G. Chincarini, S. Covino, P. D'Avanzo, A. Fernández-Soto, C. Guidorzi, F. Mannucci, R. Margutti, C.C. Thöne, L.A. Antonelli, S.D. Barthelmy, M. de Pasquale, V. D'Elia, F. Fiore, D. Fugazza, L.K. Hunt, E. Maiorano, S. Marinoni, F.E. Marshall, E. Molinari, J. Nousek, E. Pian, J.L. Racusin, L. Stella, L. Amati, G. Andreuzzi, G. Cusumano, E.E. Fenimore, P. Ferrero, P. Giommi, D. Guetta, S.T. Holland, K. Hurley, G.L. Israel, J. Mao, C.B. Markwardt, N. Masetti, C. Pagani, E. Palazzi, D.M. Palmer, S. Piranomonte, G. Tagliaferri, V. Testa, GRB090423 at a redshift of $z \sim 8.1$. Nature **461**, 1258–1260 (2009). doi:10.1038/nature08445

R. Salvaterra, S. Campana, S.D. Vergani, S. Covino, P. D'Avanzo, D. Fugazza, G. Ghirlanda, G. Ghisellini, A. Melandri, L. Nava, B. Sbarufatti, H. Flores, S. Piranomonte, G. Tagliaferri, A complete sample of bright Swift long gamma-ray bursts. I. Sample presentation, luminosity function and evolution. Astrophys. J. **749**, 68 (2012). doi:10.1088/0004-637X/749/1/68

R. Sari, Linear polarization and proper motion in the afterglow of beamed gamma-ray bursts. Astrophys. J. Lett. **524**, 43–46 (1999). doi:10.1086/312294

S. Savaglio, K. Glazebrook, D. Le Borgne, S. Juneau, R.G. Abraham, H.-W. Chen, D. Crampton, P.J. McCarthy, R.G. Carlberg, R.O. Marzke, K. Roth, I. Jørgensen, R. Murowinski, The Gemini deep survey. VII. The redshift evolution of the mass-metallicity relation. Astrophys. J. **635**, 260–279 (2005). doi:10.1086/497331

B.E. Schaefer, The Hubble diagram to redshift 6 from 69 gamma-ray bursts. Astrophys. J. **660**, 16–46 (2007). doi:10.1086/511742

B.E. Schaefer, C.L. Gerardy, P. Höflich, A. Panaitescu, R. Quimby, J. Mader, G.J. Hill, P. Kumar, J.C. Wheeler, M. Eracleous, S. Sigurdsson, P. Mészáros, B. Zhang, L. Wang, F.V. Hessman, V. Petrosian, GRB 021004: a massive progenitor star surrounded by shells. Astrophys. J. **588**, 387–399 (2003). doi:10.1086/373896

M. Schmidt, Luminosities and space densities of gamma-ray bursts. Astrophys. J. Lett. **523**, 117–120 (1999). doi:10.1086/312281

R. Schneider, A. Ferrara, P. Natarajan, K. Omukai, First stars, very massive black holes, and metals. Astrophys. J. **571**, 30–39 (2002). doi:10.1086/339917

R. Schneider, K. Omukai, A.K. Inoue, A. Ferrara, Fragmentation of star-forming clouds enriched with the first dust. Mon. Not. R. Astron. Soc. **369**, 1437–1444 (2006). doi:10.1111/j.1365-2966.2006.10391.x

I.I. Shapiro, Fourth test of general relativity. Phys. Rev. Lett. **13**, 789–791 (1964). doi:10.1103/PhysRevLett.13.789

F.H. Shu, S. Lizano, D. Galli, J. Cantó, G. Laughlin, Self-similar champagne flows in H II regions. Astrophys. J. **580**, 969–979 (2002). doi:10.1086/343859

C. Sivaram, Constraints on the photon mass and charge and test of equivalence principle from GRB 990123. Bull. Astron. Soc. India **27**, 627 (1999)

A.M. Soderberg, S.R. Kulkarni, E. Nakar, E. Berger, P.B. Cameron, D.B. Fox, D. Frail, A. Gal-Yam, R. Sari, S.B. Cenko, M. Kasliwal, R.A. Chevalier, T. Piran, P.A. Price, B.P. Schmidt, G. Pooley, D.-S. Moon, B.E. Penprase, E. Ofek, A. Rau, N. Gehrels, J.A. Nousek, D.N. Burrows, S.E. Persson, P.J. McCarthy, Relativistic ejecta from X-ray flash XRF 060218 and the rate of cosmic explosions. Nature **442**, 1014–1017 (2006). doi:10.1038/nature05087

J. Sollerman, G. Östlin, J.P.U. Fynbo, J. Hjorth, A. Fruchter, K. Pedersen, On the nature of nearby GRB/SN host galaxies. New Astron. **11**, 103–115 (2005). doi:10.1016/j.newast.2005.06.004

D.N. Spergel, L. Verde, H.V. Peiris, E. Komatsu, M.R. Nolta, C.L. Bennett, M. Halpern, G. Hinshaw, N. Jarosik, A. Kogut, M. Limon, S.S. Meyer, L. Page, G.S. Tucker, J.L. Weiland, E. Wollack, E.L. Wright, First-year Wilkinson Microwave Anisotropy Probe (WMAP) observations: determination of cosmological parameters. Astrophys. J. Suppl. Ser. **148**, 175–194 (2003). doi:10.1086/377226

R. Srianand, H. Chand, P. Petitjean, B. Aracil, Limits on the time variation of the electromagnetic fine-structure constant in the low energy limit from absorption lines in the spectra of distant quasars. Phys. Rev. Lett. **92**(12), 121302 (2004). doi:10.1103/PhysRevLett.92.121302

R. Srianand, H. Chand, P. Petitjean, B. Aracil, Srianand et al. reply. Phys. Rev. Lett. **99**(23), 239002 (2007). doi:10.1103/PhysRevLett.99.239002

R. Srianand, P. Noterdaeme, C. Ledoux, P. Petitjean, First detection of CO in a high-redshift damped Lyman-α system. Astron. Astrophys. **482**, 39–42 (2008). doi:10.1051/0004-6361:200809727

R. Srianand, N. Gupta, P. Petitjean, P. Noterdaeme, C. Ledoux, C.J. Salter, D.J. Saikia, Search for cold gas in $z > 2$ damped Lyα systems: 21-cm and H_2 absorption. Mon. Not. R. Astron. Soc. **421**, 651–665 (2012). doi:10.1111/j.1365-2966.2011.20342.x

A. Stacy, V. Bromm, A. Loeb, Rotation speed of the first stars. Mon. Not. R. Astron. Soc. **413**, 543–553 (2011). doi:10.1111/j.1365-2966.2010.18152.x

A. Stacy, T.H. Greif, V. Bromm, The first stars: mass growth under protostellar feedback. Mon. Not. R. Astron. Soc. **422**, 290–309 (2012). doi:10.1111/j.1365-2966.2012.20605.x

K.Z. Stanek, T. Matheson, P.M. Garnavich, P. Martini, P. Berlind, N. Caldwell, P. Challis, W.R. Brown, R. Schild, K. Krisciunas, M.L. Calkins, J.C. Lee, N. Hathi, R.A. Jansen, R. Windhorst, L. Echevarria, D.J. Eisenstein, B. Pindor, E.W. Olszewski, P. Harding, S.T. Holland, D. Bersier, Spectroscopic discovery of the supernova 2003dh associated with GRB 030329. Astrophys. J. Lett. **591**, 17–20 (2003). doi:10.1086/376976

K.Z. Stanek, O.Y. Gnedin, J.F. Beacom, A.P. Gould, J.A. Johnson, J.A. Kollmeier, M. Modjaz, M.H. Pinsonneault, R. Pogge, D.H. Weinberg, Protecting life in the Milky Way: metals keep the GRBs away. Acta Astron. **56**, 333–345 (2006)

K.M. Svensson, A.J. Levan, N.R. Tanvir, A.S. Fruchter, L.-G. Strolger, The host galaxies of core-collapse supernovae and gamma-ray bursts. Mon. Not. R. Astron. Soc. **405**, 57–76 (2010). doi:10.1111/j.1365-2966.2010.16442.x

W.-W. Tan, X.-F. Cao, Y.-W. Yu, Determining the luminosity function of Swift long gamma-ray bursts with pseudo-redshifts. Astrophys. J. Lett. **772**, 8 (2013). doi:10.1088/2041-8205/772/1/L8

M. Tegmark, J. Silk, M.J. Rees, A. Blanchard, T. Abel, F. Palla, How small were the first cosmological objects? Astrophys. J. **474**, 1 (1997). doi:10.1086/303434

K. Toma, T. Sakamoto, P. Mészáros, Population III gamma-ray burst afterglows: constraints on stellar masses and external medium densities. Astrophys. J. **731**, 127 (2011). doi:10.1088/0004-637X/731/2/127

T. Totani, Cosmological gamma-ray bursts and evolution of galaxies. Astrophys. J. Lett. **486**, 71–74 (1997). doi:10.1086/310853

M. Trenti, R. Perna, R. Jimenez, The luminosity and stellar mass functions of GRB host galaxies: insight into the metallicity bias. Astrophys. J. **802**, 13 (2015). doi:10.1088/0004-637X/802/2/103

J.-P. Uzan, N. Aghanim, Y. Mellier, Distance duality relation from x-ray and Sunyaev-Zel'dovich observations of clusters. Phys. Rev. D **70**(8), 083533 (2004). doi:10.1103/PhysRevD.70.083533

P. Valageas, Weak gravitational lensing effects on the determination of Ω_m and Ω_Λ from SNeIa. Astron. Astrophys. **354**, 767–786 (2000)

V. Vasileiou, A. Jacholkowska, F. Piron, J. Bolmont, C. Couturier, J. Granot, F.W. Stecker, J. Cohen-Tanugi, F. Longo, Constraints on Lorentz invariance violation from Fermi-large area telescope observations of gamma-ray bursts. Phys. Rev. D **87**(12), 122001 (2013). doi:10.1103/PhysRevD.87.122001

J.S. Vink, A. de Koter, On the metallicity dependence of Wolf-Rayet winds. Astron. Astrophys. **442**, 587–596 (2005). doi:10.1051/0004-6361:20052862

F.J. Virgili, B. Zhang, K. Nagamine, J.-H. Choi, Gamma-ray burst rate: high-redshift excess and its possible origins. Mon. Not. R. Astron. Soc. **417**, 3025–3034 (2011). doi:10.1111/j.1365-2966.2011.19459.x
M. Visser, Jerk, snap and the cosmological equation of state. Class. Quantum Gravity **21**, 2603–2615 (2004). doi:10.1088/0264-9381/21/11/006
V. Vitagliano, J.-Q. Xia, S. Liberati, M. Viel, High-redshift cosmography. J. Cosmol. Astropart. Phys. **3**, 5 (2010). doi:10.1088/1475-7516/2010/03/005
P.M. Vreeswijk, C. Ledoux, A. Smette, S.L. Ellison, A.O. Jaunsen, M.I. Andersen, A.S. Fruchter, J.P.U. Fynbo, J. Hjorth, A. Kaufer, P. Møller, P. Petitjean, S. Savaglio, R.A.M.J. Wijers, Rapid-response mode VLT/UVES spectroscopy of GRB 060418. Conclusive evidence for UV pumping from the time evolution of Fe II and Ni II excited- and metastable-level populations. Astron. Astrophys. **468**, 83–96 (2007). doi:10.1051/0004-6361:20066780
D. Wanderman, T. Piran, The luminosity function and the rate of Swift's gamma-ray bursts. Mon. Not. R. Astron. Soc. **406**, 1944–1958 (2010). doi:10.1111/j.1365-2966.2010.16787.x
D. Wanderman, T. Piran, The rate, luminosity function and time delay of non-collapsar short GRBs. Mon. Not. R. Astron. Soc. **448**, 3026–3037 (2015). doi:10.1093/mnras/stv123
Y. Wang, Model-independent distance measurements from gamma-ray bursts and constraints on dark energy. Phys. Rev. D **78**(12), 123532 (2008). doi:10.1103/PhysRevD.78.123532
F.Y. Wang, Current constraints on early dark energy and growth index using latest observations. Astron. Astrophys. **543**, 91 (2012). doi:10.1051/0004-6361/201219348
F.Y. Wang, The high-redshift star formation rate derived from gamma-ray bursts: possible origin and cosmic reionization. Astron. Astrophys. **556**, 90 (2013). doi:10.1051/0004-6361/201321623
F.-Y. Wang, Z.-G. Dai, Constraining dark energy and cosmological transition redshift with type Ia supernovae. Chin. J. Astron. Astrophys. **6**, 561–571 (2006a). doi:10.1088/1009-9271/6/5/08
F.Y. Wang, Z.G. Dai, Constraining the cosmological parameters and transition redshift with gamma-ray bursts and supernovae. Mon. Not. R. Astron. Soc. **368**, 371–378 (2006b). doi:10.1111/j.1365-2966.2006.10108.x
F.Y. Wang, Z.G. Dai, High-redshift star formation rate up to $z \sim 8.3$ derived from gamma-ray bursts and influence of background cosmology. Mon. Not. R. Astron. Soc. **400**, 10–14 (2009). doi:10.1111/j.1745-3933.2009.00751.x
F.Y. Wang, Z.G. Dai, An evolving stellar initial mass function and the gamma-ray burst redshift distribution. Astrophys. J. Lett. **727**, 34 (2011a). doi:10.1088/2041-8205/727/2/L34
F.Y. Wang, Z.G. Dai, Weak gravitational lensing effects on cosmological parameters and dark energy from gamma-ray bursts. Astron. Astrophys. **536**, 96 (2011b). doi:10.1051/0004-6361/201117517
F.Y. Wang, Z.G. Dai, Self-organized criticality in X-ray flares of gamma-ray-burst afterglows. Nat. Phys. **9**, 465–467 (2013). doi:10.1038/nphys2670
F.Y. Wang, Z.G. Dai, Estimating the uncorrelated dark energy evolution in the Planck era. Phys. Rev. D **89**(2), 023004 (2014a). doi:10.1103/PhysRevD.89.023004
F.Y. Wang, Z.G. Dai, Long GRBs are metallicity-biased tracers of star formation: evidence from host galaxies and redshift distribution. Astrophys. J. Suppl. Ser. **213**, 15 (2014b). doi:10.1088/0067-0049/213/1/15
J.S. Wang, F.Y. Wang, Cosmological model of the interaction between dark matter and dark energy. Astron. Astrophys. **564**, 137 (2014). doi:10.1051/0004-6361/201322606
F.Y. Wang, Z.G. Dai, Z.-H. Zhu, Measuring dark energy with gamma-ray bursts and other cosmological probes. Astrophys. J. **667**, 1–10 (2007). doi:10.1086/520768
F.Y. Wang, Z.G. Dai, S. Qi, Probing the cosmographic parameters to distinguish between dark energy and modified gravity models. Astron. Astrophys. **507**, 53–59 (2009). doi:10.1051/0004-6361/200911998
Y. Wang, Y.-Z. Fan, D.-M. Wei, Are GRB 090423 and similar bursts due to superconducting cosmic strings? Phys. Rev. Lett. **106**(25), 259001 (2011a). doi:10.1103/PhysRevLett.106.259001
F.-Y. Wang, S. Qi, Z.-G. Dai, The updated luminosity correlations of gamma-ray bursts and cosmological implications. Mon. Not. R. Astron. Soc. **415**, 3423–3433 (2011b). doi:10.1111/j.1365-2966.2011.18961.x
F.Y. Wang, V. Bromm, T.H. Greif, A. Stacy, Z.G. Dai, A. Loeb, K.S. Cheng, Probing pre-galactic metal enrichment with high-redshift gamma-ray bursts. Astrophys. J. **760**, 27 (2012). doi:10.1088/0004-637X/760/1/27
F.Y. Wang, S.X. Yi, Z.G. Dai, Similar radiation mechanism in gamma-ray bursts and blazars: evidence from two luminosity correlations. Astrophys. J. Lett. **786**, 8 (2014). doi:10.1088/2041-8205/786/1/L8
F.Y. Wang, Z.G. Dai, E.W. Liang, Gamma-ray burst cosmology. New Astron. Rev. **67**, 1–17 (2015). doi:10.1016/j.newar.2015.03.001
H. Wei, Observational constraints on cosmological models with the updated long gamma-ray bursts. J. Cosmol. Astropart. Phys. **8**, 20 (2010). doi:10.1088/1475-7516/2010/08/020
J.-J. Wei, X.-F. Wu, F. Melia, The gamma-ray burst Hubble diagram and its implications for cosmology. Astrophys. J. **772**, 43 (2013). doi:10.1088/0004-637X/772/1/43

J.-J. Wei, X.-F. Wu, F. Melia, D.-M. Wei, L.-L. Feng, Cosmological tests using gamma-ray bursts, the star formation rate and possible abundance evolution. Mon. Not. R. Astron. Soc. **439**, 3329–3341 (2014). doi:10.1093/mnras/stu166

J.-J. Wei, Q.-B. Ma, X.-F. Wu, Utilizing the updated gamma-ray bursts and type Ia supernovae to constrain the Cardassian expansion model and dark energy. Adv. Astron. **2015**, 20 (2015). doi:10.1155/2015/576093

R.A.M.J. Wijers, J.S. Bloom, J.S. Bagla, P. Natarajan, Gamma-ray bursts from stellar remnants—probing the universe at high redshift. Mon. Not. R. Astron. Soc. **294**, 13–17 (1998). doi:10.1046/j.1365-8711.1998.01328.x

C.M. Will, The confrontation between general relativity and experiment. Living Rev. Relativ. **9**, 3 (2006). doi:10.12942/lrr-2006-3

C.M. Will, The confrontation between general relativity and experiment. Living Rev. Relativ. **17**, 4 (2014). doi:10.12942/lrr-2014-4

C. Wolf, P. Podsiadlowski, The metallicity dependence of the long-duration gamma-ray burst rate from host galaxy luminosities. Mon. Not. R. Astron. Soc. **375**, 1049–1058 (2007). doi:10.1111/j.1365-2966.2006.11373.x

S.E. Woosley, Gamma-ray bursts from stellar mass accretion disks around black holes. Astrophys. J. **405**, 273–277 (1993). doi:10.1086/172359

S.E. Woosley, J.S. Bloom, The supernova gamma-ray burst connection. Annu. Rev. Astron. Astrophys. **44**, 507–556 (2006). doi:10.1146/annurev.astro.43.072103.150558

S.E. Woosley, A. Heger, The progenitor stars of gamma-ray bursts. Astrophys. J. **637**, 914–921 (2006). doi:10.1086/498500

S.E. Woosley, T.A. Weaver, The evolution and explosion of massive stars. II. Explosive hydrodynamics and nucleosynthesis. Astrophys. J. Suppl. Ser. **101**, 181 (1995). doi:10.1086/192237

S.-W. Wu, D. Xu, F.-W. Zhang, D.-M. Wei, Gamma-ray bursts: the isotropic-equivalent-energy function and the cosmic formation rate. Mon. Not. R. Astron. Soc. **423**, 2627–2632 (2012). doi:10.1111/j.1365-2966.2012.21068.x

J.S.B. Wyithe, A. Loeb, Reionization of hydrogen and helium by early stars and quasars. Astrophys. J. **586**, 693–708 (2003). doi:10.1086/367721

J.-Q. Xia, V. Vitagliano, S. Liberati, M. Viel, Cosmography beyond standard candles and rulers. Phys. Rev. D **85**(4), 043520 (2012). doi:10.1103/PhysRevD.85.043520

C.Y. Xu, D.M. Wei, The effect of the initial mass function on the burst rate of GRBs. Acta Astron. Sin. **49**, 387–393 (2008)

D. Xu, Z.G. Dai, E.W. Liang, Can gamma-ray bursts be used to measure cosmology? A further analysis. Astrophys. J. **633**, 603–610 (2005). doi:10.1086/466509

D. Yonetoku, T. Murakami, T. Nakamura, R. Yamazaki, A.K. Inoue, K. Ioka, Gamma-ray burst formation rate inferred from the spectral peak energy-peak luminosity relation. Astrophys. J. **609**, 935–951 (2004). doi:10.1086/421285

S.-C. Yoon, N. Langer, Evolution of rapidly rotating metal-poor massive stars towards gamma-ray bursts. Astron. Astrophys. **443**, 643–648 (2005). doi:10.1051/0004-6361:20054030

S.-C. Yoon, N. Langer, C. Norman, Single star progenitors of long gamma-ray bursts. I. Model grids and redshift dependent GRB rate. Astron. Astrophys. **460**, 199–208 (2006). doi:10.1051/0004-6361:20065912

N. Yoshida, T. Abel, L. Hernquist, N. Sugiyama, Simulations of early structure formation: primordial gas clouds. Astrophys. J. **592**, 645–663 (2003). doi:10.1086/375810

B. Yu, S. Qi, T. Lu, Gamma-ray burst luminosity relations: two-dimensional versus three-dimensional correlations. Astrophys. J. Lett. **705**, 15–19 (2009). doi:10.1088/0004-637X/705/1/L15

Y.-W. Yu, K.S. Cheng, M.C. Chu, S. Yeung, Cosmic histories of star formation and reionization: an analysis with a power-law approximation. J. Cosmol. Astropart. Phys. **7**, 23 (2012). doi:10.1088/1475-7516/2012/07/023

H. Yu, F.Y. Wang, Z.G. Dai, K.S. Cheng, An unexpectedly low-redshift excess of Swift gamma-ray burst rate. Astrophys. J. Suppl. Ser. **218**, 13 (2015). doi:10.1088/0067-0049/218/1/13

H. Yüksel, M.D. Kistler, Enhanced cosmological GRB rates and implications for cosmogenic neutrinos. Phys. Rev. D **75**(8), 083004 (2007). doi:10.1103/PhysRevD.75.083004

H. Yüksel, M.D. Kistler, J.F. Beacom, A.M. Hopkins, Revealing the high-redshift star formation rate with gamma-ray bursts. Astrophys. J. Lett. **683**, 5–8 (2008). doi:10.1086/591449

B. Zhang, Astrophysics: a burst of new ideas. Nature **444**, 1010–1011 (2006). doi:10.1038/4441010a

B. Zhang, Gamma-ray bursts in the Swift era. Chin. J. Astron. Astrophys. **7**, 1–50 (2007). doi:10.1088/1009-9271/7/1/01

B. Zhang, Gamma-ray burst prompt emission. Int. J. Mod. Phys. B **23**, 30002 (2014). doi:10.1142/S021827181430002X

S. Zhang, B.-Q. Ma, Lorentz violation from gamma-ray bursts. Astropart. Phys. **61**, 108–112 (2015). doi:10.1016/j.astropartphys.2014.04.008

B. Zhang, B.-B. Zhang, F.J. Virgili, E.-W. Liang, D.A. Kann, X.-F. Wu, D. Proga, H.-J. Lv, K. Toma, P. Mészáros, D.N. Burrows, P.W.A. Roming, N. Gehrels, Discerning the physical origins of cosmological gamma-ray bursts based on multiple observational criteria: the cases of $z = 6.7$ GRB 080913, $z = 8.2$ GRB 090423, and some short/hard GRBs. Astrophys. J. **703**, 1696–1724 (2009). doi:10.1088/0004-637X/703/2/1696

G.-B. Zhao, R.G. Crittenden, L. Pogosian, X. Zhang, Examining the evidence for dynamical dark energy. Phys. Rev. Lett. **109**(17), 171301 (2012). doi:10.1103/PhysRevLett.109.171301

Space Sci Rev (2016) 202:237–279
DOI 10.1007/s11214-016-0274-z

Perspectives on Gamma-Ray Burst Physics and Cosmology with Next Generation Facilities

Weimin Yuan[1] · Lorenzo Amati[2] · John K. Cannizzo[3] · Bertrand Cordier[4] · Neil Gehrels[5] · Giancarlo Ghirlanda[6] · Diego Götz[4] · Nicolas Produit[7] · Yulei Qiu[1] · Jianchao Sun[8] · Nial R. Tanvir[9] · Jianyan Wei[1] · Chen Zhang[1]

Received: 20 December 2015 / Accepted: 29 June 2016 / Published online: 21 July 2016

Abstract High-redshift Gamma-Ray Bursts (GRBs) beyond redshift ~6 are potentially powerful tools to probe the distant early Universe. Their detections in large numbers and at truly high redshifts call for the next generation of high-energy wide-field instruments with unprecedented sensitivity at least one order of magnitude higher than the ones currently in orbit. On the other hand, follow-up observations of the afterglows of high-redshift GRBs and identification of their host galaxies, which would be difficult for the currently operating telescopes, require new, extremely large facilities of at multi-wavelengths. This

✉ W. Yuan
wmy@nao.cas.cn

L. Amati
amati@iasfbo.inaf.it

J.K. Cannizzo
john.k.cannizzo@nasa.gov

B. Cordier
bertrand.cordier@cea.fr

N. Gehrels
neil.gehrels@nasa.gov

G. Ghirlanda
giancarlo.ghirlanda@brera.inaf.it

D. Götz
diego.gotz@cea.fr

N. Produit
Nicolas.Produit@unige.ch

Y. Qiu
qiuyl@nao.cas.cn

J. Sun
sunjc@ihep.ac.cn

N.R. Tanvir
nrt3@le.ac.uk

J. Wei
wjy@nao.cas.cn

chapter describes future experiments that are expected to advance this exciting field, both being currently built and being proposed. The legacy of Swift will be continued by SVOM, which is equipped with a set of space-based multi-wavelength instruments as well as and a ground segment including a wide angle camera and two follow-up telescopes. The established Lobster-eye X-ray focusing optics provides a promising technology for the detection of faint GRBs at very large distances, based on which the THESEUS, Einstein Probe and other mission concepts have been proposed. Follow-up observations and exploration of the reionization era will be enabled by large facilities such as SKA in the radio, the 30 m class telescopes in the optical/near-IR, and the space-borne WFIRST and JWST in the optical/near-IR/mid-IR. In addition, the X-ray and γ-ray polarization experiment POLAR is also introduced.

Keywords Gamma-ray bursts · High-redshift · Gamma-ray · X-ray · Instrumentation

1 Introduction

Understanding the structure formation and the first stars around the epoch of reionization is a major driver for the next generation of space- and ground-based astronomical facilities. Being the brightest objects in the Universe, Gamma-Ray Bursts (GRBs) beyond redshift $\sim$6 shine through the dark age as cosmic beacons and are thus potentially powerful probes of the distant early Universe. So far, seven GRBs with redshifts 6.2–9.4 have been detected with the Swift satellite in its 10-year operation. New discoveries will continue to be made by the GRB mission SVOM in the foreseeable future, which is equipped with a set of space-based multi-wavelength instruments as well as a ground segment including a wide angle camera and two follow-up telescopes. Looking far into the future, a full exploration of the early Universe requires a much larger sample ($\sim$100) and events at even higher redshifts, i.e. redshift $\sim$10 and beyond. This calls for the next generation of high-energy wide-field instruments with unprecedented sensitivity one order of magnitude or more higher than Swift/BAT. The proposed THESEUS and Einstein Probe missions (and others alike), which

C. Zhang
chzhang@nao.cas.cn

[1] Key Laboratory of Space Astronomy and Technology, National Astronomical Observatories, Chinese Academy of Sciences, Datun Rd 20A, Beijing 100012, China

[2] INAF–IASF Bologna, via P. Gobetti 101, 40129 Bologna, Italy

[3] CRESST/Joint Center for Astrophysics, Univ. of Maryland, Baltimore County, Baltimore, MD 21250, USA

[4] CEA-Irfu, Service d'Astrophysique, 91191 Gif-sur-Yvette, France

[5] Astrophysics Division, NASA/Goddard Space Flight Center, Greenbelt, MD 20771, USA

[6] Istituto Nazionale di Astrofisica INAF, Osservatorio Astronomico di Brera, Via E. Bianchi 46, 23807 Merate, LC, Italy

[7] ISDC, Chemin d' Ecogia 16, 1290 Versoix, Switzerland

[8] Institute of High-Energy Physics, Chinese Academy of Sciences, Yuquan Lu 19B, Shijingshan District, Beijing, China

[9] Dept. of Physics and Astronomy, University of Leicester, University Road, Leicester, LE1 7RH, UK

are based on novel technology of X-ray focusing, are promising concepts to achieve these goals. On the other hand, follow-up observations in multi-wavebands, particularly in the near- and mid-infrared, of the afterglows of high-redshift GRBs are key to identify their host galaxies, to measure the redshifts, and to enable detailed astrophysical and cosmological studies. This can be facilitated by the advent of the ground- and space-based large telescopes at multi-wavelengths currently being under construction, such as the 30-m class optical/near-IR telescopes, Square Kilometer Array (SKA) in the radio, and the space-borne WFIRST and JWST missions in the optical/near-IR/mid-IR. Last, complementary to the exploration of the early Universe, future extensive measurement of the X-ray and γ-ray polarization of GRBs, expected to be achieved by POLAR will open up a new dimension in understanding the radiation region geometry and radiation mechanism of GRB. This chapter describes future experiments that are expected to advance this exciting field, both currently being built and being proposed, including SVOM (Sect. 2; contributed by B. Cordier, D. Götz, Y. Qiu and J. Wei), POLAR (Sect. 3; J.C. Sun and N. Produit), WFIRST & JWST (Sect. 4; N. Gehrels and J.K. Cannizzo), TMT, GMT & E-ELT (Sect. 5; N.T. Tanvir), SKA (Sect. 6; G. Ghirlanda), Einstein Probe (Sect. 7; W. Yuan and C. Zhang), and THESEUS (Sect. 8; L. Amati).

2 SVOM

The success of the Swift mission for catching GRBs and other transients illustrates the benefit of its unique combination of space agility, fast communication with the ground, multi-wavelength observing capability, and long lifetime. These capabilities have permitted the detection of nearly 1000 GRBs and the measure of nearly 300 redshifts, have offered a new look at GRB progenitors and have led to several important discoveries, like the existence of GRBs at the epoch of the re-ionization of the universe. The aim of SVOM (Space-based multi-band astronomical Variable Objects Monitor) is to continue the exploration of the transient universe with a set of space-based multi-wavelength instruments, following the way opened by Swift. SVOM is a space mission developed in cooperation between the Chinese National Space Agency (CNSA), the Chinese Academy of Science (CAS) and the French Space Agency (CNES). The mission features a medium size satellite, a set of space and ground instruments designed to detect, locate and follow-up GRBs of all kinds, a anti-Sun pointing strategy allowing the immediate follow-up of SVOM GRBs with ground based telescopes, and a fast data transmission to the ground. The satellite (see Fig. 1) carries two wide field high energy instruments: a coded-mask gamma-ray imager called ECLAIRs, and a gamma-ray spectrometer called GRM, and two narrow field telescopes that can measure the evolution of the afterglow after a slew of the satellite: an X-ray telescope called MXT and an optical telescope called VT. The ground segment includes additional instrumentation: a wide angle optical camera (GWAC) monitoring the field of view of ECLAIRs in real time during part of the orbit, and two 1-meter class robotic follow-up telescopes (the GFTs). SVOM has some unique features: an energy threshold of ECLAIRs at 4 keV enabling the detection of faint soft GRBs (e.g. XRFs and high-redshift GRBs); a good match in sensitivity between the X-ray and optical space telescopes which permits the detection of most GRB afterglows with both telescopes; and a set of optical instruments on the ground dedicated to the mission. The mission has recently been confirmed by the Chinese and French space agencies for a launch in 2021, and it has entered in an active phase of development. We present an overview of the scientific objectives of SVOM in the next sub-section, and a brief description of the instrumental aspects of the mission in the following ones.

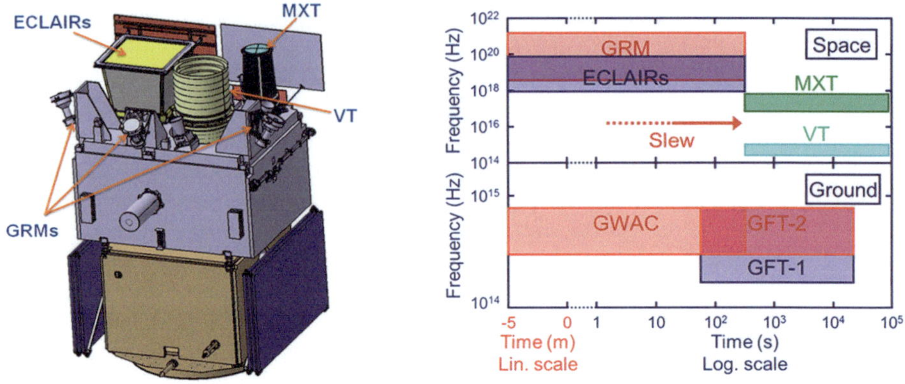

Fig. 1 Schematic showing the SVOM spacecraft with its multi-wavelength space payload. It consists of two wide-field instruments: ECLAIRs and the Gamma-Ray Monitor (GRM) for the observation of the GRB prompt emission and two narrow field instruments: the Micro channel X-ray Telescope (MXT) and the Visible Telescope (VT) for the observation of the GRB afterglow emission. *Right*: Space and ground instruments join to enable a unique coverage in time and wavelength

2.1 Scientific Objectives

The main science goal of SVOM is the study of cosmic transients detected in hard X-rays and in the optical domain. While the mission has been designed for the study of GRBs, it is also well suited for the study of other types of high-energy transients like Tidal Disruption Events (TDEs), Active Galactic Nuclei (AGNs), or galactic X-ray binaries and magnetars. For this type of sources, SVOM is both a "discovery machine", with wide field instruments that survey a significant fraction of the sky (ECLAIRs, GRM and GWAC), and a "follow-up machine", with fast pointing telescopes in space and on the ground (MXT, VT, and GFTs) that provide a multi-wavelength follow-up of different kinds of sources, with good sensitivity and a high duty cycle. The follow-up can be triggered by the satellite itself or from the ground, upon reception of a request for target of opportunity observations (ToO). In this section we concentrate on the objectives of SVOM for GRBs. Through GRB population simulations we have evaluated a GRB detection rate of 70–80 GRBs/yr for ECLAIRs and 90 GRBs/yr for GRM. One essential goal of SVOM is to provide GRBs with a redshift measurement. The redshift is required to measure the energetics of the burst and the epoch at which the GRB occurred in the history of the universe. Four elements in the design of the mission concur to facilitate the measure of the redshift for SVOM GRBs: a near anti-solar pointing during most of the orbit ensuring that SVOM GRBs can be quickly observed with ground based telescopes, a good sensitivity of the on-board optical telescope permitting the rapid identification of high-z candidates (which are not detected at visible wavelengths), Near Infra-Red follow-up on the ground to look for the afterglows of dark/high redshift GRBs, and agreements with the community to promote the optical spectroscopy of SVOM GRBs using large telescopes. With this strategy we expect to measure the redshift of more than 50% of SVOM GRBs, constructing a sample that could potentially be more representative of the intrinsic GRB population than the Swift sample. One drawback of this pointing strategy is that it avoids the galactic plane. Galactic sources can nevertheless be observed during target of opportunity pointings. In the following, we discuss a few selected topics about GRBs for which SVOM may bring decisive progress.

2.1.1 GRB Progenitors

In order to get a better understanding of the GRB phenomenon SVOM was designed to detect all kinds of GRBs, and to provide extensive multi-wavelength observations of the prompt GRB and its afterglow. In addition to classical long GRBs that will be detected by the two wide field instruments (ECLAIRs and GRM), SVOM will routinely detect short bursts with its gamma-ray monitor and soft GRBs (XRFs or highly redshifted GRBs) with ECLAIRs. The ability to detect soft GRBs with a spectral energy distribution (SED) peaking below 20–30 keV will favor the detection of X-ray flashes. As shown by BeppoSAX and HETE-2 (Heise et al. 2001; Sakamoto et al. 2004), these faint GRBs can only be detected if they are close enough and if their spectral energy distribution peaks at low energies (because faint GRBs with a SED peaking at high energies radiate too few photons to be detected). The detection of X-Ray Flashes in the local universe ($z < 0.1$) will permit detailed studies of their associated supernovae (as was the case for XRF 020903; Soderberg et al. 2005), providing crucial clues to understand the broader context of the SN-GRB connection.

2.1.2 GRB Physics

The instrument suite of SVOM will provide good multi-wavelength coverage of GRBs. For those occurring in the field of view of GWAC, the prompt emission will be measured from 1 eV to 5 MeV, with the GWACs, ECLAIRs, and the GRM. The prompt optical emission may also be observed by one GFT for GRBs lasting longer than 40 seconds. GRB afterglows will be observed with the two narrow field instruments on-board the satellite and with the ground follow-up telescopes on Earth. For some GRBs, SVOM will provide a very complete view of the phenomenon and its evolution, hopefully bringing new insight into the complex physics at work in these events. A lesson learned from Swift is that a few well observed GRBs may crucially improve our understanding of GRB physics, as was the case for GRB 130427, a bright nearby burst detected by Swift, Fermi and various optical and radio telescopes on the ground (Maselli et al. 2014; Perley et al. 2014). The physical processes at work within the jet remain not well understood even after the observation of hundreds of GRBs. Comprehensive discussions of the theoretical challenges connected with the understanding of the prompt GRB emission can be found in Kumar and Zhang (2015) and Zhang (2014). The authors show that several crucial questions connected to the physics of the ultra-relativistic jet and its interaction with the surrounding medium remain unanswered, like the nature and content of the jet (is the energy stored in the baryons or in magnetic fields?), the mechanisms of particle acceleration, the micro physics and the dominant radiation processes, the importance of the reverse shock, the role of pairs, etc. Performing multi-wavelength observations during the prompt emission and the early afterglow, SVOM will provide key observations to understand the physics of relativistic jets. GRBs detected with SVOM will also benefit from contemporaneous or follow-up observations with a novel generation of powerful instruments, like CTA (the Cerenkov Telescope Array) for very high energy photons, LSST (Large Synoptic Survey Telescope) for optical transients associated with on-axis and off-axis GRBs, and the precursors of SKA (the Square Kilometer Array) in radio.

2.1.3 Cosmology

Their extreme luminosity of GRBs permits their detection in hard X-rays up to very high redshifts, possibly beyond $z = 10$ and the spectroscopy of the optical afterglow provides the redshift of the burst and a tomographic vision of the line of sight to the burst. With Swift and the measure of about 300 redshifts, GRBs are providing new diagnostics of the distant Universe. When the signal to noise ratio (SNR) of the optical spectrum of the afterglow is sufficient, we get detailed information on the circumburst medium, on the gas and dust in the host galaxy, on the intergalactic medium and the intervening systems. With a smaller SNR, the measure of the redshift allows locating the time of the explosion in the history of the universe and reconstructing the history of the GRB formation rate, which traces the formation rate of massive stars. There is of course a special interest in very distant GRBs ($z > 5$), which provide a unique view on the young Universe, especially since they occur in galaxies which are undetectable with other methods of observation. One exciting challenge of GRB missions is the detection of GRBs resulting from the explosion of population III stars (the first generation of stars formed with pristine gas containing no metals). Such events are expected to be rare, to occur at high redshifts, to have no detectable hosts (except in absorption in the spectrum of their optical afterglow) and afterglows that are only detectable in the near infrared. They could be similar to some very long GRBs found at lower redshift (Mészáros and Rees 2010; Gendre et al. 2013). We expect to detect about 5 GRBs/yr at redshift $z > 5$ with ECLAIRs, but they will be useful only if we can measure their redshift. One difficulty is that the optical afterglows of GRBs fade very quickly, and after a few hours they are often too faint to permit measuring the redshift of the burst. In order to quickly identify high-z candidates that deserve deep spectroscopy in the NIR, we rely on the sensitivity of VT and on fast NIR follow-up telescopes on the ground. Dark GRBs, whose afterglows are not detected in the VT, are good candidates for high-z bursts, but they can also be extinct by dust in the vicinity of the source. The nature of these events (distant or extinct GRB) will be confirmed quickly with fast visible/NIR photometry from the ground, allowing the most appropriate spectroscopic follow-up. In some cases, GRBs detected by SVOM could trigger follow-up observations by the JWST (James Webb Space Telescope) for the study of the hosts.

2.2 Space Borne Instruments

In this section and the following, we give a brief description of the instruments of SVOM. In order to be both a discovery machine and a follow-up machine, SVOM features two types of instruments: space based instruments and ground based instruments. We briefly describe these instruments below.

2.2.1 ECLAIRs the Hard X-Ray Coded Mask Imaging Camera

ECLAIRs (see Fig. 2) is the instrument on-board the satellite that will detect and locate the GRBs. ECLAIRs is made of four parts: a pixellated detection plane (1024 cm^2) with its read-out electronics, a coded mask, a shield defining a field of view of 2 steradians ($89° \times 89°$), and a processing unit in charge of detecting and locating transient sources. The detection plane is made of 200 modules of 32 CdTe detectors each, for a total of 6400 detectors of size $4 \times 4 \times 1$ mm. Each module is read by a customized ASIC connected to an electronics that encodes the position, the time and the energy of each photon. One of the requirements of ECLAIRs is to reach an energy threshold of 4 keV, in order to study soft GRBs like X-Ray Flashes and highly redshifted GRBs. As shown in Fig. 2, the first modules satisfy the low energy threshold requirement (Godet et al. 2014). The coded mask is a square of side 54 cm located at a distance of 46 cm from the detection plane; it has an opening fraction of 40% and provides a localization accuracy of several arc minutes ($\sim 14'$ for a source at the

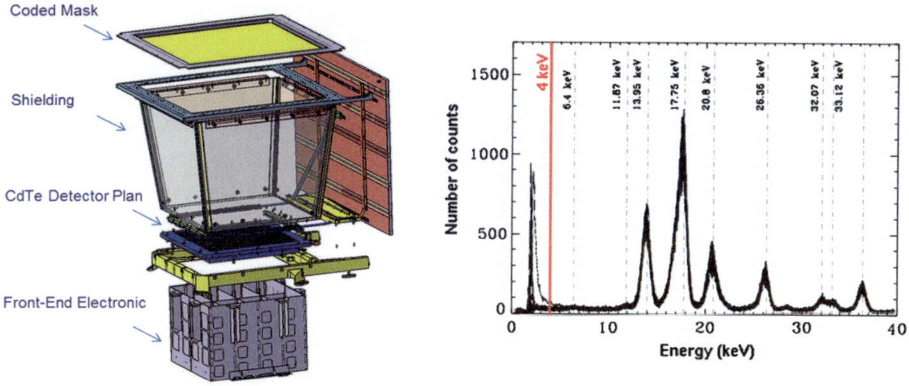

Fig. 2 Schematics showing the different sub-systems of ECLAIRs except the data processing unit in charge of the GRB detection and localization. *Right*: Laboratory spectral measurements performed on a detector module prototype (a 32 CdTe pixel matrix) with radioactive sources (241 Am). The *red vertical line* corresponds to the expected low energy threshold of the ECLAIRs camera [6]

limit of detection). The instrument features a count rate trigger and an image trigger, like Swift. These triggers are computed, from the photon data, in several energy bands and on time-scales ranging from 10 ms to several minutes. Our simulations show that ECLAIRs will detect 70–80 GRBs/yr.

2.2.2 GRM: The Gamma-Ray Monitor

GRM consists of a set of three detection modules. Each of them is made of a scintillating crystal (sodium iodide), a photomultiplier and its readout electronics. Each detector has a surface area of 200 cm^2 and a thickness of 1.5 cm. One piece of plastic scintillator in front of NaI (Tl) is used to distinguish low energy electrons from normal X-rays. The three modules are pointed at different directions to form a total field of view of 2 sr, within which rough (*of the order of 10 degrees radius*) localization of transient sources can be achieved on-board. The energy range of the GRM is 15–5000 keV, extending the energy range of ECLAIRs towards high energies to measure E_{Peak} for a large fraction of SVOM GRBs. We expect that GRM will detect >90 GRBs/yr. GRM will have a good sensitivity to short/hard GRBs, like the GBM of Fermi. GRM can generate on-board GRM-only triggers, taking use of only GRM detectors. Such triggers with localization information will be transferred to ECLAIRs for trigger enhancement on the short GRBs, and to ground facilities (e.g. GWAC, GW experiments) for joint observations. A calibration detector containing one radioactive ^{241}Am isotope is installed on the edge of each detection module, for the purpose of gain monitoring and energy calibration. In addition, a particle monitor auxiliary to GRM can generate South Atlantic Anomaly alerts and help protecting the detection modules.

2.2.3 MXT: The Microchannel X-Ray Telescope

MXT (see Fig. 3) is a light and compact focusing X-ray telescope designed to observe and measure the properties of the GRB X-ray afterglow after a slew of the satellite. The telescope will implement a novel technique to focus X-rays, based on micropore optics arranged in a lobster eye geometry. The use of micropore optics instead of classical X-ray

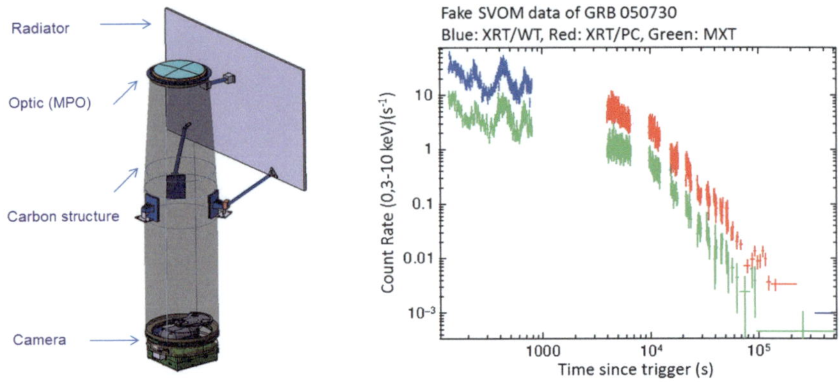

Fig. 3 Schematics showing the different sub-systems of the MXT. *Right*: XRT measured (*blue* and *red points*) versus MXT simulated (*green points*) *light curve* of GRB 050730, a GRB with a median flux in the Swift/XRT afterglow database

electro-formed mirrors permits a significant reduction of the size and weight of the telescope, fitting on a medium size satellite like SVOM. The optics has a diameter of 24 cm and a focal length of 1 meter (Götz et al. 2014). MXT will make use of the radiation hard pn-CCD detector developed for the DUO mission and adopted in a larger version for the eROSITA mission (Meidinger et al. 2014; Predehl et al. 2014). The MXT detector is made of 256 × 256 Si pixels of 75 μm side and has an expected energy resolution of 75 eV at 1 keV. MXT will be operated in the energy range of 0.2–10 keV, will have an effective area of 45 cm^2 at 1 keV, and a field of view of 64 × 64 arc minutes. Despite the smaller effective area with respect to XRT on board Swift at 1 keV (about 120 cm^2) MXT will be able to fully characterize the GRBs light curves (as shown in Fig. 3). In fact with a sensitivity of 7×10^{-13} $\mathrm{erg\,cm^{-2}\,s^{-1}}$ in 10^4 s, MXT will detect the afterglows of more than 90% of SVOM GRBs. Indeed its sensitivity is well adapted to early GRB afterglow observations, and its PSF of about 4 arcmin will allow to significantly reduce the ECLAIRs error boxes. Simulations based on the Swift-XRT database show that the localization accuracy for the MXT is $\sim 13''$ for 50% of the bursts, 10 minutes after the trigger (statistical uncertainty only).

2.2.4 *VT: The Visible Telescope*

The Visible Telescope (VT) is a dedicated optical follow-up telescope on board the SVOM satellite. Its main purpose is to detect and observe the optical afterglows of gamma-ray bursts localized by ECLAIRs. It is a Ritchey–Chretien telescope (see Fig. 4) with a diameter of 40 cm and an f-ratio of 9. Its limiting magnitude is about 22.5 (M_V) for an integration time of 300 seconds VT is designed to maximize the detection efficiency of GRB's optical afterglows. Instead of a filter wheel, a dichroic beam splitter is used to divide the light into two channels, in which the GRB afterglow can be observed simultaneously. Their wavelength ranges are from 0.4 μm to 0.65 μm (blue channel) and from 0.65 μm to 1 μm (red channel). Each channel is equipped with a $2k \times 2k$ CCD detector. While the CCD for the blue channel is a normal thinned back-illuminated one, a deep-depleted one is adopted for the red channel to obtain high sensitivity at long wavelength. The Quantum Efficiency (QE) of the red-channel CCD at 0.9 μm is over 50%, which enables VT to have the capability of detecting GRBs with the redshift larger than 6.5. The field of view of VT is

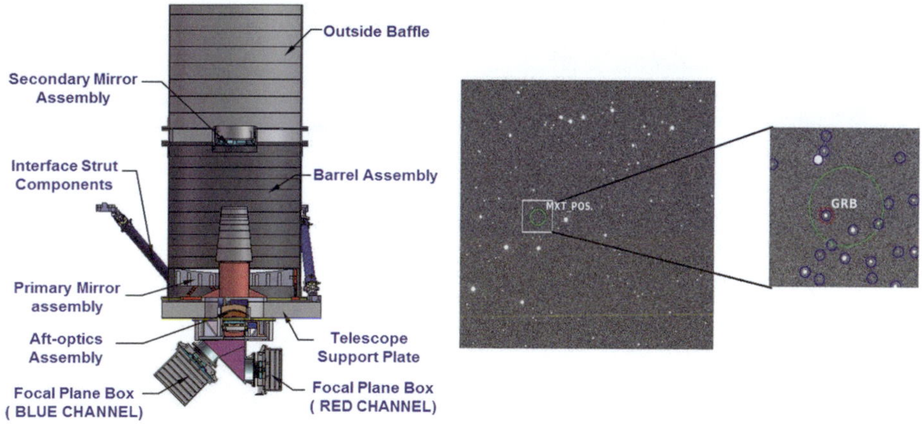

Fig. 4 Schematic showing the different sub-systems of the VT. *Right*: a simulated image of the VT and its associated finding chart

about $26' \times 26'$, which can cover the error box of ECLAIRs in most cases. Both CCDs have a pixel size of 13.5 μm × 13.5 μm, corresponding to spatial resolutions of 0.77 arc second. This ensures the GRB positioning accuracy to be greatly improved by VT from several arc minutes (ECLAIRs) and tens *of* arc-seconds (MXT) to a level of sub-arc second.

In order to promptly provide the GRB alerts with the sub-arc second accuracy, VT will do some data processing on board. After a GRB has been localized by the co-aligned MXT, lists of sources are extracted from the VT sub-images whose centerers and sizes are determined by the GRB positions and the corresponding error boxes provided by MXT. The lists are immediately downlinked to the ground through the VHF network. Then, the ground software will make finding charts with these lists (see Fig. 4) and search the optical counterparts of the GRB by comparing the lists with the existing catalogs. If a counterpart is identified, an alert will be then produced and distributed to world-wide astronomical community, which is useful for triggering large ground-based telescopes to measure the redshifts of the GRBs by spectroscopy. VT is expected to do a good job on detecting high-redshift GRBs. The confirmed high-redshift GRBs are rare in the Swift era, in contrast to a theoretical prediction of a fraction of more than 10% (Salvaterra 2015). This is probably due to the fact that for most Swift GRBs the early-time optical imaging follow-up is not deep enough for a quick identification and some faint GRBs cannot be spectroscopically observed in time by the large ground-based telescopes. This passive situation will be significantly improved by SVOM, due to the high sensitivity of VT, in particular at long wavelength, and the prompt optical-counterpart alerts. Additionally, the anti-solar pointing strategy of SVOM allows GRBs to be spectroscopically observed by large ground-based telescopes at the early time of the bursts. Consequently, more high-redshift GRBs are expected to be identified in the SVOM era. VT is also used to support the platform to achieve the required high pointing stability. A Fine Guidance Sensor (FGS) is mounted on the VT focal plane to measure relative image motions. Its images are processed in real time by a specialized data processing unit to get the centroid positions of several stars brighter than the magnitude of 15 (M_V). The results are sent at a frequency of 1 Hz to the platform to improve the pointing stability, which enables VT to have a good performance in a long exposure time.

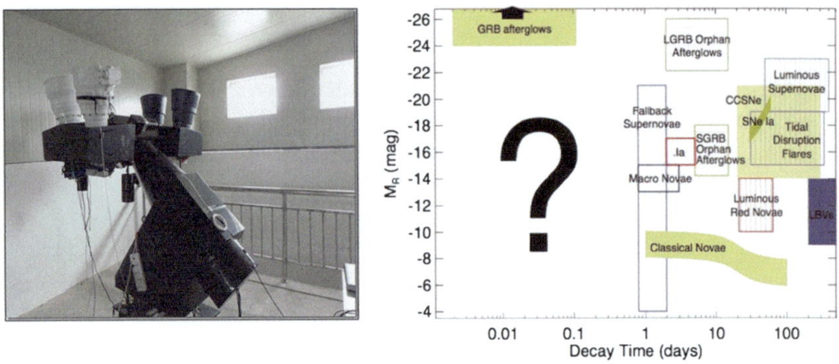

Fig. 5 First prototype of one GWAC module. The final system will be composed by 9 modules. *Right*: the figure shows the discovery space of an instrument dedicated to short time-scale optical transients (Wozniak et al. 2009)

2.3 Ground Based Instruments

The ground follow-up instruments constitute an important part of the mission. Three instruments are developed for the follow-up of SVOM GRBs: a wide angle camera that surveys a significant fraction of the sky for transients, and two robotic telescopes. In addition to these dedicated instruments, the SVOM collaboration will seek agreements with various existing telescopes or networks willing to contribute to the follow-up of SVOM GRBs.

2.3.1 GWAC: The Ground Wide Angle Camera

GWAC (see Fig. 5) provides a unique way to survey a large field of view for optical transients. The instrument will monitor 63% of ECLAIRs field of view, looking for optical transients occurring before, during and after GRBs. GWAC will also have its own trigger system, providing alerts to the world. GWAC is a complex system: the heart of the system is a set of 36 wide angle cameras with a diameter of 18 cm and a focal length of 22 cm, together these cameras cover a field of view of 5000 sq. deg. They use $4k \times 4k$ CCD detectors, sensitive in the range of wavelength 500–800 nm. These cameras reach a limiting magnitude $V = 16$ (5σ) in a typical 10 second exposure. This set of cameras is completed by two 60 cm robotic telescopes. equipped with EMCCD cameras. These telescopes will provide multicolor photometry of the transients discovered by GWAC with a temporal resolution $\leq$1 second.

2.3.2 GFTs: The Ground Follow-up Telescopes

The ground follow-up telescopes have two main goals. Firstly, they measure the photometric evolution of the optical afterglow in the first minutes after the trigger in a broad range of visible and NIR wavelengths, with a temporal resolution of few seconds. Secondly, when an afterglow is detected, they provide its position with arc second precision within 5 minutes of the trigger. Some essential features of the GFTs are their field of view, their size, and their sensitivity in the near infrared. The field of view ($\sim$30 arc minutes) enables observing quickly the entire error boxes of ECLAIRs. The size, typically 1 meter, allows the detection of all visible (i.e. non-dark) afterglows at the condition to arrive within few minutes after the trigger (Akerlof and Swan 2007; Klotz et al. 2009). Finally, the near infrared sensitivity permits the detection of high-z GRBs and GRBs extinct by dust, whose afterglow are obscured in the visible domain (Greiner et al. 2008). GFTs are especially useful for the study of the early afterglow during the slew of the satellite, and for the rapid identification of the optical afterglow in various cases: when SVOM cannot slew to the burst or when the slew is delayed due to pointing constraints, and when the optical afterglow is only visible in the NIR. One telescope is located in China at Xinglong Observatory and the other one will be located in Mexico at San Pedro Mártir.

2.4 The System

In order to facilitate measuring the redshifts of GRBs detected with ECLAIRs, the instruments of SVOM will be pointed close to the anti-solar direction. Most of the year the optical axis of the SVOM instruments will be pointed at about 45° from the anti-solar direction. This pointing is interleaved with avoidance periods during which the satellite passes away from the Sco X-1 source and the galactic plane. This strategy ensures that SVOM GRBs will be in the night hemisphere and quickly observable from the ground by large telescopes. More details on the SVOM pointing strategy, can be found in Cordier et al. (2008).

As soon as a GRB will be located, its coordinates and its main characteristics will be sent to the ground within seconds with a VHF antenna. The VHF signal will be received by one of the ∼40 ground stations distributed around the Earth below the orbit. The data will then be relayed to the Operation Center, which will send SVOM alerts to the internet via the GCN and VO Event networks (http://gcn.gsfc.nasa.gov/), and to the ground instruments GWAC and the GFTs. SVOM can also perform target of opportunity observations (with MXT and VT for instance), with a delay of few hours, which depends on the availability of uplink communication with the satellite.

SVOM will try to select high-z GRB candidates by analyzing MXT and VT data, and multi-band photometry data from GFTs. If a GRB is detected by MXT in the soft X-rays, but not by VT in the optical band, it will be selected as a candidate of high-z or an optically dark GRB. Ground telescopes with NIR capabilities are encouraged to follow up these candidates as soon as possible to try to measure their redshift. GFTs will be able to measure photometric redshifts by observing GRB afterglows with multiple filters from the visible to the NIR bands.

2.4.1 Conclusion

SVOM will be a highly versatile astronomy satellite, with built-in multi-wavelength capabilities, autonomous re-pointing and dedicated ground follow-up. With its peculiar pointing strategy and the low energy detection threshold, SVOM is expected to improve the number of GRBs detected at high redshift, and hence to contribute to the use of GRBs as probes of the young Universe.

Beyond the GRB studies emphasized here, SVOM will bring new observations about all types of high energy transients, in particular those of extragalactic origin (TDEs, AGNs, etc.).

3 POLAR—Space-Borne Gamma-Ray Bursts Compton Polarimeter

Until recent years, only a few polarization measurement results on Gamma-Ray Burst (GRB) prompt emissions have been published (Yonetoku et al. 2011), while in the same

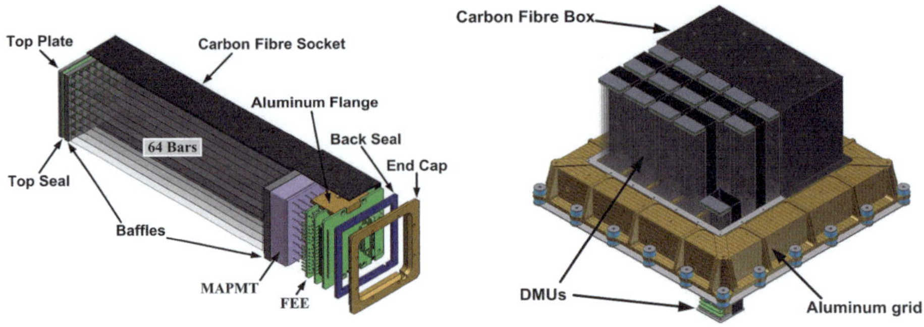

Fig. 6 Detector design of POLAR. *Left*: Detector modular unit; *Right*: polarization detector OBOX

time more and more hard X-ray/Gamma-ray polarization measurement instruments have been put into operation or under development. All these efforts will accumulate a larger observation database on GRB's polarization.

POLAR (Produit et al. 2005) is a novel space-borne Compton polarimeter dedicated for the precise measurement of the polarization of GRB's prompt emission, *which is expected to be on-board the Chinese space laboratory "Tiangong-2 (TG-2)" with a scheduled launch date in late 2016*. The future detailed measurement of the polarization of GRBs will lead to a better understanding of its radiation region geometry and emission mechanisms. POLAR will contribute greatly to two aspects on the polarization observation of GRBs. One is its high sensitivity which is minimum detectable polarization (MDP). The MDP of POLAR can be as low as $\sim$10% (Suarez 2010). The other one is its high observation statistics. $\sim$50 GRBs/year are expected to be detected by POLAR. This will provide significant results for restricting the radiation mechanism of GRBs.

3.1 Design of POLAR Detector

POLAR is composed of the polarization detector (OBOX) and electronic cabinet (IBOX). OBOX consists of 25 detector modular units (DMU). Each DMU is composed of 64 low-Z material plastic scintillator (EJ-248M) bars, read-out by a flat-panel multi-anode photomultipliers H8500 and ASIC front-end electronics. The CAD view of DMU and OBOX are shown in Fig. 6.

The linear polarization degree and polarization direction of the polarized Gamma-rays can be reconstructed by POLAR through Compton scattering effect. In order to improve the Compton scattering efficiency, the low-Z material plastic scintillator has been selected as the detection material of POLAR as it is more stable chemically and mechanically.

3.2 Detection Principle

The main interaction between the hard X-rays/Gamma-rays within the energy range 50 keV $\sim$ 500 keV and plastic scintillator material is Compton scattering. The differential cross-section is given by the Klein–Nishina formula for the polarized Compton scattering interaction process:

$$\begin{aligned}\frac{d\sigma_p}{d\Omega} &= \frac{r_0^2}{2}\varepsilon^2\left[\varepsilon+\varepsilon^{-1}-2\sin^2\theta\cos^2\eta\right]\\ &= \frac{r_0^2}{2}\varepsilon^2\left[\varepsilon+\varepsilon^{-1}-\sin^2\theta+\sin^2\theta\cos\left(2\left(\eta+\frac{\pi}{2}\right)\right)\right],\end{aligned} \qquad (1)$$

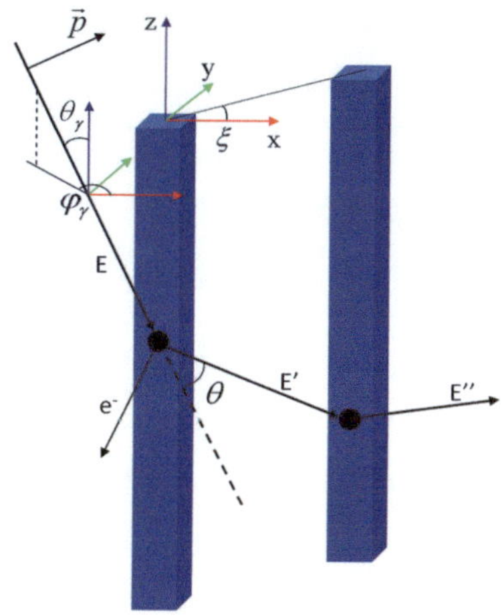

Fig. 7 Detection principle of POLAR by Compton scattering effect

where r_0 is the classic radius of an electron and $\varepsilon = E'/E$ is the ratio of the energy of the scattered photon and the energy of the incident photon, θ is the Compton scattering angle and η is the scattering azimuth angle.

An incident photon interacts with one of the 1600 PS bars in POLAR through Compton scattering effect; the recoil electron will be absorbed by the PS bar and its deposited energy will be detected by POLAR and readout by the following electronics. The scattered photon will interact with next PS bars until being absorbed or escaping the detector array. For the unpolarized Gamma-rays, the scattering azimuth angle is isotropic, while for the polarized Gamma-rays, the scattering azimuth angle distribution is related to the incident photons' polarization degree and polarization angle. Therefore, the polarization property of the incident Gamma-rays can be reconstructed by analyzing the distribution of the Compton scattering azimuth angles. Figure 7 shows the detection principle of POLAR.

In Fig. 7, $\vec{P}$ is the polarization vector of the incident Gamma-rays, θ_γ and φ_γ describe the incident photons' direction, e^- is the recoil electron, ξ is the projection of the Compton scattering azimuth angle in the detector plane.

3.3 Physics Performance Study with Monte-Carlo Method

The performance of POLAR has been studied using the Monte-Carlo simulation method. The GEANT4 toolkit is used for the simulations. The Band (Band et al. 1993) function is used to create the spectrum of an GRB. Typically, the standard GRB spectrum ($\alpha = -1.0$, $\beta = -2.5$ and $E_{peak} = 200$ keV) is used for the study. Figure 8 shows the triggered events on the 1600 detection channels of POLAR with different GRB incident angles.

For different incident angles of GRBs, the reconstructed modulation factor by POLAR can be quite different. Figure 9 shows the modulation factor measured by POLAR as a function of the incident angle of GRB when the GRB photons are 100% polarized. More simulation results are discussed else where (Suarez 2010; Xiong et al. 2009; Sun 2012).

3.4 Calibration Tests

In 2015, the full flight model of POLAR has been constructed. A series of calibration activities have been performed to verify the basic working function of the instrument, test and optimize its working parameters, study the polarization measurement performance of the

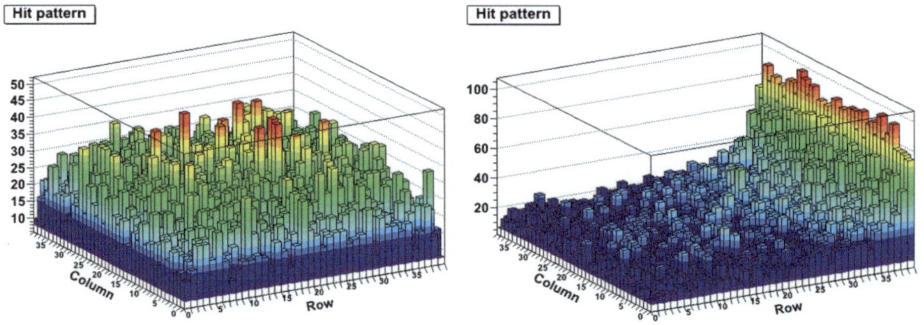

Fig. 8 Triggering pattern of POLAR detector with standard incident GRB photons. *Left*: $\theta_\gamma = 0°$, $\varphi_\gamma = 0°$; *Right*: $\theta_\gamma = 45°$, $\varphi_\gamma = 0°$

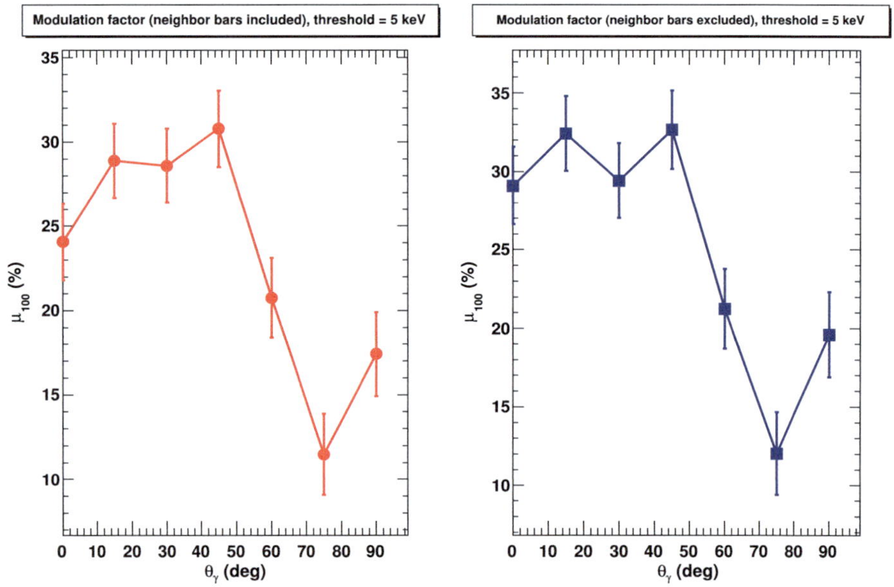

Fig. 9 Simulated modulation factor (100% polarized) as a function of the incident angle θ_γ of GRB; φ_γ is set to 0° for simplicity. *Left*: neighboring bars included; *Right*: neighboring bars excluded

detector, verify and improve the data analysis procedure and algorithm, as well as to optimize the Monte-Carlo simulation model of POLAR which will be used in the in-orbit data analysis. Among all of the calibration tests, the most important one is the polarization measurement of POLAR performed with the European Synchrotron Radiation Facility (ESRF) beam. The beamline used for the test is ID11 which can provide us with 100% polarized hard X-rays between 35 keV $\sim$ 140 keV. The beam of ID11 is horizontally polarized. The size of the beam used for POLAR calibration is $0.5 \times 0.5\ \text{mm}^2$ with original beam intensity $\sim 10^7$ phs/s, which was reduced to about $10^3 \sim 10^4$ phs/s with absorber to avoid the data pileup in the electronics.

During the ESRF beam test, four different beam energies were used, i.e., 140 keV, 110 keV, 80 keV and 60 keV. For each beam energy, three different beam incident angles (0°,

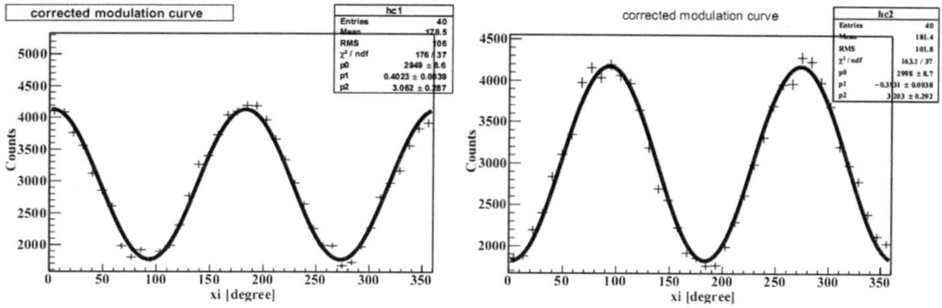

Fig. 10 Modulation curve measured during the ESRF beam tests. *Left*: the polarization direction of the beam relative to POLAR detector is 90°; *Right*: the polarization direction of the beam relative to POLAR detector is 0°

Table 1 Main technical properties of POLAR

No.	Property	Performance
1	Detector material	Plastic scintillator (EJ-248M)
2	Yearly detectable GRBs	~50
3	GRB localization accuracy	$\leq 5°$ ($F_{\text{total}} \geq 10^{-5}$ erg cm^{-2})
4	Detection energy range	50 keV ~ 500 keV
5	Field of view	$\pm 70° \times \pm 70°$
6	Minimum detectable polarization	$<10\%$ ($F_{\text{total}} \geq 3 \times 10^{-5}$ erg cm^{-2})

30° and 60°) were chosen, where 0° is on-axis, 30° and 60° are off-axis. Figure 10 shows the very preliminary analysis results of the reconstructed Compton scattering azimuth angle distributions (modulation curves) with 140 keV beam. When fitting the modulation curves, the calculated modulation factors for 90° and 0° polarized beam are $40.23\% \pm 0.0039\%$ and $39.31\% \pm 0.0038\%$, respectively. The measurement results are quite similar to the Monte-Carlo simulation result of 40.90%, which verifies the accuracy of the model for the Monte-Carlo simulations.

3.5 Summary and Discussions

For GRB prompt emissions, the most interesting energy range is about 100 keV ~ 200 keV, within which POLAR has its best performance. The Monte-Carlo simulation results show that for the GRBs with total energy fluence F_{total} larger than 3×10^{-5} erg cm^{-2}, the minimum detectable polarization (MDP) of POLAR can reach below ~10%. In addition, for strong GRBs POLAR can localize their positions with accuracy better than 5° ($F_{\text{total}} \geq 10^{-5}$ erg cm^{-2}). The main technical properties of POLAR are summarized in Table 1. The calibration test results show that POLAR performs as expected on the polarization measurement. After eliminating all kinds of system effects of the instrument during the data analysis procedure and optimizing the Monte-Carlo simulation model, the final measurement results and simulation results are quite similar. The calibration results as well as the optimized instrument working parameters will be used in the further Monte-Carlo simulations and real in-orbit operations.

POLAR has also undergone and passed the space qualification tests, such as thermal cycling, vibration, shock and thermal vacuum, etc. These tests can verify the reliability of POLAR instrument and guarantee its long-term operation and observation.

4 *WFIRST* and *JWST*

4.1 *WFIRST*

4.1.1 Instrumentation

WFIRST consists of two instruments, a Wide-Field Instrument (WFI) and a coronagraph.

The WFI will be capable of imaging and spectroscopy over thousands of square degrees and monitoring of SNe and microlensing fields. It has sensitivity in the 0.7–2 μm bandpass and a 0.28 deg^2 field-of-view (FOV), about a hundred times that of *JWST*. The WFI has 18 H4RG[1] detectors (288 Mpixels) and utilizes 6 filter imaging. The wide field instrument includes two channels, a wide field channel and an integral field unit (IFU) spectrograph channel. The wide field channel includes three mirrors and a filter/grism wheel to provide an imaging mode covering 0.76–2.0 μm and a spectroscopy mode covering 1.35–1.95 μm. The IFU channel uses an image slicer and spectrograph to provide individual spectra of each 0.15 arcsec wide strip covering the 0.6–2.0 μm over a 3.00×3.15 arcsec FOV.

The coronagraph can image ice and gas giant exoplanets as well as debris disks. It has a 400–1000 nm bandpass, $\leq 10^{-9}$ contrast, and a 100 msec inner working angle (at 400 nm). The coronagraph instrument includes an imaging mode and a spectroscopic mode to perform exoplanet direct imaging and spectroscopic characterization of planets and debris disks around nearby stars.

4.1.2 Science

WFIRST has a 2.4 m wide-field IR telescope (0.7–2 μm) and an exoplanet imaging coronagraph instrument (400–1000 nm). It will (i) carry out galaxy surveys over thousands of square degrees down to $J = 27$ AB to study dark energy (Fig. 11) weak lensing and baryon acoustic oscillations, (ii) monitor a few square degrees for dark energy SN Ia studies, (iii) perform microlensing observations of the galactic bulge for an exoplanet census, and (iv) undertake direct imaging observations of nearby exoplanets with a pathfinder coronagraph. The mission will have a robust and well-funded guest observer program for 25% of the observing time. With a <1 hr TOO response time (TBD, under study), *WFIRST* will be a powerful tool for time domain astronomy and for coordinated observations with gravitational wave experiments. Gravitational wave events produced by mergers of nearby binary neutron stars (LIGO-Virgo, cf. Kanner et al. 2012) or extragalactic supermassive black hole binaries (*LISA*) will produce electromagnetic radiation that *WFIRST* can observe.

[1] H4RG is a hybrid CMOS (Complimentary Metal-Oxide-Semiconductor) 4K×4K optical imager made by Teledyne Scientific & Imaging.

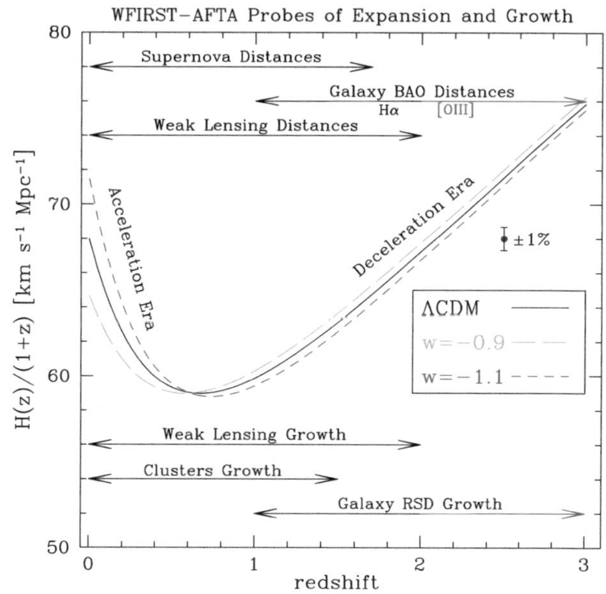

Fig. 11 Dark energy techniques used by *WFIRST* (Spergel et al. 2015)

4.2 *JWST*

4.2.1 Instrumentation

The Integrated Science Instrument Module (ISIM) will house the four main instruments that will detect light from distant stars and galaxies, and planets orbiting other stars. The ISIM will include:

(i) the Near-Infrared Camera, or NIRCam (University of Arizona),

(ii) the Near-Infrared Spectrograph, or NIRSpec (ESA, with components provided by NASA/GSFC),

(iii) the Mid-Infrared Instrument, or MIRI (European Consortium, European Space Agency (ESA), and NASA Jet Propulsion Laboratory),

and (iv) the Fine Guidance Sensor/ Near InfraRed Imager and Slitless Spectrograph, or FGS/NIRISS (Canadian Space Agency).

The 6.5 m primary mirror, a gold-coated beryllium reflector, will have a collecting area about five times that of *HST*. *JWST* will be oriented towards near-IR astronomy, but with sensitivity also in orange and red light, as well as the mid-IR. Its coverage will span 0.6–27 μm. The motivation for an emphasis on near-IR to mid-IR is three-fold: high-z objects have their visible emissions shifted into the IR, cold objects such as debris disks and planets emit most strongly in the IR, and this band has been difficult to study from the ground or by previous space missions.

JWST will operate near the Earth–Sun L2 Lagrange point, $\sim 1.5 \times 10^{11}$ m beyond the Earth. Objects near this point can orbit the Sun synchronously with the Earth, allowing the telescope to remain at a roughly constant distance, using a single sunshield to block heat and light from the Sun and Earth. This will keep the temperature of the spacecraft below 50 K which is necessary for IR observations.

Launch is scheduled for 2018 on an Ariane 5 rocket. Its nominal mission length is five years, with a goal of ten years.

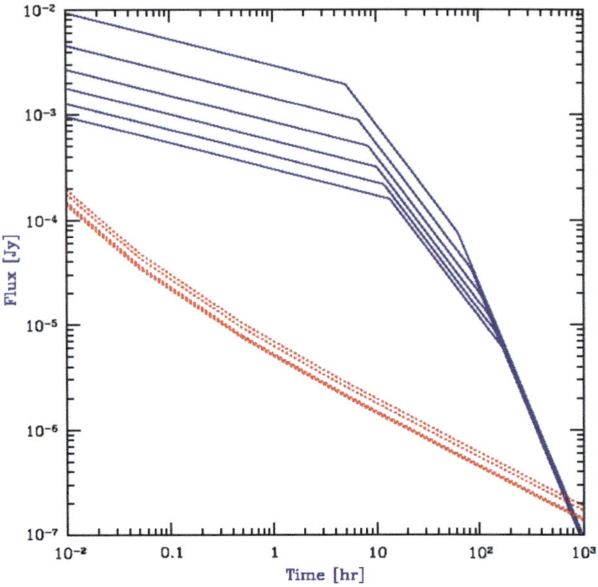

Fig. 12 Detectability of high-z GRB afterglows versus observer time $t - T0$ (Bromm and Loeb 2007; from Barkana and Loeb 2004). GRB afterglow light curves are presented at the redshifted Lyα wavelength (*solid blue curves*). *JWST*/NIRSpec detection threshold (*dotted red curves*) are shown adopting a spectral resolution $\lambda/\Delta\lambda = 5000$, a signal to noise ratio of 5 per spectral resolution element, and an exposure time of $0.2(t - T0)$. For each set of curves $z = 5, 7, 9, 11, 13$, and 15, respectively (*top* to *bottom*)

4.2.2 Science

JWST's scientific mission has four main components: (i) to search for light from the first stars and galaxies, (ii) to study the formation and evolution of galaxies, (iii) to understand the formation of stars and planetary systems, and (iv) to study planetary systems and the origins of life. This science will be facilitated by *JWST*'s near-IR capabilities.

JWST is expected to see the very first galaxies, forming just a few 100 Myr after the Big Bang. IR observations allow the study of objects and regions of space which would be obscured by gas and dust in the visible, such as the molecular clouds where stars are born, the circumstellar disks that give rise to planets, and the cores of active galaxies. Relatively cool objects emit primarily in the IR. Most objects cooler than stars are better studied in the IR, including ISM clouds, brown dwarfs, planets—both in our own and other solar systems, and comets and Kuiper belt objects. In addition, *JWST* will be a valuable tool in high-z GRB follow-up (Fig. 12) thanks to its ability to see back to early time.

4.3 Summary

JWST and *WFIRST* will bring in a new era of NIR and MIR astrophysics. *JWST* will provide very deep ($J = 29$) observations with multiple instruments. *WFIRST* will provide wide-field (1000's of square degrees) deep ($J = 27$) surveys. For both missions GRB follow-up science will be an integral part of science programs; *JWST* will have a 2 d TOO response time, while *WFIRST* will have a <1 hr response time. TOO follow-ups for both missions will be limited to few per year.

Table 2 Summary of planned 30 m class telescope characteristics

Telescope	Location	Altitude (m)	Diameter (m)	Mirror technology
GMT	Las Campanas	2550	24.5	7 monolithic mirrors, common focus
TMT	Mauna Kea	4050	30	Segmented primary
E-ELT	Cerro Armazones	3060	39.3	Segmented primary

5 The 30 m Class Ground-based Telescopes

5.1 Introduction

As of 2015, construction has begun for three giant 30 m class ground-based optical/infra-red telescopes: the Giant Magellan Telescope (GMT) at Las Campanas, Chile, the Thirty Meter Telescope (TMT) at Mauna Kea, Hawaii, and the European Extremely Large Telescope (E-ELT) at Cerro Armazones, also in Chile. Thus by 2025 it is expected astronomers will have access to facilities significantly more powerful than the existing state-of-the-art 8 m generation. Their advantage goes beyond simply enhanced light-gathering-power, since each telescope is also planned to have sophisticated capabilities for high-order adaptive optics, thus providing spatial resolutions even better than the space-based *HST* and *JWST* for many observations (e.g., Gilmozzi and Spyromilio 2007). A summary of the basic vital statistics for each telescope is given in Table 2.

A major driver for construction of these facilities is the exploration of early structure formation, during and before the epoch of reionization. This pushes towards optimization in the near infra-red, since the Gunn–Peterson trough renders the universe opaque at rest-frame wavelengths below 1215Å, which corresponds to $\lambda > 1$ μm at $z \gtrsim 7$. In recognition of this, all three telescope consortia have prioritized building of intermediate resolution, integral field spectrographs with nIR capability; namely HARMONI on the E-ELT, IRIS on the TMT and GMTIFS on the GMT. HARMONI and IRIS, indeed, have been selected as first light instruments. With spectral resolutions of $R \sim 4000$–$10\,000$, the majority of the available spectral range is between bright night-sky lines, and therefore provides low background and consequently high sensitivity.

5.2 High Redshift Transient Science

5.2.1 Science Drivers

Gamma-ray bursts are extremely bright, and hence visible in principle to very high redshifts. This, combined with their very broad spectral energy distributions, and association with massive star death, has long been recognized as making them potentially very powerful probes of the early universe (Lamb and Reichart 2000; Tanvir and Jakobsson 2007). They trace early star formation, pin-point the positions of primeval galaxies, and provide luminous backlights for absorption spectroscopy (Chornock et al. 2013; Sparre et al. 2014; Hartoog et al. 2015).

To-date, the highest spectroscopic redshift for a GRB is $z = 8.2$ for GRB 090423 (Tanvir et al. 2009), whilst a photometric redshift of $z \approx 9.4$ was found for GRB 090429B (Cucchiara et al. 2011). However, despite their high intrinsic luminosities, at these distances follow-up with the current generation of telescopes and instruments is hard for all but the most extreme events. Thus it is likely that some high-z GRBs detected by *Swift* have not been

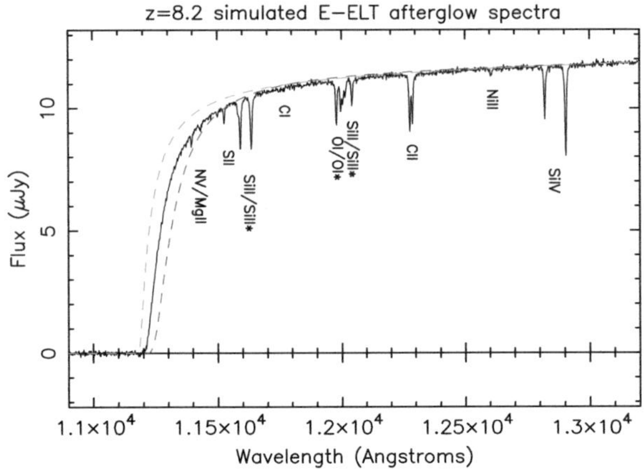

Fig. 13 Simulated spectrum (*black solid curve*) of typical GRB afterglow around the Ly-α line, as it would be seen in a 30 min exposure with E-ELT. The high signal-to-noise allows precise determination of neutral hydrogen column in both the host and intervening intergalactic medium, together with abundances for many metal species (the simulation is of 1% of Solar metallicity). Attempts to fit the damping wing with a host HI only model is shown by the *red dashed curve*, and a 100% neutral IGM only model as a *green dashed curve*; both provide poor fits. This shows that with such S/N it is possible to decompose the host HI column density and IGM neutral fraction

recognized as such, and those that have been recognized have usually lacked the high-S/N data to fully exploit their potential.

The next generation telescopes combine near infra-red optimization with exquisite point-source sensitivity, thanks to their large collecting area and AO-assisted high spatial resolution. Thus afterglows for which only redshifts could be obtained previously, will in the future be used to study abundances, neutral hydrogen, dust and molecular content in even very faint hosts. All of these properties are extremely difficult to study by other means. This potential is illustrated graphically in Fig. 13, which shows a simulated E-ELT spectrum of a typical GRB afterglow at $z = 8.2$ (in fact based on observed luminosity of GRB 090423). The measurement of HI column in both host and intergalactic medium (IGM) in a statistical sample of high-z afterglows would be of considerable importance in understanding the sources driving reionization. The former provides the distribution of opacities, and hence escape fractions for ionizing radiation, along the lines of sight to massive stars in the early universe (Chen et al. 2007; Fynbo et al. 2009), while the latter would allow us to map the progress of reionization and its variation from field to field (McQuinn et al. 2008).

Other transients that may be studied by 30 m class telescopes include supernovae of all types, particularly superluminous supernovae that, although rare, are bright enough to be studied in detail at $z > 6$. This class of event may include pair-instability supernovae, the likely end-point of many massive population III stars (Heger et al. 2003).

5.2.2 *Challenges*

Observing transients with 30 m class telescopes offers great promise, but also presents particular challenges. Some challenges are for the wider astronomical community, in particular it is obviously essential that we maintain a capability to discover transient sources to feed into down-stream follow-up. In the 2020s new sources of transient discovery should be operating, notably the Large Synoptic Survey Telescope (LSST), the advanced generation gravitational wave detectors (e.g., aLIGO, aVirgo), and the SVOM satellite. However, of these only SVOM will be capable of detecting high redshift sources, specifically gamma-ray bursts (GRBs) and related high-energy transients. Since SVOM's capabilities are broadly similar to those of *Swift* we still only expect a modest rate of high-z bursts to be detected. Another challenge for transient follow-up, particularly of GRBs, is that their rapid time-evolution means that observations should ideally be done on a time-scale of hours after trigger. With the 30 m telescopes being restricted to just two geographical regions of the Earth, this limits the potential for follow-up, and, especially if one also considers weather and technical down-time, strongly motivates the continued operation of smaller telescopes as part of a transient follow-up network.

Several other challenges are at least more within the control of the consortia building the 30 m class telescopes. First and foremost is the provision of suitable imaging and spectroscopic capability. Unlike some other areas of astronomy, wide field is not key here, but good spatial resolution and (simultaneous) wide wavelength coverage is important. Since speed is of the essence, operational efficiency is also very important, both to realize rapid responses to trigger requests (quick instrument/mode changes, active optics reconfiguring, target acquisition, etc.), and also to expedite quick-look reduction and transmission of data to observers (or their software agents) allowing further follow-up to be initiated with minimal delay. These capabilities require attention even in the design phase, for example to ensure that suitable data calibration products will exist to support the desired quick-look pipeline processing.

5.3 Host Galaxies

In addition to the transients themselves, the 30 m class telescopes will provide unprecedented insights into their host galaxies. In general, high-z galaxies found in Lyman-break surveys are compact (e.g., Curtis-Lake et al. 2014), typically a few tenths of an arcsec, and hence only crudely resolved at *HST* resolution in the nIR. GRB hosts seem to be even fainter (Tanvir et al. 2012), and probably smaller, but with spatial resolutions as good as $\sim$0.04 arcsec for E-ELT, significant detail will be resolvable, allowing us to see probe the physical conditions leading to early star formation.

5.4 Conclusions

Within ten years, if these ambitious plans come to fruition, the astronomical community will have access to three 30 m class telescopes. Thanks to their huge grasp, adaptive optics capability, and near-infrared optimization, these facilities will revolutionize our view of early structure formation in and before the era of reionization. Observations of transients and their host galaxies has a particularly important role to play, as they provide a route to studying in remarkable detail the physical nature of the very small star-forming galaxies that are thought to have dominated star formation at that time.

6 Radio Observations of GRBs and Orphan Afterglows

6.1 Radio Observations of GRBs

Radio observations of GRBs dates back to the end of the nineties when the afterglow discovery (Costa et al. 1997; van Paradijs et al. 1997) opened a new era in the field of GRBs with the follow up of their long wavelength long lasting afterglow emissions. The study of

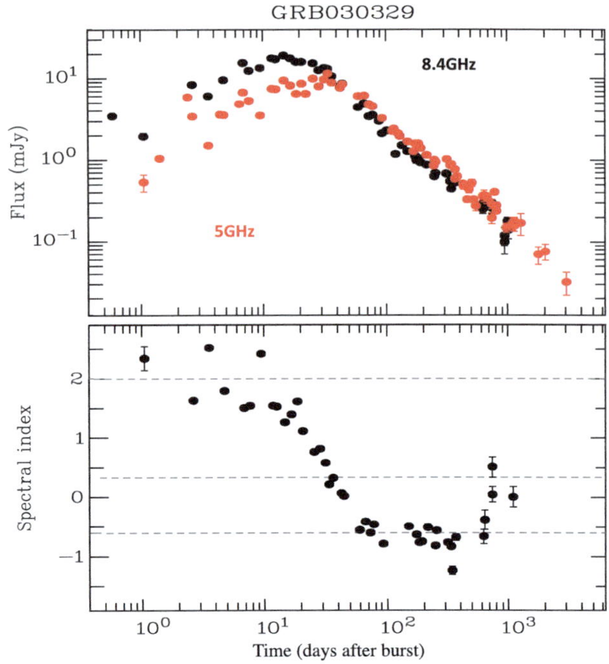

Fig. 14 Radio light curve (*top panel*) at 8.5 GHz (*black symbols*) and 5 GHz (*red symbols*) of GRB 030329. Evolution of the spectral index (computed between 5 and 8.5 GHz) as a function of time (*bottom panel*). Dashed lines in the bottom panel show reference slopes of $-2/3$, $1/3$ and 2 (from *bottom* to *top*). Figure adapted from Granot and van der Horst (2014)

X-ray and optical emission from GRBs was revolutionized, starting in 2005, by the Swift satellite (Gehrels et al. 2009) thanks to its fast (within one minute) pinpointing (with subarcmin precision) of the location of the γ-ray event. Currently, ~95% of the bursts detected by the γ-ray detector (BAT—Burst Alert Telescope—15–150 keV) are detected on board by the X-Ray Telescope (XRT; 0.1–3 keV) and ~75% of these are also detected at optical wavelengths.

Until the last two years, the detection rate of GRBs in the radio was around ~30%. This small percentage (if compared to the optical and X-ray) remained unchanged before and after the launch of Swift (Chandra and Frail 2012). The ameliorated sensitivity of the upgraded Very Large Array (i.e. JVLA) increased the detection rate to 60% in the last two years. This seems to support the idea that the small detection rate was mainly due to the sensitivity limits of past radio follow up programs (Ghirlanda et al. 2013). Most often, these detections were at the level of sub-mJy flux densities in the 0.6–660 GHz (53% at 8.5 GHz) frequency range and were obtained within dedicated observing programs with the largest radio telescopes (VLA, WSRT, ATCA, GMRT). The median peak luminosity of the detected long GRBs is 10^{31} erg s-1 Hz^{-1} at 8.4 GHz while the few detected short GRBs and SN/GRBs are a factor 100 and 10 less luminous, respectively. Regarding the distance scale of radio detected GRBs there seems to be no preference: both low redshift (e.g. GRB 980425—Kulkarni et al. (1998)) and high redshift (GRB 090423—Chandra et al. (2010)) events have been detected in the radio band.

In those events with multi epoch detections (i.e. radio light curve) a typical peak luminosity of $\sim 2 \times 10^{31}$ erg s^{-1} Hz^{-1} is reached after about 10–20 days (3–6 days in the source rest frame) since the burst trigger. Typical post peak decay index is of order unity. Figure 14 shows the radio light curve (top panel) of GRB 030329, one of the GRBs with the densest/longest monitoring in the radio band so far. The light curve shows a peak at ~15 days at 8.5 GHz and slightly later (~45 days) at 5 GHz, a typical behavior of the radio light curve. Within the standard afterglow model where the radiation is produced by synchrotron emission at the external shock (produced by the deceleration of the jet by the interstellar medium), the broad band spectral energy distribution (SED) can be described by a characteristic peak and at least three break frequencies: the self–absorption frequency ν_{sa} below which the spectrum is steep (with typical slopes $\nu^{2-5/2}$), the electron minimum frequency ν_m corresponding to the minimum energy (γ_m) of the relativistically shock accelerated electrons and the cooling frequency ν_c which corresponds to the minimum energy of the electrons that cool on dynamical timescales. The typical ordering is $\nu_{sa} < \nu_m < \nu_c$ at relatively early times (days–weeks) and $\nu_m < \nu_{sa} < \nu_c$ at late times. The evolution of these frequencies and of the peak flux with time depends on the jet dynamics. The standard afterglow model (e.g. van Eerten et al. (2012)) links the spectrum (dependent on the micro physical parameters describing the shock—i.e. electron energy index, fraction of energy in relativistic electrons and in magnetic field) and the jet dynamics (dependent on the macro physical parameters, i.e. the kinetic energy of the jet and the density of the external medium).

The radio band, therefore, is affected by ν_{sa} and ν_m and their relative position. The appearance of a peak in the radio light curve is produced by the passage across the observing band of these frequencies. In the example of GRB 030329 shown in Fig. 14 the bottom panel shows that the spectrum is self absorbed at early times (with index -2 typical of the SED below ν_{sa}) and becomes softer at later times reaching a slope -0.6 which can be used to infer the electron energy power law slope, $p \sim 2.2$ in this case. The early time spectral variation is due to interstellar scintillation. In general, in the radio band, the time of the peak increases and its brightness decreases with decreasing radio frequency.

Radio emission, when detected, can be followed on very long timescales. While the X-ray (optical) afterglow fades below the sensitivity limits (below the host galaxy flux level) within weeks since the burst, radio emission has been observed years after (e.g. GRB 970508—Frail et al. (2000); GRB 980703—Mesler and Pihlström (2013)) with GRB 030329 (Fig. 14) still detected at 5 GHz almost a decade after the γ-ray event. In this respect, radio observations are unique in probing the transition of the blast wave to the non–relativistic regime (i.e. trans-relativistic phase). In particular, during this phase estimates of the outflow parameters such as the kinetic energy (Berger et al. 2004; Frail et al. 2001, 2005; Shivvers and Berger 2011) do not depend on relativistic effects and jet aperture (if spherical symmetry is reached) differently from the same parameters estimates performed during the early relativistic phase (although radio calorimetry is not assumptions free—e.g. Eichler and Waxman (2005)).

Another unique feature of the radio band is the possibility to study the broad band SED of dark bursts (see e.g. Melandri et al. (2012) for a recent compilation of dark bursts properties): by combining multifrequency radio observations with X-ray data it is possible to model the time evolution of the SED and infer, for bursts without optical detection, an estimate of the optical extinction. One of the best cases, GRB 051022 (Rol et al. 2007), turned out to be highly extinguished ($\sim$2.3 mag in the infrared and at least 5.4 mag in the optical) possibly due to some dust clumps located along the line of sight within the host galaxy.

Polarization is a powerful diagnostic of the jet composition. In particular, the reverse shock emission (produced by the deceleration shock propagating backwards into the jet outflow) and its polarization level can constrain the magnetic field structure of the jet. Reverse shock emission in the optical band, characterized by a flare like light curve (i.e. with rise and decay steeper than the typical slopes predicted by the afterglow emission), has been observed in a handful of bursts (e.g. 080319B—Racusin et al. (2008); see Japelj et al. (2014)). Radio flares from the reverse shock has been observed only in a couple of bursts (990123—Kulkarni et al. (1999)) with the most recent case of 130427A (Laskar et al. 2013). This is a relatively nearby GRB (at $z = 0.34$) with an extraordinary energetic budget (exceeding

10^{54} erg). Its emission has been detected up to the GeV bad by Fermi and broadband SED modeling has produced different interpretations (e.g. Maselli et al. (2014)). The radio band offered the opportunity to interpret the early time radio flare as due to the reverse shock (also observed in the optical—Verstrand et al. 2013) suggesting a low density environment (Laskar et al. 2013; Perley et al. 2014). However, additional radio data (van der Horst et al. 2014) suggested also a possible alternative interpretation of a double jet component. This bursts is one of the best examples where the combination of multi wavelength data allow us to break the degeneracy among the free parameters of the standard model while it still represents a challenge for the standard model itself (Maselli et al. 2014).

Radio afterglow observations offer in principle the unique opportunity to measure the jet expansion and constrain the GRB size. This can be done (i) directly through imaging as for (the only case) GRB 030329 which was observed to expand over a couple of months thus indicating an apparent velocity 3–5 the speed of light (Taylor et al. 2004; Pihlström et al. 2007; Mesler et al. 2012) or (ii) indirectly through radio scintillation (i.e. from the measurement of the time it is quenched when the projected source size is larger than the galactic ISM inhomogeneities causing it).

At very late times, when the afterglow emission has faded below the host galaxy, radio observations allows us to derive an independent estimate of the star formation in GRB hosts thus providing an estimate of the dust obscuration. Systematic search of radio host emission provided several upper limits at the level of 10–100 μJy and only few detections at redshift $z < 1$ (Michałowski et al. 2012; Perley and Perley 2013; Perley et al. 2015). Despite the small detection rates, these studies indicate that the host star formation rates derived from the radio, which have the advantage of probing the unobscured SFR are $\ll 100\ M_{\odot}\ yr^{-1}$. One major source of uncertainty in these estimates is the unknown host radio spectrum, which requires multifrequency observations. Radio hosts observations will shed light on the class of dark GRBs ($\sim$20–30) i.e. bursts that are not detected in the Optical band (Perley et al. 2015).

6.2 The Future of Radio Observations of GRBs

We are now entering the era of advanced radio facilities. The updated JVLA, the largest radio telescopes working in interferometric mode (VLBI, VLBA, EVN) and the advent of the SKA precursors are in the reach of detecting and monitoring GRBs. In particular, the Square Kilometer Array (SKA—Carilli et al. (2003)) will be the most powerful radio facility for continuum, spectroscopy and survey studies of the radio sky. The observatory will be composed of two sites (Australia and South Africa) where thousands of different antennas (from aperture arrays to single dishes) will be operated over a wide frequency range extending to 9 GHz at the highest frequencies.

Through a population synthesis code it is possible to predict the radio flux density at the time when the radio light curve peaks for a synthetic population of GRBs (Ghirlanda et al. 2013). The assumptions of this simulations are discussed in (Ghirlanda et al. 2013, 2014, 2015). Figure 15 shows the contour levels corresponding to the distribution of the simulated population. The axes represent the flux of the radio emission at the time when the peak of the emission is reached. The different contour levels refer to four characteristic frequencies of the SKA-MID configuration.[2] The dashed lines are the 5σ sensitivity limits obtained for a 12 h exposure. Figure 15 shows that the characteristic timescale of the peak of the afterglow light curve moves to longer times at lower frequencies. Current estimates

[2] https://www.skatelescope.org/news/ska-science-book/.

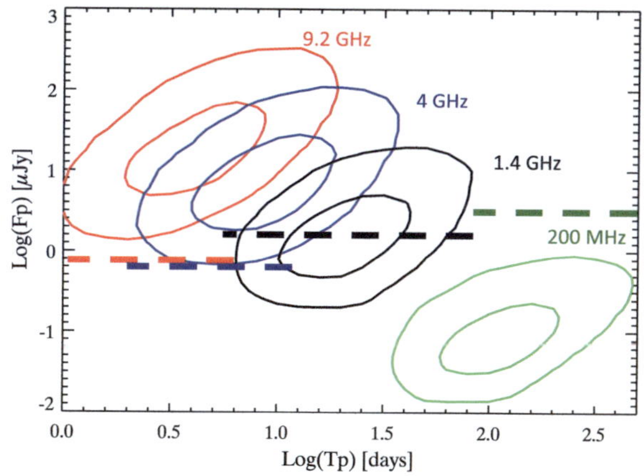

Fig. 15 Contour levels (68% and 95%) of the distribution of the population of GRBs corresponding to the time when the peak of the afterglow is reached and the corresponding flux. Four characteristic frequencies corresponding to the SKA1-MID (REF) are shown with different *colors*. The *dashed lines* show the expected continuum sensitivity limits at 5σ for a 12 h exposure

with the SKA1-MID sensitivity limits (Burlon et al. 2015) show that it will be possible to detect most of the population of GRBs at relatively large frequencies (from 4 to 9.2 GHz, blue and red curves in Fig. 15). Still the self absorption dominating the lowest frequencies will make accessible with the SKA-1 roughly half of the bursts at 1.4 GHz (black curves) and almost make it tough to detect any GRB at 200 MHz (green curves). These estimates only describe the power of the radio observations to study the radio afterglow emission of GRBs. In general the follow up campaigns of GRBs will strongly benefit from the increased sensitivities of present and forthcoming radio facilities provided that the detection rate of GRBs in the γ-ray band (i.e. the triggers) is maintained/increased with current/future GRB detectors (from Swift and Fermi now to SVOM in the future).

The high sensitivity of the SKA will make it possible to perform late time radio observations of the bursts when they have transitioned to the non–relativistic regime. We have computed (Ghirlanda et al. 2013; Burlon et al. 2015) the expected flux density at the non–relativistic transition and compared to the SKA1-MID expected limits. Only a handful of objects reach the sensitivity limit of current facilities (a few μJy at best). We foresee that the full SKA will routinely observe a significant fraction (15–25%) of the whole GRB afterglow population at late times.

One challenging aspect of GRBs in the forthcoming era of extended radio facilities and in the final SKA era will the possibility to access their parent population. We know that GRBs are jetted sources with a typical opening angle of a few degree. We detect only those bursts whose jet is closely aligned with our line of sight. This means that for burst that we detect in γ-rays there are hundreds of events that point their jet elsewhere. Depending on the jet structure these events can still be detected as slow transients in wide field surveys. Orphan afterglow properties have been studied in the literature Rossi et al. (2008), Ghirlanda et al. (2014, 2015). Despite their large number no convincing evidence of a radio orphan GRB has been reported so far. This is most likely due to the sensitivity limits of past radio surveys. Forthcoming radio surveys like the VAST/ASKAP (operating at 1.4 GHz) or the MeerKAT or EVLA (operating at 8.4 GHz) could detect 3×10^{-3} and 3×10^{-1} OA $\mathrm{deg}^{-2}\,\mathrm{yr}^{-1}$, respectively. The deeper SKA survey, reaching the μJy flux limit, could detect up to 0.2–1.5 OA $\mathrm{deg}^{-2}\,\mathrm{yr}^{-1}$ (Ghirlanda et al. 2014; Metzger et al. 2015).

The SKA era will be transformational for the study of GRBs. The SKA will access close to 100% of the GRBs detected in γ-rays and will substantially contribute to the multi wavelength follow up of these sources. Radio observations will extend the timescale of follow up to very late times when the afterglow emission at other frequencies will be already undetectable. We will get an unprecedented insight into the true energy budget of GRBs and it will become possible to probe both the macrophysics of the ambient medium (e.g. the density profile) and the microphysics of the shocks Finally, it will become possible to detect the so far elusive class of "orphan afterglows" that should appear as slowly evolving transients detectable with the SKA on weekly basis in its wide field survey.

7 Lobster-eye Micro-pore Optics for X-Ray Focusing and the Einstein Probe Mission

Time-domain astronomy will see its golden era towards the end of this decade with the advent of major wide-field facilities across the electromagnetic spectrum and in the multi-messenger realms including gravitational wave and neutrinos. In the X-ray regime, the driving science calls for new generation instruments with high sensitivity, good angular resolution (a few arc-minutes or less) and a large sky coverage (field of view of order of thousands square degrees). These requirements can be fulfilled by wide-field X-ray focusing optics—the emerging lobster-eye Micro-Pore Optics (MPO), as focusing imaging results in enormously enhanced gain in signal to noise, and thus high detecting sensitivity. The Einstein Probe (EP), which is a candidate mission of priority of the Chinese Academy of Sciences with an intended launch date around 2020, is based on this lobster-eye MPO technology. Its aim is to monitor a large fraction of the whole sky at high cadences with sensitivity in X-ray at least one order of magnitude deeper than the most sensitive all-sky monitoring type instruments ever built.

Among various types of faint transients that EP is expected to discover, high-redshift gamma-ray bursts are of particular interest since, as the most luminous objects in the Universe at their peaks, they provide a unique tool to study the early Universe beyond redshifts 6 and up to $\sim$20. During this epoch the first luminous objects as Pop-III stars (as well as Pop-II stars) are expected to form and start to re-ionize the Universe out of the dark ages (Ciardi and Ferrara 2005). Whilst synthesizing metals they quickly evolved and exploded, and thereby polluted their environment. The detection of these stars individually would be extremely difficult or almost impossible, however, even with the next generation of space observatories like JWST. The only way of observing Pop-III stars in action is to detect their explosive deaths. They have been predicted to produce a GRB-like event (Bromm and Loeb 2006; Wang et al. 2012). By using high-redshift GRBs as beacons we could identify and probe the regions where the first stars and their remnants (the first black holes) were formed.

7.1 Micro-pore Optics X-Ray Focusing Technology

X-ray focusing instruments rely on multiple smooth refection surfaces almost parallel to incident X-rays, which can be arranged in several different configurations. One is Lobster-eye optics (Angel 1979), which mimics the imaging principle of the eyes of lobsters as shown in Fig. 16. Incoming light is reflected off the walls of many tiny square pores arranged on a sphere and pointed towards the concentric spherical center. The reflection surfaces are configured orthogonal to each other without a specific optical axis, and thus the FOV can in principle subtend the entire solid angle of 4π. In contrast, the conventional Wolter-I optics (Wolter 1952) (used for Chandra, XMM-Newton, Swift/XTR, etc.) has an tubular,

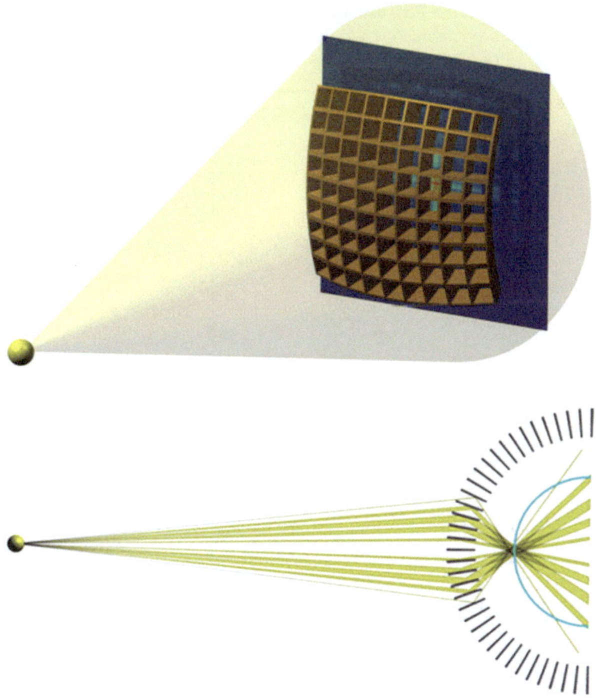

Fig. 16 *Upper panel*: Lobster eye optics: X-rays from source illuminate micro channel grid (*gold color*) and are focused on detector plane with cross-like Pointing Spreading Function (*right, blue*). *Lower panel*: Optical path in a lobster eye optics. (Credit: adopted from http://www.x-ray-optics.de/)

Fig. 17 MPO device (*left*) and a close-up under microscope (*right*). The pore size is of 20 micrometers

rotationally symmetric elliptical or parabolic reflection surface followed by a hyperbolic surface, which is impossible to achieve a FOV larger than one degree.

Lobster-eye telescopes can be constructed using novel X-ray focusing device, the so-called Micro-Pore Optics (MPO) (Fraser et al. 1993). This technology integrates millions of square pores as shown in Fig. 17 with sizes of about tens of micrometers on a very thin ($\sim$1 mm) glass piece of several square centimeters. The walls of pores are smooth enough to reflect X-ray photons up to several Kilo-eV. MPO is one of the most compact X-ray optics with the largest effective area to weight ratio. MPO is employed for the MIXS X-ray telescope onboard EAS's Mercury mission BepiColombo (Fraser et al.

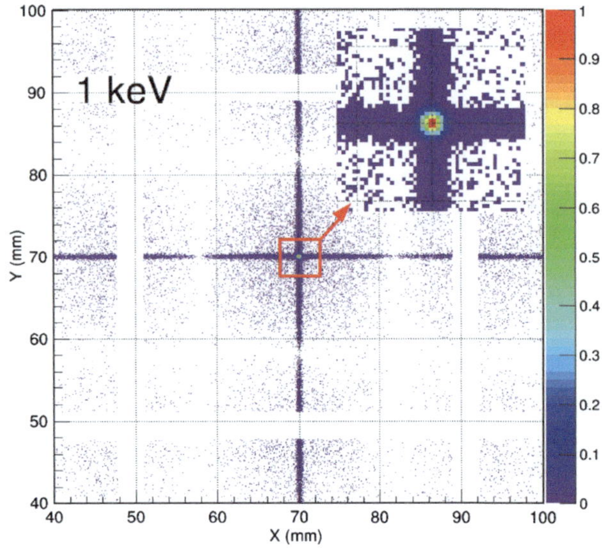

Fig. 18 Simulated PSF at 1 keV of a Lobster-eye telescope with a focal length of 375 mm

2010), where a payload mass is strictly constrained, *as well as for the MXT X-ray telescope onboard SVOM (Sect. 2)*. A lobster-eye telescope with MPO devices can achieve a huge instantaneous FOV with very light weight, which is unique for X-ray imaging telescopes.

A simulated point spread function (PSF) of a Lobster-eye telescope is shown in Fig. 18. The cross-like feature is caused by complex reflection processes of X-ray photons across the square pore. The incoming X-rays are split into four beams corresponding respectively to no reflection (straight through, forming background), single reflection (cross arms) or double reflections off adjacent walls of the pore (focal spot). The FoV of the optical arrangement indicated in Fig. 16 is only limited by the size of the optics (the number of MPO pieces) or the size of the detector. The PSF remains almost unchanged over the entire FOV without vignetting of the effective area. Such a wide-field lobster-eye telescope provides the technological basis of the next generation wide-field X-ray monitors to detect faint and short-lived phenomena like high-redshift Gamma-Ray Bursts, distant X-ray novae and tidal disruption events.

7.2 Einstein Probe

The Einstein Probe is a candidate mission of priority in the Advanced Study Phase of the Space Science Pilot Program of the Chinese Academy of Sciences (CAS), with an intended launch date around 2020. It will discover and characterize high-energy transients and monitor variable objects in the soft X-ray band with unprecedented sensitivity. Its primary scientific goals are to: (1) reveal quiescent black holes at almost all astrophysical mass scales and study how matter falls onto them by detecting transient X-ray flares, particularly stars being tidally-disrupted by otherwise dormant massive black holes at galactic centers; (2) discover the X-ray photonic counterparts of gravitational-wave transients found with the next generation of gravitational-wave detectors and precisely locate them; (3) carry out systematic and sensitive surveys of high-energy transients, to discover faint X-ray transients of various types, such as high-redshift GRBs, supernova shock breakout, and previously unknown transients.

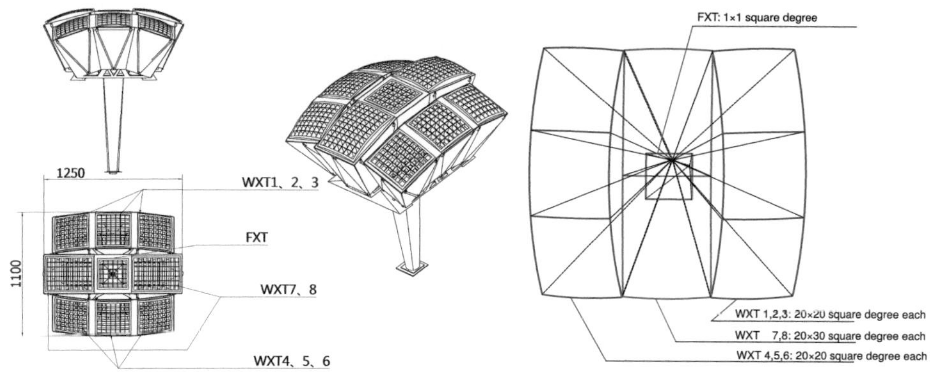

Fig. 19 (*Left*) Layout of WXT modules and FXT. (*Right*) Sketch of the field of view of WXT (not to scale). [Figures adopted from Yuan et al. (2015)]

Table 3 Specifications of WXT and FXT

Parameters	WXT	FXT
Number of modules	8	1
Field-of-view	$60° \times 60°$	$1° \times 1°$
Focal length (mm)	375	1,400
Angular resolution FWHM (arcmin)	<5	<5
Nominal pandpass (keV)	0.5–4.0	0.5–4.0
Energy resolution @1 keV	40%	100 eV
Effective area (central focus + arms) (cm^2)	8 @0.7 keV	80 @1 keV
Sensitivity ($erg\,s^{-1}\,cm^{-2}$ @1,000 s)	$\sim 1 \times 10^{-11}$	$\sim 3 \times 10^{-12}$

The payload of EP consists of a Wide-field X-ray Telescope (WXT) with a field-of-view of 60×60 deg^2 in the nominal 0.5–4 keV band, and a Follow-up X-ray telescope (FXT) with a larger light-collecting power than WXT, both based on the lobster-eye MPO technology. In addition, EP is equipped with a fast alert telemetry system, in order to trigger multi-wavelength follow-up observations from the worldwide community. The nominal mission lifetime is three years with a goal of five years. For a more detailed description of the scientific goals and instrumentation of EP, please refer to Yuan et al. (2015).

The WXT telescope consists of eight modules, with FXT mounted at the center (Fig. 19). Table 3 lists the main specifications of both WXT and FXT. The eight WXT modules make up a spherical array mosaicked by 441 concentric MPO pieces, each of 40 mm $\times$ 40 mm in size. The total FoV subtends a solid angle of 60×60 deg^2 ($\sim$1.1 steradian), which is about $1/11$ of the whole sky. WXT has a large focal plane size, 420×420 mm^2 in total. As a baseline detector, gas detectors based on GEM (Gas electron multiplier) are currently being developed. The effective area and sensitivity curve of WXT are shown in Fig. 20. FXT is a narrow field-of-view ($\sim$1 deg) telescope of the same lobster-eye MPO type with a focal length of 1.4 m, yielding a much larger effective area ($\sim$80 cm^2 at 1 keV). FXT has an aperture size of about 240 mm, which is mosaicked by 6×6 MPO pieces. FXT will employ CCD as the focal plane detector to gain better spectral performance.

The layout of the EP satellite and its FoV as well as the observing mode are illustrated in Fig. 21. The satellite will be in a circular orbit with an altitude of about 600 km and an

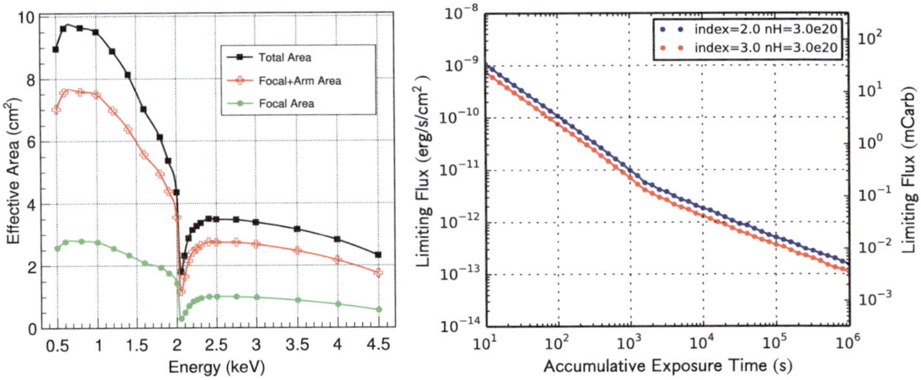

Fig. 20 (*Left*) Simulated effective area of EP/WXT with a GEM detector, for the central focal spot (*green*), central plus the cruciform arms (*red*), and total (*black*; plus unfocused X-rays as diffuse background). The MPO arrays are coated with Iridium, and have surface roughness of ~0.55 nm and the tilts of pores following a Gaussian distribution with $\sigma = 0.85$ arcmin. Xenon gas and a window of a 40 nm-thick Si_3N_4 foil coated with 30 nm-thick Aluminum are used for the detector. [Figure adopted from Zhao et al. (2014)]. (*Right*) Detecting sensitivity as a function of accumulative exposure time (assuming a source spectrum of an absorbed power-law with a photon index of $\Gamma = 2$ and 3, respectively)

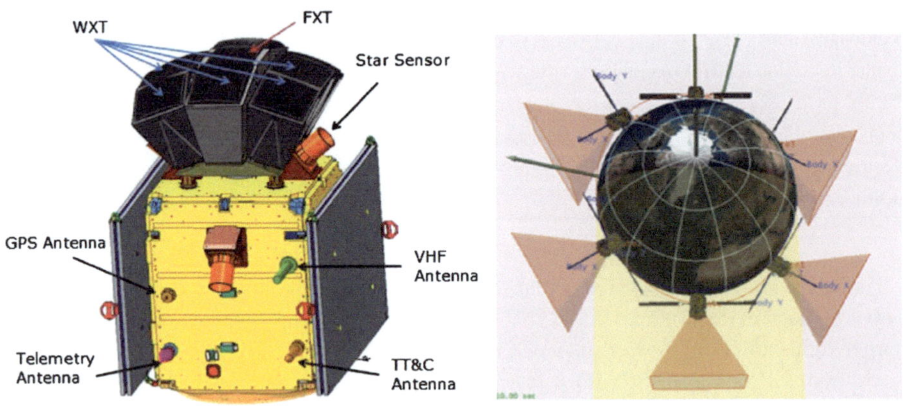

Fig. 21 (*Left*) Layout of the Einstein Probe satellite. (*Right*) Illustration of the field-of-view and pointed observations in one orbit. [Figures adopted from Yuan et al. (2015)]

inclination angle less than 30 degrees. The survey strategy of EP will be composed of a series of pointings (each of about 11-minute exposure) to mosaic the night sky in the directions avoiding the Sun. Over three orbits (~97 minutes each) almost the entire night sky will be sampled. The pointing directions are shifted by about 1 degree per day to compensate the daily movement of the Sun on the sky. In this way, the entire sky will be covered within half-a-year's operation. Once a transient source is detected with WXT and is classified and triggered by the processing and alerting system onboard, the satellite will slew to a new position to enable pointed follow-up observations of the new source with FXT. Meanwhile, WXT continues to monitor the new sky region centering the position of the transient. The alert data of transients are expected to be downlinked within one minute or so via the fast telemetry system, for which the French VHF system is considered.

7.3 Summary

The emerging lobster-eye MPO optics provides a promising technology to realize X-ray focusing, thus the improvement of detecting sensitivity, while retaining large field-of-view of over thousands of square degrees. Its application to X-ray all-sky monitors is expected to revolutionize the high-energy time-domain astrophysics. The Einstein Probe mission, which is based on this technology, will attempt to address some of the key questions in astrophysics and cosmology by capturing faint flashes of X-ray radiation produced by energetic events within a cosmic horizon far beyond the reach of any current and previous missions. Its scientific impact will span a wide range of research fields in astrophysics and fundamental physics, from stars, compact objects in our and nearby galaxies, black holes, supernovae, GRBs, galaxies to cosmology. In particularly, EP has the capability of detecting high-redshift GRBs at $z \sim 10$ and beyond. It is expected that EP will detect 10–30 GRBs at $z > 6$ (3–9 GRBs at $z > 8$) per year (Wu et al., in preparation), and will thus deliver a considerable sample of high-redshift GRBs with which to study the early Universe within its nominal lifetime of three years.

8 THESEUS

8.1 Gamma-Ray Bursts as Probes of the Early Universe

The study of the Universe before and during the epoch of reionization represents one of the major themes for the next generation of space and ground based observational facilities. Many questions about the first phases of structure formation in the early Universe will still be open in the late 2020s: when and how did first stars/galaxies form? What are their properties? When and how fast was the Universe enriched with metals? How did reionization proceed?

Because of their huge luminosities, mostly emitted in the X and gamma-rays, their redshift distribution extending at least to $z \sim 10$ and their association with explosive death of massive stars and star forming regions, Gamma-Ray Bursts (GRBs) are unique and powerful tools for investigating the early Universe: SFR evolution, physics of re-ionization, galaxies metallicity evolution and luminosity function, first generation (pop III) stars.

The European community played a fundamental role in the enormous progress in the field of GRBs in the last 15–20 years (BeppoSAX, HETE-2, Swift, AGILE, Fermi, plus enormous efforts in optical IR and radio follow-up). In 2012, two European proposals for ESA Call for Small mission dedicated to GRBs and all-sky monitoring: GAME (led by Italy, SDD-based cameras + CZT-based camera + scintillator based detectors) and A-STAR (led by UK, lobster-eye telescopes + CdTe detectors). Following all this unique experience and efforts, a White Paper on GRBs as probes of the early Universe submitted in response to ESA Call for science theme for next L2/L3 missions (Amati et al. 2013) was very well considered by ESA.

The THESEUS proposal described below is a result and follow-on of all this unique experience and efforts by the European GRB community, reinforcing also the collaboration with extra-European (e.g., USA) GRB communities and opening itself to tight collaboration with the cosmology and transients astrophysical communities.

8.2 The THESEUS Mission Concept

The Transient High Energy Sky and Early Universe Surveyor (THESEUS) is a mission concept *that will be submitted to ESA in 2016 by a large international collaboration in response to the Call for next M5 mission* within the Cosmic Vision Program. The primary scientific goals of the mission are linked to the following Cosmic Vision themes: 4.1 Early Universe, 4.2 The Universe taking shape, and 4.3 The evolving violent Universe.

As detailed below, the main goal of THESEUS is fully exploiting GRBs as cosmological probes, thus providing a fundamental and unique step forward in our understanding of the early Universe. More in general, THESEUS would vastly increase the discovery space of several classes of high energy transient phenomena over the entire cosmic history.

This is achieved via a unique payload providing an unprecedented combination of: (i) wide and deep sky monitoring in a broad energy band (0.3 keV–20 MeV); (ii) focusing capabilities in the soft X-ray band granting large grasp and high angular resolution; and (iii) on board near-IR capabilities for immediate transient identification and first redshift estimate.

8.3 Scientific Goals

The main scientific goals of THESUES can be summarized as follows.

(a) Exploring the Early Universe (cosmic dawn and reionization era) by unveiling the Gamma-Ray Burst (GRBs) population in the first billion years, namely to perform unprecedented studies of the star formation history up to $z \sim 8$–9 and more; to detect and study the primordial (pop III) star population: when did the first stars form and how did the earliest pop III and pop II stars influence their environments?; to investigate the re-ionization epoch, the interstellar medium (ISM) and intergalactic medium (IGM) up to $z \sim 8$–9 and more: how did re-ionization proceed as a function of environment, and was radiation from massive stars its primary driver? How did cosmic chemical evolution proceed as a function of time and environment?; to investigate the properties of the early galaxies and what was the galaxies global star formation in the re-ionization era, and to investigate the dark energy properties and evolution. The complementarity of THESEUS under this respect with other "cosmology" mission investigating the CMB or the large scale structure of the Universe is illustrated in Fig. 22.

(b) Performing an unprecedented deep survey of the soft X-ray transient Universe in order to: Fill the present gap in the discovery space of new classes of transients events, thus providing unexpected phenomena and discoveries; Provide a fundamental step forward in the comprehension of the physics of various classes of Galactic and extra-Galactic transients, like, e.g.: tidal disruption events TDE, magnetars/SGRs, SN shock break-out, Soft X-ray Transients SFXTS, thermonuclear bursts from accreting neutron stars, Novae, dwarf novae, stellar flares, AGNs/Blazars; Provide real time trigger and accurate (∼1 arcmin within a few seconds; ∼1 arcsec within a few minutes) location of (long/short) GRBs and high-energy transients for follow-up with next-generation optical (EELT), IR (JWST), radio (SKA), X-rays (ATHENA), TeV (CTA) or neutrino telescopes and identify electromagnetic counterpart of detections by next generation gravitational wave detectors.

Additional science. By satisfying the requirements coming from the above main science drivers, the THESEUS payload will also automatically be capable to perform excellent secondary and observatory science, e.g.: unprecedented insights in the physics and progenitors of GRBs and their connection with peculiar core-collapse SNe; substantially increased detection rate and characterization of subenergetic GRBs and X-Ray Flashes; IR survey and

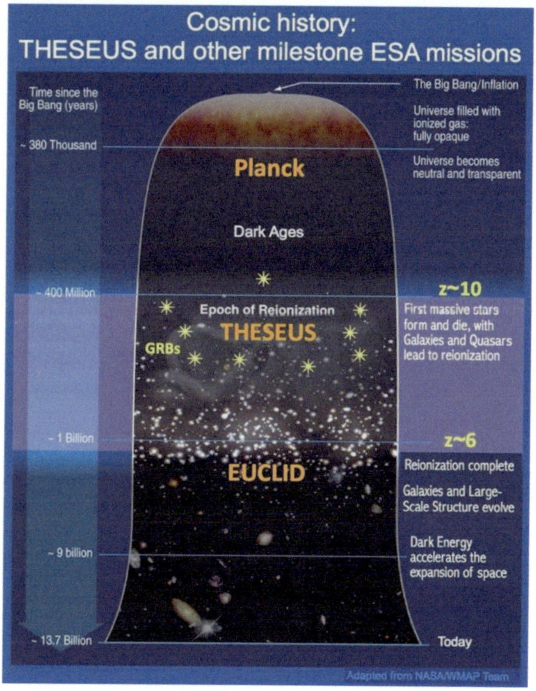

Fig. 22 Example of the complementarity and uniqueness of the THESEUS mission w/r to other cosmological measurements. (Credits: Caltech/NASA and the THESEUS Collaboration)

guest observer possibilities, thus allowing a strong community involvement; survey capabilities of transient phenomena similar to the Large Synoptic Survey Telescope (LSST) in the optical: a remarkable scientific synergy can be anticipated.

8.4 Payload and Mission Profile

The scientific goals which come from a full exploration of the early Universe requires the detection of a factor ten more GRBs (about 100) in the first billion years of the Universe ($z > 6$), in the 3 years prime mission life time of THESEUS. Such a requirement is well beyond the capabilities of current and near future GRB detectors (Swift/BAT, the most sensitive one, has detected only very few GRBs above $z = 6$ in 10 years). As supported by intensive simulations performed by us and other works in the literature, the needed substantial increase of high-z GRBs requires both an increase of $\sim$1 order of magnitude in sensitivity and an extension of the detector passband down to the soft X-rays (0.5–1 keV). Such capabilities must be provided over a broad field of view ($\sim$1 sr) with a source location accuracy $<2'$, in order to allow efficient counterpart detection, on-board spectroscopy and redshift measurement and optical and IR follow-up observations. Such performances can best be obtained by including in the payload a monitor based on the lobster-eye telescope technique, capable of focusing soft X-rays in the 0.3–6 keV energy band over a large FOV. Such instrumentation has been under development for several years at the University of Leicester, has an high TRL level (e.g., BepiColombo) and can perform an all-sky survey in the soft X-rays with an unprecedented combination of FOV, source location accuracy ($<1'$) and sensitivity thus addressing both main science goals of the mission. An onboard infrared telescope of the 0.5–1 m class is also needed, together with spacecraft fast slewing capability (e.g., 10–20 deg/min), in order to provide prompt identification of the GRB optical/IR counterpart, refinement of the position down to a few arcs (thus enabling follow-up with the largest ground and space observatories), on-board redshift determination and spectroscopy of the counterpart and of the host galaxy. The telescope may also be used for multiple observatory and survey science goals. Finally, the inclusion in the payload of a broad field of view hard X-ray detection system covering the same survey FOV as the lobster-eye telescopes and extending the energy band from few keV up to several MeV will increase significantly the capabilities of the mission. As the lobster-eye telescopes can be triggered by several classes of transient phenomena (e.g., flare stars, X-ray bursts, etc), the hard X-ray detection system provides an efficient means to identify true GRBs and detect other transient sources (e.g., short GRBs). The joint data from the three instruments will characterize transients in terms of luminosity, spectra and timing properties over a broad energy band, thus getting fundamental insights into their physics.

Based on the above, the THESEUS payload consists of the instruments shortly described below.

Soft X-ray Imager (SXI): a set of Lobster Eye (0.3–6 keV) telescopes covering a total FOV of 1 sr field with 0.5–1 arcmin source location accuracy. Each module is a focusing wide field lobster eye telescope based on the optical principles described in previous sections. The optics aperture is 290 × 290 mm^2 formed by an array of 7 × 7 square pore Micro Channel Plates (MCPs). The MCPs are 40 × 40 mm^2 and are mounted on a spherical frame with radius of curvature 600 mm (2 times the focal length of 300 mm). The open aperture provided by each plate is 38 × 38 mm^2; the outer dimension of the optics frame is 320 × 320 mm^2. The focal plane of each SXI module is a spherical surface of radius of curvature 600 mm situated at a distance of 300 mm (the focal length) from the optics aperture. The detectors for each module comprise a 2 × 2 array of large format CCDs baselined to be supplied by e2v technologies (UK) Ltd. Each CCD has an active area of 61 × 61 mm^2; the detectors are tilted to approximate to the spherical focal surface.

InfraRed Telescope (IRT): a 70 cm class near-infrared (up to 2 microns) telescope (IRT) with imaging and moderate spectral capabilities. The telescope (optics and tube assembly) will be made of SiC, a material that has been used in other space missions (such as Gaia, Herschel, Sentinel 2 and SPICA). Simulations using a 0.7 m aperture Cassegrain space borne NIR telescope (with a 0.23 m secondary mirror), using a Teledyne Hawaii-2RG 2048 × 2048 pixels detector (18 micron/pix, resulting in 0.3 arcsec/pix plate scale) show that, for a 22.5 (H) point like source in a single 300 s exposure one could expect a SNR of 6. In order to achieve such performances the telescope needs to be cooled at 240 (+/−3) K by passive means, conductive and radiative insulations. Regarding the instrument, the optics box needs to be cooled to 190 (+/−5) K to by a two stage cooler for which the first stage will cool the optics and the second stage (the cold end) will cool the IR detector itself to 95 (+/−10) K: this allows the detector dark current to be kept at an acceptable level. The mechanical envelope of IRT is a cylinder with 80 cm diameter and 180 cm height. A sunshield is placed on top of the telescope baffle for IRT straylight protection. The thermal hardware is composed by a pulse tube cooling the Detector and FEE electronics and a set of thermal straps extracting the heat from the electronic boxes and camera optics coupled to a radiator located on the spacecraft structure. The overall telescope mass is 112.6 kg and the total power supply is 95 W.

X-Gamma-rays Spectrometer (XGS): non-imaging spectrometer (XGS) based on SDD + CsI, covering the same FOV as the Lobster telescope extending its energy band up to 20 MeV. The XGS consists of 25 modules made of scintillator bars optically coupled to an array of Silicon Drift Detectors (SDD) PhotoDiodes (PD) tightly packaged to each other. Both SDD-PDs and scintillator detect X- and gamma-rays. The top SDD-PD, fac-

Table 4 Main characteristics of the THESEUS/SXI instrument (Credits: P. O'Brien, J. Osborne, D. Willingale and the THESEUS Collaboration)

Energy band (keV)	0.3–5
Telescope type:	Lobster eye
Optics aperture (mm^2)	290×290
Optics configuration	7×7 square pore MCPs
MCP size (mm^2)	40×40
Focal length (mm)	300
Focal plane shape	spherical
Focal plane detectors	CCD array
Size of each CCD (mm^2)	61×61
Pixel size (μm)	30
Pixel number	1024×1024
Number of CCDs	4
Field of view (square deg)	542.8
Angular accuracy (best, worst) (arcsec)	$(<10, 105)$

Table 5 Main characteristics of the THESEUS/XGS instrument (Credits: F. Fuschino, C. Labanti, M. Marisaldi and the THESEUS Collaboration)

Energy band (keV)	2–20 000
Detection principle	Siswitch
Low-energy detector	Silicon Drift Detector
High energy detector	CsI(Tl)
Separation of energy losses in Si and CsI (Tl)	Pulse shape analysis
Number of modules	25
Size of each module (mm^3)	$130 \times 130 \times 100$
Lateral passive shielding/module	0.5 mm W
Slat collimator/module	0.5 mm W
Overall collimator FOV (FWHM)	1.8 sr
Average useful area in the SXI FOV (cm^2)	~2000

ing the X-/gamma-ray entrance window, is operated both as X-ray detector for low energy X-ray photons interacting in Silicon and as a read-out system of the scintillation light resulting from X-/gamma-ray interactions in the scintillator. The bottom SDD-PD at the other extreme of the crystal bar operates only as a read-out system for the scintillations. The discrimination between energy losses in Si and CsI is based on the different shape of charge pulses resulting from X-ray interactions in Si or from the collection of the scintillation light thanks to their different timing properties (Marisaldi et al. 2005). Each bar is made of scintillating crystal $5 \times 5 \times 30$ mm^3 in size. Each extreme of the bar is covered with a PD for the read-out of the scintillation light, while the other sides of the bar are wrapped with a light reflecting material convoying the scintillation light towards the PDs. The scintillator material is CsI (Tl) peaking its light emission at about 560 nm. The PD is realized with the technique of Silicon Drift Detectors with an active area of 5×5 mm^2 so matching the scintillator cross section.

The main instruments characteristics are summarized in Tables 4, 5 and 6, and a possible payload accommodation sketch is shown in Fig. 23. Sensitivity curves of the SXI and XGS are jointly shown in Fig. 24. Examples of expected performances in terms of GRB detection rate as a function of redshift and rate of different classes of transients are shown in Fig. 25 and Table 7, respectively.

Table 6 Main characteristics of the THESEUS/IRT instrument (Credits: A.J. Castro-Tirado, V. Reglero, D. Gotz and the THESEUS Collaboration)

Telescope type:	Cassegrain		
Primary & Secondary size:	700 mm & 230 mm		
Material:	SiC (for both optics and optical tube assembly)		
Detector type:	Teledyne Hawaii-2RG 2048 × 2048 pixels (18 μm each)		
Imaging plate scale	$0''.3$/pixel		
Field of view:	$10' \times 10'$	$6' \times 6'$	$2''.1 \times 2''.1$
Resolution ($\lambda/\delta\lambda$)	2–3 (imaging)	20 (low-res)	1700 (high-res) with $0''.4$ slit
Sensitivity (AB mag):	H = 22.5 (30 s)	H = 20.8 (300 s)	H = 19.3 (1800 s)
Filters:	ZYJH	Prism	VPH grating
Wavelength range (μm):	0.7–1.8 (imaging)	0.7–1.8 (low-res)	0.7–1.8 (high-res)
Total envelope size (mm):	800∅ × 1800		
Power (W):	95		
Mass (kg):	112.6		

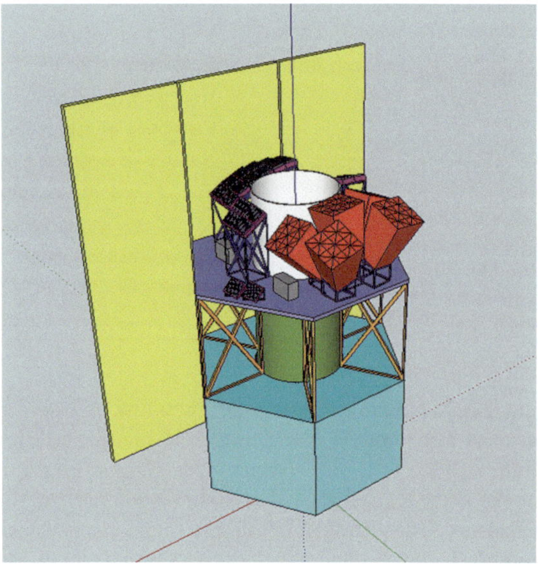

Fig. 23 Schematic view of the payload and satellite in THESEUS: the central *green* and *white* tube contains the IRT detectors, optics and buffle; the *red* modules are the SXI; the *black–pink* modules are the XGS. (Credits: F. Fuschino, C. Labanti and the THESEUS Collaboration)

8.5 Mission Profile and Consortium

The proposed mission profile includes an onboard data handling (OBDH) system capable of detecting, identifying and localizing likely transients in the SXI and XGS FOV; the capability of promptly (within a few tens of seconds at most) transmitting to ground the trigger time and position of GRBs (and other transients of interest); and a spacecraft slewing capability of 30/min. The baseline launcher/orbit configuration is a launch with Vega to a low inclination low Earth orbit (LEO, ~600 km, <5), which has the unique advantages of granting a low and stable background level in the high-energy instruments, allowing the exploitation of the Earths magnetic field for spacecraft fast slewing and facilitating the prompt transmission of transients trigger and positions to ground. The basic observing strategy ASI antenna in Malindi and, as an option, the Brazilian antenna in Alcantara were proposed

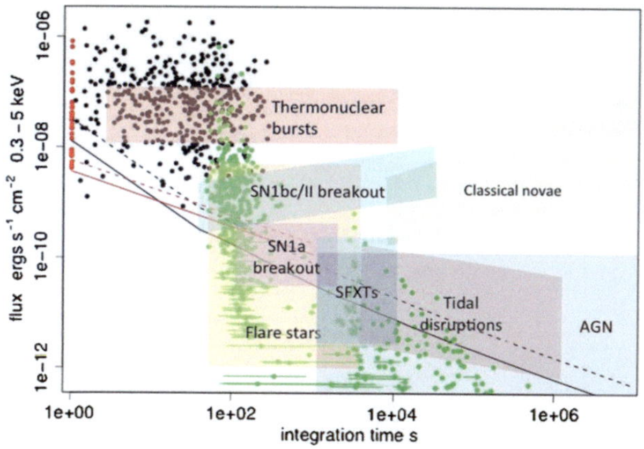

Fig. 24 Sensitivity of the SXI (*black curves*) and XGS (*red*) vs. integration time. The *solid curves* assume a source column density of 5×10^{20} cm^{-2} (i.e. well out of the Galactic plane and very little intrinsic absorption). The *dotted curves* assume a source column density of 10^{22} cm^{-2} (significant intrinsic absorption). The *black dots* are the peak fluxes for Swift BAT GRBs plotted against T90/2. The flux in the soft band 0.3–10 keV was estimated using the T90 BAT spectral fit including the absorption from the XRT spectral fit. The *red dots* are those GRBs for which T90/2 is less than 1 second. The *green dots* are the initial fluxes and times since trigger at the start of the Swift XRT GRB light-curves. The *horizontal lines* indicate the duration of the first time bin in the XRT light-curve. The various *shaded regions* illustrate variability and flux regions for different types of transients and variable sources. (Credits: D. Willingale, P. O'Brien, J. Osborne and the THESEUS Collaboration)

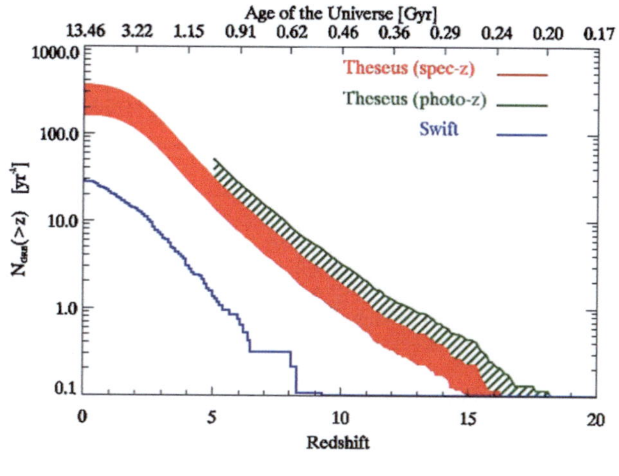

Fig. 25 The annual rate of GRBs predicted for THESEUS SXI (*red*) compared to Swift (*blue*). The *upper scale* shows the age of the Universe. For Swift the actual number of known redshifts is approximately one third that plotted and none were determined on board (the *blue curve* has been linearly scaled upwards to match the total Swift trigger rate). For THESEUS the *red region* uses the simulations from Ghirlanda et al. (2015) and adopts the instrument sensitivity for the SXI. (Credits: G. Ghirlanda, R. Salvaterra and the THESEUS Collaboration)

as ground stations. The basic observing strategy is based on alternating anti-sun and $\sim$ polar pointing, a compromise between maximum sky coverage, optimization of follow-up observations from the ground, instruments requirements (thermal, etc.). Considered prompt downlink options include: NASA/TDRSS, ESA/EDRS, WHF network, IRIDIUM network, ORBCOMM. MOC and SOC were proposed to be managed by ESA, while the SDC was proposed to be managed by ASI (ASDC).

The total payload mass of THESEUS, including all contingencies, was estimated to be $\sim$350 kg, and the total spacecraft dry mass about 1000 kg (power about 230 W and 800 W,

Table 7 Theseus detection rates for different astrophysical transients and variables (Credits: J. Osborne, P. O'Brien, D. Willingale and the THESEUS Collaboration)

Transient type	SXI Rate
GW sources	0.03–33 yr^{-1}
Magnetars	40 day^{-1}
SN shock breakout	4 yr^{-1}
TDE	50 yr^{-1}
AGN+Blazars	350 day^{-1}
Thermonuclear bursts	35 day^{-1}
Novae	250 yr^{-1}
Dwarf novae	30 day^{-1}
SFXTs	1000 yr^{-1}
Stellar flares	400 yr^{-1}
Stellar super flares	200 yr^{-1}

respectively). The foreseen telemetry budget is fully compatible with the capabilities of the X-band, which will be the standard for next ESA M missions.

The THESEUS payload consortium for the ESA/M4 proposal is organized as follow. The SXI will be responsibility of the UK (led by the University of Leicester), the XGS will be responsibility of Italy (led by INAF, the University of Ferrara and INFN), and the IRT will be the responsibility of Spain (a consortium led by IAA-CSIC including UV, INTA, UGR and UMA) for the camera and ESA for the optics. The core consortium includes also Poland (led by CBK) and Denmark (led by DTU) for payload data handling hardware and software. Czech Republic (led by CTU) is also involved for contributions to the SXI, as well as Slovenia (SPACE-SI) for communications. France (CNES, CEA) is available to provide the SVOM network of VHF antennae and associated control center. A junior international contribution is foreseen by USA (NASA) for the TDRSS system and contributions to the XGS and IRT camera. Finally, Hungary and Ireland have declared its interest in investigating their contributions to different payload components during assessment phase.

Acknowledgements The authors are grateful to the International Space Science Institute of Beijing (ISSI-Beijing), its executive director Prof. M. Falanga and all the staff for hosting and funding the workshop "Gamma-Ray Bursts: a tool to Explore the Young Universe" held in Beijing from April 10 to 15 2015. W. Yuan and C. Zhang thank R. Willingale, J.P. Osborne and P. O'Brien for their contribution to the EP project and acknowledge support of the "Strategic Priority Research Program on Space Science" (Grant number No. XDA04061100) and "Gravitational Wave Universe Research" (No. XDB23000000) of the Chinese Academy of Sciences. B. Cordier and D. Götz acknowledge financial support of the UnivEarthS Labex program at Sorbonne Paris Cité (ANR-10-LABX-0023 and ANR-11-IDEX-0005-02). Sun acknowledges support from the 973 program 2014CB845802 and NSFC 11503028.

References

C.W. Akerlof, H.F. Swan, An estimation of the gamma-ray burst afterglow apparent optical brightness distribution function. Astrophys. J. **671**, 1868–1876 (2007). doi:10.1086/523081

L. Amati, J.-L. Atteia, L. Balazs, S. Basa, J. Becker Tjus, D.F. Bersier, M. Boer, S. Campana, B. Ciardi, S. Covino, F. Daigne, M. Feroci, A. Ferrara, F. Frontera, J.P.U. Fynbo, G. Ghirlanda, G. Ghisellini, S. Glover, J. Greiner, D. Gotz, L. Hanlon, J. Hjorth, R. Hudec, U. Katz, S. Khochfar, R. Klessen, M. Kowalski, A.J. Levan, S. McBreen, A. Mesinger, R. Mochkovitch, P. O'Brien, J.P. Osborne, P. Petitjean, O. Reimer, E. Resconi, S. Rosswog, F. Ryde, R. Salvaterra, S. Savaglio, R. Schneider, G. Tagliaferri, N.R. Tanvir, A. van der Horst, Light from the cosmic frontier: gamma-ray bursts. ArXiv e-prints (2013)

J.R.P. Angel, Lobster eyes as X-ray telescopes. Astrophys. J. **233**, 364–373 (1979). doi:10.1086/157397

D. Band, J. Matteson, L. Ford, B. Schaefer, D. Palmer, B. Teegarden, T. Cline, M. Briggs, W. Paciesas, G. Pendleton, G. Fishman, C. Kouveliotou, C. Meegan, R. Wilson, P. Lestrade, BATSE observations of gamma-ray burst spectra. I—Spectral diversity. Astrophys. J. **413**, 281–292 (1993). doi:10.1086/172995

E. Berger, S.R. Kulkarni, D.A. Frail, The nonrelativistic evolution of GRBs 980703 and 970508: beaming-independent calorimetry. Astrophys. J. **612**, 966–973 (2004). doi:10.1086/422809

V. Bromm, A. Loeb, High-redshift gamma-ray bursts from population III progenitors. Astrophys. J. **642**, 382–388 (2006). doi:10.1086/500799

D. Burlon, G. Ghirlanda, A. van der Horst, T. Murphy, R.A.M.J. Wijers, B. Gaensler, G. Ghisellini, I. Prandoni, The SKA view of gamma-ray bursts, in *Advancing Astrophysics with the Square Kilometre Array, (AASKA14)*, vol. 52 (2015)

C.L. Carilli, R.J. Ivison, D.A. Frail, Variability of submillijansky radio sources. Astrophys. J. **590**, 192–196 (2003)

P. Chandra, D.A. Frail, A radio-selected sample of gamma-ray burst afterglows. Astrophys. J. **746**, 156 (2012). doi:10.1088/0004-637X/746/2/156

P. Chandra, D.A. Frail, D. Fox, S. Kulkarni, E. Berger, S.B. Cenko, D.C.-J. Bock, F. Harrison, M. Kasliwal, Discovery of radio afterglow from the most distant cosmic explosion. Astrophys. J. Lett. **712**, 31–35 (2010). doi:10.1088/2041-8205/712/1/L31

H.-W. Chen, J.X. Prochaska, E. Ramirez-Ruiz, J.S. Bloom, M. Dessauges-Zavadsky, R.J. Foley, On the absence of wind signatures in GRB afterglow spectra: constraints on the Wolf–Rayet winds of GRB progenitors. Astrophys. J. **663**, 420–436 (2007). doi:10.1086/518111

R. Chornock, E. Berger, D.B. Fox, R. Lunnan, M.R. Drout, W.-f. Fong, T. Laskar, K.C. Roth, GRB 130606A as a probe of the intergalactic medium and the interstellar medium in a star-forming galaxy in the first Gyr after the Big Bang. Astrophys. J. **774**, 26 (2013). doi:10.1088/0004-637X/774/1/26

B. Ciardi, A. Ferrara, The first cosmic structures and their effects. Space Sci. Rev. **116**, 625–705 (2005). doi:10.1007/s11214-005-3592-0

B. Cordier, F. Desclaux, J. Foliard, S. Schanne, SVOM pointing strategy: how to optimize the redshift measurements? in *American Institute of Physics Conference Series*, ed. by M. Galassi, D. Palmer, E. Fenimore American Institute of Physics Conference Series, vol. 1000, 2008, pp. 585–588. doi:10.1063/1.2943538

E. Costa, F. Frontera, J. Heise, M. Feroci, J. in't Zand, F. Fiore, M.N. Cinti, D. Dal Fiume, L. Nicastro, Discovery of an X-ray afterglow associated with the γ-ray burst of 28 February 1997. Nature **387**, 783–785 (1997)

A. Cucchiara, A.J. Levan, D.B. Fox, N.R. Tanvir, T.N. Ukwatta, E. Berger, T. Krühler, A. Küpcü Yoldaş, X.F. Wu, K. Toma, J. Greiner, F.E. Olivares, A. Rowlinson, L. Amati, T. Sakamoto, K. Roth, A. Stephens, A. Fritz, J.P.U. Fynbo, J. Hjorth, D. Malesani, P. Jakobsson, K. Wiersema, P.T. O'Brien, A.M. Soderberg, R.J. Foley, A.S. Fruchter, J. Rhoads, R.E. Rutledge, B.P. Schmidt, M.A. Dopita, P. Podsiadlowski, R. Willingale, C. Wolf, S.R. Kulkarni, P. D'Avanzo, A photometric redshift of $z \sim 9.4$ for GRB 090429B. Astrophys. J. **736**, 7 (2011). doi:10.1088/0004-637X/736/1/7

E. Curtis-Lake, R.J. McLure, J.S. Dunlop, A.B. Rogers, T. Targett, A. Dekel, R.S. Ellis, S.M. Faber, H.C. Ferguson, N.A. Grogin, K.-H. Huang, D.D. Kocevski, A.M. Koekemoer, K. Lai, E. Mármol-Queraltó, B.E. Robertson, No evidence for evolution in the typical rest-frame UV sizes or the incidence of morphological disturbance amongst L_* galaxies at $4 < z < 8$. ArXiv e-prints (2014)

D. Eichler, E. Waxman, The efficiency of electron acceleration in collisionless shocks and gamma-ray burst energetics. Astrophys. J. **627**, 861–867 (2005). doi:10.1086/430596

D.A. Frail, E. Waxman, S.R. Kulkarni, A 450 day light curve of the radio afterglow of GRB 970508: fireball calorimetry. Astrophys. J. **537**, 191–204 (2000). doi:10.1086/309024

D.A. Frail, S.R. Kulkarni, R. Sari, S.G. Djorgovski, J.S. Bloom, T.J. Galama, D.E. Reichart, E. Berger, F.A. Harrison, P.A. Price, S.A. Yost, A. Diercks, R.W. Goodrich, F. Chaffee, Beaming in gamma-ray bursts: evidence for a standard energy reservoir. Astrophys. J. Lett. **562**, 55–58 (2001)

D.A. Frail, A.M. Soderberg, S.R. Kulkarni, E. Berger, S. Yost, D.W. Fox, F.A. Harrison, Accurate calorimetry of GRB 030329. Astrophys. J. **619**, 994–998 (2005). doi:10.1086/426680

G.W. Fraser, A.N. Brunton, J.E. Lees, J.F. Pearson, W.B. Feller, X-ray focusing using square-pore microchannel plates first observation of cruxiform image structure. Nucl. Instrum. Methods Phys. Res., Sect. A, Accel. Spectrom. Detect. Assoc. Equip. **324**, 404–407 (1993). doi:10.1016/0168-9002(93)91003-6

G.W. Fraser, J.D. Carpenter, D.A. Rothery, J.F. Pearson, A. Martindale, J. Huovelin, J. Treis, M. Anand, M. Anttila, M. Ashcroft, J. Benkoff, P. Bland, A. Bowyer, A. Bradley, J. Bridges, C. Brown, C. Bulloch, E.J. Bunce, U. Christensen, M. Evans, R. Fairbend, M. Feasey, F. Giannini, S. Hermann, M. Hesse, M. Hilchenbach, T. Jorden, K. Joy, M. Kaipiainen, I. Kitchingman, P. Lechner, G. Lutz, A. Malkki, K. Muinonen, J. Näränen, P. Portin, M. Prydderch, J.S. Juan, E. Sclater, E. Schyns, T.J. Stevenson, L. Strüder, M. Syrjasuo, D. Talboys, P. Thomas,

C. Whitford, S. Whitehead, The mercury imaging x-ray spectrometer (mixs) on bepicolombo. Planet. Space Sci. **58**(1–2), 79–95 (2010). doi:10.1016/j.pss.2009.05.004. Comprehensive science investigations of Mercury: the scientific goals of the joint ESA/JAXA mission BepiColombo. http://www.sciencedirect.com/science/article/pii/S0032063309001482

J.P.U. Fynbo, P. Jakobsson, J.X. Prochaska, D. Malesani, C. Ledoux, A. de Ugarte Postigo, M. Nardini, P.M. Vreeswijk, K. Wiersema, J. Hjorth, J. Sollerman, H.-W. Chen, C.C. Thöne, G. Björnsson, J.S. Bloom, A.J. Castro-Tirado, L. Christensen, A. De Cia, A.S. Fruchter, J. Gorosabel, J.F. Graham, A.O. Jaunsen, B.L. Jensen, D.A. Kann, C. Kouveliotou, A.J. Levan, J. Maund, N. Masetti, B. Milvang-Jensen, E. Palazzi, D.A. Perley, E. Pian, E. Rol, P. Schady, R.L.C. Starling, N.R. Tanvir, D.J. Watson, D. Xu, T. Augusteijn, F. Grundahl, J. Telting, P.-O. Quirion, Low-resolution spectroscopy of gamma-ray burst optical afterglows: biases in the Swift sample and characterization of the absorbers. Astrophys. J. Suppl. Ser. **185**, 526–573 (2009). doi:10.1088/0067-0049/185/2/526

N. Gehrels, E. Ramirez-Ruiz, D.B. Fox, Gamma-ray bursts in the Swift era. Annu. Rev. Astron. Astrophys. **47**, 567–617 (2009). doi:10.1146/annurev.astro.46.060407.145147

B. Gendre, G. Stratta, J.L. Atteia, S. Basa, M. Boër, D.M. Coward, S. Cutini, V. D'Elia, E.J. Howell, A. Klotz, L. Piro, The ultra-long gamma-ray burst 111209A: the collapse of a blue supergiant? Astrophys. J. **766**, 30 (2013). doi:10.1088/0004-637X/766/1/30

G. Ghirlanda, R. Salvaterra, D. Burlon, S. Campana, A. Melandri, M.G. Bernardini, S. Covino, P. D'Avanzo, V. D'Elia, G. Ghisellini, L. Nava, I. Prandoni, L. Sironi, G. Tagliaferri, S.D. Vergani, A. Wolter, Radio afterglows of a complete sample of bright Swift GRBs: predictions from present days to the SKA era. Mon. Not. R. Astron. Soc. **435**, 2543–2551 (2013)

G. Ghirlanda, D. Burlon, G. Ghisellini, R. Salvaterra, M.G. Bernardini, S. Campana, S. Covino, P. D'Avanzo, V. D'Elia, A. Melandri, T. Murphy, L. Nava, S.D. Vergani, G. Tagliaferri, GRB orphan afterglows in present and future radio transient surveys. Publ. Astron. Soc. Aust. **31**, 22 (2014)

G. Ghirlanda, R. Salvaterra, S. Campana, S.D. Vergani, J. Japelj, M.G. Bernardini, D. Burlon, P. D'Avanzo, A. Melandri, A. Gomboc, F. Nappo, R. Paladini, A. Pescalli, O.S. Salafia, G. Tagliaferri, Unveiling the population of orphan γ-ray bursts. Astron. Astrophys. **578**, 71 (2015). doi:10.1051/0004-6361/201526112

R. Gilmozzi, J. Spyromilio, The European Extremely Large Telescope (E-ELT). Messenger **127**, 11 (2007)

O. Godet, G. Nasser, J.-. Atteia, B. Cordier, P. Mandrou, D. Barret, H. Triou, R. Pons, C. Amoros, S. Bordon, O. Gevin, F. Gonzalez, D. Götz, A. Gros, B. Houret, C. Lachaud, K. Lacombe, W. Marty, K. Mercier, D. Rambaud, P. Ramon, G. Rouaix, S. Schanne, V. Waegebaert, The X-/Gamma-ray camera ECLAIRs for the Gamma-ray burst mission SVOM, in *Society of Photo-Optical Instrumentation Engineers (SPIE) Conference Series*. Society of Photo-Optical Instrumentation Engineers (SPIE) Conference Series, vol. 9144, 2014, p. 24. doi:10.1117/12.2055507

D. Götz, J. Osborne, B. Cordier, J. Paul, P. Evans, A. Beardmore, A. Martindale, R. Willingale, P. O'Brien, S. Basa, C. Rossin, O. Godet, N. Webb, J. Greiner, K. Nandra, N. Meidinger, E. Perinati, A. Santangelo, K. Mercier, F. Gonzalez, The microchannel x-ray telescope for the gamma-ray burst mission SVOM, in *Society of Photo-Optical Instrumentation Engineers (SPIE) Conference Series*. Society of Photo-Optical Instrumentation Engineers (SPIE) Conference Series, vol. 9144, 2014, p. 23. doi:10.1117/12.2054898

J. Granot, A.J. van der Horst, Gamma-ray burst jets and their radio observations. Publ. Astron. Soc. Aust. **31**, 8 (2014). doi:10.1017/pasa.2013.44

J. Greiner, W. Bornemann, C. Clemens, M. Deuter, G. Hasinger, M. Honsberg, H. Huber, S. Huber, M. Krauss, T. Krühler, A. Küpcü Yoldaş, H. Mayer-Hasselwander, B. Mican, N. Primak, F. Schrey, I. Steiner, G. Szokoly, C.C. Thöne, A. Yoldaş, S. Klose, U. Laux, J. Winkler, GROND—a 7-channel imager. Publ. Astron. Soc. Pac. **120**, 405–424 (2008). doi:10.1086/587032

O.E. Hartoog, D. Malesani, J.P.U. Fynbo, T. Goto, T. Krühler, P.M. Vreeswijk, A. De Cia, D. Xu, P. Møller, S. Covino, V. D'Elia, H. Flores, P. Goldoni, J. Hjorth, P. Jakobsson, J.-K. Krogager, L. Kaper, C. Ledoux, A.J. Levan, B. Milvang-Jensen, J. Sollerman, M. Sparre, G. Tagliaferri, N.R. Tanvir, A. de Ugarte Postigo, S.D. Vergani, K. Wiersema, J. Datson, R. Salinas, K. Mikkelsen, N. Aghanim, VLT/X-Shooter spectroscopy of the afterglow of the Swift GRB 130606A. chemical abundances and reionisation at $z \sim 6$. Astron. Astrophys. **580**, 139 (2015). doi:10.1051/0004-6361/201425001

A. Heger, C.L. Fryer, S.E. Woosley, N. Langer, D.H. Hartmann, How massive single stars end their life. Astrophys. J. **591**, 288–300 (2003). doi:10.1086/375341

J. Heise, J.I. Zand, R.M. Kippen, P.M. Woods, X-ray flashes and X-ray rich gamma ray bursts, in *Gamma-Ray Bursts in the Afterglow Era*, ed. by E. Costa, F. Frontera, J. Hjorth, 2001, p. 16. doi:10.1007/10853853_4

J. Japelj, D. Kopač, S. Kobayashi, R. Harrison, C. Guidorzi, F.J. Virgili, C.G. Mundell, A. Melandri, A. Gomboc, Phenomenology of reverse-shock emission in the optical afterglows of gamma-ray bursts. Astrophys. J. **785**, 84 (2014). doi:10.1088/0004-637X/785/2/84

A. Klotz, M. Boër, J.L. Atteia, B. Gendre, Early optical observations of gamma-ray bursts by the TAROT telescopes: period 2001–2008. Astron. J. **137**, 4100–4108 (2009). doi:10.1088/0004-6256/137/5/4100

S.R. Kulkarni, D.A. Frail, M.H. Wieringa, R.D. Ekers, E.M. Sadler, R.M. Wark, J.L. Higdon, E.S. Phinney, J.S. Bloom, Radio emission from the unusual supernova 1998bw and its association with the γ-ray burst of 25 April 1998. Nature **395**, 663–669 (1998). doi:10.1038/27139

S.R. Kulkarni, D.A. Frail, R. Sari, G.H. Moriarty-Schieven, D.S. Shepherd, P. Udomprasert, A.C.S. Readhead, J.S. Bloom, M. Feroci, E. Costa, Discovery of a radio flare from GRB 990123. Astrophys. J. Lett. **522**, 97–100 (1999). doi:10.1086/312227

P. Kumar, B. Zhang, The physics of gamma-ray bursts and relativistic jets. Phys. Rep. **561**, 1–109 (2015). doi:10.1016/j.physrep.2014.09.008

D.Q. Lamb, D.E. Reichart, Gamma-ray bursts as a probe of the very high redshift universe. Astrophys. J. **536**, 1–18 (2000). doi:10.1086/308918

T. Laskar, E. Berger, B.A. Zauderer, R. Margutti, A.M. Soderberg, S. Chakraborti, R. Lunnan, R. Chornock, P. Chandra, A. Ray, A reverse shock in GRB 130427A. Astrophys. J. **776**, 119 (2013). doi:10.1088/0004-637X/776/2/119

A. Maselli, A. Melandri, L. Nava, C.G. Mundell, N. Kawai, S. Campana, S. Covino, J.R. Cummings, G. Cusumano, P.A. Evans, G. Ghirlanda, G. Ghisellini, C. Guidorzi, S. Kobayashi, P. Kuin, V. La Parola, V. Mangano, S. Oates, T. Sakamoto, M. Serino, F. Virgili, B.-B. Zhang, S. Barthelmy, A. Beardmore, M.G. Bernardini, D. Bersier, D. Burrows, G. Calderone, M. Capalbi, J. Chiang, P. D'Avanzo, V. D'Elia, M. De Pasquale, D. Fugazza, N. Gehrels, A. Gomboc, R. Harrison, H. Hanayama, J. Japelj, J. Kennea, D. Kopac, C. Kouveliotou, D. Kuroda, A. Levan, D. Malesani, F. Marshall, J. Nousek, P. O'Brien, J.P. Osborne, C. Pagani, K.L. Page, M. Page, M. Perri, T. Pritchard, P. Romano, Y. Saito, B. Sbarufatti, R. Salvaterra, I. Steele, N. Tanvir, G. Vianello, B. Weigand, K. Wiersema, Y. Yatsu, T. Yoshii, G. Tagliaferri, GRB 130427A: a nearby ordinary monster. Science **343**, 48–51 (2014). doi:10.1126/science.1242279

M. McQuinn, A. Lidz, M. Zaldarriaga, L. Hernquist, S. Dutta, Probing the neutral fraction of the IGM with GRBs during the epoch of reionization. Mon. Not. R. Astron. Soc. **388**, 1101–1110 (2008). doi:10.1111/j.1365-2966.2008.13271.x

N. Meidinger, R. Andritschke, W. Bornemann, D. Coutinho, V. Emberger, O. Hälker, W. Kink, B. Mican, S. Müller, D. Pietschner, P. Predehl, J. Reiffers, Report on the eROSITA camera system, in *Society of Photo-Optical Instrumentation Engineers (SPIE) Conference Series*. Society of Photo-Optical Instrumentation Engineers (SPIE) Conference Series, vol. 9144, 2014, p. 1. doi:10.1117/12.2055703

A. Melandri, B. Sbarufatti, P. D'Avanzo, R. Salvaterra, S. Campana, S. Covino, S.D. Vergani, L. Nava, G. Ghisellini, G. Ghirlanda, D. Fugazza, V. Mangano, M. Capalbi, G. Tagliaferri, The dark bursts population in a complete sample of bright Swift long gamma-ray bursts. Mon. Not. R. Astron. Soc. **421**, 1265–1272 (2012). doi:10.1111/j.1365-2966.2011.20398.x

R.A. Mesler, Y.M. Pihlström, Calorimetry of GRB 030329: simultaneous model fitting to the broadband radio afterglow and the observed image expansion rate. Astrophys. J. **774**, 77 (2013). doi:10.1088/0004-637X/774/1/77

R.A. Mesler, Y.M. Pihlström, G.B. Taylor, J. Granot, VLBI and archival VLA and WSRT observations of the GRB 030329 radio afterglow. Astrophys. J. **759**, 4 (2012). doi:10.1088/0004-637X/759/1/4

P. Mészáros, M.J. Rees, Population III gamma-ray bursts. Astrophys. J. **715**, 967–971 (2010). doi:10.1088/0004-637X/715/2/967

B.D. Metzger, P.K.G. Williams, E. Berger, Extragalactic transients in the era of wide-field radio surveys. I. Detection rates and light curve characteristics. ArXiv e-prints (2015)

M.J. Michałowski, A. Kamble, J. Hjorth, D. Malesani, R.F. Reinfrank, L. Bonavera, J.M. Castro Cerón, E. Ibar, J.S. Dunlop, J.P.U. Fynbo, M.A. Garrett, P. Jakobsson, D.L. Kaplan, T. Krühler, A.J. Levan, M. Massardi, S. Pal, J. Sollerman, N.R. Tanvir, A.J. van der Horst, D. Watson, K. Wiersema, The optically unbiased GRB host (TOUGH) survey. VI. Radio observations at $z >\sim 1$ and consistency with typical star-forming galaxies. Astrophys. J. **755**, 85 (2012). doi:10.1088/0004-637X/755/2/85

D.A. Perley, R.A. Perley, Radio constraints on heavily obscured star formation. Astrophys. J. **778**, 172 (2013). doi:10.1088/0004-637X/778/2/172

D.A. Perley, S.B. Cenko, A. Corsi, N.R. Tanvir, A.J. Levan, D.A. Kann, E. Sonbas, K. Wiersema, W. Zheng, X.-H. Zhao, J.-M. Bai, M. Bremer, A.J. Castro-Tirado, L. Chang, K.I. Clubb, D. Frail, A. Fruchter, E. Göğüş, J. Greiner, T. Güver, A. Horesh, A.V. Filippenko, S. Klose, J. Mao, A.N. Morgan, A.S. Pozanenko, S. Schmidl, B. Stecklum, M. Tanga, A.A. Volnova, A.E. Volvach, J.-G. Wang, J.-M. Winters, Y.-X. Xin, The afterglow of GRB 130427A from 1 to 10^{16} GHz. Astrophys. J. **781**, 37 (2014). doi:10.1088/0004-637X/781/1/37

D.A. Perley, R.A. Perley, J. Hjorth, M.J. Michałowski, S.B. Cenko, P. Jakobsson, T. Krühler, A.J. Levan, D. Malesani, N.R. Tanvir, Connecting GRBs and ULIRGs: a sensitive, unbiased survey for radio emission from gamma-ray burst host galaxies at $0.0 < z < 2.5$. Astrophys. J. **801**, 102 (2015). doi:10.1088/0004-637X/801/2/102

Y.M. Pihlström, G.B. Taylor, J. Granot, S. Doeleman, Stirring the embers: high-sensitivity VLBI observations of GRB 030329. Astrophys. J. **664**, 411–415 (2007). doi:10.1086/518955

P. Predehl, R. Andritschke, W. Becker, W. Bornemann, H. Bräuninger, H. Brunner, T. Boller, V. Burwitz, W. Burkert, N. Clerc, E. Churazov, D. Coutinho, K. Dennerl, J. Eder, V. Emberger, T. Eraerds, M.J. Freyberg, P. Friedrich, M. Fürmetz, A. Georgakakis, C. Grossberger, F. Haberl, O. Hälker, G. Hartner, G. Hasinger, J. Hoelzl, H. Huber, A. von Kienlin, W. Kink, I. Kreykenbohm, G. Lamer, I. Lomakin, I. Lapchov, L. Lovisari, N. Meidinger, A. Merloni, B. Mican, J. Mohr, S. Müller, K. Nandra, F. Pacaud, M.N. Pavlinsky, E. Perinati, E. Pfeffermann, D. Pietschner, J. Reiffers, T. Reiprich, J. Robrade, M. Salvato, A.E. Santangelo, M. Sasaki, H. Scheuerle, C. Schmid, J. Schmitt, A.D. Schwope, R. Sunyaev, C. Tenzer, L. Tiedemann, W. Xu, V. Yaroshenko, S. Walther, M. Wille, J. Wilms, Y.-Y. Zhang, EROSITA on SRG, in *Society of Photo-Optical Instrumentation Engineers (SPIE) Conference Series*. Society of Photo-Optical Instrumentation Engineers (SPIE) Conference Series, vol. 9144, 2014, p. 1. doi:10.1117/12.2055426

N. Produit, F. Barao, S. Deluit, W. Hajdas, C. Leluc, M. Pohl, D. Rapin, J.-P. Vialle, R. Walter, C. Wigger, POLAR, a compact detector for gamma-ray bursts photon polarization measurements. Nucl. Instrum. Methods Phys. Res., Sect. A, Accel. Spectrom. Detect. Assoc. Equip. **550**, 616–625 (2005). doi:10.1016/j.nima.2005.05.066

J.L. Racusin, S.V. Karpov, M. Sokolowski, J. Granot, X.F. Wu, V. Pal'Shin, S. Covino, A.J. van der Horst, S.R. Oates, P. Schady, R.J. Smith, J. Cummings, R.L.C. Starling, L.W. Piotrowski, B. Zhang, P.A. Evans, S.T. Holland, K. Malek, M.T. Page, L. Vetere, R. Margutti, C. Guidorzi, A.P. Kamble, P.A. Curran, A. Beardmore, C. Kouveliotou, L. Mankiewicz, A. Melandri, P.T. O'Brien, K.L. Page, T. Piran, N.R. Tanvir, G. Wrochna, R.L. Aptekar, S. Barthelmy, C. Bartolini, G.M. Beskin, S. Bondar, M. Bremer, S. Campana, A. Castro-Tirado, A. Cucchiara, M. Cwiok, P. D'Avanzo, V. D'Elia, M. Della Valle, A. de Ugarte Postigo, W. Dominik, A. Falcone, F. Fiore, D.B. Fox, D.D. Frederiks, A.S. Fruchter, D. Fugazza, M.A. Garrett, N. Gehrels, S. Golenetskii, A. Gomboc, J. Gorosabel, G. Greco, A. Guarnieri, S. Immler, M. Jelinek, G. Kasprowicz, V. La Parola, A.J. Levan, V. Mangano, E.P. Mazets, E. Molinari, A. Moretti, K. Nawrocki, P.P. Oleynik, J.P. Osborne, C. Pagani, S.B. Pandey, Z. Paragi, M. Perri, A. Piccioni, E. Ramirez-Ruiz, P.W.A. Roming, I.A. Steele, R.G. Strom, V. Testa, G. Tosti, M.V. Ulanov, K. Wiersema, R.A.M.J. Wijers, J.M. Winters, A.F. Zarnecki, F. Zerbi, P. Mészáros, G. Chincarini, D.N. Burrows, Broadband observations of the naked-eye γ-ray burst GRB080319B. Nature **455**, 183–188 (2008). doi:10.1038/nature07270

E. Rol, A. van der Horst, K. Wiersema, S.K. Patel, A. Levan, M. Nysewander, C. Kouveliotou, R.A.M.J. Wijers, N. Tanvir, D. Reichart, A.S. Fruchter, J. Graham, J.-E. Ovaldsen, A.O. Jaunsen, P. Jonker, W. van Ham, J. Hjorth, R.L.C. Starling, P.T. O'Brien, J. Fynbo, D.N. Burrows, R. Strom, GRB 051022: physical parameters and extinction of a prototype dark burst. Astrophys. J. **669**, 1098–1106 (2007). doi:10.1086/521336

E.M. Rossi, R. Perna, F. Daigne, 'orphan' afterglows in the universal structured jet model for γ-ray bursts. Mon. Not. R. Astron. Soc. **390**, 675–682 (2008)

T. Sakamoto, D.Q. Lamb, C. Graziani, T.Q. Donaghy, M. Suzuki, G. Ricker, J.-L. Atteia, N. Kawai, A. Yoshida, Y. Shirasaki, T. Tamagawa, K. Torii, M. Matsuoka, E.E. Fenimore, M. Galassi, T. Tavenner, J. Doty, R. Vanderspek, G.B. Crew, J. Villasenor, N. Butler, G. Prigozhin, J.G. Jernigan, C. Barraud, M. Boer, J.-P. Dezalay, J.-F. Olive, K. Hurley, A. Levine, G. Monnelly, F. Martel, E. Morgan, S.E. Woosley, T. Cline, J. Braga, R. Manchanda, G. Pizzichini, K. Takagishi, M. Yamauchi, High energy transient explorer 2 observations of the extremely soft X-ray flash XRF 020903. Astrophys. J. **602**, 875–885 (2004). doi:10.1086/381232

R. Salvaterra, High redshift gamma-ray bursts. J. High Energy Astrophys. **7**, 35–43 (2015). doi:10.1016/j.jheap.2015.03.001

I. Shivvers, E. Berger, A beaming-independent estimate of the energy distribution of long gamma-ray bursts: initial results and future prospects. Astrophys. J. **734**, 58 (2011). doi:10.1088/0004-637X/734/1/58

A.M. Soderberg, S.R. Kulkarni, D.B. Fox, E. Berger, P.A. Price, S.B. Cenko, D.A. Howell, A. Gal-Yam, D.C. Leonard, D.A. Frail, D. Moon, R.A. Chevalier, M. Hamuy, K.C. Hurley, D. Kelson, K. Koviak, W. Krzeminski, P. Kumar, A. MacFadyen, P.J. McCarthy, H.S. Park, B.A. Peterson, M.M. Phillips, M. Rauch, M. Roth, B.P. Schmidt, S. Shectman, An HST search for supernovae accompanying X-ray flashes. Astrophys. J. **627**, 877–887 (2005). doi:10.1086/430405

M. Sparre, O.E. Hartoog, T. Krühler, J.P.U. Fynbo, D.J. Watson, K. Wiersema, V. D'Elia, T. Zafar, P.M.J. Afonso, S. Covino, A. de Ugarte Postigo, H. Flores, P. Goldoni, J. Greiner, J. Hjorth, P. Jakobsson, L. Kaper, S. Klose, A.J. Levan, D. Malesani, B. Milvang-Jensen, M. Nardini, S. Piranomonte, J. Sollerman, R. Sánchez-Ramírez, S. Schulze, N.R. Tanvir, S.D. Vergani, R.A.M.J. Wijers, The metallicity and dust content of a redshift 5 gamma-ray burst host galaxy. Astrophys. J. **785**, 150 (2014). doi:10.1088/0004-637X/785/2/150

G.W. Suarez, PhD thesis, University of Geneva, 2010

J. Sun, PhD thesis, Graduate University of Chinese Academy of Sciences, 2012

N.R. Tanvir, P. Jakobsson, Observations of GRBs at high redshift. Philos. Trans. R. Soc. Lond. Ser. A, Math. Phys. Sci. **365**, 1377–1384 (2007). doi:10.1098/rsta.2006.1992

N.R. Tanvir, D.B. Fox, A.J. Levan, E. Berger, K. Wiersema, J.P.U. Fynbo, A. Cucchiara, T. Krühler, N. Gehrels, J.S. Bloom, J. Greiner, P.A. Evans, E. Rol, F. Olivares, J. Hjorth, P. Jakobsson, J. Farihi, R. Willingale, R.L.C. Starling, S.B. Cenko, D. Perley, J.R. Maund, J. Duke, R.A.M.J. Wijers, A.J. Adamson, A. Allan, M.N. Bremer, D.N. Burrows, A.J. Castro-Tirado, B. Cavanagh, A. de Ugarte Postigo, M.A. Dopita, T.A. Fatkhullin, A.S. Fruchter, R.J. Foley, J. Gorosabel, J. Kennea, T. Kerr, S. Klose, H.A. Krimm, V.N. Komarova, S.R. Kulkarni, A.S. Moskvitin, C.G. Mundell, T. Naylor, K. Page, B.E. Penprase, M. Perri, P. Podsiadlowski, K. Roth, R.E. Rutledge, T. Sakamoto, P. Schady, B.P. Schmidt, A.M. Soderberg, J. Sollerman, A.W. Stephens, G. Stratta, T.N. Ukwatta, D. Watson, E. Westra, T. Wold, C. Wolf, A γ-ray burst at a redshift of $z \sim 8.2$. Nature **461**, 1254–1257 (2009). doi:10.1038/nature08459

N.R. Tanvir, A.J. Levan, A.S. Fruchter, J.P.U. Fynbo, J. Hjorth, K. Wiersema, M.N. Bremer, J. Rhoads, P. Jakobsson, P.T. O'Brien, E.R. Stanway, D. Bersier, P. Natarajan, J. Greiner, D. Watson, A.J. Castro-Tirado, R.A.M.J. Wijers, R.L.C. Starling, K. Misra, J.F. Graham, C. Kouveliotou, Star formation in the early universe: beyond the tip of the iceberg. Astrophys. J. **754**, 46 (2012). doi:10.1088/0004-637X/754/1/46

G.B. Taylor, D.A. Frail, E. Berger, S.R. Kulkarni, The angular size and proper motion of the afterglow of GRB 030329. Astrophys. J. Lett. **609**, 1–4 (2004). doi:10.1086/422554

A.J. van der Horst, Z. Paragi, A.G. de Bruyn, J. Granot, C. Kouveliotou, K. Wiersema, R.L.C. Starling, P.A. Curran, R.A.M.J. Wijers, A. Rowlinson, G.A. Anderson, R.P. Fender, J. Yang, R.G. Strom, A comprehensive radio view of the extremely bright gamma-ray burst 130427A. Mon. Not. R. Astron. Soc. **444**, 3151–3163 (2014). doi:10.1093/mnras/stu1664

H. van Eerten, A. van der Horst, A. MacFadyen, Gamma-ray burst afterglow broadband fitting based directly on hydrodynamics simulations. Astrophys. J. **749**, 44 (2012)

J. van Paradijs, P.J. Groot, T. Galama, C. Kouveliotou, R.G. Strom, J. Telting, R.G.M. Rutten, G.J. Fishman, Transient optical emission from the error box of the γ-ray burst of 28 February 1997. Nature **386**, 686–689 (1997)

F.Y. Wang, V. Bromm, T.H. Greif, A. Stacy, Z.G. Dai, A. Loeb, K.S. Cheng, Probing pre-galactic metal enrichment with high-redshift gamma-ray bursts. Astrophys. J. **760**, 27 (2012). doi:10.1088/0004-637X/760/1/27

H. Wolter, Spiegelsysteme streifenden Einfalls als abbildende Optiken für Röntgenstrahlen. Ann. Phys. **445**(1-2), 94–114 (1952). doi:10.1002/andp.19524450108

P.R. Wozniak, S.R. Kulkarni, J. Bloom, N. Brandt, D. Fox, A. Gal-Yam, N. Gehrels, L. Jones, M. Kasliwal, C. Kouveliotou, E. Nakar, A. Rau, M. Strauss, W.T. Vestrand, Explosive transients in the distant universe, in *astro2010: The Astronomy and Astrophysics Decadal Survey*. ArXiv Astrophysics e-prints, vol. 2010 (2009), p. 326

S. Xiong, N. Produit, B. Wu, Expected performance of a hard X-ray polarimeter (POLAR) by Monte Carlo simulation. Nucl. Instrum. Methods Phys. Res., Sect. A, Accel. Spectrom. Detect. Assoc. Equip. **606**, 552–559 (2009). doi:10.1016/j.nima.2009.04.033

D. Yonetoku, T. Murakami, S. Gunji, T. Mihara, K. Toma, T. Sakashita, Y. Morihara, T. Takahashi, N. Toukairin, H. Fujimoto, Y. Kodama, S. Kubo, IKAROS Demonstration Team, Detection of gamma-ray polarization in prompt emission of GRB 100826A. Astrophys. J. Lett. **743**, 30 (2011). doi:10.1088/2041-8205/743/2/L30

W. Yuan, C. Zhang, H. Feng, S.N. Zhang, Z.X. Ling, D.H. Zhao, J. Deng, Y.L. Qiu, J.P. Osborne, P. O'Brien, R. Willingale, J.S. Lapington, Einstein probe—a small mission to monitor and explore the dynamic X-ray universe, in *Proceedings of Swift: 10 Years of Discovery*, SWIFT 10, 2–5 December 2014, La Sapienza University, Rome, Italy (2015), p. 6. arXiv:1506.07735

B. Zhang, Gamma-ray burst prompt emission. Int. J. Mod. Phys. D **23**, 30002 (2014). doi:10.1142/S021827181430002X

D. Zhao, C. Zhang, W. Yuan, R. Willingale, Z. Ling, H. Feng, H. Li, J. Ji, W. Wang, S. Zhang, Ray tracing simulations for the wide-field x-ray telescope of the Einstein probe mission based on Geant4 and XRTG4, in *Society of Photo-Optical Instrumentation Engineers (SPIE) Conference Series*. Society of Photo-Optical Instrumentation Engineers (SPIE) Conference Series, vol. 9144, 2014, p. 4. doi:10.1117/12.2055434

MIX
Papier aus verantwortungsvollen Quellen
Paper from responsible sources
FSC® C105338

If you have any concerns about our products,
you can contact us on
ProductSafety@springernature.com

In case Publisher is established outside the EU,
the EU authorized representative is:
Springer Nature Customer Service Center GmbH
Europaplatz 3, 69115 Heidelberg, Germany

Printed by Libri Plureos GmbH
in Hamburg, Germany